METHODS IN MOLECULAR BIOLOGY™

For further volumes:
http://www.springer.com/series/7651

Strain Engineering

Methods and Protocols

Edited by

James A. Williams

Nature Technology Corporation, Lincoln, NE, USA

Humana Press

Editor
James A. Williams, Ph.D
Nature Technology Corporation
Lincoln, NE
USA
jwilliams@natx.com

ISSN 1064-3745 e-ISSN 1940-6029
ISBN 978-1-4939-5846-7 ISBN 978-1-61779-197-0 (eBook)
DOI 10.1007/978-1-61779-197-0
Springer New York Dordrecht Heidelberg London

Printed on acid-free paper

Humana Press is part of Springer Science+Business Media (www.springer.com)

Preface

Microbial strain engineering is used to improve production of bioproducts. Classical strain engineering is performed by repeated cycles of random mutation and selection. These methods have greatly contributed to strain improvement, but have serious drawbacks. Uncharacterized "non-specific" secondary deleterious mutations will be introduced into the genome during each mutagenesis cycle, and accumulate in the selected strain. Classical methods also do not allow the introduction of new genetic material and are not suitable for complex strain development applications such as metabolic engineering of organisms to enable cell-based conversion of biomass into biofuels.

For complex strain engineering projects such as metabolic engineering for biofuels production, a starting point "chassis" organism must be selected. This may be a commonly used industrial organism such as *Escherichia coli* or *Sacharromyces cerevisiae*. While these industrial organisms are not inherently adapted for production of biofuels, new genes and functions can be rapidly imported using existing comprehensive strain engineering toolkits. Many of these methods draw upon the fully annotated genome sequences of *E. coli* and *S. cerevisiae* that ushered in a new age of rationale design-based strain engineering.

Alternatively, a native organism with existing biochemical pathways and production potential for biofuels is selected as the chassis. However, native strains often are not adapted for industrial fermentation and lack existing molecular biology tools necessary for efficient strain engineering.

Recently, fully annotated genome sequences of many important native microbial organisms have become publically available as a resource for researchers. The availability of these genomic resources will enable adaptation of *E. coli* or *S. cerevisiae*-based rationale design strain engineering methods to native organisms.

In this book, powerful new genetic engineering-based strain engineering methods are presented for rational modification of a variety of model organisms. These methods are particularly powerful when utilized to manipulate microbes for which sequenced and annotated genomes are available. Collectively, these methods systematically introduce genome alterations in a precise manner, allowing creation of novel strains carrying only desired genome alterations.

In *Section 1*, *E. coli*-based bacterial strain engineering strategies are reviewed. State-of-the-art methods for targeted gene knockout are presented, as well as their sequential application for scarless genome modification. Methods for random gene knockout by transposon mutagenesis are also described.

Cutting edge methods for identification of adaptation-selected genes are presented in chapters describing genome engineering using oligonucleotide-mediated targeted gene replacement and microarray-based genetic footprinting of random transposon libraries.

Methods to optimize synthetic operons for metabolic engineering applications are described. Methods for introduction of genes and operons into the bacterial chromosome are presented in a chapter on integration plasmid-based chromosomal expression of native and foreign genes.

Strategies to assemble combinations of tagged integration plasmids, gene knockouts, or knockout collections (e.g., Keio collection) are discussed in a chapter on high-throughput double mutant assembly via conjugation. Protocols to assemble multiply modified strains are provided in a chapter on P1 transduction.

In *Section 2*, analogous microbial engineering strategies for eukaryotic cells are presented, using the yeast *S. cerevisiae* as a model. This section also includes chapters describing creation and phenotypic trait selection with signature-tagged barcoded mutant collections and libraries of mutant transcription factors; these methodologies have application in a wide range of microorganisms.

In *Section 3*, examples of the proliferative adaptations of these base technologies to strain engineer industrially important prokaryotic or eukaryotic microbial systems are presented. Introductory chapters on transformation and broad host range plasmid vectors provide design guidance to develop robust methods for the critical first step of efficiently introducing functional DNA into new microbes. This effort is guided by identification in the annotated genome of genes whose products are detrimental to efficient transformation, for example, restriction endonucleases and secreted nonspecific nucleases. Targeted elimination or neutralization of these genes improves broad host range plasmid transformation. In the case of fungi, nonhomologous recombination genes are also identified and eliminated, to facilitate development of targeted homologous recombination-based methods. This subsection then describes methods for applied strain engineering of microbial organisms (prokaryotic and eukaryotic) with bioenergy potential for which sequenced and annotated genomes are available. Once basic DNA transformation, replicating plasmids, and homologous recombination-based chromosome integration methods in new organisms are available, other techniques described in Sections 1 and 2 can be adapted. For example, to facilitate application of the *E. coli* integration plasmid technology described in Chapter 8, phage integration sites can be integrated into the genome at a permissive site by homologous recombination, and the corresponding phage integrase supplied on a broad host range plasmid.

Written for: Molecular and cellular biologists, molecular geneticists, bioengineers, and microbiologists working in academia, pharmaceutical and biotechnology that perform microbial strain engineering.

Lincoln, NE, USA ***James A. Williams***

Contents

Contributors

DANIEL E. AGNEW • *Department of Chemical and Biological Engineering, University of Wisconsin-Madison, Madison, WI, USA*

HAL S. ALPER • *Department of Chemical Engineering, The University of Texas at Austin, Austin, TX, USA*

KATSUTOSHI ARA • *Biological Science Laboratories, Kao Corporation, Tochigi, Japan*

AUDREY L. ATKIN • *School of Biological Sciences, University of Nebraska – Lincoln, Lincoln, NE, USA*

MOHAN BABU • *Banting and Best Department of Medical Research, University of Toronto, Toronto, ON, Canada*

ARVIND A. BHAGWAT • *Environmental Microbial and Food Safety Laboratory, U.S. Department of Agriculture, Beltsville, MD, USA; Division Environmental Microbial & Food Safety Laboratory, Organization USDA-ARS, Beltsville, MD, USA*

MEDHA BHAGWAT • *NIH Library, Office of Research Services, National Institutes of Health, Bethesda, MD, USA*

NIKË BHARUCHA • *Department of Molecular, Cellular, and Developmental Biology, Life Sciences Institute, University of Michigan, Ann Arbor, MI, USA*

PAUL BLUM • *School of Biological Sciences, University of Nebraska, Lincoln, NE, USA*

TRYGVE BRAUTASET • *Department of Biotechnology, SINTEF Materials and Chemistry, Trondheim, Norway*

STEPHEN T. CARTMAN • *Clostridia Research Group, BBSRC Sustainable Bioenergy Centre, School of Molecular Medical Sciences, Centre for Biomolecular Sciences, The University of Nottingham, Nottingham, UK*

YUN-PENG CHAO • *Department of Chemical Engineering, Feng Chia University, Taichung, Taiwan*

PO TING CHEN • *Department of Biotechnology, Southern Taiwan University, Tainan, Taiwan*

SHAN-YU CHEN • *Graduate School of Biotechnology and Bioengineering, Yuan Ze University, Taoyuan, Taiwan*

CHUNG-JEN CHIANG • *Department of Medical Laboratory Science and Biotechnology, China Medical University, Taichung, Taiwan*

CLARE M. COOKSLEY • *Clostridia Research Group, BBSRC Sustainable Bioenergy Centre, School of Molecular Medical Sciences, Centre for Biomolecular Sciences, The University of Nottingham, Nottingham, UK*

KARL DANA • *School of Biological Sciences, University of Nebraska, Lincoln, NE, USA*

ELIE J. DINER • *Biomolecular Science and Engineering Program, University of California, Santa Barbara, Santa Barbara, CA, USA*

GRACE L. DOUGLAS • *Department of Food, Bioprocessing & Nutrition Sciences, North Carolina State University, Raleigh, NC, USA*

ANDREW EMILI • *Department of Molecular Genetics, Donelly Centre for Cellular and Biomolecular Research (CCBR),University of Toronto, Toronto, ON, Canada*

ALLA GAGARINOVA • *Department of Molecular Genetics, University of Toronto, Toronto, ON, Canada*
FERNANDO GARZA-SÁNCHEZ • *Department of Molecular, Cellular and Developmental Biology, University of California, Santa Barbara, Santa Barbara, CA, USA*
YONG JUN GOH • *Department of Food, Bioprocessing & Nutrition Sciences, North Carolina State University, Raleigh, NC, USA*
JACK GREENBLATT • *Banting and Best Department of Medical Research, University of Toronto, Terrence Donnelly Center for Cellular and Biomolecular Research, 160 College Street Toronto, ON, Canada; Department of Molecular Genetics, University of Toronto, 1 King's College Circle, Toronto, ON, Canada*
CHRISTOPHER S. HAYES • *Biomolecular Science and Engineering Program, Department of Molecular, Cellular and Developmental Biology, University of California, Santa Barbara, Santa Barbara, CA, USA*
JOHN T. HEAP • *Clostridia Research Group, BBSRC Sustainable Bioenergy Centre, School of Molecular Medical Sciences, Centre for Biomolecular Sciences, The University of Nottingham, Nottingham, UK*
JOHANNES H. HEGEMANN • *Heinrich-Heine-Universität, Lehrstuhl für Funktionelle Genomforschung der Mikroorganismen, Düsseldorf, Germany*
SVEN BORIS HEICK • *Heinrich-Heine-Universität, Lehrstuhl für Funktionelle Genomforschung der Mikroorganismen, Düsseldorf, Germany*
LES M. HOFFMAN • *Epicentre Biotechnologies, an Illumina company, Madison, WI, USA*
ALISON K. HOTTES • *Department of Molecular Biology, Lewis-Sigler Institute for Integrative Genomics, Princeton University, Princeton, NJ, USA*
MASAYUKI INUI • *Research Institute of Innovative Technology for the Earth (RITE), Kizugawa-Shi, Kyoto, Japan*
ANDREAS KAUFMANN • *LMC RISC, ETH Zürich, HPM F16, Zürich, Switzerland*
SUN CHANG KIM • *Department of Biological Sciences, Korea Advanced Institute of Science and Technology (KAIST), Daejeon, South Korea*
KATSUHIKO KITAMOTO • *Department of Biotechnology, The University of Tokyo, Tokyo, Japan*
TODD R. KLAENHAMMER • *Department of Food, Bioprocessing & Nutrition Sciences, North Carolina State University, Raleigh, NC, USA*
MICHAEL KNOP • *Cell Biology and Biophysics, ZMBH, Univeristät Heidelberg, Heidelberg, Germany*
SARAH A. KUEHNE • *Clostridia Research Group, BBSRC Sustainable Bioenergy Centre, School of Molecular Medical Sciences, Centre for Biomolecular Sciences, The University of Nottingham, Nottingham, UK*
ANUJ KUMAR • *Department of Molecular, Cellular, and Developmental Biology, Life Sciences Institute, University of Michigan, Ann Arbor, MI, USA*
RAHMI LALE • *Department of Biotechnology, Norwegian University of Science and Technology (NTNU), Trondheim, Norway*
AMANDA M. LANZA • *Department of Chemical Engineering, The University of Texas at Austin, Austin, TX, USA*
JUN HYOUNG LEE • *Department of Biological Sciences, Korea Advanced Institute of Science and Technology (KAIST), Daedeok Science Town, Daejeon, South Korea*

YUKARI MAEZATO • *School of Biological Sciences, University of Nebraska, Lincoln, NE, USA*
JUN-ICHI MARUYAMA • *Department of Biotechnology, The University of Tokyo, Tokyo, Japan*
NIGEL P. MINTON • *Clostridia Research Group, BBSRC Sustainable Bioenergy Centre, School of Molecular Medical Sciences, Centre for Biomolecular Sciences, University Park, The University of Nottingham, Nottingham, UK*
SEAN D. MOORE • *Burnett School of Biomedical Sciences, College of Medicine, University of Central Florida, Orlando, FL, USA*
TAKUYA MORIMOTO • *Graduate School of Information Science, Nara Institute of Science and Technology, Takayama, Ikoma, Nara, Japan*
KENAN C. MURPHY • *Department of Microbial and Physiological systems, University of Massachusetts Medical School, Worcester, MA, USA*
COREY NISLOW • *Director, Donnelly Sequencing Center, The Donnelly Centre, Toronto, Canada*
NAOTAKE OGASAWARA • *Graduate School of Information Science, Nara Institute of Science and Technology, Takayama, Ikoma, Nara, Japan*
JULIA OH • *Genetics and Molecular Biology Branch, National Human Genome Research Institute, NIH, Bethesda, MD, USA*
KATSUYA OZAKI • *Biological Science Laboratories, Kao Corporation, Tochigi, Japan*
KATHERINE M. PAPPAS • *Department of Genetics & Biotechnology, Faculty of Biology, University of Athens, Athens, Greece*
BRIAN F. PFLEGER • *Department of Chemical and Biological Engineering, University of Wisconsin-Madison, Madison, WI, USA*
KENNETH E. RUDD • *Department of Biochemistry and Molecular Biology, Miller School of Medicine, University of Miami, Miami, FL, USA*
NANDITA SARKAR • *System Biosciences, Mountain View, CA, USA*
BONG HYUN SUNG • *Industrial Biotechnology and Bioenergy Research Center, Korea Research Institute of Bioscience & Biotechnology (KRIBB), Daejeon, South Korea*
NOBUAKI SUZUKI • *Research Institute of Innovative Technology for the Earth (RITE), Kizugawa-Shi, Kyoto, Japan*
TOHRU SUZUKI • *The United Graduate School of Agricultural Science, Gifu University, Gifu, Gifu Prefecture, Japan*
HENDRIK SZURMANT • *Department of Molecular and Experimental Medicine, The Scripps Research Institute, La Jolla, CA, USA*
SAEED TAVAZOIE • *245 Carl Icahn Laboratory, Washington Road, Princeton, NJ, USA*
SVEIN VALLA • *Department of Biotechnology, Norwegian University of Science and Technology (NTNU), Trondheim, Norway*
KARIN VAN DIJK • *Biology Department, Creighton University, Omaha, NE, USA*
ADAM C. WILSON • *Department of Biology, Georgia State University, Atlanta,GA, USA*
TAO XU • *Department of Molecular, Cellular, and Developmental Biology, Life Sciences Institute, University of Michigan, Ann Arbor, MI, USA*
KAZUMASA YASUI • *The United Graduate School of Agricultural Science, Gifu University, Gifu, Gifu Prefecture, Japan*
HIDEAKI YUKAWA • *Research Institute of Innovative Technology for the Earth (RITE), Kizugawa-Shi, Kyoto, Japan*
JINDAN ZHOU • *Department of Electrical and Computer Engineering, University of Miami, Coral Gables, FL, USA*

Part I

E. coli

Chapter 1

Bacterial Genome Reengineering

Jindan Zhou and Kenneth E. Rudd

Abstract

The web application PrimerPair at ecogene.org generates large sets of paired DNA sequences surrounding all protein and RNA genes of *Escherichia coli* K-12. Many DNA fragments, which these primers amplify, can be used to implement a genome reengineering strategy using complementary in vitro cloning and in vivo recombineering. The integration of a primer design tool with a model organism database increases the level of quality control. Computer-assisted design of gene primer pairs relies upon having highly accurate genomic DNA sequence information that exactly matches the DNA of the cells being used in the laboratory to ensure predictable DNA hybridizations. It is equally crucial to have confidence that the predicted start codons define the locations of genes accurately. Annotations in the EcoGene database are queried by PrimerPair to eliminate pseudogenes, IS elements, and other problematic genes before the design process starts. These projects progressively familiarize users with the EcoGene content, scope, and application interfaces that are useful for genome reengineering projects.

The first protocol leads to the design of a pair of primer sequences that were used to clone and express a single gene. The N-terminal protein sequence was experimentally verified and the protein was detected in the periplasm. This is followed by instructions to design PCR primer pairs for cloning gene fragments encoding 50 periplasmic proteins without their signal peptides. The design process begins with the user simply designating one pair of forward and reverse primer endpoint positions relative to all start and stop codon positions. The gene name, genomic coordinates, and primer DNA sequences are reported to the user. When making chromosomal deletions, the integrity of the provisional primer design is checked to see whether it will generate any unwanted double deletions with adjacent genes. The bad designs are recalculated and replacement primers are provided alongside the requested primers. A list of all genes with overlaps includes those expressed from the translational coupling motifs 5′-UGAUG-3′ and 5′-AUGA-3′. Rigid alignments of the 893 ribosome binding sites (RBSs) linked to the AUG codons of this coupled subset are assessed for information content using WebLogo 3.0. These specialized logos are missing the G at the prominent information peak position normally seen in the rigid alignment of all genes. This novel GHOLE motif was apparently masked by the normal RBSs in two previously published rigid alignments. We propose a model constraining the distance between the ATG and the RBS, obviating the need for a flexible linker model to reveal a Shine–Dalgarno-like sequence.

Key words: Polymerase chain reaction, Genetic engineering, *Escherichia coli*, Internet, Software, Databases, Quality control, Annotation, Bioinformatics, Microbiology

James A. Williams (ed.), *Strain Engineering: Methods and Protocols*, Methods in Molecular Biology, vol. 765, DOI 10.1007/978-1-61779-197-0_1, © Springer Science+Business Media, LLC 2011

1. Introduction

Reinterpretation, redesign, and repetition of published experiments demonstrate progress in any field of active scientific investigation. Misinterpretations fomented by the misannotation of gene starts has led to incorrect models to explain the effects of some novel *Escherichia coli secM* (1) and *mak* (2) genetic regulatory mutations. One of us (K.E.R.) helped reinterpret the *secM* results based on a revised *secM* start codon prediction, which was subsequently verified experimentally (3). The original interpretation of the *mak*-up mutation as being in the *mak* coding region has been clarified by verification of an internal GTG codon as the true initiation codon placing the *mak*-up mutation in the promoter region (4). The unpublished Mak N-terminal protein sequence can be found on the *mak* GenePage as a personal communication. A third example suggests that a plasmid construct used to characterize a predicted DNA binding protein requires reengineering; a pseudogene fragment was almost certainly cloned instead of an intact allele (5). The findings that YkgA does not crosstalk with the *mar/sox/rob* regulon, possibly because it is lacking its activator domain, can be interpreted as inconclusive until an intact clone of *ykgA* is reengineered.

EcoGene is concerned with the genome sequence annotation quality control issues of accuracy, comprehensiveness, and timeliness. The choices for translation start codons have been under extensive manual review for many years by the EcoGene curator (K.E.R.) and over 800 revisions have been made. EcoGene pioneered methods for the comprehensive annotation of bacterial pseudogenes, a difficult and inconsistent annotation process currently being reconsidered at Genbank. EcoGene is home to the Verified Set, a curated compilation of published N-terminal protein sequences that is used to document 900 start codons, fMet cleavages, and type I signal peptide cleavages (6). EcoGene has many curated compilations including lipoproteins (7) and small proteins (8, 9). EcoGene has thousands of up-to-date online bibliographies linked to GenePages and TopicPages. Over 500 TopicPages organize a wide variety of linked gene sets that can be retrieved as FASTA libraries using Boolean logic queries.

Based upon new results, further revisions can be made. This information can include better start site predictions based on alignments to new homologous sequences. Or an alternate start codon can also be identified and revised including mass spectrometry or protein sequencing. The primer design functions in EcoGene will automatically use the latest revised gene intervals as curators at EcoGene and other databases collect, correct, and interpret biological publications. We plan to further develop ecogene.org to allow the import of DNA sequence and annotations from existing Genbank genome records. When this is completed, the bulk primer design function can be used to design deletion and cloning primers for all the genes of any bacterial genome, including the warnings about double deletions. Although most bacterial Genbank records are not being actively updated, the corresponding RefSeq genome records are updated to some extent. However, before bulk primer design is attempted for another organism, the annotation should be reviewed and corrected as necessary.

Recent developments with the *E. coli* K-12 Keio mutant collection (10–12) and the ASKA clone libraries (13–15) illustrate that bacterial genome engineering can be a reiterative process refining and extending existing genomic biotechnology resource collections (GBRCs). This process involves (1) experimental and computational error detection and documentation within GBRCs, (2) remediation of GBRCs by the reuse, repair, and replacement of prior components, and (3) GBRC expansion in scope, content, reliability, and applicability. We describe this cyclic improvement over time as genome reengineering, involving redesign using bioinformatics followed by laboratory remanufacture and redistribution.

The *E. coli* K-12 ASKA and Keio GBRCs are used in thousands of laboratories; their in-house reengineering started early and continues (16). Middle generation GBRCs with the core set of ASKA cloned intervals and Keio deletion alleles are available from other laboratories. For example, the ASKA clone inserts have been moved into Gateway entry clones (14). The EcoGene laboratory has moved the entire Keio mutant collection into a clean genetic background using P1 outcrosses in 24-well culture dishes. We also use *re*-recombineering as an alternative to P1 transduction for transferring Keio alleles into our MG1655(Seq) *rph+* recipient strain to avoid co-transduction of closely linked mutations. We have tested, added, subtracted, replaced, and rescued hundreds of Keio mutant alleles (see Note 1).

Bacterial genome reengineering and genome sequence annotation are codependent processes that develop optimally when integrated as interdisciplinary systems biology. Reengineering is important because it is critical that current and future GBRCs be as reliable as possible to best serve as foundation resources for their use in laboratory experiments and modeling. Genome reengineering is a quality control system that can accelerate postgenomic research, and be part of a current world trend toward broader interdisciplinary experimental networks, sustainable highly integrated web servers with strong user contributions, and more open access to the top scientific journals.

2. Materials

Data sources and web applications used in our protocols are noted.

2.1. EcoGene Is in a Network of Genome Annotation Databases

EcoGene has a shared maintenance agreement for Genbank U00096 with Guy Plunkett and Fred Blattner who completed the *E. coli K-12* MG16555(Seq) genome sequence used to seed EcoGene (6, 17). A subset of regularly updated EcoGene database tables are transmitted to NCBI monthly and processed into annotation updates (Klimke, W., Tatusova, T., Fedorov, B. and K.E.R., unpublished). EcoGene curators process information in newly released *E. coli* journal articles as source material for most of the EcoGene/Genbank annotation updates.

Colleagues at NCBI (http://www.ncbi.nlm.nih.gov), UniProtKB/Swiss-Prot (http://www.uniprot.org), ASAP (http://asap.ahabs.wisc.edu/asap/home.php), EcoCyc (http://ecocyc.org), RegulonDB (http://regulondb.ccg.unam.mx), the Coli Genetic Stock Center (http://cgsc2.biology.yale.edu/index.php), GenoBase (http://ecoli.aist-nara.ac.jp/), GenExpDB (http://chase.ou.edu/oubcf), PrFEcT (www.prfect.org), EchoBASE (http://ecoli-york.org), EcoliWiki (http://ecoliwiki.net), CyberCell (http://redpoll.pharmacy.ualberta.ca/CCDB), GenomeAtlas (http://www.cbs.dtu.dk/staff/dave/TIGRconf3.html), and EcoliHub (http://ecolihub.org) also curate the biomedical literature and at times provide EcoGene with important contributions for updating the EcoGene/Genbank annotations. RegulonDB is the source for the transcription factor binding sites depicted in EcoGene, with permission. COMBREX (combrex.org) is promoting biochemical functional investigations and fostering bioinformatics and experimentalist collaborations for *E. coli* and other microbes. EcoGene fosters database integration and its GenePages have links to many websites with information about *E. coli* K-12. The vector maps and sequences of pET28a, pET15b, and pET26b Novagen pET vectors can be accessed at the Merck Chemicals website http://www.merck-chemicals.com/life-science-research/pet/c_2tOb.s1OkacAAAE-jWhl9.zLX. The SMS server provided by Dr. Paul Stothard at the University of Alberta, Canada, is used to obtain the reverse complements of DNA sequences (http://www.ualberta.ca/~stothard/javascript/rev_comp.html).

2.2. Re-recombineering Away from Host Mutations

Re-recombineering is a useful alternative to phage P1 for moving mutations from strain to strain without moving linked markers and without bringing live phage into the laboratory. The Coli Genetic Stock Center database is an excellent source for strain genotypes and pedigrees. We *re*-recombineered genes linked to these Keio collection host mutations, listed in the CGSC genotype: CGSC#: 7636 (BW25113), F-, Δ(*araD-araB*)567, Δ*lacZ*4787(::*rrnB*-3), λ-, *rph*-1, Δ(*rhaD-rhaB*)568, *hsdR*514.

2.3. Signal Peptide and Restriction Site Cleavages

Type I signal predictions were performed using the SignalP web server at http://www.cbs.dtu.dk/services/SignalP for all *E. coli* K-12 proteins and were individually inspected to differentiate uncleaved signal anchors from signal cleavage sites (18). The previous EcoGene, UniProtKB/Swiss-Prot, EchoLOCATION (19), PRED-TAT (20), and TatP (21) signal peptide predictions were compared in order to assemble the EcoGene compilation of proven and predicted type I signal protein cleavage sites to use as a reliable PrimerPair resource. A similar methodology was previously used to create EcoGene's curated compilation of lipoprotein type II signal peptide cleavage sites (7). REBASE (http://rebase.neb.com) is our source of information about the restriction enzyme names and DNA sequence site specificities present in EcoGene (7). Predictions are supplanted if experimental cleavages are present in the Verified Set, which has been used for modeling methionine aminopeptidase cleavage site specificities (22). The PrimerPair primers can be further analyzed using Frank Collart's Express Primer tool (http://tools.bio.anl.gov/bioJAVA/jsp/ExpressPrimerTool) (see Note 2).

The Periplasmic Protein Design tool automatically excludes signal peptide codons during primer design guided by manually adjusted SignalP 3.0 predictions (18).

2.4. Reengineering Deletions in the Keio Collection

The adjacent gene deletions in the Keio collection are compared to the current EcoGene annotations using a supplementary table of deletion interval genome coordinates (10) and an in-house application (J.Z. and K.E.R., unpublished results). See GenoBase (http://ecoli.aist-nara.ac.jp) for more information about the Keio and ASKA GBRCs (see Note 1).

2.5. Logos of the GHOLE RBSs Associated with Translational Coupling

WebLogo 3 (http://weblogo.threeplusone.com/create.cgi) is used to create sequence logos using an alignment and conserved pattern detection algorithm based on information theory (see Note 3).

3. Methods

All protocols start with: Open a web browser and go to the EcoGene home page http://www.ecogene.org.

3.1. Designing and Redesigning a Pair of HiuH Expression Clone PCR Primers

This *hiuH* (*yedX*) primer pair was used in our laboratory to amplify a PCR fragment with 5′ NcoI and 3′ XhoI restriction sites for directional cloning into pET28a to construct pYedX-His. We demonstrated that the over-expressed HiuH periplasmic protein was present in both processed and unprocessed forms, with the vast majority of the protein present as an insoluble precursor (R. Mitchell, N. Hus, and K.E.R., Fig. 1, unpublished results).

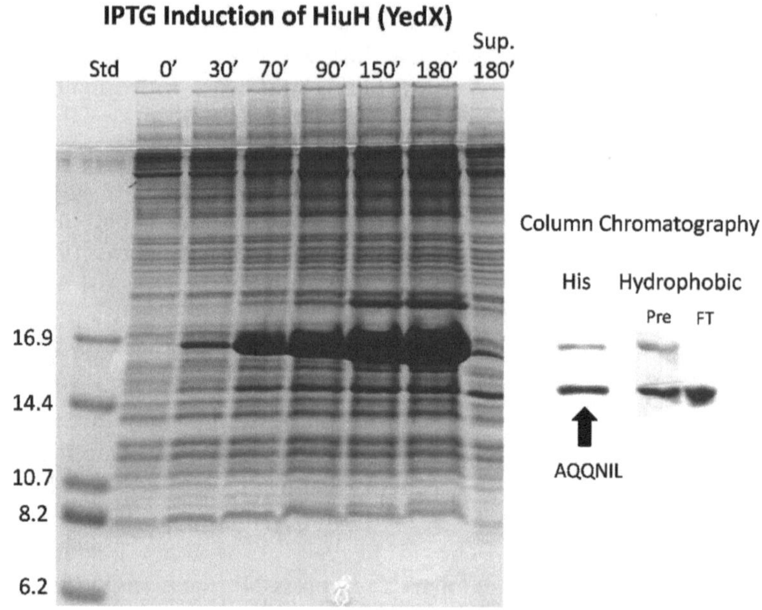

Fig. 1. A Tris-Tricine PAGE gel depicts IPTG-induced increases in HiuH(YedX) expression from pYedX-His with time. Very little HiuH comes down with the pellet. Both cleaved and uncleaved-signal forms of C-tagged HiuH were purified on a nickel affinity column, as designated above the N-terminal protein sequence. The Pre and FT columns show that hydrophobic interaction chromatography removed uncleaved HiuH, but crystallization still failed.

We sequenced the N terminus of the soluble processed form of HiuH to verify that the signal peptide is cleaved after residue 23. The poor recovery of soluble mature HiuH protein that is depicted was unsuitable for HiuH structural determination, so the *hiuH* primers were redesigned to eliminate the 23 *hiuH* N-terminal signal peptide codons. This new primer pair was used to construct a pET15b-derived clone called pHis-YedXΔN, which produces large amounts of soluble homogeneous HiuH protein that was used to solve the crystal structure (Zuo, Y., Ballanco, J., Shah, J., Wang, Y., Rivera, S., Ragan, T.J., Hernandez, G., Nelersa, C.M., Mitchell, R., Rudd, K.E., and Malhotra, A., unpublished, 2006; PDB 2IGL). The successful primer redesign used to make this reengineered clone is recapitulated in the last steps of this procedure.

1. Enter "hiuH" into the Gene Search window and select submit to link to the *hiuH* GenePage. We first design a PCR primer pair to clone the amplicon of full length *hiuH* gene into the pET28a expression vector NcoI (**ccatgg**) 5′ and XhoI (**ctcgag**) 3′ cloning sites creating a C-terminal hexa-histidine affinity label (His-tag).
2. Click "DNA Sequence" to go to the *hiuH* DNA sequence page. Click "Coordinates" to add both local and genomic numbering scales to the DNA sequence.

3. Copy the 20 bases following but not including the *hiuH* ATG start codon into a file. Add gcgcgcgc**ccatgg***gc* to the 5′ end to get the 36 base start(fwd) PCR primer sequence 5′-gcgcgcgc-**ccatgg***gc*TTAAAGCGTTATTTAGTACT-3′. The extra *gc* bases between the NcoI site and the *hiuH* sequence keep the vector NcoI site ATG in frame with the rest of the *hiuH* ORF, replacing the native ATG codon with ATGGGC encoding Met-Gly. The gcgcgcgc end spacer preceding the NcoI site can be almost any sequence, should be at least four bases, and is used to preserve the NcoI cut site in the amplicon.
4. Copy the 20 bases immediately preceding but not including the *hiuH* TAA stop codon to get 5′-ATTCAACCTATCGT GGCAGT-3′. Reverse complement the DNA sequence and add the end spacer and XhoI restriction site sequence gcgcgcgc**ctcgag** to get the 34 base stop(rev) primer sequence 5′- gcgcgcgc**ctcgag**ACTGCCACGATAGGTTGAAT-3′. No extra bases between the cloning site and *hiuH* are needed since the XhoI site is already in frame with the His-tag and stop codons of the vector.
5. Go back to the *hiuH* GenePage to manually inspect for internal cloning sites in *hiuH*. Click the SitesMap button to reveal the restriction maps. Click the SelectSites button to select up to seven restriction enzyme sites to view.
6. Enter "DpnI, NcoI, XhoI" in the Sites entry box of the Restriction Sites Selection pop-up window and Submit. DpnI GATC sites are present in most genes, forcing a SitesMap to be depicted on the GenePage regardless of which other enzymes selected. Click the magnifying glass icon next to the CloseSites button up to three times to get a closer look. It appears there are no XhoI or NcoI sites in *hiuH*, but there is one DpnI site in *hiuH* and another one just past the stop codon.
7. Set values of 100 bp in both Upstream and Downstream DNA Sequence entry windows and select DNA Sequence, then select the SITES button to reveal two GATC sequences, one in the coding sequence, and another located 20 bp past the *hiuH* stop codon. The Sites Positions pop-up window lists the restriction site positions relative to the start codon are given. Return to the *hiuH* GenePage.
8. One can calculate primer Tm values, predict secondary structure, and check for primer dimers, but we routinely obtain the precise PCR amplicons we target without checking. Generally, we only have nonoptimal design alternatives. If the primers fail to produce amplicons under one set of PCR reaction conditions, one can look more closely at the DNA properties of the primers and usually find conditions that will allow amplification. This completes the design of the first pair of *hiuH* cloning primers.

9. The redesigned *hiuH* primers will incorporate 5′ NdeI **catatg** and 3′ BamHI **ggattc** sites for cloning into pET15b to attach a thrombin-cleavable N-terminal His-tag to the mature HiuH protein. Set values of 20 bp in both Upstream and Downstream DNA Sequence entry windows and select DNA Sequence.
10. Starting at the 24th codon triplet GCA, copy 20 bases and add **catatg** to the 5′ end to add the NdeI site, then add the gcgcgcgc end spacer to get the 34 base *hiuH* start(fwd) primer sequence 5′-gcgcgcgc**catatg**GCACAACAAAACATTCTTAG-3′.
11. Copy the last 20 bases of the *hiuH* gene including the native stop codon TAA to be utilized. Get the reverse complement of the DNA sequence and add the BamHI site and end spacer to get the 34 base stop(rev) primer sequence 5′-gcgcgcgc **ggattc**TTAACTGCCACGATAGGTTGAA-3′. This completes the redesign of the *hiuH* cloning primer pair.

3.2. Designing Primers for Re-recombineering the Keio lacY784::kan Cassette

The parent BW25113 strain in which the Keio alleles were constructed contains several pre-existing mutations including Δ*lacZ4787* (10). We constructed a *lacZ*::kan deletion strain KRE10345 and use its genomic DNA as our universal *kan* cassette template to create dozens of new deletions with no background colonies (see Note 1). P1 transduction using the Keio *lacY784*::kan as donor would co-transduce the adjacent Δ*lacZ4787::rrnB-3* mutation highly. *Re*-recombineering primers can cleanly and economically amplify and transfer a Keio allele from a genomic DNA template without having to use the de novo recombineering 70-mers. This *re*-recombineering example utilizes a pair of 20-mers starting 30 bp away from the *lacY* gene borders for the 50 bp of homology needed for an efficient phage lambda recombinase reaction (23).

1. Enter "lacY" into the Gene Search window and go to the *lacY* GenePage.
2. Click "DNA Sequence" to go to the *lacY* DNA sequence page and select "Coordinates" to add both local and genomic numbering scales to the DNA sequence.
3. Set values of 50 bp in both Upstream and Downstream DNA Sequence entry windows and select DNA Sequence.
4. Copy the first 20 bp of DNA to obtain the start (fwd) primer sequence 5′-AATAACCGGGCAGGCCATGT-3′.
5. Copy and reverse complement the last 20 bp to get the stop(rev) primer sequence 5′-ATGATATGTTGGTCGGATAA-3′. This completes the design of the *lacY* *re*-recombineering primers.

These are bioinformatics methods to design primers for laboratory experiments. Our laboratory experiments using these primers are referred to but the laboratory protocols used, e.g., restriction enzyme digestion, DNA ligation, and plasmid transformation, are not the subject of these methods.

3.3. Using PrimerPair for the Batchwise Design of PCR Primers for Gene Cloning

1. Click on the EcoTopics button to go to the EcoTopics Search page.
2. Click the radio button Title Only.
3. Enter the search term “periplasmic” and hit Search.
4. Click the link to go to the “Periplasmic binding proteins for ABC transporters” TopicPage.
5. Click the Genes button on the TopicPage to get to the Gene Search Results page with 52 genes as shown in Fig. 2, a composite EcoGene 2.0 figure that it also depicts sample GenePage maps and features, as well as overlapping genes for later procedures.
6. Click the PrimerPair button to get to the PrimerPair Design Page (Fig. 3). Three types of genes will be filtered out as unsuitable for PrimerPair: pseudogenes, IS element transposase ORFs, and extensively overlapping genes. One pseudogene is eliminated from the binding proteins gene set, leaving 51 gene PCR amplicon primers to design.
7. Retain the default settings: *cloning*, *protein-only* genes, and *20*-mer primer lengths.
8. Since all of the selected genes have proven or predicted signal peptides, a radio button option “Offset to exclude signal peptides” automatically appears in the cloning section that allows for the design of cloning primers to amplify normally exported proteins without their native signal peptides. Choose this option by selecting the radio button. This automatically disables the “start offset” entry window and overrides it with a different offset length corresponding to each proven or predicted signal peptide.
9. Click the “stop offset” radio button labeled inside in the cloning subsection and enter a stop offset value of “3” in the data entry window since a C-terminal hexa-histidine affinity tag and stop codon from the pET28a vector will be utilized.
10. Terminal restriction site and end spacer sequences are added by selecting “Your sequence” in the cloning Add-ons section. In the end spacer entry window for the start (fwd) primers, enter **gcgcgcgcgc** and enter the NcoI site sequence **ccatgg** in the restriction site entry window. Likewise, enter **gcgcgcgcgc** as end spacer and the XhoI site sequence **ctcgag** as the stop (rev) primer add-ons.
11. Do a test run of PrimerPair by selecting Download Data to check for restriction sites contained within the EcoGene sequences of the PCR amplicons. Save the tab-delimited text output file to your computer. Open the file in a text editor or

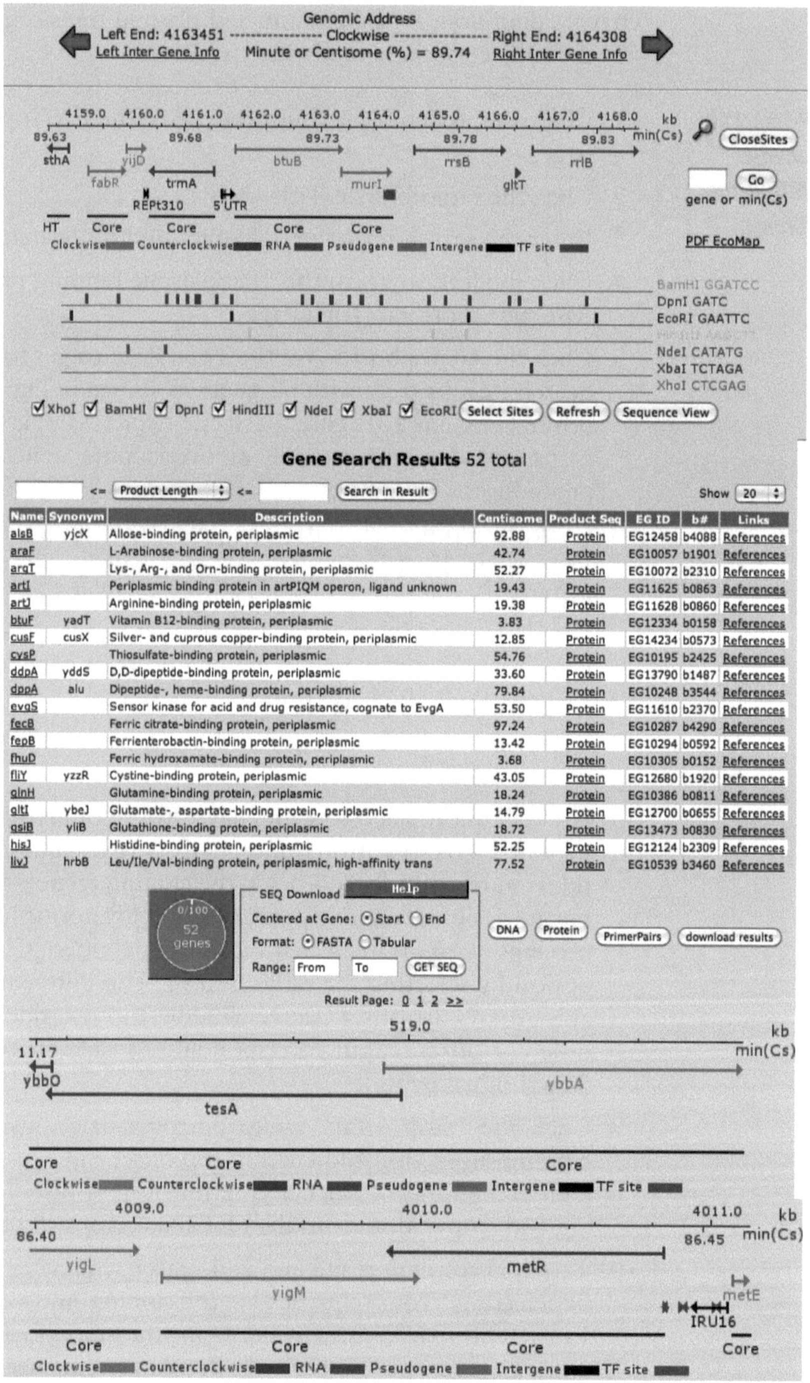

Name	Synonym	Description	Centisome	Product Seq	EG ID	b#	Links
alsB	yjcX	Allose-binding protein, periplasmic	92.88	Protein	EG12458	b4088	References
araF		L-Arabinose-binding protein, periplasmic	42.74	Protein	EG10057	b1901	References
argT		Lys-, Arg-, and Orn-binding protein, periplasmic	52.27	Protein	EG10072	b2310	References
artI		Periplasmic binding protein in artPIQM operon, ligand unknown	19.43	Protein	EG11625	b0863	References
artJ		Arginine-binding protein, periplasmic	19.38	Protein	EG11628	b0860	References
btuF	yadT	Vitamin B12-binding protein, periplasmic	3.83	Protein	EG12334	b0158	References
cusF	cusX	Silver- and cuprous copper-binding protein, periplasmic	12.85	Protein	EG14234	b0573	References
cysP		Thiosulfate-binding protein, periplasmic	54.76	Protein	EG10195	b2425	References
ddpA	yddS	D,D-dipeptide-binding protein, periplasmic	33.60	Protein	EG13790	b1487	References
dppA	alu	Dipeptide-, heme-binding protein, periplasmic	79.84	Protein	EG10248	b3544	References
evgS		Sensor kinase for acid and drug resistance, cognate to EvgA	53.50	Protein	EG11610	b2370	References
fecB		Ferric citrate-binding protein, periplasmic	97.24	Protein	EG10287	b4290	References
fepB		Ferrienterobactin-binding protein, periplasmic	13.42	Protein	EG10294	b0592	References
fhuD		Ferric hydroxamate-binding protein, periplasmic	3.68	Protein	EG10305	b0152	References
fliY	yzzR	Cystine-binding protein, periplasmic	43.05	Protein	EG12680	b1920	References
glnH		Glutamine-binding protein, periplasmic	18.24	Protein	EG10386	b0811	References
gltI	ybeJ	Glutamate-, aspartate-binding protein, periplasmic	14.79	Protein	EG12700	b0655	References
gsiB	yliB	Glutathione-binding protein, periplasmic	18.72	Protein	EG13473	b0830	References
hisJ		Histidine-binding protein, periplasmic	52.25	Protein	EG12124	b2309	References
livJ	hrbB	Leu/Ile/Val-binding protein, periplasmic, high-affinity trans	77.52	Protein	EG10539	b3460	References

Fig. 2. The EcoGene 2.0 interface. The lower portion of the *murI* GenePage shows the *btuB-murI* start–stop overlap and the restriction site maps. Below the maps is the GeneSearch Results page for the periplasmic proteins and the SEQ Download section. At the bottom of the composite figure are the gene maps for the start–start overlap pair *tesA-ybbA* and the stop–stop overlap pair *yigM-metR*.

spreadsheet program. The last two columns list the number of internal restriction sites matching the sequences entered in the restriction site entry window, identifying the genes with internal cloning restriction sites. Among the 51 known and predicted periplasmic binding protein genes, *cysP* and *malE* each have one internal NcoI site and *ugpB* and *evgS* each have one internal XhoI site, as shown in Fig. 3.

12. Extra bases can be added after the restriction sites to preserve the open reading frames. Add **gc** to the start restriction site add-on entry window so the add-on sequence is now **ccatg-ggc**. This will disable the check for internal restriction sites that was done during the previous test run because they are not recognized by PrimerPair as a restriction site. This start add-on will now add Met-Gly to the mature protein sequences, which will be retained in the cytoplasm. The number of bases

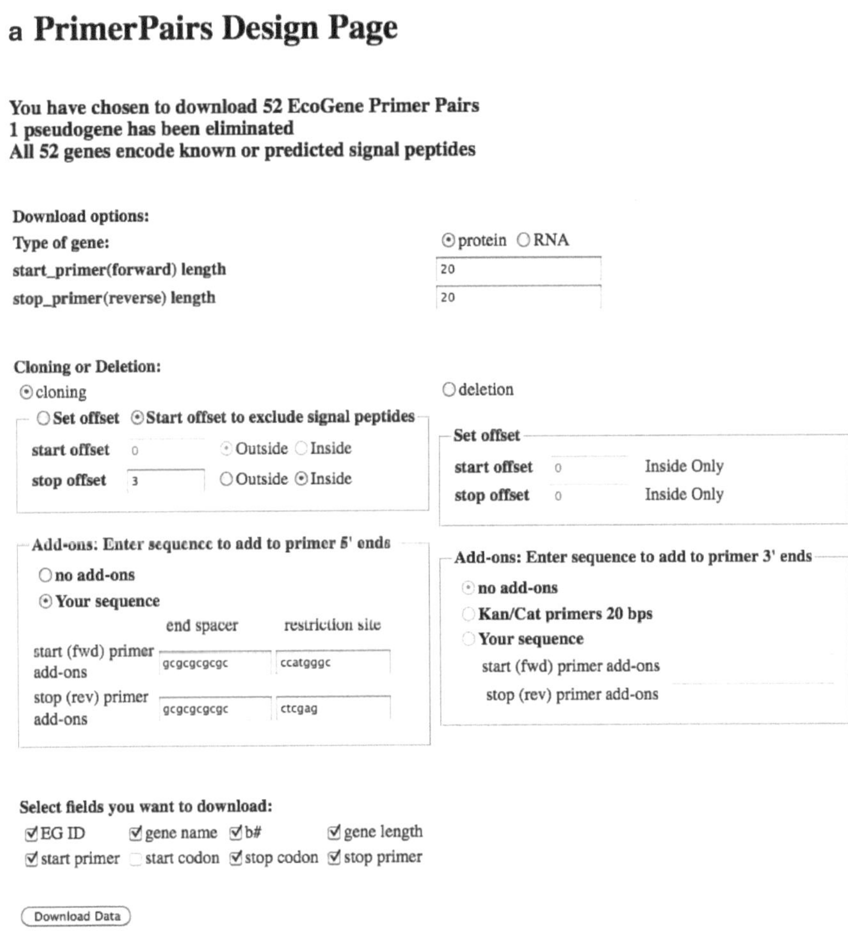

Fig. 3. The PrimerPairs Design Page (**a**) and a clones report (**b**). These settings will create 51 primer pairs for the cloning of periplasmic solute binding proteins into the XhoI and NcoI sites of vector pE28a. The actual 20-mers are depicted. The number of XhoI and NcoI sites predicted to be in the amplicons is denoted in the last two columns.

b

Primers Info

EG_ID	gene	b#	primer_add_on_start_primer(fwd)	mature_length	primer_add_on_stop_primer(rev)	stop_codon(rev)	start restriction sites	stop restriction sites
EG10057	araF	b1901	gcgcgcgcgcccatggGAGAACCTGAAGCTCGGTTT	921	gcgcgcgcgcctcgagCTTACCGCCTAAACCTTTTT	TTA	0	0
EG10072	argT	b2310	gcgcgcgcgcccatggGCGCTACCGGAGACGGTACG	717	gcgcgcgcgcctcgagGTCACCGTAGACATTAAAGT	TCA	0	0
EG10195	cysP	b2425	gcgcgcgcgcccatggACGGAACTGCTGAACAGTTC	942	gcgcgcgcgcctcgagGTTACGCCCCGCCGCTAACA	TCA	1	0
EG10248	dppA	b3544	gcgcgcgcgcccatggAAAACTCTGGTTTATTGCTC	1524	gcgcgcgcgcctcgagTTCGATAGAGACGTTTTCGA	TTA	0	0
EG10287	fecB	b4290	gcgcgcgcgcccatggGCCACGGTTCAGGACGAACA	840	gcgcgcgcgcctcgagTTTCACAACGGTAAGCGGCT	TCA	0	0
EG10294	fepB	b0592	gcgcgcgcgcccatggGCTGACTGGCCGCGTCAGAT	879	gcgcgcgcgcctcgagAAACAGCGCCTTAAGCCTAT	TTA	0	0
EG10305	fhuD	b0152	gcgcgcgcgcccatggGCGGCTATTGATCCCAATCG	801	gcgcgcgcgcctcgagCGCTTTACCTCCGATGGCGT	TCA	0	0
EG10386	glnH	b0811	gcgcgcgcgcccatggGCGGATAAAAAATTAGTTGT	681	gcgcgcgcgcctcgagTTTCGGTTCAGTACCGAACC	TTA	0	0
EG10539	livJ	b3460	gcgcgcgcgcccatggGAAGATATTAAAGTCGCGGT	1035	gcgcgcgcgcctcgagCTTCGCATCGGTCGCCGTGC	TTA	0	0
EG10540	livK	b3458	gcgcgcgcgcccatggGACGATATTAAAGTCGCCGT	1041	gcgcgcgcgcctcgagCTTGGCTGCCGTGGATGAAC	TCA	0	0
EG10554	malE	b4034	gcgcgcgcgcccatggAAAATCGAAGAAGGTAAACT	1113	gcgcgcgcgcctcgagCTTGGTGATACGAGTCTGCG	TTA	1	0
EG10593	mglB	b2150	gcgcgcgcgcccatggGCTGATACTCGCATTGGTGT	930	gcgcgcgcgcctcgagTTTCTTGCTGAATTCAGCCA	TTA	0	0
EG10674	oppA	b1243	gcgcgcgcgcccatggGCTGATGTACCCGCAGGCGT	1554	gcgcgcgcgcctcgagGTGCTTCACAATGTACATAT	TTA	0	0
EG10714	phnD	b4105	gcgcgcgcgcccatggGAAGAGCAGGAAAAGGCGTT	939	gcgcgcgcgcctcgagCTGCACCGCTTTACTCACCG	TTA	0	0
EG10734	pstS	b3728	gcgcgcgcgcccatggGAAGCAAGCCTGACAGGTGC	966	gcgcgcgcgcctcgagGTACAGCGGCTTACCGCTAC	TTA	0	0
EG10752	potD	b1123	gcgcgcgcgcccatggGATGACAACAACACGCTGTA	978	gcgcgcgcgcctcgagACGTCCTGCTTTCAGCTTCT	TTA	0	0
EG10773	proX	b2679	gcgcgcgcgcccatggGCCGATCTGCCGGGCAAAGG	930	gcgcgcgcgcctcgagCTTCTGCGCTGCCAGCGCCT	TTA	0	0
EG10815	rbsB	b3751	gcgcgcgcgcccatggAAAGACACCATCGCGCTGGT	816	gcgcgcgcgcctcgagCTGCTTAACAACCAGTTTCA	CTA	0	0
EG10929	sbp	b3917	gcgcgcgcgcccatggAAGGATATTCAGCTTCTTAA	933	gcgcgcgcgcctcgagGCGTTTGCTGATCTGATCGA	TCA	0	0
EG11047	ugpB	b3453	gcgcgcgcgcccatggGTGACGACCATTCCGTTCTG	1248	gcgcgcgcgcctcgagAGACTTCGTCGATTTCTCAA	TTA	0	1
EG11373	mltF	b2558	gcgcgcgcgcccatggCTCTGGCCATCCATTCCCTG	1494	gcgcgcgcgcctcgagATTTTGTTTCTCTTCACTCC	TTA	0	0
EG11574	thiB	b0068	gcgcgcgcgcccatggAAACCCGTTCTGACTGTTTA	930	gcgcgcgcgcctcgagACGGCTGACGGCGCGTTGCC	TTA	0	0
EG11610	evgS	b2370	gcgcgcgcgcccatggGACGAAGATTACATCGAATA	3531	gcgcgcgcgcctcgagGTCATTTTTCTGACAGAAAA	TTA	0	1
EG11625	artI	b0863	gcgcgcgcgcccatggGCCGAAACCATTCGTTTTGC	675	gcgcgcgcgcctcgagCTTCTGGAACCATTTGTTGT	TTA	0	0
EG11628	artJ	b0860	gcgcgcgcgcccatggGCAGAGAAAATCAATTTTGG	675	gcgcgcgcgcctcgagCTGTGGGAACCACTGGTCAC	TTA	0	0
EG11629	potF	b0854	gcgcgcgcgcccatggGCTGAACAAAAAACACTCCA	1035	gcgcgcgcgcctcgagTTTTCCGCTCTTCACTTTGG	TTA	0	0
EG12012	osmF	b2131	gcgcgcgcgcccatggGCTTCCCCCGTTAAAGTCGG	849	gcgcgcgcgcctcgagCTTCGTCCACCCTTTTTGTT	TTA	0	0
EG12037	yejA	b2177	gcgcgcgcgcccatggCAGGCTATCAAGGAAAGCTA	1758	gcgcgcgcgcctcgagCTCTCCCTGTTTGCTGGCGG	CTA	0	0
EG12075	nikA	b3476	gcgcgcgcgcccatggGCTGCACCAGATGAAATCAC	1509	gcgcgcgcgcctcgagAGGTTTCACCGGTTTAATCT	TTA	0	0
EG12124	hisJ	b2309	gcgcgcgcgcccatggGCGATTCCGCAAAACATCCG	717	gcgcgcgcgcctcgagGCCACCATAAACATCAAAAT	TTA	0	0
EG12334	btuF	b0158	gcgcgcgcgcccatggGCGCCGCGCGTCATCACGCT	735	gcgcgcgcgcctcgagATCTACCTGTGAAAGCGCAT	CTA	0	0
EG12427	modA	b0763	gcgcgcgcgcccatggGATGAAGGGAAAATCACGGT	702	gcgcgcgcgcctcgagCTTGATTGTAAATCCGTAAC	TTA	0	0
EG12458	alsB	b4088	gcgcgcgcgcccatggGCCGCCGAATATGCTGTCGT	867	gcgcgcgcgcctcgagTTGAGTGACCAGGATTGAAT	TTA	0	0
EG12517	ytfQ	b4227	gcgcgcgcgcccatggGCTCCATTAACCGTTGGATT	894	gcgcgcgcgcctcgagATACCCCATATTTTTCTTCT	TCA	0	0
EG12616	torT	b0994	gcgcgcgcgcccatggGCTGATAACCTGTTGCGCTG	975	gcgcgcgcgcctcgagTTTCTTAGCCGCTGATGTGT	TTA	0	0
EG12618	lptA	b3200	gcgcgcgcgcccatggGTAACCGGAGACACTGATCA	477	gcgcgcgcgcctcgagATTACCCTTCTTCTGTGCCG	TTA	0	0
EG12680	**fliY**	b1920	gcgcgcgcgcccatggGATGAAGGTCTGCTTAATAA	714	gcgcgcgcgcctcgagTTTGGTCACATCAGCACCAA	TTA	0	0
EG12700	gltI	b0655	gcgcgcgcgcccatggGATGACGCCGCCCCGGCAGC	843	gcgcgcgcgcctcgagGTTCAGTGCCTTGTCATTCG	TTA	0	0
EG12798	mlaC	b3192	gcgcgcgcgcccatggGCAGACCAGACCAATCCGTA	573	gcgcgcgcgcctcgagTTTTTTCTCTTCCAGAGTGA	TTA	0	0
EG13021	ygiS	b3020	gcgcgcgcgcccatggGCTGACGTTCCCGCCAACAC	1548	gcgcgcgcgcctcgagATGTGCCTTGATATACAACT	TCA	0	0
EG13300	tauA	b0365	gcgcgcgcgcccatggGTGAACGTCACCGTGGCGTA	897	gcgcgcgcgcctcgagTTGCACGAAGCGCGAGGTAA	TTA	0	0
EG13376	mppA	b1329	gcgcgcgcgcccatggGCAGAAGTTCCGAGCGGCAC	1548	gcgcgcgcgcctcgagATGCTTCACAATATACATAG	TCA	0	0
EG13467	yphF	b2548	gcgcgcgcgcccatggGCGGAAAAAGAAATGACCAT	906	gcgcgcgcgcctcgagGGGCAGACCATCAACGTGCG	TTA	0	0
EG13473	gsiB	b0830	gcgcgcgcgcccatggGCCAAAGATGTGGTGGTGGC	1461	gcgcgcgcgcctcgagTTGCAAATCCGCGTCTTCAA	TTA	0	0
EG13707	ssuA	b0936	gcgcgcgcgcccatggGCAGAATCCTCGCCTGAAGC	897	gcgcgcgcgcctcgagTAATTGTTTTCCTTCCAGTT	TCA	0	0
EG13762	ydcS	b1440	gcgcgcgcgcccatggGCCGAACCGCCTACCAATTT	1080	gcgcgcgcgcctcgagGCGACCGCCCATAATGGCAA	TTA	0	0
EG13790	ddpA	b1487	gcgcgcgcgcccatggGCCGTACCAAAAGATATGCT	1476	gcgcgcgcgcctcgagTTTACTCATGGTATTGATAT	TTA	0	0
EG13911	ycjN	b1310	gcgcgcgcgcccatggTGTAAAGAAGAAAATAAAAC	1233	gcgcgcgcgcctcgagGTGCTGTTCGATCAGTTCAT	TTA	0	0
EG14234	cusF	b0573	gcgcgcgcgcccatggAACGAACATCATCATGAAAC	267	gcgcgcgcgcctcgagCTGGCTGACTTTAATATCCT	TTA	0	0
EG20252	xylF	b3566	gcgcgcgcgcccatggAAAGAAGTCAAAATAGGTAT	924	gcgcgcgcgcctcgagCAGCTCGCTCTCTTTGTGGA	TTA	0	0
EG20254	sapA	b1294	gcgcgcgcgcccatggGCGCCTGAATCTCCCCCGCA	1581	gcgcgcgcgcctcgagTGGTTTTTTCACCTCATCCT	TCA	0	0

Fig. 3. (continued)

to add to the primers to preserve the reading frame must be determined for each cloning vector restriction site used. The Leu-Glu codons of the XhoI cloning site in pET28a are already in frame with the six vector His-tag codons followed by a TGA stop codon, adding LGHHHHHH to the end of the target proteins, so no extra bases need to be added between the stop restriction site sequence add-on and the gene sequence. Click "Download Data" to produce the final set of Primer Pairs and save a tab-delimited text file to your local computer. Remember to remove and redesign the four genes with internal cut sites (see Note 4). Before ordering the primers, select a subset of the primer sequences and check the sequences to make sure they match the expected sequences.

3.4. Using PrimerPair to Redesign Deletions and Minimize Adjacent Gene Damage

1. Go to the EcoSearch page.
2. Select Protein in the Product Type menu.
3. Use the Product Size windows to enter 1 as the Minimum and 10,000 as the Maximum Product Length values and select Gene Query to retrieve all the protein-coding genes.

4. The Gene Search Results page lists 4,274 genes. Click the PrimerPair button to go to the PrimerPair Design Page (Fig. 4). One hundred seventy-three pseudogenes and 17 IS element transposase genes are filtered out at this stage, as noted at the top of the Design Page.
5. In the Download options section, keep the protein default selection for Type of Gene and change both primer lengths to 50.
6. In the Cloning or Deletion section, select the deletion radio button.
7. In the Add-ons section, select Kan/Cat primers 20 bps for amplifying from a chromosomal cassette.
8. Enter values for a start inside offset of 3 and a stop inside offset of 21.

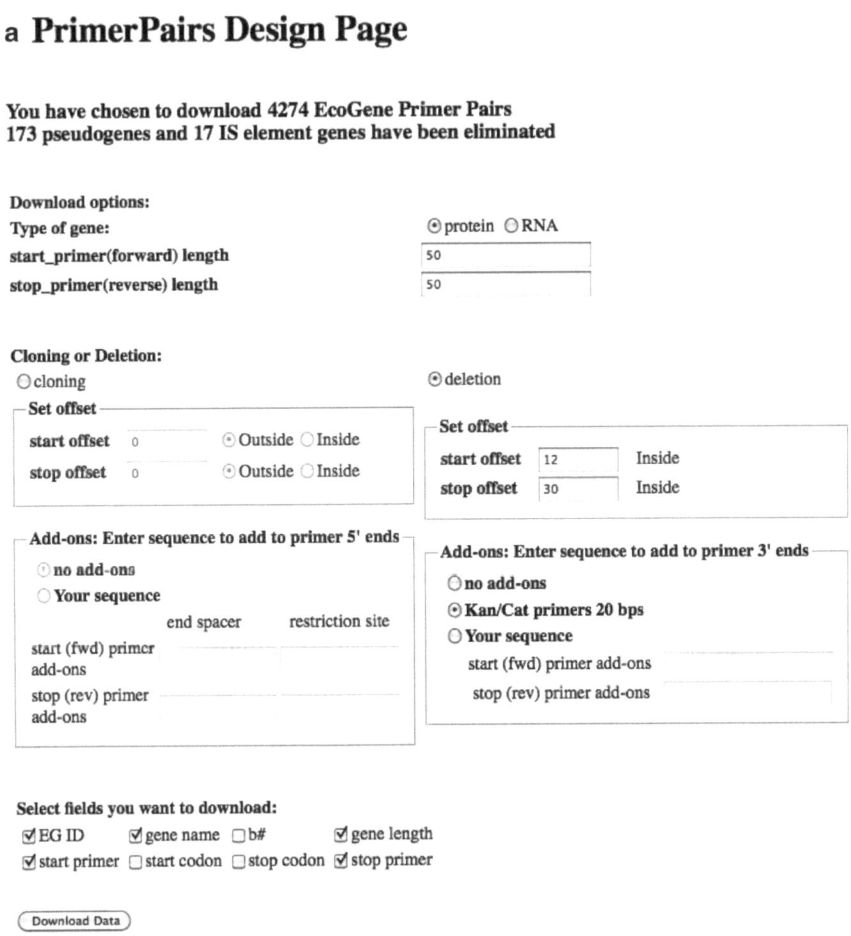

Fig. 4. The PrimerPairs Design Page (**a**) and a deletion report (**b**). These settings will create >4,000 deletion primer pairs that will leave four N-terminal codons and ten C-terminal codons intact as a proposed optimal setting. The first 50 bases target the PCR amplicons to the chromosome and the last 20 bases prime the kanamycin cassette to make the PCR amplicons. The last two columns contain the replacement primers. The actual 70-mers are not depicted.

b

Primers Info					Double Deletions			All non-overlapping primers	
EG_ID	gene	primer_add_on_start_primer(fwd)	deletion/gene_length	primer_add_on_stop_primer(rev)	Gene Affected	5' or 3'	Overlap	Start primer(fwd)	Stop primer (rev)
EG10126	btuB	70mer	1803/1845	70mer	murI	5'	26	70mer	70mer
EG11542	tesA	70mer	585/627	70mer	ybbA	5'	21	70mer	70mer
EG11657	ybbA	70mer	645/687	70mer	tesA	5'	21	70mer	70mer
EG12778	yraM	70mer	1995/2037	70mer	yraN	5'	13	70mer	70mer
EG13271	panE	70mer	870/912	70mer	yajL	5'	8	70mer	70mer
EG10862	rnpA	70mer	318/360	70mer	yidD	5'	7	70mer	70mer
EG11910	**pflD**	70mer	2256/2298	70mer	**pflC**	5'	5	70mer	70mer
EG11414	holD	70mer	372/414	70mer	rimI	5'	2	70mer	70mer
EG12806	lptC	70mer	534/576	70mer	lptA	5'	2	70mer	70mer
EG13562	yagW	70mer	1602/1644	70mer	yagV	5'	2	70mer	70mer
EG13646	dpiB	70mer	1617/1659	70mer	dpiA	5'	2	70mer	70mer
EG14021	yebS	70mer	1242/1284	70mer	yebT	5'	2	70mer	70mer
EG10591	metR	70mer	912/954	70mer	yigM	3'	83	70mer	70mer
EG11471	yigM	70mer	858/900	70mer	metR	3'	83	70mer	70mer
EG11204	murI	70mer	816/858	70mer	btuB	3'	44	70mer	70mer
EG14247	ymfI	70mer	300/342	70mer	ymfJ	3'	33	70mer	70mer
EG14248	ymfJ	70mer	267/309	70mer	ymfI	3'	33	70mer	70mer
EG12779	yraN	70mer	354/396	70mer	yraM	3'	31	70mer	70mer
EG13272	yajL	70mer	549/591	70mer	panE	3'	26	70mer	70mer
EG11348	yidD	70mer	216/258	70mer	rnpA	3'	25	70mer	70mer
EG11911	**pflC**	70mer	837/879	70mer	**pflD**	3'	23	70mer	70mer
EG10850	rimI	70mer	405/447	70mer	holD	3'	20	70mer	70mer
EG12618	lptA	70mer	516/558	70mer	lptC	3'	20	70mer	70mer
EG13544	dpiA	70mer	639/681	70mer	dpiB	3'	20	70mer	70mer
EG13561	yagV	70mer	669/711	70mer	yagW	3'	20	70mer	70mer
EG14022	yebT	70mer	2592/2634	70mer	yebS	3'	20	70mer	70mer
EG13848	clcB	70mer	1215/1257	70mer	ynfK	3'	18	70mer	70mer
EG13849	ynfK	70mer	654/696	70mer	clcB	3'	18	70mer	70mer
EG11605	smg	70mer	432/474	70mer	smf	3'	17	70mer	70mer
EG11757	yjeE	70mer	420/462	70mer	yjeF	3'	17	70mer	70mer
EG12599	yjjW	70mer	822/864	70mer	yjjI	3'	17	70mer	70mer
EG13532	ybdM	70mer	588/630	70mer	ybdN	3'	16	70mer	70mer
EG14005	ynjC	70mer	1494/1536	70mer	ynjB	3'	16	70mer	70mer
EG20257	nrdE	70mer	2103/2145	70mer	nrdI	3'	16	70mer	70mer
EG11721	yidZ	70mer	918/960	70mer	mdtL	3'	14	70mer	70mer
EG11751	otsA	70mer	1383/1425	70mer	otsB	3'	14	70mer	70mer
EG13498	yeaL	70mer	405/447	70mer	yeaM	3'	14	70mer	70mer
EG13499	yeaM	70mer	780/822	70mer	yeaL	3'	14	70mer	70mer
EG13539	citG	70mer	837/879	70mer	citX	3'	14	70mer	70mer
EG13572	wcaD	70mer	1176/1218	70mer	wcaC	3'	14	70mer	70mer
EG13963	sufD	70mer	1230/1272	70mer	sufC	3'	14	70mer	70mer
EG14190	eutQ	70mer	660/702	70mer	eutP	3'	14	70mer	70mer
EG11573	thiP	70mer	1569/1611	70mer	thiB	3'	13	70mer	70mer
EG12178	dsbD	70mer	1656/1698	70mer	cutA	3'	13	70mer	70mer
EG10704	pgpA	70mer	477/519	70mer	thiL	3'	11	70mer	70mer
EG12276	xylH	70mer	1140/1182	70mer	xylG	3'	11	70mer	70mer
EG13125	ygcR	70mer	738/780	70mer	ygcS	3'	11	70mer	70mer
EG13993	cho	70mer	846/888	70mer	ves	3'	11	70mer	70mer
EG13994	ves	70mer	534/576	70mer	cho	3'	11	70mer	70mer
EG11958	alsC	70mer	939/981	70mer	alsA	3'	10	70mer	70mer
EG13673	ybhQ	70mer	369/411	70mer	ybhR	3'	9	70mer	70mer

Fig. 4. (continued)

9. Do test run of PrimerPair by selecting Download Data to check for adjacent-gene deletions. Deselect the "b#," start and stop codon fields as unnecessary. This may take a few minutes to process. Name and save the tab-delimited text output file to your computer, then open it in a spreadsheet program. The output file will resemble that in Fig. 4 including the two overlap check columns indicating if the adjacent gene would be deleted at its 5′ or 3′ end and how many bp would be deleted. The output file also notes 30 overlapping genes that were filtered out at this step because they are in a built-in PrimerPair exception list.
10. The test output indicates that 500/4,065 of these primers would delete one or more basepairs from an adjacent gene when the start codon, the last six sense codons and the stop codon are deleted from each target gene, start and stop inside offset of 3 and 21, respectively (see Fig. 4). The last two columns of the test run output file contain 500 automatic replacement primer pairs in addition to the good primers. This redesigned primer pair set can guide the reengineering

of a completed, corrected mutant collection. First one can systematically evaluate the 3/21 offset strategy, establishing optimization parameters to guide genome reengineering strategies. One can systematically increase the deletion offsets to identify initial offset values that minimize the formation of double deletions. In this way, one can minimize the number of primer pairs that need to be automatically adjusted for better standardization.

11. Plot distribution histograms of the 32 adjacent-gene 5′ deletion lengths and the 468 adjacent-gene 3′ deletion lengths separately as shown in Fig. 5 and inset. Use the spreadsheet data from the sorted Overlap column, as in Fig. 4. The lengths of the 32 5′ deletions vary from 1 to 35 bp and 250/468 of the 3′ deletions remove only the last bp of the stop codons.

12. Perform additional PrimerPair test runs in order to independently vary the start and stop offsets using start/stop inside offset values of 0/0, 3/0, 0/21, 6/21, 9/21, 12/21, 3/30, 6/30, 9/30, and 12/30. Open the text output files in a spreadsheet program. Sort the output files with a primary sort on the column labeled "5′ or 3′" and a secondary sort on the Overlap column. Count the number of adjacent genes with 5′ or 3′ deletions for each start and stop setting. These data are summarized in Table 1.

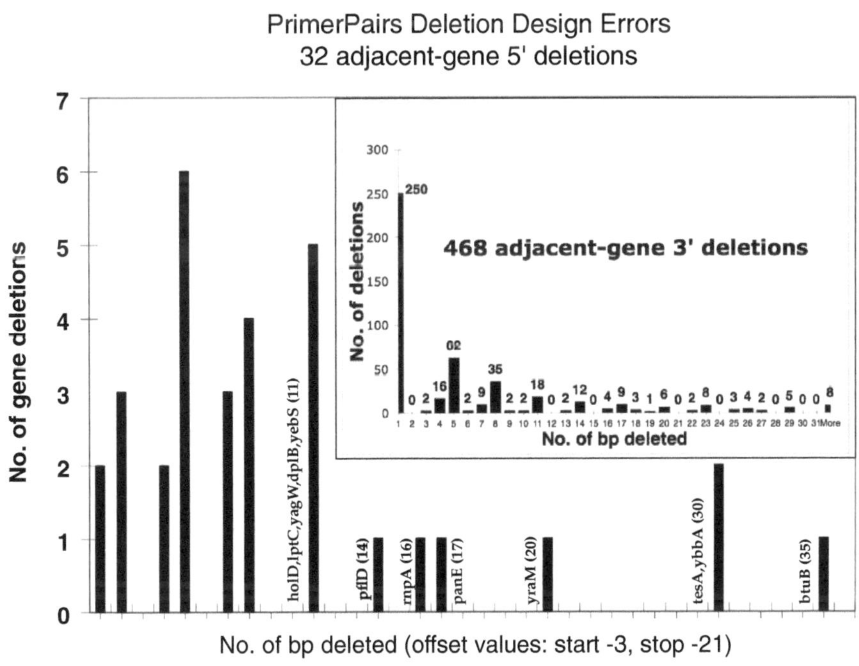

Fig. 5. Size distributions of adjacent 5′ and 3′ deletions. At the common inside offset settings of −3 for starts and −21 for stops, 500 unwanted neighboring deletions are made. The peaks in the 3′ deletions depicted in the inset are periodic with a descending cycle of three. The overlapping regions are referred to as OLEs in the text.

Table 1
PrimerPair test runs to independently vary the start and stop offsets

Start/stop offsets	3′ deletions	5′ deletions	Totals
0/0	741	597	1,338
3/0	585	597	1,182
12/0	223	596	819
0/21	622	37	659
0/30	609	15	624
3/21	468	32	500
6/21	218	32	250
9/21	142	32	174
12/21	103	32	135
3/30	450	12	462
6/30	201	12	213
9/30	125	12	137
12/30	88	12	100

13. The optimization in Table 1 indicates that the primary benefit of the inside offsets is to minimize the collateral deletion of partially overlapping genes. It is also important to avoid the deletion of ribosome binding sites (RBSs) preceding the start codons of the downstream partners of tightly coupled genes in operons. For this reason, a stop offset of 21 bp is routinely used. A 21-bp offset also eliminates 94% of the adjacent gene 5′ deletions (560/597). Start codons are very sensitive to deletion as they are generally null mutations that should be avoided (see Note 1).

3.5. Detecting the GHOLE Motifs: Revealing Noise Hidden by Too Much Signal

The shape of the histogram of the 468 adjacent gene 3′ deletions in Fig. 5 deserves further analysis. The 3-bp cycle oscillating decay might be explained in part by the avoidance of in-frame deletions lengths ($3n$), although the $3n+1$ positions are even lower. The initial peak of 250 1 bp deletions created using the 3/21 offsets are explained as the abundant ATGA translational coupling motifs. When the offsets are set to 0, the PrimerPair error report contains a list of all the overlapping intervals and their lengths. We refer to these as OLEs (overlapping little ends) and number them by their length. Since Fig. 5 has an offset of 3, OLE4 (ATGA) has length of 1 and OLEs 1, 2, and 3 are missing.

OLE1 is the name given by us for the TGATG 1 bp overlap translational coupling motif for this analysis. This procedure is how these motifs were originally identified and visualized. A detailed analysis of OLE and GHOLE motifs will be presented elsewhere.

1. The sorted output used to generate the data in Table 1 is mined for the unique identifiers (EG ids) to retrieve gene sets. Cut-and-paste and collect all the EG ids in separate text files for OLEs 1, 4, 8, 11,14, and 17 from a table like the one in Fig. 3 for both 3′ and 5′ OLEs. These are all the peaks in Fig. 3, but the real OLEs are three bases longer in your 0 offset PrimerPair test run spreadsheet values.
2. Use EcoSearch to separately upload each of the text files of EG ids and select Gene Query to get a Gene Search Results page similar to the one depicted in Fig. 2 for the periplasmic proteins.
3. Under the gene descriptions there is the SEQ Download box that allows one to download FASTA library files centered around either the start or stop codon positions, similar to PrimerPair. This download box was designed to examine at gene regulatory regions but was also used for primer design prior to PrimerPair.
4. Twelve gene sets with 5′ and 3′ overlap genes for all six over-represented OLEs have already been collected. The OLEs are in the start codon regions of the target genes if it is noted that they delete the 3′ ends of adjacent genes. Likewise, the OLEs are in the stop codon regions of the target genes if the deletion affects 5′ end adjacent genes. PrimerPair does not report the adjacent gene identifiers, just the target gene whose deletion causes the adjacent deletion. Name the files OLE1-5 and OLE1-3, etc, and retrieve the gene records in EcoSearch with a text file upload. The non-OLEs are the starts (stops) on the other side of each OLE stop (start) region. They are also systematically collected so files do not get mixed up and to use as controls if desired. There are duplicates across your lists from genes that are coupled at both ends and these should be taken out, but we use them as is since they are so few.
5. In the SEQ Download box for each gene set, leave the default FASTA format set as is and choose Start or End according to the 5′-stop, 3′-start rule. Set the range from –20 to 20. Adjust the intervals so they will all line up on the ATG at position 21, as in Fig. 6 by setting the range at –24 and 16 for the 5′ stop codon linked OLE1-5s.
6. Go to WebLogo 3 (http://weblogo.threeplusone.com/create.cgi) and upload your FASTA library files (see Note 3). After some trial and error you should be able to assemble the gallery of sequence logos depicted in Fig. 6.

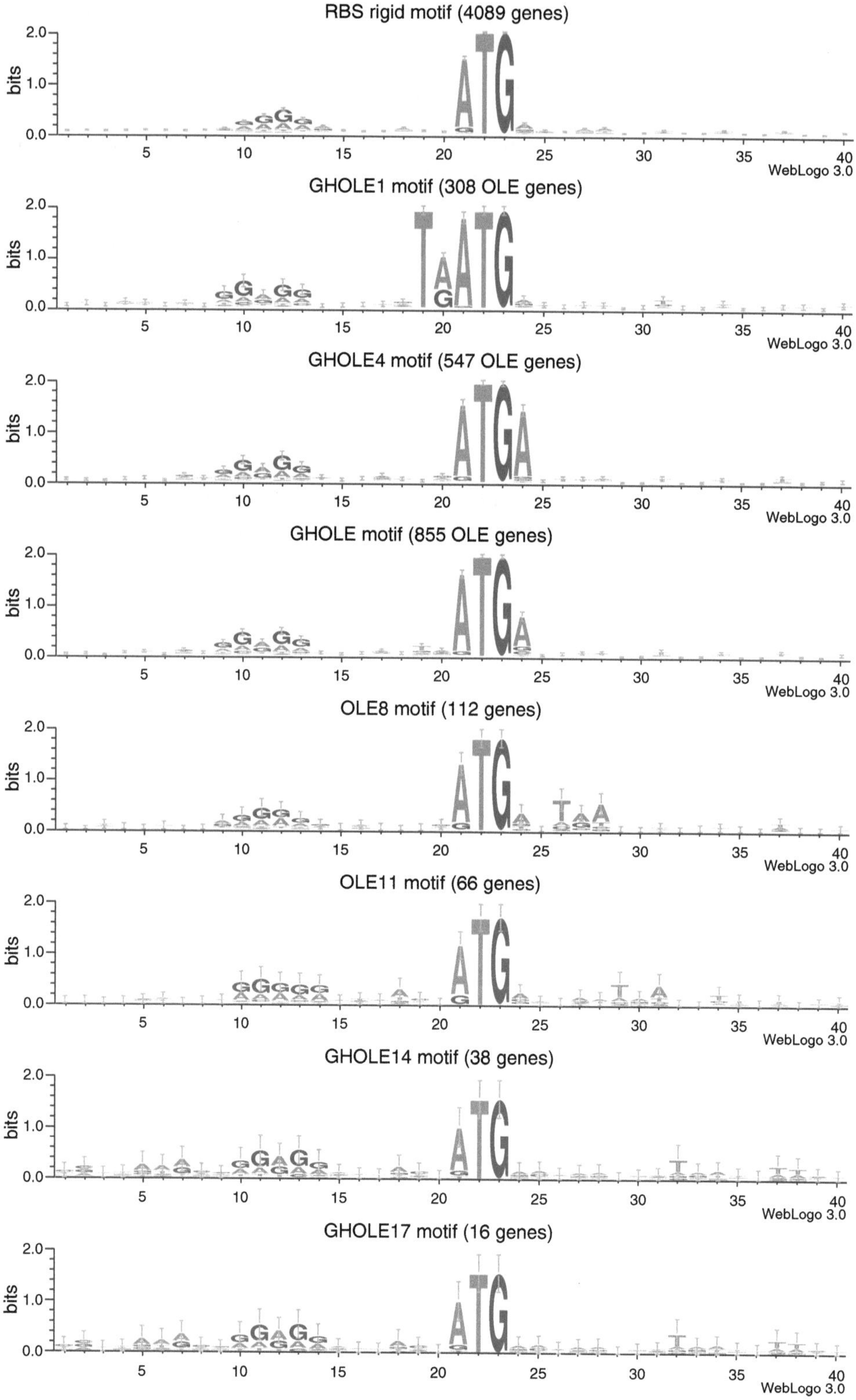

Fig. 6. Sequence logos for GHOLEs 1, 4, 14, and 17. WebLogo 3.0 was used to graphically represent the information content measured in bits. Two bits of information are contained in invariant residues like the TG of the start codons. OLE8 and OLE11 are not GHOLEs, but OLE11 is much flatter than the control rigid model. No gaps are allowed in the fixed alignments used to generate rigid RBS models.

7. Add a control logo for all *E. coli* gene RBS regions using fixed alignments to create rigid RBS models (24). Get all *E. coli* protein-coding genes as in steps 1–3 of Subheading 3.4. SEQ Download does not have a pseudogene prefilter like PrimerPair does, but there is a simple two-step procedure to get rid of pseudogenes.
8. On the Gene Search Results with all genes, select download results to get to the table download page.
9. You can select protein-only here if you forgot to earlier, but retain the exclude pseudogenes option. Deselect gene name from the field selection box and download all the intact protein gene ids. Now you can upload that file in EcoSearch and download −20 to 20 ATG regions to get a FASTA library free of pseudogenes to make the control logo at the top of Fig. 6.
10. When using simple blocked no-gapping rigid alignments like those here, a G-rich bump is in front of RBSs due to a variable gap between the Shine–Dalgarno (SD) region and the ATG initiation region (IR) that conceal the SD sequence; flexible alignments reveal the full anti-SD sequence (24). But Fig. 6 suggests there may be two types of slightly shifted RBS motifs utilized, the normal one revealed by our flexible alignments and another one, closer and rigid, revealed in Fig. 6. GGAGG appears out of the lump as a loss of information content specifically for G, so A (really U) becomes the top base when the OLE subsets are used.
11. We name these novel, rigid, closer-to-the-ATG GGAGG motifs GHOLEs for Gaplessly Hovering near Overlapping Little Ends. Strikingly they disappear and then reappear as they move one base even closer to the IR as the OLEs get longer (Fig. 6). This may have to do with by applying tension on a ribosome waiting at an OLE1 or OLE2 to get on the coupled gene, which only occurs if de novo translation is blocked or weak. The GGAGG is a strong RBS, and being both strong and close may jerk the ribosome into proceeding with the coupling. Many of the OLEs appear to have extensive RBSs mixed in with weak RBSs. It is very heterogenous group of RBSs. All the more unexpected that a rigid GGAGG model emerging from a gapless alignment of OLE RBSs was observed.
12. Next would be a refinement step where the OLE datasets are cleaned up to get a better signal (24). Alignment programs and inspection are used to locate the OLEs that are not start–stop and remove them. Examples of stop–stop and start–start overlaps are at the bottom of Fig. 2. It will also be very interesting to see what emerges from using gapped alignments and flexible RBS modeling to continue to see what emerges from OLE and other gene subsets to help deconvolute multiple modes of translation initiation. We note that the GGAGG

motif emerges due to a dramatic loss of the information content of single position specifically for G. The A that replaces it does not rise in information content, so it is a case of a single noisy base position being lost in the presence of too much signal strength. However, further work must be done to determine the biological significance, if any, of GHOLEs.

4. Notes

1. The EcoGene laboratory has done extensive work validating and reengineering the Keio collection (Dague, D., Kaya, Y, Jones, K.L., and Rudd, K.E., unpublished results). The collection has been transferred into MG1655(Seq) *rph*+ by P1 transduction. The *rph*-1 frameshift mutation (25) in MG1655(Seq) was cleanly repaired and provided by Don Court. MG1655(Seq) (CGSC# 7740) is the MG1655 strain that was sequenced right after it picked up the IS1H insertion at *flhDC* causing hypermotility. MG1655 and the Keio parent strain BW25113 are unstable, poorly motile strains that revert to hypermotility easily, however, MG1655(Seq) is stabilized (26). We sampled 94 Keio deletion strains, two from each tray, and found that 40% of the collection has either IS1- or IS5-induced hypermotility. Another quarter of the Keio collection has unmapped hypermotility mutations, some of which we have mapped near *ompR*. The Keio collection is not isogenic. We separated over 100 slow-growing deletions from their unlinked unknown suppressors, e.g., we showed a Keio *rluD*::kan strain has a *prfB* suppressor (27). We constructed over 100 missing deletions de novo and *re*-recombineered the ones close to all the BW25113 host mutations. Our P1 restricted outcrosses followed by PCR test identified a similar set of essential gene deletions complemented by tandem duplications in the collection as previously reported (11). More than 500 double mutations delete small parts of the adjacent genes' 3′ ends; so far we have demonstrated that *bioCDF*, *hisCI*, *purK*, and *fliGH* are still functional. However, there are more than 30 adjacent gene 5′ deletions that need to be reengineered.
2. The Periplasmic Protein Design tool automatically excludes signal peptide codons during primer design guided by manually adjusted SignalP 3.0 predictions. Similar to PrimerPairs, the Express Primer tool can accept user-specified sites and end spacers. In addition, the Express Primer tool can add its own restriction enzyme recognition sites, i.e., AvrII, SpeI, or XbaI, and can recognize sites already appended to the input primers. Dr. Frank Collart is sharing information with EcoGene as

part of a COMBREX-funded project (see Subheading 2.1) and has already cloned all the *E. coli* periplasmic solute-binding proteins using the Periplasmic Protein Design tool that is why this example is presented (F. Collart, personal communication).

3. WebLogo 3.0 is an easy way to make sequence logos. It can accept either the FASTA or tabular DNA sequence formats that the SEQ Download box on the Gene Search Results provides, and it allows the user to designate a subinterval for the logo. Make certain all the lines in the alignments have the same number of characters. After some trial and error at the default settings, change to a PDF setting and select download. The rigid sequence logos for all 1,089 intact RBSs depicted in Fig. 6 closely resemble the ones that Tom Schneider and K.E.R. published in 1992, updated in our analysis of gapped alignments (24), and in collaboration with Gisela Storz, small ORF RBSs (9).
4. For genes that have internal NcoI or XhoI sites, compatible 4 bp sticky ends may be generated using a primer containing an adjacent AarI Type IIS restriction enzyme site (5′-CACCTGC$(N)_{4/8}$-3′) or alternative Type IIS enzyme site if the gene contains an internal AarI site. Alternatively, if XhoI is present in the gene pick an alternative enzyme present in the pET28a multiple cloning site, taking care to maintain the reading frame such that the protein is produced in frame with the C-terminal His-tag. If NcoI is present in the gene, use an alternative enzyme site in the cloning site, taking care to ensure the encoded protein is in frame with the N-terminal His-tag.

Acknowledgments

We thank Guy Plunkett (ASAP), Mary Berlyn (CGSC), Ingrid Keseler (EcoCyc), Bill Klimke (NBCI), Tatiana Tatusova (NBCI), Boris Fedorov (NCBI), and Andrea Auchincloss (UniProtKB/Swiss-Prot) for collaborating on EcoGene/Genbank updates. We thank Don Court for providing us with a scarless *rph+* derivative of MG1655(Seq) that we distribute as KRE10000. We thank Frank Collart, Brian Miller, Arun Malhotra, Yuhong Zuo, past EcoGene laboratory members Yusuf Kaya, Kristi Jones, Rick Mitchell, Nir Hus, and current member Darryl Dague for communicating unpublished results. K.E.R. thanks Tom Schneider for introducing him to bits and thanks Bobby Baum for discussions about errors in publications. We thank Barry Wanner and Mike Gribskov for hosting a mirror EcoGene site at Purdue. We thank Julio Collado-Vides for granting permission to use the RegulonDB TFBS sites in EcoGene. We thank Rich Roberts, Martin Steffen, and Simon Kasif of COMBREX for their support.

We acknowledge Frank Collart's Express Primer and Periplasmic Protein Design tools as the inspiration for PrimerPairs. This work was supported by NIH grants R01-GM58560 and by a COMBREX sub-award from NIH RC2-GM92602.

References

1. Oliver D., Norman J., Sarker S. (1998) Regulation of *Escherichia coli* secA by cellular protein secretion proficiency requires an intact gene X signal sequence and an active translocon. *J Bacteriol* **180**, 5240–5242.
2. Sproul A. A., Lambourne L. T., Jean-Jacques D. J., Kornberg H. L. (2001) Genetic control of manno(fructo)kinase activity in *Escherichia coli. Proc Natl Acad Sci U S A* **98**, 15257–15259.
3. Sarker S., Rudd K. E., Oliver D. (2000) Revised translation start site for *secM* defines an atypical signal peptide that regulates *Escherichia coli secA* expression. *J Bacteriol* **182**, 5592–5595.
4. Miller B.G., Raines R.T. (2004) Identifying latent enzyme activities: substrate ambiguity within modern bacteria sugar kinases. *Biochemistry* **43**, 6387–6392.
5. Martin R. G., Gillette W. K., Rosner J. L. (2000) The *ykgA* gene of *Escherichia coli. Mol Microbiol* **37**, 978–979.
6. Rudd K. E. (2000) EcoGene: a genome sequence database for *Escherichia coli* K-12. *Nucleic Acids Res* **28**, 60–64.
7. Gonnet P., Rudd K. E., Lisacek F. (2004) Fine-tuning the prediction of sequences cleaved by signal peptidase II: a curated set of proven and predicted lipoproteins of *Escherichia coli* K-12. *Proteomics* **4**, 1597–1613.
8. Rudd K. E., Humphery-Smith I., Wasinger V. C., Bairoch A. (1998) Low molecular weight proteins: a challenge for post-genomic research. *Electrophoresis* **19**, 536–544.
9. Hemm M. R., Paul B. J., Schneider T. D., Storz G., Rudd K. E. (2008) Small membrane proteins found by comparative genomics and ribosome binding site models. *Mol Microbiol* **70**, 1487–1501.
10. Baba T., Ara T., Hasegawa M., Takai Y., Okumura Y., Baba M., Datsenko K. A., Tomita M., Wanner B. L., Mori H. (2006) Construction of *Escherichia coli* K-12 in-frame, single-gene knockout mutants: the Keio collection. *Mol Syst Biol* **2**, 2006 0008.
11. Yamamoto N., Nakahigashi K., Nakamichi T., Yoshino M., Takai Y., Touda Y., Furubayashi A., Kinjyo S., Dose H., Hasegawa M., Datsenko K. A., Nakayashiki T., Tomita M., Wanner B. L., Mori H. (2009) Update on the Keio collection of *Escherichia coli* single-gene deletion mutants. *Mol Syst Biol* **5**, 335.
12. Babu M., Musso G., Diaz-Mejia J. J., Butland G., Greenblatt J. F., Emili A. (2009) Systems-level approaches for identifying and analyzing genetic interaction networks in *Escherichia coli* and extensions to other prokaryotes. *Mol Biosyst* **5**, 1439–1455.
13. Kitagawa M., Ara T., Arifuzzaman M., Ioka-Nakamichi T., Inamoto E., Toyonaga H., Mori H. (2005) Complete set of ORF clones of *Escherichia coli* ASKA library (a complete set of *E. coli* K-12 ORF archive): unique resources for biological research. *DNA Res* **12**, 291–299.
14. Rajagopala S. V., Yamamoto N., Zweifel A. E., Nakamichi T., Huang H. K., Mendez-Rios J. D., Franca-Koh J., Boorgula M. P., Fujita K., Suzuki K., Hu J. C., Wanner B. L., Mori H., Uetz P. (2010) The *Escherichia coli* K-12 ORFeome: a resource for comparative molecular microbiology. *BMC Genomics* **11**, 470.
15. Desai K. (2010) Recruitment of genes and enzymes conferring resistance to the nonnatural toxin bromoacetate. *Proc Natl Acad Sci U S A* **107**, 17968–17973.
16. Butland G., Babu M., Diaz-Mejia J. J., Bohdana F., Phanse S., Gold B., Yang W., Li J., Gagarinova A. G., Pogoutse O., Mori H., Wanner B. L., Lo H., Wasniewski J., Christopolous C., Ali M., Venn P., Safavi-Naini A., Sourour N., Caron S., Choi J. Y., Laigle L., Nazarians-Armavil A., Deshpande A., Joe S., Datsenko K. A., Yamamoto N., Andrews B. J., Boone C., Ding H., Sheikh B., Moreno-Hagelseib G., Greenblatt J. F., Emili A. (2008) eSGA: *E. coli* synthetic genetic array analysis. *Nat Methods* **5**, 789–795.
17. Blattner F. R., Plunkett G., 3rd, Bloch C. A., Perna N. T., Burland V., Riley M., Collado-Vides J., Glasner J. D., Rode C. K., Mayhew G. F., Gregor J., Davis N. W., Kirkpatrick H. A., Goeden M. A., Rose D. J., Mau B., Shao Y. (1997) The complete genome sequence of *Escherichia coli* K-12. *Science* **277**, 1453–1462.
18. Bendtsen J. D., Nielsen H., von Heijne G., Brunak S. (2004) Improved prediction of signal peptides: SignalP 3.0. *J Mol Biol* **340**, 783–795.

19. Horler R. S., Butcher A., Papangelopoulos N., Ashton P. D., Thomas G. H. (2009) EchoLOCATION: an in silico analysis of the subcellular locations of *Escherichia coli* proteins and comparison with experimentally derived locations. *Bioinformatics* **25**, 163–166.
20. Bagos P. G., Nikolaou E. P., Liakopoulos T. D., Tsirigos K. D. (2010) Combined prediction of Tat and Sec signal peptides with Hidden Markov Models. *Bioinformatics Epub.*
21. Bendtsen J. D., Nielsen H., Widdick D., Palmer T., Brunak S. (2005) Prediction of twin-arginine signal peptides. *BMC Bioinformatics* **6**, 167.
22. Frottin F., Martinez A., Peynot P., Mitra S., Holz R. C., Giglione C., Meinnel T. (2006) The proteomics of N-terminal methionine cleavage. *Mol Cell Proteomics* **5**, 2336–2349.
23. Thomason L., Court D. L., Bubunenko M., Costantino N., Wilson H., Datta S., Oppenheim A. (2007) Recombineering: genetic engineering in bacteria using homologous recombination. *Curr Protoc Mol Biol Chapter 1*, Unit 1 16.
24. Shultzaberger R. K., Bucheimer R. E., Rudd K. E., Schneider T. D. (2001) Anatomy of *Escherichia coli* ribosome binding sites. *J Mol Biol* **313**, 215–228.
25. Jensen K. F. (1993) The *Escherichia coli* K-12 "wild types" W3110 and MG1655 have an rph frameshift mutation that leads to pyrimidine starvation due to low *pyrE* expression levels. *J Bacteriol* **175**, 3401–3407.
26. Barker C. S., Pruss B. M., Matsumura P. (2004) Increased motility of *Escherichia coli* by insertion sequence element integration into the regulatory region of the flhD operon. *J Bacteriol* **186**, 7529–7537.
27. Ejby M., Sorensen M. A., Pedersen S. (2007) Pseudouridylation of helix 69 of 23S rRNA is necessary for an effective translation termination. *Proc Natl Acad Sci U S A* **104**, 19410–19415.

Chapter 2

Targeted Chromosomal Gene Knockout Using PCR Fragments

Kenan C. Murphy

Abstract

The development of recombineering technology has converged to a point that virtually any type of genetic modification can be made in the *Escherichia coli* chromosome. The most straightforward modification is a chromosomal gene knockout, which is done by electroporation of a PCR fragment that contains a selectable drug marker flanked by 50 bp of target DNA. The phage λ Red recombination system expressed in vivo from a plasmid promotes deletion of the gene of interest at high efficiency. The combination of this technology with site-specific recombination systems of Cre and Flp has enabled genetic engineers to construct a variety of marked and precise gene knockouts in a variety of microbial chromosomes. The basic protocols for designing PCR substrates for recombineering, generating recombineering-proficient electrocompetent strains of *E. coli*, and for selection and verification of recombinant clones are described.

Key words: Recombineering, Lambda red, Gene replacement, Strain development, Electroporation, Phage lambda, Beta, Exo, Gam, PCR

1. Introduction

The precise deletion of a gene of interest in the *Escherichia coli* chromosome is a central step to understanding gene function or to remove undesirable byproducts for strain engineering purposes. Classically, this has been done by random mutagenesis, or by integrating nonreplicating plasmids containing an altered target gene, with the hope of being able to generate a resolution event that excises the wild-type copy leaving the modified (deleted) copy of the gene in the chromosome. These processes were often time consuming and/or unsuccessful at generating gene knockouts.

James A. Williams (ed.), *Strain Engineering: Methods and Protocols*, Methods in Molecular Biology, vol. 765,
DOI 10.1007/978-1-61779-197-0_2,

In the last decade, a new approach has evolved that takes advantage of the recombination proficiency of the bacteriophage λ Red recombination system (identified by *re*combination *d*efective phage mutants) and of the *rac* prophage RecET system; the process has been referred to as "recombineering" (from *recombinational* engin*eering*) (1–7). The key to successful use of this system is that the Red system consists of only two genes (*exo* and *beta*) that initiate a recombination event that requires only limited amounts of homology to the target gene (~40–50 bp). The λ Exo protein is a processive 5′–3′ dsDNA exonuclease that binds to dsDNA ends and degrades the 5′ strand at the site of entry, leaving 3′ ssDNA tails (8, 9). The λ Beta protein, which binds to the ssDNA generated by λ Exo, is a member of a class of proteins known as single-stranded DNA annealing proteins (SSAPs) that share a common ring-like quaternary structure, promote annealing of ssDNA in vitro, and stimulate DNA recombination events in vivo (10–15). The Red functions are assisted by the λ *gam* gene, which encodes an inhibitor of the host RecBCD enzyme, a destructive dsDNA exonuclease that would otherwise compete with the Red functions for dsDNA ends (16–18). In recombineering events, it is thought that the action of the Red genes in vivo produces either a long ssDNA intermediate bound by Beta, or a linear dsDNA molecule that has Beta bound to 3′ ssDNA overhangs on either end of the substrate (5, 19). In both models, the replication fork is the likely target for the Red-generated intermediates (19, 20). These interactions might occur via annealing of the ssDNA intermediate to the lagging strand template of a replication fork, or by consecutive interactions of each end of the dsDNA intermediate with two independent replication forks.

The procedure presented here describes a simple straightforward method for generating a gene knockout in *E. coli*. An *E. coli* strain of choice, containing a plasmid that overexpresses the λ *exo*, *beta*, and *gam* genes, is electroporated with a PCR product that contains a drug marker flanked by 50 bases of homology to the target gene (or region) to be deleted. The endpoints of the deletion are dictated by sequences within the PCR primers. Following electroporation, the cells are grown out and plated on antibiotic-selection plates. Gene knockouts can be easily obtained in one day, are verified by PCR analyses, and can be transferred into clean genetic backgrounds by P1 transduction (if so desired).

2. Materials

2.1. Reagents

1. LB medium: 10 g tryptone, 5 g yeast extract, 5 g NaCl, 1 ml 1 M NaOH. Mix components in 1 l of distilled water and sterilize by autoclaving for 30 min; store at room temperature.

For LB plates, add 15 g agar, autoclave as above, cool for 30 min at room temperature, add antibiotics as needed, and pour into 100 mm × 15 mm petri plates using 25–30 ml per plate.

2. Electroporation washing buffer: 10% glycerol. Dilute 100 ml of glycerol in 900 ml distilled deionized water, autoclave 500 ml in two 1-l flasks, and store at 4°C (6 months).
3. Ampicillin. The stock solution is dissolved at 10 mg/ml in 90% ethanol and stored at −20°C (1 year). Use between 25 and 100 μg/ml in LB plates for growing Amp^R gene replacements; use at 100 μg/ml for growing cells containing pKM208 in culture.
4. Chloramphenicol. The stock solution is dissolved at 20 mg/ml in 90% ethanol and stored at −20°C (1 year). Use at a concentration of 15 μg/ml in LB plates for selecting Cam^R gene replacements.
5. Kanamycin monosulfate. The stock solution is dissolved at 20 mg/ml in water and stored at 4°C (1 month). Use at 20 μg/ml in LB plates for selecting Kan^R gene replacements.
6. Tetracycline. The stock solution is dissolved at 10 mg/ml in 90% ethanol and stored at −20°C (1 year). Use at 3–10 μg/ml in LB plates for selecting Tet^R gene replacements.
7. Isopropylthiogalactopyranoside (IPTG) – Added to cell cultures for induction of the *red* and *gam* functions from pKM208. Dissolve 238 mg of IPTG powder into 10 ml deionized H_2O; filter sterilize, and store at −20°C (6 months).
8. Agarose. Use at 0.75–1.5% for analysis of PCR products.
9. Pfu-Ultra II Fusion HS DNA polymerase (Stratagene, 600670-51). Enzyme used for generating PCR recombineering substrates.
10. Taq DNA polymerase. Enzyme used for colony PCR to check structure of recombinant clones.
11. QIAprep Spin Miniprep kit (Qiagen, 27106). Used for the isolation of plasmids from 5 ml of culture.
12. QIAquick PCR purification kit (Qiagen, 28104). Used for the purification of PCR products to be used as substrates for recombineering.
13. pJW168–Amp^R, pSC101-derived, Cre recombinase expressing plasmid (21) (Lucigen, 42200-1).
14. EB (elution buffer): 10 mM Tris–HCl, pH 8.5.
15. PBS: Dissolve 8 g NaCl, 0.2 g KCl, 1.44 g Na_2HPO_4, and 0.24 g KH_2PO_4 in 800 ml of distilled water. Adjust pH to 7.4 with HCl; add water to 1 l and autoclave.
16. dNTPs: 2.5 mM each of dATP, dCTP, dGTP, dTTP.

Table 1
Annealing sequences for drug cassettes

Antibiotic (cassette length)	Gene(s)	Primer pair (5′ to 3′)[a]	Drug concentration (μg/ml)
Kanamycin (944 bp)	Tn*903* (*aph*) *type I*	CACGTTGTGTCTCAAAATCTC TACAACCAATTAACCAATTCTG	20
Kanamycin (949 bp)	Tn5 (*aph*) type II	TATGGACAGCAAGCGAACCG TCAGAAGAACTCGTCAAGAAG	20
Chloramphenicol (822 bp)	Tn*9 cat*	TGAGACGTTGATCGGCACGT ATTCAGGCGTAGCACCAGGC	15
Ampicillin (975 bp)	Tn*3 bla*	CGCGGAACCCCTATTTGTTT GGTCTGACAGTTACCAATGC	50
Tetracycline (1,996 bp)	Tn*10 tetRA*	CTCGACATCTTGGTTACCGT CGCGGAATAACATCATTTGG	7
Gentamicin (616 bp)	Tn*1696 aacC*	CGAATCCATGTGGGAGTTTA TTAGGTGGCGGTACTTGGGT	10

[a]These sequences should be placed on the 3′ ends of the primers used to generate the recombineering substrate

17. Dimethyl sulfoxide (DMSO), molecular biology grade.
18. Sterile distilled water.
19. Primers (as defined in Table 1).
20. Recombineering plasmid: Plasmid pKM208 expresses the λ *red* and *gam* functions under control of the Ptac promoter (6) and can be obtained from *addgene.com*. The plasmid contains a temperature-sensitive origin of replication (cells containing the plasmid should be grown at 30°C). The plasmid also contains the *lacI* repressor under control of its own promoter (to keep *red* and *gam* expression turned off in the uninduced state), and the *bla* gene, which confers resistance to ampicillin (see Note 1).

2.2. Equipment

1. Thermocycler (e.g., Minicycler PTC-200, MJ Research).
2. Two incubators set at 30°C and 37°C for growth of recombineering strains and recombinant colonies, respectively.
3. Two shaking water baths set at 30°C and 42°C for growth of *E. coli* recombineering cultures.

4. Spectrophotometer and cuvettes for measuring optical densities of bacterial cultures.
5. Biorad Gene Pulser Xcell Electroporation system (#165-2660) or BioRad MicroPulser Electroporator (#165-2100).
6. Electroporation cuvettes – sterile, 0.1 cm gap, package of 50 (Bio-Rad, 165-2089).
7. Centrifugation tubes – 40 ml (Nalgene, Oak Ridge Centrifuge Tubes, 3119-0050).
8. Pipets to deliver up to 1 ml (P-1000), 200 μl (P-200), or 20 μl (P-20) of liquid or culture.

3. Methods

3.1. Preparation of Targeting Substrate by PCR

1. The standard targeting substrate for recombineering is a PCR product that contains a drug marker flanked by upstream and downstream regions of the target site. Primers for the PCR are typically 70 bases in length and are designed so that 20 bases on the 3′ ends will anneal to and amplify a drug cassette of one's choice. See Table 1 for sequences and templates used for amplifying a variety of drug cassettes used in recombineering. The 50 bases on the 5′ ends of the primers contain the upstream sequence and the reverse complement of the downstream sequence, respectively of the target site (see Fig. 1).
2. A high fidelity polymerase such as Pfu UltraII fusion polymerase should be used to generate the targeting substrate. Alternative High Fidelity polymerases for this step include Platinum High Fidelity Taq polymerase (Invitrogen 11304-011), or Roche Expand High Fidelity polymerase (Roche, 04-738-250-001).
3. Prepare PCR reaction as follows: 31 μl sterile distilled water, 5 μl 10× PCR buffer (supplied by manufacturer), 5 μl 2.5 mM dNTPs, 2 μl primer A (20 μM), 2 μl primer B (20 μM), 2 μl DMSO, 2 μl template DNA (~10 ng), 1 μl High Fidelity Pfu UltraII fusion polymerase (see Fig. 1 and Note 2).
4. Perform standard PCR. We typically use the following program for 0.8–1-kb amplicons; (step 1) 95°C, 1 min; (step 2) 94°C, 30 s; (step 3) 58°C, 30 s; (step 4) 72°C, 1 min; (step 5) repeat last three steps 29 times; (step 6) 72°C, 5 min; (step 7) hold at 4°C. The extension times (step 4) should be increased for products expected to be longer than 1 kb, though check the elongation properties of the polymerase as reported by the manufacturer.
5. When completed, load 3 μl of the PCR on a 0.75% agarose gel to check for correct size and purity of the recombination

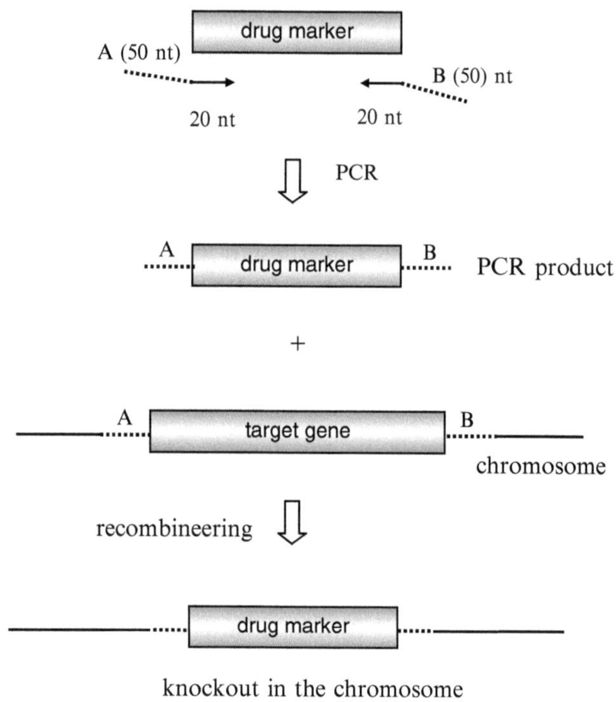

Fig. 1. Generation of recombineering substrate by PCR. The first primer contains sequence from its 5′ ends that is identical to the upstream region of the target gene (*dotted line* marked A). The second primer contains from its 5′ end the reverse complement of the sequence in the downstream region of the target gene (*dotted line* marked B). The last 20 bases of the primers anneal to and amplify the drug marker (see Table 1 for these sequences). The product of this PCR is ~1–2 kb amplicon (depending on the drug marker) which contains 50 base pair ends that are homologous to the target region. After filter cleaning and elution in a low salt buffer (EB) or water, the PCR product is electroporated into recombineering-proficient *E. coli* cells. After a growth period, the recombinant is selected on an antibiotic-selection plate.

substrate. If present as a single species, clean the PCR product with PCR-quick clean kit (Qiagen) or similar type of PCR purification kit. Elute the DNA in 30–50 μl of EB buffer or deionized water (see Note 3). If side products are present, gel-purify the recombineering substrate on a 0.75% agarose gel. If the recombineering substrate is not found, repeat PCR with 2–4°C decrease in annealing temperature and/or remove DMSO from the PCR. If band still not present (and known PCR control is working), redesign and order new primers (see Note 4).

3.2. Preparation of Recombineering-Proficient Electrocompetent E. coli Cells

1. Transform the *E. coli* strain of interest with Red-recombineering plasmid pKM208 (Amp^R). Plate transformation at 30°C on LB plates containing 100 μg/ml ampicillin overnight (see Note 5). Inoculate a fresh colony into 5 ml LB containing 100 μg/ml ampicillin and roll overnight at 30°C.

2. In a 125 ml flask, inoculate 20 ml of LB containing 100 μg/ml ampicillin with 100 μl of the 5 ml overnight culture containing pKM208. Grow cells in a shaking water bath with aeration at 30°C to an OD of 0.2 (~10^7 cells/ml) and add 200 μl of 0.1 M IPTG (final concentration is 1 mM). Continue to grow cells at 30°C.
3. At an OD between 0.4 and 0.6 (~10^8 cells/ml), place culture in a water bath prewarmed to 42°C. Aerate by shaking for an additional 15 min (see Note 6).
4. Place culture in an ice-water bath and swirl moderately for 10 min.
5. Pour culture into prechilled centrifugation tubes (Nalgene, Oak Ridge Centrifuge tubes, 3119-0050) and collect cells by centrifugation at 3,800 × *g* in SS-34 rotor. Alternatively, use sterile 50 ml Falcon tubes in swinging bucket bench top centrifuge at 3,800 × *g*. Handle tubes gently so as not to disturb the cell pellet. Pour off supernatant slowly and resuspend the cells in 2 ml of ice-cold 10% glycerol. Resuspend the cells with P-1000 pipet by gently pipeting cells back and forth (easier done in this smaller volume). Add 18 ml of ice-cold 10% glycerol, mix culture by inverting tube four to five times, and recentrifuge.
6. Resuspend the cells in 1 ml ice-cold 10% glycerol and transfer to a prechilled 1.5 ml Eppendorf tube. Spin cells in refrigerated microcentrifuge at 10k for 1 min at 4°C. Gently pour off supernatant and remove last ~200 μl with P-200 pipet, being careful as not to disturb pellet. *Repeat this step once more* (see Note 7).
7. Resuspend the pellet in 100–150 μl of ice-cold 10% glycerol with P-200 pipet by gently pipeting back and forth. Make sure no clumps are present. Place cells on ice and use within 30 min (see Note 8). This amount of cells is good for two to three trials using 50 μl of electrocompetent cells per electroporation. If more samples need to be done, the process can be scaled up by growing more cells in additional 125 ml flasks (see Note 9).

3.3. Electroporation of Recombineering-Proficient Cells with PCR Fragments

1. Prechill the electroporation cuvettes (0.1 cm) by placing in an ice-water bath for 10 min. In a prechilled sterile Eppendorf tube, mix 50 μl of electrocompetent cells with 0.1–0.5 μg of PCR substrate. Ideally, use 1–3 μl of DNA per 50 μl of electrocompetent cells. Do not exceed 5 μl of DNA per 50 μl of cells as this amount of substrate increases the possibility of arcing. Arcing occurs when the charge is dissipated as a spark outside the electroporation chamber, and no pulse is detected by the electroporation device (see Note 10).
2. Assemble the Gene Pulser II to Pulse Controller II (Bio-Rad). Select preset protocol for transformation of *E. coli* cells using

0.1 cm cuvette. If using alternate electroporation set-up, set voltage to 1,800 V, use 25 μF capacitance and 200 Ω resistance.

3. Transfer the DNA-cell mixture to a prechilled cuvette, replace the cap, and incubate on ice for 1 min. Quickly (but thoroughly) dry the cuvette with miniwipes, place the cuvette into the electroporation chamber, and release charge. The time constant should be close to 5 ms. A value much less than 5 ms for the time constant indicates impurities (i.e., salt) in the DNA sample or electrocompetent cell preparation.
4. Using the P-1000 pipet, immediately add 0.5 ml of LB to cuvette. Pipet back and forth a few times and transfer cells to 2.5 ml LB in sterile culture tube. It is not necessary to include ampicillin or IPTG in the outgrowth medium, as the Red and Gam proteins are already at their optimal levels.
5. Perform appropriate controls (see Note 11).

3.4. Outgrowth and Selection of Recombinants

1. The electroporated cells are further grown by rolling or shaking for 90–120 min at 37°C. This is an important step as it allows the cells to recover from the electrophoretic shock and express adequate amounts of the drug resistance marker gene prior to exposure to the selection plate.
2. After outgrowth, spread 0.2 and 0.5 ml aliquots of the culture on LB antibiotic-selection plates. Incubate the plates at 37°C overnight (no need to grow at 30°C, as Red-expression is no longer desired). Also plate 100 μl of 10^{-4} and 10^{-5} dilutions of the culture on LB plates to determine the total number of cells present. Percent recombineering frequency can be expressed as the fraction of drug-resistant colonies divided by total cell titer × 100. This number is often normalized to the number of recombinants per 10^8 of viable cells (see Note 12).
3. Allow the rest of the culture to grow overnight at 37°C. If no colonies appear on the plates after overnight growth, spread the rest of the culture on additional drug selection plates and incubate at 37°C overnight. Some recombinants take longer to appear than others.
4. Use drug concentrations in the plates that will select for the drug marker at single copy in the chromosome. These concentrations are lower relative to the same markers present on multicopy plasmids. Drug concentrations in the selection plates we have employed include the following: chloramphenicol, 10–15 μg/ml; kanamycin, 20 μg/ml; tetracycline, 3–7 μg/ml; gentamycin, 10 μg/ml; and ampicillin, 25–50 μg/ml.
5. If no colonies are found on the drug-selection plates, try troubleshooting (see Note 13).

3.5. Verification of Recombinants and Curing of Red-Producing Plasmid pKM208

1. Restreak candidate gene knockout strains on to fresh antibiotic-selection plates and incubate at 37°C overnight. Spontaneous mutants arising on the drug plates typically do not restreak as well on these plates as true gene replacement candidates.
2. Colony PCR can be used to verify the structure of the recombinant. A high-fidelity polymerase is not required (or recommended) for these PCRs. Use a standard Taq polymerase, which works well for colony PCRs. One should use primers that are positioned ~100 bp upstream and downstream of the sequences used for targeting the gene replacement, as well as primers reading out of the drug marker cassette. These primers (see Fig. 2) can be used to verify the 5′ junction of the knockout (primers 1 and 2), the 3′ end of the knockout (primers 3 and 4), as well as any overall differences in size of the gene replacement (primers 1 and 4). A third set of primers should be used to amplify a 500–700-bp region of the target gene or region, which should appear when wild-type cells are used as a template by colony PCR, but absent when the recombinant cells are used (see Note 14).
3. Design primers #2, #3, #5, and #6 (see Fig. 2) to give PCR products in the 500–700 bp range. These products are easy to generate by PCR and can be readily distinguished from PCR artifacts that might occur at 300 bp and below. It is also

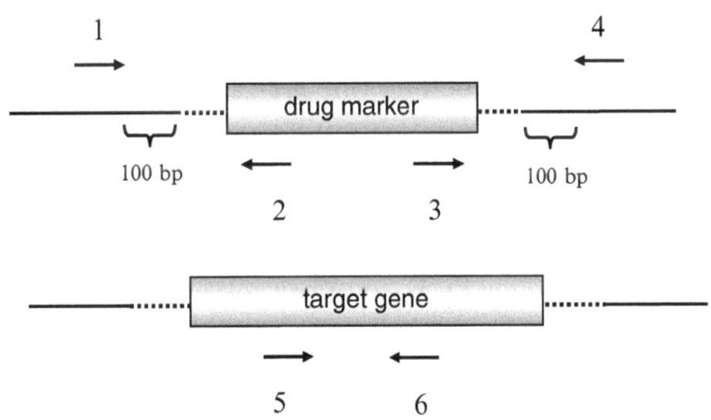

Fig. 2. Primers for gene knockout verification. To verify the 5′ junction, use a primer containing sequences ~100 bp upstream of the sequence used to generate the PCR recombineering substrate (primer #1) and a primer in the in the drug cassette reading leftward (primer #2). To verify the 3′ junction, use a primer containing sequences ~100 bp downstream of the sequence used to generate the PCR recombineering substrate (primer #4) and a primer in the in the drug cassette reading rightward (primer #3). If the size of the gene or region deleted is different from that of the drug cassette, then a PCR using primers #1 and #4 will generate a band diagnostic for the knockout. If the size of the parental and recombinant PCR product is the same, then restriction analysis can usually be used to reveal the presence of the knockout. Finally, a PCR to verify the absence of the wild-type locus in the recombinant should be performed using primers #5 and #6 (see Note 10).

a good idea to run a computer simulation of the PCR before ordering the primers, to avoid the generation of primer dimers that might interfere with detection of the diagnostic band. Amplify is a free Mac software program that can be used to simulate and test PCR reactions in silico (http://engels.genetics.wisc.edu/amplify). Alternatively, for Vector NTI program users, check primers with Oligo analyses programs Thermodynamic Properties and Oligo Duplexes.

4. The λ Red + Gam producing plasmid pKM208 contains a temperature-sensitive origin of replication, where optimal growth occurs at 30°C and restrictive growth occurs at 42°C. Thus, the recombinants can be cured of pKM208 following construction of the knockout by growth of the cells at 42°C. In some cases, streaking out two consecutive times at 42°C is required for promoting loss of the plasmid. Verification of plasmid loss can be found by sensitivity to ampicillin (100 μg/ml), followed by electrophoresis of 10 μl of a minilysate of the cell culture and noting the absence of pKM208 (8,731 bp).
5. If no recombinants are found at this point, perform troubleshooting (see Note 13).

3.6. Generation of Unmarked Gene Knockouts

A procedure for generating a gene knockout and removing the antibiotic resistance takes advantage of the phage P1 Cre-mediated site-specific recombination system (22). The *loxP* sequence (ATAA CTTCGTATA$(N)_8$TATACGAAGTTAT) is a target sites for the Cre recombinase (23). A Cre-promoted recombination event will delete the DNA between directly repeated two *loxP* sites, leaving behind one *loxP* site in the recombinant (24). The use of the Cre-*loxP* system for creating unmarked gene knockouts was developed by Sauer and Henderson (25). The removal of the drug marker after Red-mediated gene deletion is done in a similar manner as described above, with two exceptions. First, the drug marker in the PCR template plasmid should be flanked by *loxP* site-specific recombination sites. Secondly, after recovery of the marked gene deletion, a plasmid expressing the P1 Cre recombinase (pJW168) can be used to delete the drug marker from the chromosome (21). This plasmid, like pKM208, contains a temperature-sensitive origin of replication and can be easily evicted. This system is easy to employ, occurs at high frequency, and allows multiple alterations of the chromosome to occur without the need for multiple drug markers. The only concern is that there is a scar left over (the *loxP* sequence in place of drug marker). Repeated use of this procedure could leave multiple scars in the chromosome, which themselves might become substrates of unintended Cre-promoted recombination.

1. Generate a PCR recombineering substrate as described above in Subheading 3.1, but use as a template drug marker that is flanked by *loxP* target sites (21).

2. After selection, verification, and curing of the recombinant strain of the Red-producing Amp^R pKM208 plasmid, the cell is transformed with Amp^R pJW168 that expresses the Cre recombinase (21). A colony is picked and grown overnight in LB containing 100 μg/ml ampicillin at 30°C in the presence of IPTG (to induce *cre*).
3. The overnight culture is diluted 10^{-4}, 10^{-5}, and 10^{-6}-fold in PBS (or LB) and portions of the dilutions (100–200 μl) are spread on LB plates.
4. Single colonies are streaked as short patches (~0.5 cm) first on to LB plates containing the antibiotic encoded by the evicted drug marker, and then on LB plates with no drugs. This screen identifies recombinant clones that have lost the antibiotic drug cassette by Cre-mediated excision. This step is usually very efficient and drug-free recombinants are easily found.

4. Notes

1. Plasmids have been described that express the λ *red* and *gam* genes from the Ptac promoter (pKM208 – www.addgene.com) (6), the P_{BAD} promoter (pKD46 – http://cgsc.biology.yale.edu) (3), or the phage lambda P_L promoter (pSIM6 – court@ncifcrf.gov) (26). The protocol presented above describes the use of pKM208, where expression of the *red* and *gam* genes is induced by the addition of IPTG. The protocol is the same when using these other Red and Gam-producing plasmids, with the exception of the induction steps: for pKD46, *red* and *gam* are induced by the addition of 10 mM arabinose; for pSIM6, *red* and *gam* are induced by a 15-min incubation at 42°C. All these plasmids carry the same temperature-sensitive origin of replication and the *bla* gene conferring resistance to ampicillin. Options for recombineering plasmids containing drug markers other than ampicillin are available from D. Court (26) htpp://web.ncifcrf.gov/research/brb/recombineering-information.aspx; and at addgene.com (Murphy lab). To use *bla* as a gene knockout marker (Table 1), an alternative Red-producing plasmid containing a different drug marker is needed.
2. The choice of template used for generating the recombineering substrate is crucial. Intact circular plasmids should not be used as templates. While they are used at low amounts in a typical PCR (~10 ng), the template plasmid will still be present in a purified PCR product and will transform *E. coli* at high efficiency giving rise to false-positive recombinants on antibiotic-selecting plates. To prevent these false positives from arising, one can (1) gel purify the PCR product, (2)

treat the PCR product with *Dpn*I, which will digest the template plasmid but not the unmethylated PCR amplicon, (3) perform colony PCR with a strain containing the drug marker in the chromosome, (4) use drug markers cloned into conditionally replicating vectors such as R6K *oriRγ* origin vectors that require engineered pir+ host strains that provide the *trans*-acting Π protein for replication (3), or (5) use as a PCR template, a gel-purified fragment of the marker-containing plasmid that is free of its origin of replication. The last option is quite useful, as 1 μg of this fragment can serve as a successful template for 100 PCRs.

3. In some cases (e.g., when dealing with enteropathogenic strains of *E. coli*), the use of higher concentrations of the PCR substrate will give a better chance of recovering a recombinant. To this end, the 50 μl of cleaned PCR product can be concentrated by ethanol precipitation and resuspended in 10 μl of EB (10 mM Tris–HCl, pH 8.0). To do this, dilute 50 μl of DNA to 350 μl with precipitation buffer (20 mM Tris–HCl, 10 mM NaCl, 2 mM EDTA, 0.5M ammonium acetate, pH 6.5), add 3 μl of 10 mg/ml of tRNA (as carrier), and fill the 1.5 ml Eppendorf tube with ethanol. Vortex the mixture well, freeze at −20°C for 30 min, and spin out the precipitate at high speed in a microcentrifuge for 5 min. Remove the supernatant, dry the pellet with one wash of cold ethanol, let the pellet dry, and resuspend the DNA in 10 μl of EB. We have found that samples prepared in this way allow higher amounts of DNA to be electroporated without causing sparking (i.e., arcing, no pulse delivered to sample due to dissipation of the charge outside the cuvette, usually the result of residual salt in the sample).
4. The lack of PCR products (in general) is usually indicative of problems with one or more components of the reaction, or errors in the cycling program. But remember, the primer annealing sequences in Table 1 and their templates have been used repeatedly in successful PCRs, so a problem in generating a substrate for gene replacement (with all control reactions with known reagents working properly) most likely indicates problems with one of the primers. If so, do not spend much effort in trying to optimize the PCR, just order new primers.
5. If no transformants with pKM208 are found, try plating cells on decreasing concentrations of ampicillin (25–50 μg/ml). Once established, cells containing the plasmid should be grown in LB containing 100 μg/ml ampicillin.
6. This heat shock step is optional. It has proved useful for obtaining recombinants in pathogenic strains such as enterohemorrhagic *E. coli* and enteropathogenic *E. coli*. In *E. coli* K-12, the stimulation due to the heat shock is variable depending on the loci being deleted. The reason for this observation is not known.

7. Multiple glycerol washes are necessary to thoroughly remove salts from the cell preparation to increase resistance thus preventing arcing during electroporation.
8. For long-term storage, flash freeze the samples by swirling in an dry ice-ethanol bath, then store the cells at −80°C. The fold-less transformation is variable (depending on the initial competence), but generally expect about a five- to tenfold drop in transformation efficiency over a 6-month period. This step is useful when the total number of recombinants is not critical. However, when high transformation efficiencies are required, one should use freshly prepared cells.
9. The electroporation can also be scaled down to 25 μL of cells (just above the minimum volume required for a 0.1 cm cuvette) to allow processing of more samples.
10. If a spark occurs, chances are that the sample did not receive the appropriate charge to generate pores in the membrane to promote DNA uptake. However, we have seen examples where a spark has been observed, and upon plating the cells, recombinants were in fact recovered.
11. Perform appropriate controls. The most important control is to electroporate the host cells with no DNA. Spreading of the cells for this "blank" on antibiotic-selecting plates should give no colonies. The presence of colonies is indicative of host cell line contamination. Sometimes, this control gives rise to small colonies on the drug plates indicative of spontaneous resistance. These colonies generally should be fewer in number (relative to plates that were spread with cells containing DNA) and should not grow well upon restreaking onto fresh antibiotic-selection plates. For a positive control, knock out a gene that has been done before (*lacZ* for example), just in case there is something peculiar about the knockout being attempted.
12. When comparing the recombination rates of different strains, it is advisable to include a small amount (10–50 ng) of an intact plasmid as an electroporation control. This plasmid can be mixed directly with the PCR substrate and co-electroporated into *E. coli*. Even the same cell preparation can exhibit various transformation efficiencies when electroporated side-by-side on the same day. The plasmid should possess a different drug marker relative to the Red-producing plasmid and the recombination substrate, and the recombineering frequencies are reported as recombinants per competent cell (recombinant titer/plasmid transformant titer). Recombineering with linear dsDNA substrates is usually on the order of 10^{-4} to 10^{-5} per viable cell. One can typically expect 50 ng of an intact plasmid to transform about 10% of the cell population following electroporation. Thus, the range of recombinant titer/plasmid titer is expected to be 10^{-3} to 10^{-4}. However, these numbers

can vary depending on the purity of the DNA samples and the electrocompetence of the cells. In addition, while recombineering with small homology substrates (50 bp flanks) works in a variety of strains that are deficient for host recombination (e.g., *recA* strains), the total number of recombinants may be reduced due to lower strain viability, relative to wild type, following electroporation.

13. Troubleshooting. If no colonies or recombinant clones are found, examine this list of possible reasons/solutions:

    No Colonies

    (a) Design the primers so that the drug marker reads in the same direction as neighboring genes. If one direction does not work, try the other.

    (b) Clean PCR substrate by ethanol precipitation (see Note 3).

    (c) Problem with PCR product. Generate more or order new oligos.

    (d) Make sure cells are electrocompetent by transforming with an intact plasmid (e.g., pBR322). One should obtain at least 10^7 transformants per microgram of DNA.

    (e) Measure total numbers of survivors on LB plates. Less than 10^6 cells/ml following electroporation indicates that the cells were not grown to high enough density, were lost during centrifugation steps, or are not surviving the electroporation shock. In this last case, check for salt contamination in PCR sample or in the washed cell preparation.

    (f) Increase cell outgrowth postelectroporation to a longer period of time (2 h or more), or even overnight.

    (g) Recombineering strain was grown at 37°C, a temperature too high to maintain the recombineering plasmid (pKM208 requires growth at 30°C).

    (h) Make minilysate preparation from recombineering strain; verify the presence of pKM208 (8731 bp).

    (i) Forgot to add inducer IPTG, or added it too late.

    Colonies obtained but not recombinant targeted knockout.

    (j) Make sure the PCR substrate is free of intact plasmid (see Note 2).

    (k) Check negative control electroporation without DNA (see Note 11) to ensure cell line is not contaminated with plasmid.

14. Verification of the absence of the wild-type loci by PCR analysis is important, as one can (on occasion) find PCR products representative of the replaced target gene (including junctions between the 3′ and 5′ regions of the drug marker and

adjacent chromosomal regions of the target gene), but still find an intact target gene present on the chromosome. This anomalous event might happen when recombineering occurs in a strain that is transiently duplicated for the targeted loci, thus allowing both deleted and wild-type versions of the gene to be present in the same chromosome. Such events can mistakenly identify an essential gene as nonessential.

References

1. Murphy K. C. (1998) Use of bacteriophage lambda recombination functions to promote gene replacement in Escherichia coli. *J Bacteriol* **180**, 2063–2071.
2. Zhang Y., Buchholz F., Muyrers J. P., and Stewart, A. F. (1998) A new logic for DNA engineering using recombination in *Escherichia coli*. *Nat Genet* **20**, 123–128.
3. Datsenko K. A., and Wanner B. L. (2000) One-step inactivation of chromosomal genes in *Escherichia coli* K-12 using PCR products. *Proc Natl Acad Sci U S A* **97**, 6640–6645.
4. Yu D., Ellis H. M., Lee E. C., Jenkins N. A., Copeland N. G., and Court, D. L. (2000) An efficient recombination system for chromosome engineering in *Escherichia coli*. *Proc Natl Acad Sci U S A* **97**, 5978–5983.
5. Court D. L., Sawitzke J. A., and Thomason L. C. (2002) Genetic engineering using homologous recombination. *Annu. Rev. Genet.* **36**, 361–388.
6. Murphy K. C., and Campellone K. G. (2003) Lambda Red-mediated recombinogenic engineering of enterohemorrhagic and enteropathogenic E. coli. *BMC Mol Biol* **4**, 11.
7. Sawitzke J. A., Thomason L. C., Costantino N., Bubunenko M., Datta S., and Court D. L. (2007) Recombineering: in vivo genetic engineering in E. coli, S. enterica, and beyond. *Methods Enzymol* **421**, 171–199.
8. Little J. W. (1967) An exonuclease induced by bacteriophage lambda. II. Nature of the enzymatic reaction. *J Biol Chem* **242**, 679–686.
9. Sriprakash K. S., Lundh N., Huh M.-O., and Radding C. M. (1975) The specificity of lambda exonuclease. Interactions with single-stranded DNA. *J Biol Chem* **250**, 5438–5445.
10. Echols H., and Gingery R. (1968) Mutants of bacteriophage (lambda) defective in vegetative genetic recombination. *J Mol Biol* **34**, 239–249.
11. Signer E. R., and Weil J. (1968) Recombination in bacteriophage lambda. I. Mutants deficient in general recombination. *J Mol Biol* **34**, 261–271.
12. Kmiec E., and Holloman W. K. (1981) Beta protein of bacteriophage` lambda promotes renaturation of DNA. *J Biol Chem* **256**, 12636–12639.
13. Muniyappa K., and Radding C. M. (1986) The homologous recombination system of phage lambda. Pairing activities of beta protein. *J Biol Chem* **261**, 7472–7478.
14. Passy S. I., Yu X., Li Z., Radding C. M., and Egelman E. H. (1999) Rings and filaments of beta protein from bacteriophage lambda suggest a superfamily of recombination proteins. *Proc Natl Acad Sci U S A* **96**, 4279–4284.
15. Iyer L. M., Koonin E. V., and Aravind L. (2002) Classification and evolutionary history of the single-strand annealing proteins, RecT, Redbeta, ERF and RAD52. *BMC Genomics.* **3**, 8.
16. Karu A. E., Sakaki Y., Echols H., and Linn S. (1975) The gamma protein specified by bacteriophage gamma. Structure and inhibitory activity for the recBC enzyme of Escherichia coli. *J Biol Chem* **250**, 7377–7387.
17. Murphy K. C. (1991) Lambda Gam protein inhibits the helicase and chi-stimulated recombination activities of Escherichia coli RecBCD enzyme. *J Bacteriol* **173**, 5808–5821.
18. Murphy K. C. (2007) The lambda Gam protein inhibits RecBCD binding to dsDNA ends. *J Mol Biol* **371**, 19–24.
19. Ellis H. M., Yu D., DiTizio T., and Court D. L. (2001) High efficiency mutagenesis, repair, and engineering of chromosomal DNA using single-stranded oligonucleotides. *Proc Natl Acad Sci U S A* **98**, 6742–6746.
20. Poteete A. R. (2008) Involvement of DNA replication in phage lambda Red-mediated homologous recombination. *Mol Microbiol* **68**, 66–74.
21. Palmeros B., Wild J., Szybalski W., Le Borgne S., Hernandez-Chavez G., Gosset G., Valle F., and Bolivar F. (2000) A family of removable cassettes designed to obtain antibiotic-resistance-free genomic modifications of Escherichia coli and other bacteria. *Gene* **247**, 255–264.

22. Grindley N. D., Whiteson K. L., and Rice P. A. (2006) Mechanisms of site-specific recombination. *Annu Rev Biochem* **75**, 567–605.
23. Hoess R. H., and Abremski K. (1984) Interaction of the bacteriophage P1 recombinase Cre with the recombining site loxP. *Proc Natl Acad Sci U S A* **81**, 1026–1029.
24. Hamilton D. L., and Abremski K. (1984) Site-specific recombination by the bacteriophage P1 lox-Cre system. Cre-mediated synapsis of two lox sites. *J Mol Biol* **178**, 481–486.
25. Sauer B., and Henderson N. (1988) Site-specific DNA recombination in mammalian cells by the Cre recombinase of bacteriophage P1. *Proc Natl Acad Sci U S A* **85**, 5166–5170.
26. Datta S., Costantino N., and Court D. L. (2006) A set of recombineering plasmids for gram-negative bacteria. *Gene* **379**, 109–115.

Chapter 3

Scarless Chromosomal Gene Knockout Methods

Bong Hyun Sung, Jun Hyoung Lee, and Sun Chang Kim

Abstract

An improved and rapid genomic engineering method has been developed for the construction of custom-designed microorganisms by scarless chromosomal gene knockouts. This method, which can be performed in 2 days, permits restructuring of the *Escherichia coli* genome via scarless deletion of selected genomic regions. The deletion process is mediated by a special plasmid, pREDI, which carries two independent inducible promoters: (1) an arabinose-inducible promoter that drives expression of λ-RED recombination proteins, which carry out the replacement of a target genomic region with a marker-containing linear DNA cassette, and (2) a rhamnose-inducible promoter that drives expression of I-*Sce*I endonuclease, which accomplishes deletion of the introduced marker by double-strand breakage – mediated intramolecular recombination. This genomic deletion is performed simply by changing the carbon source in the bacterial growth medium from arabinose to rhamnose. The efficiencies of targeted region replacement and deletion of the inserted linear DNA cassette are nearly 70 and 100%, respectively. This rapid and efficient procedure can be adapted for use in generating a variety of genome modifications.

Key words: pREDI, Scarless deletion, λ-Red system, I-*Sce*I, *sacB*/sucrose, Rhamnose and arabinose induction system

1. Introduction

The complete genome sequences of a rapidly growing number of bacterial strains have provided a wealth of information on the molecular structure and organization of myriad genes and open reading frames. This vast amount of information has been used in the construction of microorganisms with restructured, custom-designed genomes. One of the most common approaches for the restruction of a microbial genome to create custom-designed microorganisms is sequence-specific deletion or insertion of target genes or DNA sequences. For the precise modification of a genome, various methods have been developed based on RecA-dependent homologous recombination (1–3).

James A. Williams (ed.), *Strain Engineering: Methods and Protocols*, Methods in Molecular Biology, vol. 765,
DOI 10.1007/978-1-61779-197-0_3,

In addition to the RecA-dependent homologous recombination system in microbes, the λ-Red or RecET recombination system has also been exploited for the modification of large DNA constructs, including bacterial chromosomes and BAC clones (4–8). In these recombination events, selection markers are necessary to confirm the insertion or deletion of targeted regions. But the inserted selection markers prevent further modifications of the genome. To avoid having residual selection markers or foreign DNA sequences within the engineered chromosomes after genome modification, the *Flp* recombinase target (*FRT*) and the *loxP*-mediated site-specific recombination systems have been used for the precise excision of selection markers with the corresponding recombinase (*Flp* or *Cre*, respectively) (5, 9–12). However, even with these site-specific recombination systems, at least one copy of an *FRT* site or a *loxP* site remains after excision of the selective markers, which limits the repeated use of these procedures (13, 14).

Therefore, a more efficient method to delete target genes or genomic regions without leaving selection markers or foreign DNA sequences behind has been developed. This procedure involves the use of the intron-encoded homing endonuclease enzyme I-*Sce*I as a counter-selection tool, which introduces a double-stranded break (DSB) in the genome (15–17). This DSB is a potent substrate for a microbial host recombination system that can repair the break by homologous recombination within the regions of sequence homology that flank the ends of the break. With the help of the host DSB-mediated repair system, several scarless modifications have been introduced into BAC clones and into the genomes of Gram-negative bacteria, such as *Escherichia coli* and *Salmonella typhimurim* (3, 7, 18–22).

Although the above methods have been used successfully to produce scarless modifications in genomes, several drawbacks remain. For example, these methods are time-consuming and labor-intensive, taking more than a week to delete a single targeted region, because of the repeated plasmid transformation and curing required for each deletion step (6, 7). Here, we describe a highly efficient and rapid single plasmid genomic engineering procedure that allows researchers to perform scarless deletion of a selected genomic region in 2 days.

2. Materials

1. *E. coli* strains: DH5α, MG1655 (23), and recombination-proficient *E. coli* strain.
2. Plasmids pSCI and pSKI (24).
3. Plasmid pREDI (24; Fig. 1a).

4. SOC medium: 2% bacto-tryptone, 0.5% bacto-yeast extract, 0.05% NaCl (pH 7.0). After sterilization by autoclaving, add sterile glucose and $MgCl_2$ to achieve 20 and 10 mM final concentration, respectively.
5. LB (Luria-Bertani) medium: 1% bacto-tryptone, 0.5% bacto-yeast extract, and 0.5% NaCl (pH 7.0), sterilized by autoclaving.
6. LB plates: LB medium supplemented before autoclaving with 1.5% bacto-agar.
7. 50 mg/mL Ampicillin (Ap): Dissolve 0.5 g ampicillin sodium salt in 10 mL double-distilled water to make 50 mg/mL stock solution. Store at −20°C. Use at a final concentration of 50 μg/mL.
8. 50 mg/mL Kanamycin (Km): Dissolve 0.5 g kanamycin sulfate in 10 mL double-distilled water. Store at −20°C. Use at a final concentration of 25 μg/mL.
9. 34 mg/mL Chloramphenicol (Cm): Dissolve 0.34 g of chloramphenicol in 10 mL 100% ethanol. Store at −20°C. Use at a final concentration of 17 μg/mL.
10. LB Ap Km plates: LB plates supplemented after autoclaving with AP (to 50 μg/mL) and Km (to 25 μg/mL).
11. LB Ap Cm plates: LB plates supplemented after autoclaving with AP (to 50 μg/mL) and Cm (to 17 μg/mL).
12. 1 M L-(+)-Arabinose: Dissolve 1.50 g L-(+)-arabinose in 10 mL double-distilled water and filter across a sterile syringe filter with 0.22 μm pore size. Use at a final concentration of 10 mM.
13. 1 M L-Rhamnose: Dissolve 1.82 g L-rhamnose monohydrate in 10 mL double-distilled water and filter across a sterile syringe filter with 0.22 μm pore size. Use at a final concentration of 10 mM.
14. 50% Sucrose: Dissolve 5.5 g sucrose in 10 mL double-distilled water and filter across the filter with 0.22 μm pore size. Use at a final concentration of 5%.
15. LB Rhamnose Sucrose plates: LB plates supplemented after autoclaving with rhamnose (to 10 mM) and sucrose (to 5%).
16. LB Sucrose plates: LB plates supplemented after autoclaving with sucrose (to 5%).
17. 10% glycerol: Dissolve 10 g in 90 ml double-distilled water and filter across a sterile syringe filter with 0.22 μm pore size. Store at 4°C.
18. TE buffer (10 mM Tris–HCl, 1 mM EDTA, pH 7.5): Dissolve 1.21 g Tris(hydoxymethyl) aminomethane and 0.292 g ethylenediaminetetraacetic acid (EDTA) in 1 L double-distilled water. Adjust to pH 7.5 with HCl.

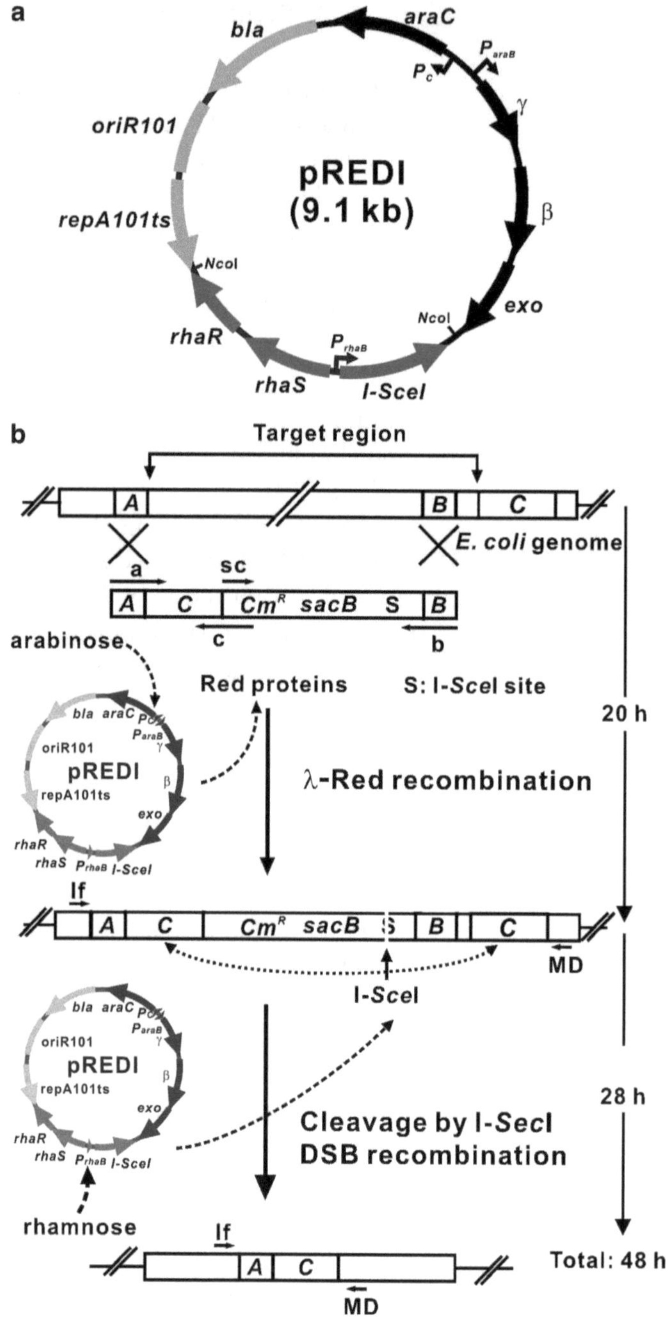

Fig. 1. Description of rapid scarless chromosomal gene knockout methods with pREDI. (**a**) Plasmid pREDI provides (1) arabinose-inducible (promoter = P_{araB}) λ-Red recombinase function (*gam* (γ), *bet* (β), and *exo*) necessary for the replacement of a target genomic region with a linear DNA cassette, and (2) rhamnose-inducible (promoter = P_{rhaB}) I-*Sce*I expression required for DSB-mediated scarless deletion. (**b**) Schematic of the scarless deletion system with pREDI. To delete the *E. coli* chromosomal targeted region between homology boxes A and C, a linear DNA cassette containing a positive selective marker (*Cm*^{R}), a negative selective marker (*sacB*), an I-*Sce*I endonuclease recognition site (S), and three homology boxes (A, B, and C) is generated by recombinant PCR using pSCI and the *E. coli* genome as templates. Recombinant PCR used primers a (forward primer that include 50-nt homology extension (**a**) and 20-nt priming sequence for the homology region C), c (reverse primer that include 20-nt reverse complement sequence of primer sc

19. Qiagen Gel Extraction kit (Qiagen, Hilden, Germany).
20. Gel electrophoresis solutions and reagents: agarose, ethidium bromide (EtBr: 500 μg/mL), and Tris–Borate–EDTA (TBE) electrophoresis buffer (45 mM Tris–borate and 1 mM EDTA).
21. DNA polymerases for polymerase chain reaction (PCR) and buffers supplied by the manufacturers.
22. Oligonucleotide primers (Genotech, Daejeon, Korea or equivalent).
23. Thermal Cycler.
24. Gene Pulser system (Bio-Rad, Herculus, CA).
25. Spectrophotometer.
26. Gel electrophoresis apparatus and equipment.

3. Methods

The methods described below outline the construction of a cassette for chromosomal gene knockout; scarless deletion of a genomic region; simultaneous deletion of two separated regions; and scarless deletion of a genomic region that contains an essential gene(s).

3.1. Construction of a Cassette for Chromosomal Genes Knockout

To delete the selected target region of an *E. coli* genome, which is housed between homology boxes *A* and *C* (see Fig. 1b), a 3.5-kb deletion cassette fragment (*A*-*C*-Cm^R-*sacB*-I-*Sce*I-*B*, see Fig. 1b) that contains three homology regions (*A*, *B*, and *C*, see Fig. 1b), a positive selection marker (Cm^R), a negative selection marker (*sacB*), and an I-*Sce*I endonuclease recognition site is constructed by recombinant PCR as follows.

Fig. 1. (continued) (5′-TAATTTCGATAAGCCAGATC-3′) and 20-nt priming sequence for the homology regions C), sc (20-nt forward primer specific to pSCI, 5′-GATCTGGCT TATCGAAATTA-3′), and b (reverse primer that include 50-nt homology extension (**b**) and 20-nt priming sequence (5′-GCATGCCTGCAGGTCGACTC -3′) for pSCI as template). The linear DNA cassette is electroporated into pREDI-containing *E. coli* cells, where the cassette can replace a target genomic segment with the help of the λ-Red proteins (Red proteins) encoded by pREDI. Next, to remove the introduced selection markers, expression of the pREDI-encoded I-*Sce*I endonuclease is induced by changing the carbon source in the medium from 10 mM arabinose to 10 mM rhamnose. As a result, the chromosome is cleaved at the I-*Sce*I endonuclease recognition site (S) present on the integrated DNA cassette, inducing the DSB repair function. Then, the DSB-mediated intramolecular recombination between the two homology arms (box C) results in the removal of the inserted deletion cassette, producing a clean, scarless deletion.

1. Amplify a 3.0-kb DNA fragment that contains a Cm^R, a *sacB*, an I-*Sce*I endonuclease recognition site, and the 50-bp homology region *B* by PCR from plasmid pSCI with 25 pmol each of primers forward (sc) and reverse (b) in a total volume of 50 μL following the manufacturer's instruction (see Fig. 1b). Run 30 amplification cycles in a thermocycler with parameters, 30 s at 94°C, 30 s at 58°C or lower (depending on the primers), and 3 min at 72°C (see Notes 1 and 2).
2. Amplify a 0.5-kb homology fragment that contains homology regions *A* and *C* from the genomic DNA of MG1655 with the forward (a) and reverse (c) primers shown in Fig. 1b following the manufacturer's instruction (the amplified 0.5-kb fragment contained a short, 20-bp flanking sequence on its 3′-end that overlapped with the 5′-end of the 3.0-kb fragment in step 1).
3. Mix 10 ng of the 3.0-kb and 0.5-kb of each PCR product with 25 pmol each of primers forward (a) and reverse (b) in a total volume of 50 μL and perform second round of PCR. Run 30 amplification cycles of 30 s at 94°C, 30 s at 56°C, and 3 min at 72°C.
4. Purify the resulting 3.5-kb linear DNA fragment (*A*-*C*-Cm^R-*sacB*-I-*Sce*I-*B*) with the Qiagen Gel Extraction kit.
5. If a Km^R, rather than a Cm^R, deletion cassette is desired, a scarless deletion cassette (*A*-*C*-Km^R-*sacB*-I-*Sce*I-*B*) is generated as described above except use of the plasmid pSKI for the selection marker Km^R instead of the plasmid pSCI (the primers sc and b can also be used for the amplification of the cassette Km^R-*sacB*-I-*Sce*I-*B*).

3.2. Scarless Deletion of a Genomic Region

The deletion process is mediated by a special plasmid, pREDI, which carries two independent inducible promoters: (1) an arabinose-inducible promoter that drives expression of λ -RED recombination proteins, which carry out the replacement of a target genomic region with the marker (Cm^R/Km^R-*sac*B-I-*Sce*I)-containing linear DNA cassette generated in Subheading 3.1 and (2) a rhamnose-inducible promoter that drives the expression of I-*Sce*I endonuclease, which accomplishes the deletion of the introduced marker by DSB-mediated intramolecular recombination (see Note 3).

3.2.1. Replacement of the Target Genomic Region with the Deletion Cassette

1. Grow the target *E. coli* cell line harboring pREDI at 30°C in 100 mL of LB medium supplemented with Ap and L-arabinose for the preparation of the electro-competent cells. Harvest the cells at early log phase (OD_{600} = 0.4) by centrifugation at 2,500 × *g* for 10 min, wash three times with ice-cold 10% glycerol, and resuspend in 400 μL of 10% glycerol.
2. Electroporate the appropriate scarless deletion cassette (400–600 ng) from Subheading 3.1 into 50 μL of the

electro-competent *E. coli* cells harboring pREDI at 2.5 kV, 25 μF, and 200 Ω.

3. Add 1 mL of SOC medium to the shocked *E. coli* cells, incubate at 30°C for 1 h with agitation, then sediment cells by a brief spin in a microcentrifuge, spread them onto LB plates containing Ap and either Cm or Km as appropriate, and incubated at 30°C for an additional 12 h.
4. Verify the correct replacement of the target genomic region with the scarless deletion cassette by colony PCR using a pair of primers (If and MD in Fig. 1b) that flanks the endpoints of the targeted region (Touch the colony with a sterile toothpick, drop the toothpick in an Eppendorf tube containing 20 μL TE buffer, vortex briefly, and use 3 μL of the cell suspension as a PCR template) (see Note 4).

3.2.2. Deletion of the Selection Markers by DSB-Mediated Homologous Recombination

1. Grow the recombinant strains from Subheading 3.2.1 to $OD_{600} = 0.4$ at 30°C in 3 L of LB medium containing Ap and rhamnose, then dilute tenfold into 3 mL of fresh LB medium containing Ap, rhamnose, and sucrose and grow to $OD_{600} = 0.4$ at 30°C (see Note 5).
2. Spread the cells on LB plates containing Ap, rhamnose, and sucrose after three rounds of serial culture with tenfold dilution (see Note 6). Grow overnight at 30°C.
3. Screen colonies for scarless deletion mutants (colonies that are sucrose-resistant and either Cm- or Km-sensitive) by replica plating the recombinants on LB plates containing either Cm or Km vs. LB plates containing sucrose.
4. Verify the excision of the selection markers by colony PCR using a pair of specific primers that flanks the endpoints of the genomic target region (primers If and MD in Fig. 1b) (see Note 7).

3.3. Simultaneous Deletion of Two Separate Regions

To delete simultaneously two targeted regions that are not adjacent to each other (*A*-*C* and *A*′-*C*′) from the microbial genome, two scarless deletion cassettes, *A*-*C*-*Cm*^{R}-*sacB*-I-*Sce*I-*B* (C1) for deletion of the first target genomic region (*A*-*C*), and *A*′-*C*′-*Km*^{R}-*sacB*-I-*Sce*I-*B*′ (K1) for deletion of the second target genomic region (*A*′-*C*′), are constructed (Fig. 2).

1. Generate two scarless deletion cassettes, *A*-*C*-*Cm*^{R}-*sacB*-I-*Sce*I-*B* (C1) and *A*′-*C*′-*Km*^{R}-*sacB*-I-*Sce*I-*B*’ (K1) as described in Subheading 3.1.
2. Replace sequentially the two targeted regions (*A*-*C* and *A*′-*C*′ in Fig. 2) with scarless deletion cassettes C1 and K1, respectively, as described in Subheading 3.2.1.
3. Check correct replacement of both targeted regions with the corresponding scarless deletion cassettes (C1 and K1) by PCR using primers If1/Ir1 and If2/Ir2, respectively.

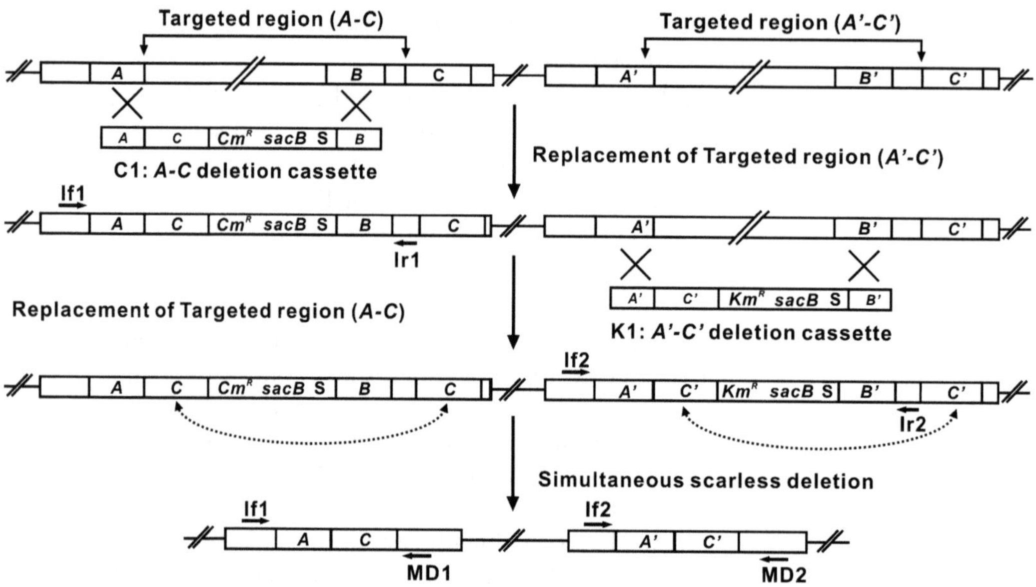

Fig. 2. Simultaneous deletion of two nonadjacent genomic targeted regions. To simultaneously delete two separate genomic regions (*A–C* and *A′–C′*), two linear DNA cassettes are constructed: (1) *A-C-Cm^R-sacB*-I-*Sce*I-*B* (C1), for deletion of the first target genomic region (between *A* and *C*), and (2) *A′-C′-Km^R-sacB*-I-*Sce*I-*B′* (K1), for deletion of the second target genomic region (between *A′* and *C′*). The *A–C* genomic region is replaced with deletion cassette C1, generating *E. coli* deletion strain Δ*A-C*::C1. Then, the *A′C′* genomic region is replaced with the deletion cassette K1, producing *E. coli* deletion strain Δ*A-C*::C1 Δ*A′-C′*::K1. The subsequent expression of the I-*Sce*I endonuclease in the double-replaced strain results in the simultaneous removal of the integrated DNA cassettes, generating the *E. coli* Δ*A-C* Δ*A′-C′* scarless double-deletion strain. Scarless deletion of the two targeted regions is confirmed by PCR using two pairs of primers (If1/MD1 and If2/MD2) specific to both ends of the targeted regions. PCR primers are indicated with *arrows*.

4. Excise the inserted selection markers from the recombinant strains by I-*Sce*I-mediated DSB repair as described in Subheading 3.2.2 and select the scarless deletion mutants (colonies that are sucrose-resistant and both Cm- and Km-sensitive) by replica plating the recombinants on LB plates containing both Cm and Km vs. LB plates containing sucrose.
5. Verify the excision of the inserted scarless deletion cassettes by PCR using a pair of specific primers (If1/MD1 and If2/MD2; Fig. 2) that flanks the endpoints of each targeted region.

3.4. Scarless Deletion of a Genomic Region that Containing an Essential Gene(s)

To delete the targeted region of the *E. coli* genome (*A–C*) that contains the essential gene (*E*), a scarless deletion cassette that houses the *E* gene as described below is prepared and used to delete the targeted *E. coli* genomic region (Fig. 3; see Note 8).

1. Construct a scarless deletion cassette (*A-E-C-Cm^R-sacB*-I-*Sce*I-*B* (E1)) and integrate it into the *E. coli* genome as described in Subheading 3.2, and verify the correct replacement of the targeted region by E1 by colony PCR with primers IE1-f and IE1-r.

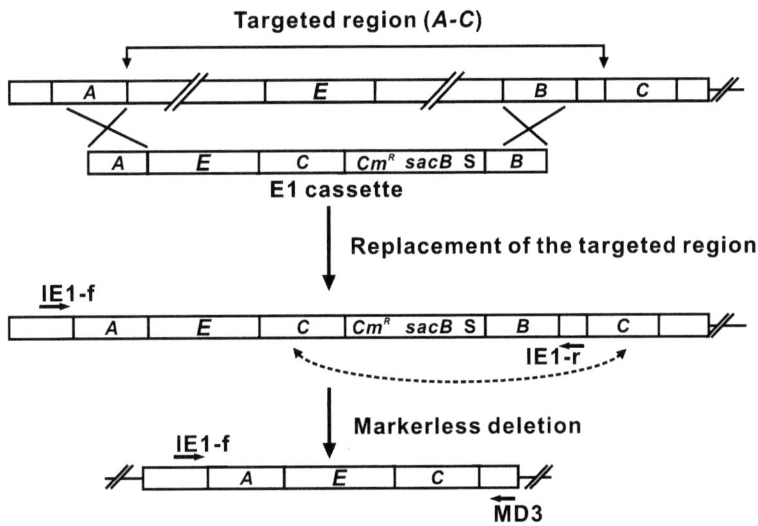

Fig. 3. Deletion of an *E. coli* genomic region that contains an essential gene. Deletion of the *E. coli* target genomic region that contains the essential gene *E* is performed with a pREDI-containing strain of *E. coli*. To delete the targeted region (between *A* and *C*), that contains the essential gene *E*, the linear DNA cassette *A-E-C-Cm^R-sacB*-I-*Sce*I-*B* (E1) is generated and used to replace the selected genomic targeted region. Scarless deletion of the introduced selection markers is carried out as described in Fig. 2. Correct replacement of the genomic targeted region and complete removal of the inserted deletion cassette (E1) are confirmed by PCR using two pairs of primers IE1-f and IE1-r, and IE1-f and MD3, respectively. All PCR primers are indicated with *arrows*.

2. Excise the inserted selection markers from the recombinant strains by I-*Sce*I-mediated DSB repair, and verify the excision of the inserted scarless deletion cassette by PCR with colonies that are Cm-sensitive and sucrose-resistant as template using a pair of specific primers that flanks the endpoints of the targeted region (IE1-f and MD3 in Fig. 3) (see Notes 9 and 10).

4. Notes

1. It has been reported that the Bet protein encoded by β gene in λ-Red recombination system binds stably to DNA strand 36 bases long (25). Therefore, DNA homologies as short as 40–60-bp on the ends of linear DNA cassette are proficient for the efficient replacement of target genomic regions.
2. The sequences of the sc primer (5′- GATCTGGCTTATC GAAATTA -3′) and 3′ end of the b primer (5′-GCATGCCT GCAGGTCGACTC -3′) are same for pSCI or pSKI vector regions.
3. For scarless deletion of a specific region of an *E. coli* genome, a two-step procedure using two different plasmids has typically been employed by researchers. Step 1 includes the transformation

of a microbe with the first plasmid for the targeting of a selected gene/genomic region and then curing of the first plasmid from the cells. Step 2 involves retransformation of the microbe with a second plasmid for the scarless deletion of the selection markers introduced in step 1, followed by curing of the second plasmid (6, 7, 19, 22). This procedure is time-consuming and labor-intensive. The one new plasmid scarless deletion system described herein is rapid and efficient and thus represents an improvement over the currently used technique.

4. It is possible that the replaced genomic regions are reintegrated into another location in the genome after λ Red recombination. Therefore, complete removal of the deletion regions should be confirmed at every step by PCRs with primers specific to the internal sites of all deleted regions.

5. The *sacB/*sucrose counter-selection procedure eliminates cells with the genomes not digested by the I-*Sce*I endonuclease, increasing the selection efficiency of the scarless deletion mutants.

6. To further improve the selection efficiency of the scarless deletion mutants, serial culture with the appropriate dilution is needed (3). One round of serial culture with tenfold dilution in the selective medium showed less than 50% selection efficiency of the correct deletion mutants. However, the selection efficiency of the correct deletion mutants was close to 100% with three rounds of serial culture with tenfold dilution in the selective medium. Therefore, our overall efficiency of scarless deletion of a targeted region was much higher than those of the previous procedures (7, 19, 22).

7. To examine the cleavage efficiency of I-*Sce*I expressed from pREDI in *E. coli*, we transformed pREDI-containing *E. coli* with pSCI, a plasmid that contains an I-*Sce*I endonuclease recognition site and a Cm^R gene. The transformants were grown at 30°C for 12 h in LB liquid medium supplemented with 10 mM rhamnose and Ap, and the resulting cells were spread on LB plates containing Ap. The cleavage efficiency of I-*Sce*I was estimated by replica plating 200 colonies on LB plates with Ap vs. LB plates with Ap and Cm. The fraction of surviving colonies on LB with Ap and Cm was lower than 5%, suggesting that more than 95% of the pSCI plasmids were cleaved by I-*Sce*I expressed from pREDI in the presence of rhamnose.

8. With appropriate modification of the scarless deletion cassette, this system can be adapted for a variety of genome modification. These include the introduction of point mutations and the insertion of genes or sequences into the genomes of *E. coli* and other Gram-negative bacterial species.

9. We observed no significant correlation between the efficiency of replacement and the size of the targeted genomic region.

The overall efficiency of the scarless deletion process ranged from 70 to 100%, and that of scarless deletion of the genomic regions containing the essential gene(s) was 9–12.5%.

10. Scarless deletion of target genomic regions that contained an essential gene(s) is not as efficient as that of nonessential targeted regions. In addition, replacement of a targeted region with two essential genes is less efficient than replacement of a targeted region that harbored only one essential gene. This is because the essential gene(s) in the scarless deletion cassette serves as a substrate for homologous recombination rather than the short 50-bp homology arms, which results in incorrect replacement of the targeted region, decreasing the deletion efficiency.

Acknowledgments

This work was supported in part by grants from 21C Frontier Program of Microbial Genomics and Applications (MG08-0204-1-0) from the Ministry of Education, Science and Technology and by grants from the Korea Science and Engineering Foundation (2008-0060733) and the Conversing Research Center Program through the National Research Foundation of Korea (2009-0082332).

References

1. Hamilton C.M., Aldea M., Washburn B.K., Babitzke P. and Kushner S.R. (1989) New method for generating deletions and gene replacements in *Escherichia coli*. *J Bacteriol*, **171**, 4617–4622.
2. Link A.J., Phillips D. and Church G.M. (1997) Methods for generating precise deletions and insertions in the genome of wild-type *Escherichia coli*: application to open reading frame characterization. *J Bacteriol*, **179**, 6228 6237.
3. Posfai G., Kolisnychenko V., Bereczki Z. and Blattner F.R. (1999) Markerless gene replacement in *Escherichia coli* stimulated by a double-strand break in the chromosome. *Nucleic Acids Res*, **27**, 4409–4415.
4. Murphy K.C. (1998) Use of bacteriophage lambda recombination functions to promote gene replacement in *Escherichia coli*. *J Bacteriol*, **180**, 2063–2071.
5. Zhang Y., Buchholz F., Muyrers J.P. and Stewart A.F. (1998) A new logic for DNA engineering using recombination in *Escherichia coli*. *Nat Genet*, **20**, 123–128.
6. Hashimoto M., Ichimura T., Mizoguchi H., Tanaka K., Fujimitsu K., Keyamura K., Ote T., Yamakawa T., Yamazaki Y., Mori H. *et al.* (2005) Cell size and nucleoid organization of engineered *Escherichia coli* cells with a reduced genome. *Mol Microbiol*, **55**, 137–149.
7. Kolisnychenko V., Plunkett G., 3rd, Herring C.D., Feher T., Posfai J., Blattner F.R. and Posfai G. (2002) Engineering a reduced *Escherichia coli* genome. *Genome Res*, **12**, 640–647.
8. Posfai G., Plunkett G., 3rd, Feher T., Frisch D., Keil G.M., Umenhoffer K., Kolisnychenko V., Stahl B., Sharma S.S., de Arruda M. *et al.* (2006) Emergent properties of reduced-genome *Escherichia coli*. *Science*, **312**, 1044–1046.
9. Copeland N.G., Jenkins N.A. and Court D.L. (2001) Recombineering: a powerful new tool for mouse functional genomics. *Nat Rev Genet*, **2**, 769–779.
10. Datsenko K.A. and Wanner B.L. (2000) One-step inactivation of chromosomal genes in *Escherichia coli* K-12 using PCR products. *Proc Natl Acad Sci U S A*, **97**, 6640–6645.

11. Lee E.C., Yu D., Martinez de Velasco J., Tessarollo L., Swing D.A., Court D.L., Jenkins N.A. and Copeland N.G. (2001) A highly efficient *Escherichia coli*-based chromosome engineering system adapted for recombinogenic targeting and subcloning of BAC DNA. *Genomics*, **73**, 56–65.
12. Oppenheim A.B., Rattray A.J., Bubunenko M., Thomason L.C. and Court D.L. (2004) *In vivo* recombineering of bacteriophage lambda by PCR fragments and single-strand oligonucleotides. *Virology*, **319**, 185–189.
13. Court D.L., Sawitzke J.A. and Thomason L.C. (2002) Genetic engineering using homologous recombination. *Annu Rev Genet*, **36**, 361–388.
14. Yu B.J., Sung B.H., Koob M.D., Lee C.H., Lee J.H., Lee W.S., Kim M.S. and Kim S.C. (2002) Minimization of the *Escherichia coli* genome using a Tn5-targeted Cre/loxP excision system. *Nat Biotechnol*, **20**, 1018–1023.
15. Choulika A., Perrin A., Dujon B. and Nicolas J.F. (1995) Induction of homologous recombination in mammalian chromosomes by using the I-SceI system of *Saccharomyces cerevisiae*. *Mol Cell Biol*, **15**, 1968–1973.
16. Rong Y.S., Titen S.W., Xie H.B., Golic M.M., Bastiani M., Bandyopadhyay P., Olivera B.M., Brodsky M., Rubin G.M. and Golic K.G. (2002) Targeted mutagenesis by homologous recombination in *D. melanogaster*. *Genes Dev*, **16**, 1568–1581.
17. Schmidt-Puchta W., Orel N., Kyryk A. and Puchta H. (2004) Intrachromosomal homologous recombination in *Arabidopsis thaliana*. *Methods Mol Biol*, **262**, 25–34.
18. Cox M.M., Layton S.L., Jiang T., Cole K., Hargis B.M., Berghman L.R., Bottje W.G. and Kwon Y.M. (2007) Scarless and site-directed mutagenesis in *Salmonella enteritidis* chromosome. *BMC Biotechnol*, 7, 59.
19. Jamsai D., Orford M., Nefedov M., Fucharoen S., Williamson R. and Ioannou P.A. (2003) Targeted modification of a human beta-globin locus BAC clone using GET Recombination and an I-Scei counterselection cassette. *Genomics*, **82**, 68–77.
20. Kang Y., Durfee T., Glasner J.D., Qiu Y., Frisch D., Winterberg K.M. and Blattner F.R. (2004) Systematic mutagenesis of the *Escherichia coli* genome. *J Bacteriol*, **186**, 4921–4930.
21. Sung B.H., Lee C.H., Yu B.J., Lee J.H., Lee J.Y., Kim M.S., Blattner F.R. and Kim S.C. (2006) Development of a biofilm production-deficient *Escherichia coli* strain as a host for biotechnological applications. *Appl Environ Microbiol*, **72**, 3336–3342.
22. Tischer B.K., von Einem J., Kaufer B. and Osterrieder N. (2006) Two-step red-mediated recombination for versatile high-efficiency markerless DNA manipulation in *Escherichia coli*. *Biotechniques*, **40**, 191–197.
23. Blattner F.R., Plunkett G., 3rd, Bloch C.A., Perna N.T., Burland V., Riley M., Collado-Vides J., Glasner J.D., Rode C.K., Mayhew G.F. *et al.* (1997) The complete genome sequence of *Escherichia coli* K-12. *Science*, **277**, 1453–1462.
24. Yu B.J., Kang K.H., Lee J.H., Sung B.H., Kim M.S. and Kim S.C. (2008) Rapid and efficient construction of markerless deletions in the *Escherichia coli* genome. *Nucleic Acids Res*, **36**, e84.
25. Yu D., Ellis H.M., Lee E.C., Jenkins N.A., Copeland N.G. and Court D.L. (2000) An efficient recombination system for chromosome engineering in *Escherichia coli*. *Proc Natl Acad Sci U S A*, **97**, 5978–5983.

Chapter 4

Random Chromosomal Gene Disruption In Vivo Using Transposomes

Les M. Hoffman

Abstract

Strain engineering of bacteria has been accomplished by many methods where mobile DNA elements (transposons) are inserted into the genomic DNA of a host organism. This chapter addresses engineering with transposable elements complexed with transposase enzyme. In traditional techniques, transposon and transposase are introduced as distinct entities. The method of mobilization into cells is often unique for each class of DNA element, and for each organism. The discovery of pre-formed transposon/transposase complexes (transposomes) that can be electroporated into living cells opens a new gateway to strain mutagenesis. Described are the preparation of electrocompetent bacterial cells and their transformation with transposomes. Once within the cell, the transposome is equipped to randomly insert its DNA into chromosomes without needing additional components. Ocr, a T7 phage protein that inhibits the host restriction of electroporated DNAs, will also be discussed as an adjunct reagent that can widen the applicability of transposomes. The transposomes used in most of the applications are commercially available, but also described is the process of making custom transposon DNAs and transposomes. The techniques are not limited to bacterial strain engineering per se and may be adapted for single-cell eukaryotes as well.

Key words: In vivo transposition, Transposome, Ocr protein, Electroporation

1. Introduction

Transposons are DNA elements that, with the assistance of transposases, can move from one genetic locus to another. Bacterial transposons of the *Tn* class are used extensively as research tools in molecular biology. They contain terminal inverted repeats and encode a transposase that excises the element from a donor site and rejoins it to DNA at a second location through a "cut-and-paste" mechanism. The molecular mechanisms of the transposon

James A. Williams (ed.), *Strain Engineering: Methods and Protocols*, Methods in Molecular Biology, vol. 765,
DOI 10.1007/978-1-61779-197-0_4, © Springer Science+Business Media, LLC 2011

Tn5 have been well characterized owing to the development of an in vitro transposition system and a hyperactive mutant of the Tn5 transposase (1).

Hyperactive Tn5 transposase complexes specifically with DNAs having inverted repeat ends, in this case chimeras of outer end (OE) and inner end (IE) Tn5 sequences (2, 3), called mosaic ends (MEs). MEs are the only sequences required for transposase binding, and any sequence between MEs becomes a transposon. The hyperactive transposase is around three orders of magnitude more efficient than wild type, and enables efficient in vivo reactions. The so-called synaptic complexes or transposomes contain transposons whose ends are brought together by the dimerization of transposase. In the absence of divalent cations the transposome is stable and catalytically inert. Once within the cytoplasm, however, magnesium ions activate transposition into cellular DNA by a cut-and-paste mechanism (2, 4).

The transposome system largely eliminates the bacterial host barriers for in vivo transposition. Host-encoded DNA restriction systems still exist, but ways to overcome them will be covered in this chapter. The Tn5 system was previously carried on plasmids (sometimes with one each for transposon and transposase), into species other than *Escherichia coli*. Transposome complexes have now been electroporated into a wide spectrum of bacterial cells in which the DNA was integrated directly into genomic (or episomal) DNA. The Tn5-based transposon inserts randomly and can create knockouts in nonessential genes. Mutagenic strain engineering is thus possible without conjugations between bacterial strains and without using "suicide" vectors (unable to replicate within the host). The lack of a transposon-borne transposase gene prevents later "hopping" of the inserted element, locking it in place.

Transposons have been artificially introduced into genomic DNA by several strategies. Prior to using synaptic complexes, the method of choice was transformation with suicide vectors encoding both transposable element and a transposase. Because the suicide plasmid does not replicate within the host, the transposon's selectable marker functions only after integration into the chromosome. Phage infection may also be used to mutagenize bacterial chromosomes by transposition. Bacteria are infected with a phage lambda derivative that is unable to either replicate or form lysogens, and carries a transposon. The transposon is maintained only if the transposable element has been incorporated into the chromosome or into a replicating episome. Both of the above methods have the disadvantage of using transposons encoding a transposase, which may cause instability of the transposon within the chromosome. The system described herein stabilizes transposable elements because the transposase is complexed outside the cell with transposon DNA and does not survive cell division.

Synaptic complexes may be introduced into many bacterial species (Table 1 and references therein) by electroporation, allowing subsequent integration of transposon DNA (2, 4). The efficiency of insertion varies with the species and strain of bacterium, but it is usually great enough to produce a library of knockout mutations for screening (5).

Table 1
Transposome-engineered species and strains

Gram-negative bacterium	References
Acinetobacter baumannii	Dorsey, C.W. et al. (2002) *Appl. Environ. Microbiol.* **68**, 6353 Tomaras, A.P. et. al. (2003) *Microbiology* **149**, 3473
Acinetobacter sp. ADP1	Lee, H. et al. (2003) *Antimicrob. Agents Chemother.* **47**, 1267
Actinobacillus pleuropneumonia	Godbout, M. et al. (2002) ASM 102nd General Meeting, abstr. B-148
Aeromonas hydrophila	Lambert, J. and McGarey, Jr., D.J. (2002) ASM 102nd General Meeting, abstr. B-378 Sarkis, Ted. et. al. (2003) Abstract: Georgia Journal of Science Meeting Florian-Frick, W. et al. (2009) J. Bacteriol. 191, 4750–4757
Agrobacterium tumefacien	Hoffman, L. et al. (2003) *Epicentre Forum* 9(2), 8
Afipia felis	Schueller, C. et al. (2009) *FEMS Micro.Lett.* **302**, 203
Bartonella henselae	Riess, T. et. al. (2003) *Gene* **313**, 103
Bdellovibrio bacteriovorus	Park, J.Y. et. al. Abstract: University of Oklahoma Health Sci. Ctr
Burkeholderia glumae	Nakata, P.A. (2002) *Plant Science* **162**, 267
Burkeholderia vietnamiensis	Menard, A. et al. (2007) *Environ. Microbiol.* **9**, 1176
Campylobacter jejuni	Lin, J. et al. (2002) *Antimicrob. Agents Chemother.* **46**, 2124
Enterobacter cloacae	Patel, E.S. et. al. (2005) Abstract 59th Annual Eastern Colleges Science Conference
Escherichia coli	Goryshin, I.Y. et al. (2000) *Nature Biotech.* **18**, 97 O'Reilly, E.K. and Kreuzer, K.N. (2004) *J. Bacteriol.* **186**, 7149 Winterberg, K.M. et. al. (2004) *Appl. and Environ. Microbiol.* **71**, 451
Escherichia coli (Shiga toxin-producing)	Lu, Y. et al. (2006) *Infect. and Immun.* **74**, 5747
Francisella tularensis	Qin, R. et al. (2006) *BMC Microbiol.* **6**, 69 Kawula, T.H. et al. (2004) *Appl. Environ. Microbiol.* **70**, 6901
Francisella novicida	Gallagher, L.A. et al. (2007) *Proc. Nat. Acad. Sci. USA* **104**, 1009
Haemophilus ducreyi	Post, D.M.B. et. al. (2005) *Infect. Immun.* **73**, 6727
Gluconacetobacter diazotrophicus	Rouws, L. et. al. (2008) *Arch. Microbiol.* (Ausgabe 00006/2008)

(continued)

Table 1 (continued)

Gram-negative bacterium	References
Moraxella catarrhalis	Holm, M.M. et al. (2003) *Infect. Immun.* **71**, 4977 Luke, N.R. et. al. (2003) *Infect. Immun.* **71**, 6426 Pearson, M.M. et. al. (2006) *Infect. Immun.* **74**, 1588
Morganella morganii	Ruzin, A. et al. (2005) *Antimicrob. Agents Chemother.* **49**, 791
Myxobacterium angiococcus	Sandmann, A. (2004) Dissertation: University of Braunschweig
Neisseria gonorrhoeae	Clark, V. and Spence, J. (2002) *Epicentre Forum* 9(2), 6
Pantoea stewartii	Minogue, T.D. et al. (2003) ASM 103nd General Meeting, abstr. H-134
Proteus mirabilis	Visalli, M.A. et al. (2003) *Antimicrob. Agents Chemother.* **47**, 665
Proteus vulgaris	Goryshin, I.Y. et al. (2000) *Nature Biotech.* **18**, 97
Pseudomonas sp. BW11M1	De los Santos, P.E. et. al. (2005) *FEMS Microb. Letters* **244**, 243
Pseudomonas sp. MMSS-8	Hoffman, L.M. et al. (2000) *Genetica* **108**, 19
Pseudomonas aeruginosa	Filiatrault, M. et al. (2006) *Infec. Immun.* **74,** 4237 Weagley, C. and Karkhoff-Schweizer, R., unpublished results; Sriramulu, D.D. et. al. (2005) *J. Med. Microbiol.* **54**, 667
Pseudomonas putida	Regenhardt, D. (2003) Dissertation: University of Braunschweig
Pseudomonas syringae	Bretz, J. et al. (2002) *Mol. Microbiol.* **45**, 397
Rhodopseudomonas palustres	Oda, Y. et. al. (2005) *J. Bacteriol.* **187**, 7784
Rickettsia monacensis	Baldridge, G.D. et. al. (2005) *Appl. and Environ. Microbiol.* **71**, 2095
Rickettsia prowazekii	Qin, A. et al. (2004) *Appl. Environ. Microbiol.* **70**, 2816 Tucker, A.M. et. al. (2005) *Ann. N.Y. Acad. Sci.* **1063**, 35
Rubrivivax gelatinosus	Vanzin, G.F. et al. (2002) *Proc. U.S. DOE Hydrogen Program Rev., Natl. Renewable* *Energy Laboratory*, CP-610-32405
Salmonella enterica	Clavijo, R.I. et al. (2006) *Appl. Environ. Microbiol.* **72,** 1055 Anriany, Y. et al. (2006) *Appl. Environ. Microbiol.* **72**, 5002 Hu, W.S. et. al. (2005) *Antimicrob. Agents and Chemother.* **49**, 3955
Salmonella typhimurium	Goryshin, I.Y. et al. (2000) *Nature Biotech.* **18**, 97 Jordan, D. et. al. (2004) *J. Appl. Microbiol.* **97,** 1054
Serratia marcesens	Su, L.H. et. al. (2005) Abstract: 15th ECCMID
Shigella boydii	Agle, M.E., unpublished results
Silicibacter sp. TM1040	Miller, T.R. (2004) Dissertation: University of Maryland Biotechnology Institute
Silicibacter pomeroyi	Buchan, A. et al. (2003) ASM 103nd General Meeting, abstr. N-304 Howard, E. and Henriksen, J. et al. (2006) *Science* **314**, 649–652 Burgmann, H. et al. (2007) *Environ. Microbiology* **9,** 2742

(continued)

Table 1 (continued)

Gram-negative bacterium	References
Stenotrophomonas maltophilia	Huang, T.-P. et al. (2006) *J. Bacteriol.* **188**, 3116
Xanthomonas campestris	Qian, W. et. al. (2005) *Genome Research* **15**, 757 Sun, Q. et. al. (2003) *FEMS Microbiol. Letters* **226**, 145
Xanthomonas citri	Levano-Garcia, J. et al. (2005) *BioTechniques* **38**, 225
Xanthomonas oryzae	Furutani, A. et al. (2003) *J. Bacteriol.* **186**, 1374 Sun, Q. et. al. (2003) *FEMS Microbiol. Letters* **226,** 145 Tsuge, S. et. al. (2004) *Phytopathology* **94**, 478
Xylella fastidiosa	Guilhabert, M.R. et al. (2001) *Mol. Plant Microbe Interact.***14**, 70 Koide, T. et. al. (2004) *Current Microbiol.* **48**, 247
Zymomonas mobilis	Zhang, M. (2007) US Patent 7,223,575

Gram-positive bacterium	References
Bacillus subtilis	Bertram, R. et al. (2005) *Nucl. Acids Res.* **33**, e153
Clavibacter michiganensis subsp. *Sepedonicus*	Ishimaru, C.A. et. al. (2005) Update: Colorado AES projects
Clostridium perfringens	Vidal, J. et al. (2009) *PLoS ONE* **4**(7), e6232
Corynebacterium diphtheriae	Oram, D.M. et al. (2002) *J. Bacteriol.* **184**, 5723
Corynebacterium glutamicum	Kawaguchi, H. et. al. (2006) *Appl. Environ. Microbiol.* **72**, 3418 Suzuki, N. et. al. (2006) *Appl. Environ. Microbiol.* **72**, 3750
Corynebacterium matruchotii	Takayama, K. et al. (2003) *Biochem. J.* **373**, 465 Wang, C. et. al. (2006) *Biochem Biophys Res Commun.* **340**, 953
Lactobaciilus casei	Ito M., et al. (2010) *J. Appl. Microbiol.* **109**, 657
Mycobacterium avium	Laurent, J.-P. et al. (2003) *J. Bacteriol.* **185**, 5003 Cangelosi, G.A. et al. (2006) *Antimicrob. Agents Chemother.* **50**, 461 Philalay, J.S. et. al. (2004) *Antimicrob. Agents Chemother.* **48**, 3412
Mycobacterium bovis (BCG)	Stewart, G.R. et. al. (2005) *PLoS Pathol.***1**, e33
Mycobacterium smegmatis	Derbyshire, K.M. et al. (2000) *Epicentre Forum* 7(2), 1 Chen, C.K. et. al. (2002) *Microbiol.* **48**, 289 Flores, A.R. et al. (2005) *J. Bacteriol.* **187**, 1892 Maus, C. E. et al. (2005) *Antimicrob. Agents Chemother.* **49**,571
Mycobacterium tuberculosis	Maus, C.E. et. al. (2005) *Antimicrob. Agents Chemother.* **49**, 571
Mycobacterium ulcerans	Alford, T.D. and Small, P.L.C. (2002) ASM 102nd General Meeting, abstr. U-29
Rhodococcus equi	Mangan, M.W. and Meijer, W.G. (2001) *FEMS Microbiol. Lett.* **205**, 243 Miranda-CasoLuengo, R. et. al. (2005) *J. Bacteriol.* **187**, 3438
Rhodococcus sp. 124	Rao, S. (2003) *BUG J.* **6**, 151

(continued)

Table 1 (continued)

Gram-negative bacterium	References
Rhodococcus erythropolis	Tanaka, Y. et al. (2002) *Arch. Microbiol.* **178**, 351
Rhodococcus rhodochrous	Fernandes, P.J. et al. (2001) *Microbiology* **147**, 2529
Spiroplasma citri	Mutaqin, K.H. (2005) Dissertation: Oklahoma State University
Streptococcus pyogenes	Cho, K.H. and Caparon, M.G. (2004) ASM 104th General Meeting, abstr. B-316
Thiomicrospira crunogena	Dobrinski, K.P. et. al. (2006) ASM 106th Annual General Meeting abstr.
Other microorganisms	
Saccharomyces cerevisiae	Goryshin, I.Y. et al. (2000) *Nature Biotechnol.* **18,** 97
Trypanosoma brucei	Shi, H. et al. (2002) *Mol. Biochem. Parasitol.* **121**, 14

There are several methods to find the locations of Tn5-derived element inserting in mutagenized chromosomes. Kirby (6) describes the use of rescue cloning, in which genomic DNA is restricted and ligated to produce rescue plasmids from the vicinity of transposons. A conditional origin of replication within the transposon (R6Kγ in the case of EZ-Tn5™ constructs) allows replication and the antibiotic resistance gene is used for selection. Genomic DNA can also be directly sequenced with primers directed from the ends of the transposable element (4).

Evolution has produced many ways for foreign DNA to be prevented from integration into genomes, and phages have similarly developed their own methods to circumvent detection and destruction. The first protein to be produced during infection of *E. coli* by bacteriophage T7 is "overcome classical restriction" (ocr), the product of gene 0.3 (7). Interestingly, this phage-encoded protein mimics DNA and acts as a molecular decoy to draw Type I restriction endonucleases away from nonmodified phage DNA (8). Ocr crystallographic structure reveals a protein that resembles B-form DNA and whose dimer has a bend of 33.6° (9). Ocr inhibits host restriction long enough to allow the foreign DNA to attain methylation and protection against restriction.

Researchers at Epicentre Biotechnologies discovered that ocr protein could be electroporated into bacterial cells and can enhance co-introduced transposome efficiency. In taxa with well-characterized restriction-modification systems, ocr makes a dramatic difference in in vivo transposition or transformation results. Ocr (TypeOne™ Inhibitor, Epicentre) dramatically improved plasmid or fosmid transformation efficiencies when the host was restriction-positive for a site contained in the episomal DNA

Table 2
TypeOne restriction inhibitor effects on transformation efficiencies

Strain	Type I R-M system	TypeOne™ inhibitor	Type of DNA or transposome™	Recombinants per microgram DNA
S. typhimurium LT2	StyL TIII	–	pUC19 (100 pg)	3.0×10^{6}
S. typhimurium LT2	StyL TIII	+	pUC19 (100 pg)	3.0×10^{8}
S. typhimurium LB5000	None	–	pUC19 (100 pg)	2.0×10^{10}
E. coli MG1655	EcoKl	–	48 Kb fosmid (50 ng)	3.0×10^{3}
E. coli MG1655	EcoKl	+	48 Kb fosmid (50 ng)	1.4×10^{6}
S. typhimurium LT2	StyL TIII	–	EZ::TN™ <R6Kγori/KAN-2>Tnp Transposome™ (1 μl)	1.3×10^{4}
S. typhimurium LT2	StyL TIII	+	EZ::TN™ <R6Kγori/KAN-2>Tnp Transposome™ (1 μl)	1.0×10^{6}
A. tumefaciens	None	–	EZ::TN™ <R6Kγori/KAN-2>Tnp Transposome™ (1 μl)	2.2×10^{5}
A. tumefaciens	None	+	EZ::TN™ <R6Kγori/KAN-2>Tnp Transposome™ (1 μl)	1.3×10^{5}

(Table 2, ref. 10). Fosmid transformations of wild-type *E. coli* were over 450-fold more efficient after the addition of the inhibitor to electroporations (Table 2). The EZ::TN™ <R6Kγori/KAN-2>Tnp Transposome™ contains six recognition sites for the type I *Salmonella typhimurium* StyL TIII nuclease. When this transposome was electroporated into *S. typhimurium* LT2 together with ocr, the number of clones with transposon insertions was increased by 75-fold (Table 2). The addition of TypeOne Inhibitor did not change insertion efficiency when there was no restriction activity in the cell, as shown for *Agrobacterium tumefaciens* (see Note 1).

Transposomes are more universally applicable for bacteria with type I restriction/modification systems when ocr protein is electrophoresed along with synaptic complexes. Because the type I restriction status of many strains is unknown, it may be useful to electroporate transposomes with and without ocr to test its effects. Whether ocr is effective in a specific cell type may be difficult to determine, but the phage protein can be included prophylactically during in vivo transposome mutagenesis. Type I restriction-modification systems (R/M Type I) are common in bacteria, but it is difficult to predict whether a particular strain's R/M system will affect transposomes without knowing the sequence specificity of the restriction enzyme and the sequence of the transposon.

How random are the in vivo gene interruptions by Tn5 transposomes? Southern blots of genomic DNA show that at the resolution level of agarose gels, the transposition is random (11). Kang et al. (12) sequenced 1,960 Tn5 transposition sites in *E. coli* genes and concluded that there may be a slight preference for guanosines at the insertion sites, but no other bias was observed. Other data, including those of Reznikoff et al. (13), imply that regions of middling GC content are slightly favored over areas of high or low GC.

Table 2 lists organisms whose genomes have been electroporated and mutagenized with Tn5 transposomes, and citations for each species. The list includes gram-negative and gram-positive bacteria and several eukaryotes.

I chose three organisms to highlight for in vivo transposition techniques: the model organism workhorse *E. coli*, a gram-positive pathogen; *Clostridium perfringens*; and the marine bacterium *Silicibacter* sp. TM1040. The *Clostridia* are low GC gram-positive bacteria and are common in gastrointestinal tracts. They live in virtually all of the anaerobic habitats of nature where organic compounds are found, including soils, aquatic sediments, and the intestinal tracts of animals (14). *C. perfringens* is the most genetically tractable of the pathogenic *Clostridia*, and its virulence and physiology are well studied. *Silicibacter* sp. TM1040 is a good example of a well-characterized bacterium from marine environments (15), a representative of the *Roseobacter* clade of *Alphaproteobacteria*. These bacteria are highly adapted to form symbioses with unicellular eukaryotic phytoplankton, and may be crucial to the health of corals and to ocean-atmosphere sulfur flux (16).

No one transposome mutagenesis method can be universal, but within these general bacterial classifications there are commonalities, and the methods described can be adapted for many species. The largest factors for success with transposome technologies may be obtaining efficient electroporation and preventing host restriction systems from degrading transposon DNAs. Other helpful hints for strain engineering with transposomes are found at the website http://www.epibio.com/guides/helpful%20hints%20for%20using%20transposomes.pdf. This site is periodically updated with new suggestions.

2. Materials

1. Pre-formed Transposome: The EZ-Tn5™ <R6K ori/KAN-2> Transposome™ Kit (Epicentre Biotechnologies, Madison, WI). It is the formulation of choice for most applications, but other transposomes can be formed from custom transposable elements (see Subheading 3.3).

Table 3
EZ-Tn5 pMOD vectors

EZ-Tn5™ transposon construction vectors	*ori* that is located on vector *outside* of the ME sequences	*ori* that is located within the ME sequences
pMOD™-2<MCS>	colE1	None
pMOD™-3<R6Kγ*ori*/MCS>	colE1	R6Kγ*ori*
pMOD™-4<MCS>	R6Kγ*ori*	None
pMOD™-5<R6Kγ*ori*/MCS>	None	R6Kγ*ori*
pMOD™-6<KAN-2/MCS>	colE1	None

2. EZ-Tn5™ Transposase (Epicentre Biotechnologies).
3. Restriction Inhibitor: TypeOne Restriction Inhibitor (Epicentre Biotechnologies).
4. Bacterial Cells: Specific bacterial strains, often more amenable to electroporation than the wild type, are available from ATCC or from research laboratories worldwide (see Table 1).
5. Plasmids for Building Custom Transposons: see Table 3 for commercially available plasmid vectors.
6. 10 mg/ml erythromycin: Prepare in ethanol. Store refrigerated.
7. 50 mg/ml kanamycin: Prepare in distilled water and filter sterilize. Store refrigerated.
8. 50 mg/ml ampicillin: Prepare in distilled water and filter sterilize. Store refrigerated.
9. Brain Heart Infusion Plates +40 μg/ml erythromycin: For each liter of medium mix 250 g calf brain infusion, 200 g beef heart infusion, 10 g proteose peptone, 5 g sodium chloride, 2.5 g disodium phosphate, 2 g dextrose, and 15 g agar. Autoclave for 15 min at 121°C. When the medium is cooled add 4 ml/l of 10 mg/ml erythromycin.
10. LB Medium: For 1 l of LB, 10 g tryptone, 5 g yeast extract, and 10 g NaCl are added. Autoclave for 15 min at 121°C.
11. LB Agar plates: LB medium with 1.5% Bacto agar added before autoclaving.
12. LB Agar plates containing 50 μg/ml ampicillin and 40 μg/ml erythromycin: LB agar with ampicillin (1 ml of 50 mg/l stock) and erythromycin (4 ml of 10 mg/ml stock) added per liter after autoclaving and cooling.
13. LB Agar plates containing 30 μg/ml kanamycin: LB agar with kanamycin (0.6 ml of 50 mg/l stock) added per liter after autoclaving and cooling.

14. No Salt LB medium: For 1 l of LB, 10 g tryptone and 5 g yeast extract are added. Autoclave for 15 min at 121°C.
15. TGY Medium: To make 1 l of TGY, mix 5 g yeast extract, 10 g peptone, and 2 g glucose. Autoclave for 15 min at 121°C.
16. HIASW Medium: Add 25 g heart infusion broth (Difco) plus 15 g Instant Ocean sea salts (Aquarium Systems, Mentor, OH) to 1 l with deionized water. Autoclave for 15 min at 121°C.
17. HIASW agar + 50 μg/ml kanamycin plates: HIASW medium with 1.5% Bacto agar added before autoclaving. When the medium is cooled add 1 ml/l of 50 mg/ml kanamycin stock.
18. Marine Broth Medium: Suspend 55 g of the medium (Carl Roth, Karlsruhe, DE) in 1 l of distilled water. Heat to boiling, agitate frequently until completely dissolved. Sterilize at 121°C for 15 min.
19. TE Buffer: TE is 10 mM Tris–HCl, pH 7.5, 1 mM EDTA.
20. Electroporation Solution: The solution consists of 10% PEG 8000 in distilled water, autoclaved for 15 min at 121°C.
21. PCR Primers for erythromycin gene amplification: The primers erm-Fwd-*Eco*RI (5′-AAGGGAATTCCTAAAAATTTGTAAT TAAGAAGGAGT) and erm-Rev-*Hind*III (5′-AAGGAAG CTTCCAAATTTACAAAAGCGACTCATA) can be obtained from Integrated DNA Technologies (Coralville, IA).
22. 10% Glycerol: The solution consists of 10% glycerol in distilled water, autoclaved for 15 min at 121°C.
23. Vector p-MOD2: The plasmid is available from Epicentre Biotechnologies.
24. Vector pJIR751: The shuttle plasmid is available from ATCC (Manassas, VA).
25. GELase™ Agarose Gel-Digesting Preparation: The enzyme and buffer are available from Epicentre Biotechnologies.
26. Plasmid Miniprep Kit: The Zyppy™ Plasmid Miniprep Kit (Zymo Research, Orange, CA) is used according to the manufacturer's instructions.
27. DNA Clean & Concentrator™-5: Spin column kits are obtained from Zymo Research and used according to directions.
28. Restriction Enzymes *Eco*RI, *Hind*III, *Pvu*II, *Psh*AI, and *Eco*RV and DNA ligase. These are available from New England Biolabs (Ipswich, MA).

29. Precast Agarose Gels: Pre-made gels may be obtained from Sigma (St. Louis, MO).
30. Electroporator: Gene Pulser with a Pulse Controller and 0.1 and 0.2 cm gap electroporation cuvettes (Bio-Rad Laboratories, Richmond, CA) or equivalent (e.g., Eppendorf Multiporator).
31. MasterPure DNA Purification Kit (Epicentre Biotechnologies).
32. DNA Clean & Concentrator™-5 spin column (Zymo Research).
33. Electrocompetent EC100D *pir+* or *pir*-116 *E. coli* (Epicentre Biotechnologies).
34. KAN-2 FP-1 and R6KAN-2 RP-1 sequencing primers (Epicentre Biotechnologies).

3. Methods

Newer techniques, for example, electroporating gram-positive species with transposomes, are listed after the standard method for gram-negative electrocompetent cells such as *E. coli*. Due to the extensive number of microorganisms successfully mutated in vivo with transposomes, it would be impractical to list methodologies for each species. The individual articles in Table 1 are recommended as references regarding specific species.

3.1. Preparation of Electrocompetent E. coli Cells

1. Streak for single colonies from –70°C glycerol stock onto a plate of appropriate medium. Start a 50 ml culture of the organism in no salt LB broth at 37°C, shaking at 200 rpm overnight. From the overnight culture, use 25 ml inoculum into 1 l of no salt LB broth (prewarmed to 37°). Grow at 37°C, shaking at approximately 200 rpm, to A_{600} = 0.6–0.75. Chill on ice immediately.
2. Spin culture 10,000×*g* 10 min and resuspend in 200 ml of ice-cold 10% glycerol.
3. Spin 10,000×*g* 10 min and resuspend in 150 ml cold 10% glycerol.
4. Spin 10,000×*g* 10 min and resuspend in 100 ml cold 10% glycerol.
5. Spin 10,000×*g* 10 min and decant, removing most of the 10% glycerol.
6. To pellet add 1–2 ml 10% glycerol. Resuspend gently with 1 ml pipettor. Dilute an aliquot of the cells 1:300 (10 µl to 3 ml) in 10% glycerol. Its A_{600} should be between 0.7 and 0.85, which indicates an A_{600} = 200–250 of the undiluted cells. If the cell

concentration is low, they can be pelleted in a microcentrifuge at 10,000×*g* for 5 min and brought to the desired volume.

7. Aliquot 110 μl cells into prechilled 1.5 ml microcentrifuge tubes. Freeze at −70° (see Note 2).

3.2. Electroporation of E. coli Cells

1. Thaw electrocompetent cells on ice and aliquot 50 μl per sample into 500 μl microcentrifuge tubes on ice.
2. Add the transposome in 1 μl of TE buffer.
3. Add sample to a sterile 2 mm gap electroporation cuvette. Electroporate at 2.5 kV for *E. coli* if using a Multiporator (Eppendorf). Optimal settings for other instruments may vary; with the Bio-Rad Gene Pulser, use 2.5 kV for a 0.2-cm gap cuvette and 1.8 kV for a 0.1-cm gap cuvette (25–80 μl).
4. Add approximately 0.3 ml of LB broth to cell and rinse cells from the cuvette with a 1 ml pipettor. Add cells to remainder of 1 ml of LB broth and shake at 370 rpm for 30 min to 1 h.
5. Plate 10–100 μl of the outgrowth on the appropriate selective plates (see Note 3).

3.3. Construction of an Erythromycin Resistance Tn5 Transposome for Clostridium sp. (17)

The pMOD series of plasmids was designed to allow creation of constructs containing custom antibiotic resistance gene cassettes, promoters, etc., that are not offered commercially. All contain multiple cloning sites flanked by transposon MEs that are in turn flanked by *Pvu*II/*PshA*I restriction sites. After constructing the appropriate transposon it can be easily excised with *Pvu*II or *PshA*I (see Notes 4 and 5).

Table 3 lists features of the current pMOD vectors that are available. Transposons conferring kanamycin or trimethoprim (DHFR) resistance are available in pre-formed transposomes. Tetracycline resistance transposons in an in vitro transposition kit can be adapted for cell electroporation by the addition of EZ-Tn5 Transposase (see step 7 in Subheading 3.3).

1. Plasmid pMOD-2 is digested with *Eco*RI and *Hind*III (New England Biolabs). The restriction nucleases are inactivated by heating 15 min at 70°C.
2. The erythromycin resistance gene of the *E. coli–C. perfringens* shuttle vector pJIR751 is amplified by PCR. The primers are erm-Fwd-*Eco*RI and erm-Rev-*Hind*III.
3. The PCR product is resolved by agarose gel electrophoresis and purified using GELase and digested with *Eco*RI and *Hind*III.
4. The linear vector pMOD-2 and the purified PCR of the erythromycin resistance gene are ligated by standard techniques.

5. The ligation is transformed into an electrocompetent *E. coli* such as DH5α, and the transformants are selected on LB plates containing ampicillin and erythromycin.
6. The resulting plasmid is digested with *Pvu*II (see Note 4) and the approximately 900 bp transposon segment is purified by agarose gel electrophoresis.
7. After quantifying the amount of transposon DNA, it is reacted with EZ-Tn5 Transposase as follows: 2 μl of transposome DNA (100 ng/μl in TE buffer) are mixed with 4 μl EZ-Tn5 transposase and 2 μl of glycerol.
8. After 30 min at room temperature, the transposome is stably formed and can be stored at –20°C.

3.4. Preparation of Electrocompetent C. perfringens and Cell Electroporation (17) (See Notes 2 and 3)

1. Late-exponential-phase cultures ($A_{600}=1.2$) of strain 13 *C. perfringens* grown anaerobically in TGY medium at 37°C are harvested and washed with electroporation solution (see steps 1–5 in Subheading 3.1).
2. The cell pellets are suspended in 1/20 of a volume of electroporation solution, and 0.4 ml of the cell suspension are mixed with 1 μl of the transposome. The cells are incubated on ice for 5 min.
3. Electroporate in a 0.2-cm gap electroporation cuvette as previously described in step 3 in Subheading 3.2 with a Gene Pulser set at 1,500 V, 25 μF, and 200 Ω, and a pulse delivery time between 30 and 40 ms.
4. Electroporated transposome-containing bacteria are grown in 3 ml of prewarmed TGY medium for 3 h at 37° and plated onto brain heart infusion agar plates with erythromycin (40 μg/ml). Plates are incubated at 37°C for 18 h under anaerobic conditions.

3.5. Preparation of Electrocompetent Silicibacter sp. Cells (18) (See Notes 2 and 3)

1. TM1040 cells are grown to an A_{578} of 0.5 in Marine Broth (MB) medium with shaking at 200 rpm at 30°C.
2. Ten milliliter cultures are centrifuged for 15 min at 3,200×*g*. Pelleted cells are washed five times with 10 ml cold 10% (v/v) glycerol. Then, the cell pellet is resuspended in 400 μl 10% (v/v) glycerol.
3. 50 μl aliquots are frozen in liquid nitrogen and stored at –80°C.

3.6. Co-Electroporation of ocr Protein into Electrocompetent Silicibacter sp. Cells (15, 16) (See Note 1)

1. A 65-μl sample of electrocompetent cells of TM1040 is mixed with 25 ng of the transposome, and 1 μl of TypeOne™ Restriction Inhibitor, and the bacteria are electroporated in a 0.2-cm gap electroporation cuvette at 2.5 kV/cm, 400 Ω, and 25 μF using a Bio-Rad GenePulser.

2. The cells are suspended in 1 ml of prewarmed HIASW broth and incubated at 30°C with shaking for 2 h.
3. After incubation, 100-μl samples of the culture are spread on HIASW agar containing kanamycin and incubated for 48 h at 30°C.

3.7. Location of Transposon <R6Kγori/KAN-2> Insertion Sites in E. coli by Ligation Capture (19) (See Note 6)

1. Transposon <R6Kγori/KAN-2> mutants of *E. coli* are grown individually in 2 ml LB medium in the presence of 50 μg/ml kanamycin overnight at 37°C and 200 rpm.
2. 1.5 ml of each is used to prepare chromosomal DNA using the MasterPure DNA Purification Kit.
3. 5 μl of each of the chromosomal DNA preps is digested with *Eco*RV in a final volume of 20 μl overnight at 37°C. The *Eco*RV is heat inactivated at 80°C for 20 min.
4. The 20 μl digest is then re-ligated in 100 μl final volume using T4 DNA ligase for 48 h at 4°C.
5. Each re-ligation is then cleaned up using a Zymo DNA Clean & Concentrator™-5 spin column and eluted in 20 μl water.
6. 2 μl of re-ligated, *Eco*RV-digested chromosomal DNA from each mutant is electroporated into 40 μl of electrocompetent EC100D *pir+* cells using a 1-mm cuvette, 100 Ω, 25 μF, and 20 kV/cm, outgrowth is in 1 ml LB medium at 37°C for 1 h.
7. Each of the transformations is then plated out on LB agar supplemented with 30 μg/ml kanamycin and incubated overnight at 37°C.
8. From each plate where colonies grew one colony is picked and grown in 5 ml L-Broth plus 30 μg/ml kanamycin at 37°C overnight.
9. Minipreps of plasmid are prepared and 10 μl plasmid DNA is sequenced using transposon-specific primers KAN-2 FP-1 and R6KAN-2 RP-1.

4. Notes

1. TypeOne Restriction Inhibitor can be added to electroporations as a protective measure, even when there is no information about the host bacteria's restriction/modification system. The protein will not inhibit in vitro transposon insertions in the absence of any type 1 restriction endonucleases (Table 2).

2. The stability of frozen competent cells can vary widely between strains and species of bacteria. It may be helpful to test the efficiency of transformation with a compatible plasmid over a course of time. Freshly prepared cells will nearly always outperform frozen preparations.
3. Transformed cell growth conditions after electroporation can be critical for obtaining good transposition efficiencies. The duration, temperature, and medium used for outgrowth can be tested for their effects on colony counts on antibiotic plates.
4. The pMOD series of transposon construction vectors have both *Pvu*II and *PshA*I (BoxI) restriction sites flanking the transposon ends (mosaic ends or MEs). *PshA*I cuts less frequently than *Pvu*II, in many DNAs and may be used to excise transposomes from the vector in cases where the transposon has an internal *Pvu*II site.
5. Plasmids containing transposons for custom transposomes require complete restriction digestions. Any remaining uncut plasmid is transformed and may replicate in the host cell to confer antibiotic resistance. Even transposon fragments cut from an agarose gels can be contaminated with small amounts of uncut plasmid, leading to background colonies if the plasmid can replicate in the host.
6. The EZ-Tn5 <R6Kγ ori/KAN-2> Transposome, or a custom transposome with the same replication origin, is the construct of choice for experiments in which the insertion sites need to be determined. The conditional origin of replication in the transposon is operative in EC100D *pir+* or *pir*-116 *E. coli*, which express a protein essential for the R6Kγ origin to function.

Acknowledgments

The assistance, advice, and amicability of the staff at Epicentre Biotechnologies has been appreciated, and contributed much to the development and dispersal of transposome technologies throughout the bioresearch community. Thanks to Dr. Fred Hyde for curating the list of transposome publications. Without decades of transposable element research in the Bill Reznikoff group at the University of Wisconsin-Madison, none of these technologies would have come to fruition. I also thank all the researchers who enabled this chapter by using transposome technology to modify strains.

References

1. Goryshin I. Y. and Reznikoff W. S. (1998) Tn5 *in vitro* transposition. *J. Biol. Chem.* **273**, 7367–7374.
2. Goryshin I. Y., Jendrisak J., Hoffman L. M., Meis R., and ReznikoffW.S. (2000) Insertional transposon mutagenesis by electroporation of released Tn5 transposition complexes. Nat. Biotechnol. **18**, 97–100.
3. Zhou M., Bhasin A., and Reznikoff W. R. (1998) Molecular genetic analysis of transposase-end DNA sequence recognition: cooperativity of three adjacent base-pairs in specific interaction with a mutant Tn5 transposase. J. Mol. Biol. **276,** 913–925.
4. Hoffman L. M., Jendrisak J. J., Meis R.J., Goryshin I. Y., and Reznikoff W. S.(2000) Transposome insertional mutagenesis and direct sequencing of microbial genomes. *Genetica* **108**, 19–24.
5. Hoffman L. M. and Jendrisak J. J. (2002) Transposomes: a system for identifying genes involved in bacterial pathogenesis. *Methods Enzymol.* **358**,128–40.
6. Kirby J. R. (2007) *In Vivo* Mutagenesis Using EZ-Tn5™. *Methods Enzymol.* **421**, 17–21.
7. Krüger D. H.,Schroeder C.,Hansen S.,and Rosenthal H.A. (1977). Active protection by bacteriophages T3 and T7 against *E. coli* B- and K-specific restriction of their DNA. *Mol Gen Genet.* **153,** 99–106.
8. Atanasiu C.,Byron O.,McMiken H.,Sturrock S.S.,and Dryden D.T.(2001) Characterization of the structure of ocr, the gene 0.3 protein of bacteriophage T7. *Nucleic Acids Res.* **29**, 3059–68.
9. Walkinshaw M.D., Taylor P., Sturrock S. S., Atanasiu C., Berge T., Henderson R. M., et al. (2002) Structure of ocr from bacteriophage T7, a protein that mimics B-form DNA. *Mol. Cell* **9**, 187–194.
10. Hoffman L. M., Haskins D. J., and Jendrisak J. J. (2003) TypeOne™ Inhibitor Improves Transformation Efficiencies by Blocking Type I Restriction and Modification Systems *In Vivo. Epicentre Forum* **9** (2), 8.
11. Fernandes P. J., Powell J. A., and Archer J. A. (2001) Construction of *Rhodococcus* random mutagenesis libraries using Tn5 transposition complexes. *Microbiology* **147**, 2529–2536.
12. KangY.,Durfee T.,Glasner J.D.,Qiu Y.,Frisch D.,Winterberg K.M., and Blattner F.R. (2004) Systematic mutagenesis of the *Escherichia coli* genome. *J. Bacteriol.* **186**, 4921–30.
13. Reznikoff W. S.,Goryshin I.Y.,and Jendrisak J. J. (2004). Tn5 as a molecular genetics tool: *In vitro* transposition and the coupling of *in vitro* technologies with *in vivo* transposition. *Methods Mol Biol.* **260**, 83–96.
14. Leser T. D.,Amenuvor J. Z.,Jensen T. K.,Lindecrona R.H.,Boye M.,and Møller K. (2002) Culture-independent analysis of gut bacteria: the pig gastrointestinal tract microbiota revisited. *Appl. Environ. Microbiol.* **68**, 673–690.
15. Belas R., Horikawa E., Aizawa S., and Suvanasuthi R. (2009) Genetic determinants of *Silicibacter* sp. TM1040 motility. *J. Bacteriol.* **191**, 4502–4512.
16. Howard E. C., Henriksen J. R., Buchan A., Reisch C. R., Bürgmann H., Welsh R. et al. (2006) Bacterial Taxa Limit Sulfur Flux from the Ocean. *Science* **314**, 649–652.
17. Vidal J. E., Chen J., Li J. and McClane B.A. (2009) Use of an EZ-Tn5-based random mutagenesis system to identify a novel toxin regulatory locus in *Clostridium perfringens* strain 13. *PLoS One* **14**, e6232.
18. Piekarski T., Buchholz I., Drepper T., Schobert M., Wagner-Doebler I., Tielen P., and Jahn D. (2009) Genetic tools for the investigation of *Roseobacter* clade bacteria. *BMC Microbiol.* **9**, 265.
19. Chaudhuri R.R., Peters S.E., Pleasance S.J., Northen H., Willers C., Paterson G.K., et al.(2009) Comprehensive Identification of *Salmonella enterica* Serovar Typhimurium Genes Required for Infection of BALB/c Mice. *PLoS Pathog.* **5**, e1000529.

Chapter 5

Genome Engineering Using Targeted Oligonucleotide Libraries and Functional Selection

Elie J. Diner, Fernando Garza-Sánchez, and Christopher S. Hayes

Abstract

The λ phage Red proteins greatly enhance homologous recombination in *Escherichia coli*. Red-mediated recombination or "recombineering" can be used to construct targeted gene deletions as well as to introduce point mutations into the genome. Here, we describe our method for scanning mutagenesis using recombineered oligonucleotide libraries. This approach entails randomization of specific codons within a target gene, followed by functional selection to isolate mutants. Oligonucleotide library mutagenesis has generated hundreds of novel antibiotic resistance mutations in genes encoding ribosomal proteins, and should be applicable to other systems for which functional selections exist.

Key words: Antibiotic resistance, Electroporation, Oligonucleotide, Recombineering, Ribosomal proteins, Spectinomycin

1. Introduction

Red-mediated recombination, or "recombineering," exploits the bacteriophage λ Red proteins (Gam, Exo, and Beta), which promote homologous recombination in *Escherichia coli*. The Gam protein inhibits the *E. coli* RecBCD nuclease, thereby prolonging the half-life of transformed linear duplex DNA. The 5′–3′ exonuclease activity of Exo generates single-stranded DNA, which then anneals to complementary regions on the chromosome through the activity of Beta (1, 2). Red-mediated recombination is quite efficient, occurring between DNAs with as little as 50 nucleotides of homology. This technology has had a tremendous impact on the pace of molecular genetic research in *E. coli*, facilitating the construction of an ordered collection of single-gene knock-out mutants (3). Recombineering has been extended to other γ-proteobacteria (*Shigella flexneri*, *Salmonella typhimurium*, *Pseudomonas syringae*)

James A. Williams (ed.), *Strain Engineering: Methods and Protocols*, Methods in Molecular Biology, vol. 765, DOI 10.1007/978-1-61779-197-0_5,

and to *Mycobacteria* species (4, 5), and this approach could be applicable to many other systems.

Court and colleagues have shown that recombineering may also be used to introduce single-stranded oligonucleotides directly into the *E. coli* chromosome (6). This recombination requires only the Beta protein, which facilitates hybridization of oligonucleotides to single-stranded regions of DNA that are exposed during replication. Consistent with this model, oligonucleotides that anneal to the lagging strand template DNA recombine more efficiently than those complementary to the leading strand (7). Presumably, the hybridized oligonucleotide is ligated to newly replicated DNA, resulting in its incorporation into the genome. Some oligonucleotide recombination efficiencies can be quite high, allowing the engineered mutations to be isolated by screening. Recombination efficiencies can be further enhanced by recombineering into methyl mismatch repair (MMR) deficient mutants, which are unable to correct mismatched base pairs introduced by the recombined oligonucleotide (7) (see Note 1). Thus, oligonucleotide recombineering is a remarkably powerful molecular genetic tool capable of introducing single nucleotide mutations into the genome without the cumbersome rounds of positive and negative selection required for standard allelic exchange.

We have exploited Red-mediated recombineering to perform scanning mutagenesis using synthetic oligonucleotide libraries. Oligonucleotide libraries containing one or more randomized codons are introduced into Red-expressing cells by electroporation, and the transformed cells are then subjected to functional selections to isolate mutants. This approach requires foreknowledge that mutagenesis of a particular target gene will result in a selectable (or screenable) phenotype, and is therefore best suited to characterized systems. We have used oligonucleotide libraries to mutagenize the genes encoding ribosomal proteins L4, L22, and S12, resulting in the identification of novel macrolide and aminoglycoside antibiotic resistance mutations (8, 9). One powerful feature of this approach is that unusual missense mutations, such as Lys (AAR) to Phe (TTY), can be isolated readily, whereas these mutations are virtually impossible to obtain using standard chemical mutagenesis.

Here, we apply oligonucleotide library mutagenesis to the *E. coli rpsE* gene, which encodes ribosomal protein S5. Mutations that alter loop 2 (residues 21–31) in S5 have been shown to confer spectinomycin resistance to *E. coli*, *Bacillus subtilis*, *Streptomyces roseosporus*, and *Pasteurella multocida* (10–16). Based on these observations, we randomized the codons corresponding to Val25 and Lys26 using recombineered oligonucleotide libraries and selected the mutagenized cells for resistance to spectinomycin. These genetic selections resulted in the isolation of 20 distinct spectinomycin-resistance mutations that encode 14 different S5 protein variants, 13 of which have not been previously described. In principle, this methodology may be used to mutagenize any gene for which a strong functional selection exists.

2. Materials

2.1. Preparation of Electrocompetent E. coli Cells

1. Luria-Bertani (LB) medium: 10 g NaCl, 10 g tryptone, 5 g yeast extract, 1 mL of 1N NaOH per liter. Autoclave to sterilize.
2. Antibiotic stock solutions: 75 mg/mL ampicillin in 70% ethanol; 33 mg/mL chloramphenicol in 70% ethanol; 50 mg/mL spectinomycin in water.
3. Carbohydrate stock solutions: 20% L-arabinose in water; 40% D-glucose in water.
4. Temperature-controlled environmental rotary shaker (set at 30°C).
5. Temperature-controlled shaking water bath (set at 42°C).
6. NANOpure water: Thermo-Barnstead, or equivalent 18 MΩ-cm source sterilized by autoclave, and chilled to 0°C in an ice-water bath.
7. Sorvall RC5B superspeed centrifuge and SS-34 rotor (or equivalent) centrifuge/rotor combination.
8. Polypropylene 50 mL Oak Ridge centrifuge tubes (Nalgene). Autoclave to sterilize.
9. Microcentrifuge chilled to 4–10°C.
10. Precooled microcentrifuge tubes.
11. Glass 10 mL pipette. Autoclave to sterilize.

2.2. Electroporation and Selection

1. Mutagenic single-stranded oligonucleotides: 60–100 bp nucleotides (IDT, Coralville, IA or equivalent), dissolved at 0.5–2.5 μM in sterilized water. For the results presented here, we used oligonucleotides: **rpsE(V25X)**, 5′ – CTG TGA AGG AGA AAA TAC GAC CAC CTT **TNN N**GG TTT TAG ATA CGC GGT TTA CCG CGA TCA – 3′; and **rpsE(K26X)**, 5′ – GAG CTG TGA AGG AGA AAA TAC GAC CAC **CNN N**AA CGG TTT TAG ATA CGC GGT TTA CCG CGA – 3′, where **N** indicates an equimolar mixture of all four bases.
2. Electroporation cuvettes, 1.0 mm (VWR). Chill on ice before use.
3. Electroporator: Bio-Rad Micropulser (Bio-Rad, Hercules, CA) or equivalent.
4. Sterile circular membranes, 82 mm diameter. Colony/Plaque Screen nylon (PALL-Perkin-Elmer), or Protran BA 85 nitrocellulose (Whatman).

5. LB-agar petri plates (100 × 15 mm): 10 g NaCl, 10 g tryptone, 5 g yeast extract, 1 mL 1N NaOH, 10 g agar-agar per 1 L.
6. LB-agar + antibiotic: Add appropriate antibiotic (e.g., 25 μg/mL spectinomycin) before pouring.
7. Stainless steel forceps. Sterilized with 95% ethanol followed by flaming with a Bunsen burner.
8. Sterile water (see Subheading 2.1 step 6).

2.3. Isolation of Antibiotic Resistant Clones and MAMA-PCR Screening

1. LB-agar plates (150 × 15 mm) containing appropriate antibiotic (e.g., 25 μg/mL spectinomycin) added before pouring.
2. Polymerase chain reaction (PCR) buffer (10×): 100 mM KCl, 60 mM $(NH_4)_2SO_4$, 20 mM $MgSO_4$, 200 mM Tris–HCl (pH 8.9), 1% Triton X-100.
3. Deoxynucleotide solution (2 mM each of dATP, dCTP, dGTP, and dTTP) (New England Biolabs, Beverly, MA).
4. Forward and reverse primers dissolved at 50 μM in sterilized water.
5. Thermocycler and appropriate PCR tubes. We use an old Perkin-Elmer 480 thermocycler with 0.65 mL microfuge tubes (Thermo-Fisher Scientific).
6. Gel loading buffer (3×): 50% glycerol, 0.01% xylene cyanol, 0.01% bromophenol blue.
7. TAE gel running buffer (50×): 2 M Tris-base, 0.1 M glacial acetic acid, 50 mM EDTA.
8. DNA grade agarose (EMD, Gibbstown, NJ or equivalent).
9. Ethidium bromide (100×): 5 mg/mL in water. Ethidium bromide is a potent mutagen and is light sensitive. Store in a dark container or wrap container in tin foil.

3. Methods

The following protocols use Red protein expression plasmids generated in the laboratories of Don Court (pSIM5 and pSIM6) and Barry Wanner (pKD46) (17, 18). Plasmids, pSIM5 and pSIM6 express the Red proteins under control of the native λ phage *p*L promoter, which is induced by heat-shock using the temperature-sensitive cI857 repressor (2, 18). Plasmid pKD46 allows L-arabinose-inducible Red protein expression under control of the *araBAD* promoter (17). In brief, the procedure involves induction of Red protein expression, preparation of electrocompetent cells, transformation with single-stranded oligonucleotide, and selection for recombinants. Mutagenic oligonucleotides should be between 60 and 100 nucleotides in length, and designed to anneal to the lagging strand DNA template during replication (2, 6). Ideally, 30–40 nucleotides of perfect homology should flank the mismatched nucleotides. However, we typically use 60-mer oligonucleotide libraries in which a single codon is randomized, reducing the homologous regions to only 28–29 nucleotides on either side of the randomized positions.

We have applied the mismatch amplification mutation assay-polymerase chain reaction (MAMA-PCR) to screen antibiotic resistant transformants for mutations in the target codon (19). This screen exploits the inability of Taq polymerase to extend a primer whose 3′ end does not anneal to template DNA. Primers are designed that will support efficient PCR from the wild-type gene, but not from genes containing mutations in the target codon (see Fig. 1). Conversely, primers can be designed to recognize specific mutations, thereby producing PCR product from mutant but not wild-type cells. In principle, MAMA-PCR can also be used to screen for targeted mutations in the absence of a functional selection.

3.1. Preparation of Electrocompetent *E. coli* Cells

1. Grow *E. coli* carrying plasmid pSIM5 (with 33 μg/mL chloramphenicol) or pSIM6 (with 150 μg/mL ampicillin) in LB medium overnight at 30°C. Dilute the overnight culture to OD_{600} = 0.05 in 35 mL of fresh LB media containing the appropriate antibiotic. If using plasmid pKD46 (ampicillin resistant), resuspend cells into two separate 17 mL LB-ampicillin cultures, one supplemented with 0.2% L-arabinose to induce Red protein expression, and the other supplemented with 0.2% D-glucose as an uninduced control. Grow cells at 30°C with aeration for approximately 2 h.
2. Once the cell density reaches OD_{600} ~0.5, transfer 17 mL of the culture to a 125-mL baffled Erlenmeyer flask and incubate in a shaking water bath at 42°C for 15 min (this heat shock will induce Red protein expression from plasmids pSIM5 and pSIM6). The remainder of the culture should be left as an uninduced control at 30°C. If using cells carrying plasmid pKD46, proceed to step 3 without heat shock.
3. Immediately plunge culture flasks into an ice-water bath and incubate with steady shaking for 5 min. Keep cells on ice for the remainder of the electroporation procedure.
4. Transfer each culture to a precooled Oak Ridge tube and collect cells by centrifugation at 6,000 × *g* for 7 min at 4°C in a Sorvall RC4 centrifuge using an SS-34 rotor. Alternatively, cell harvest and washing may be performed in prepackaged sterile 50 mL falcon tubes (e.g., Becton Dickinson) using a compatible swinging bucket bench top centrifuge.
5. Decant the supernatant and carefully resuspend the cell pellet in 1 mL of ice-cold, sterile water. After resuspension add

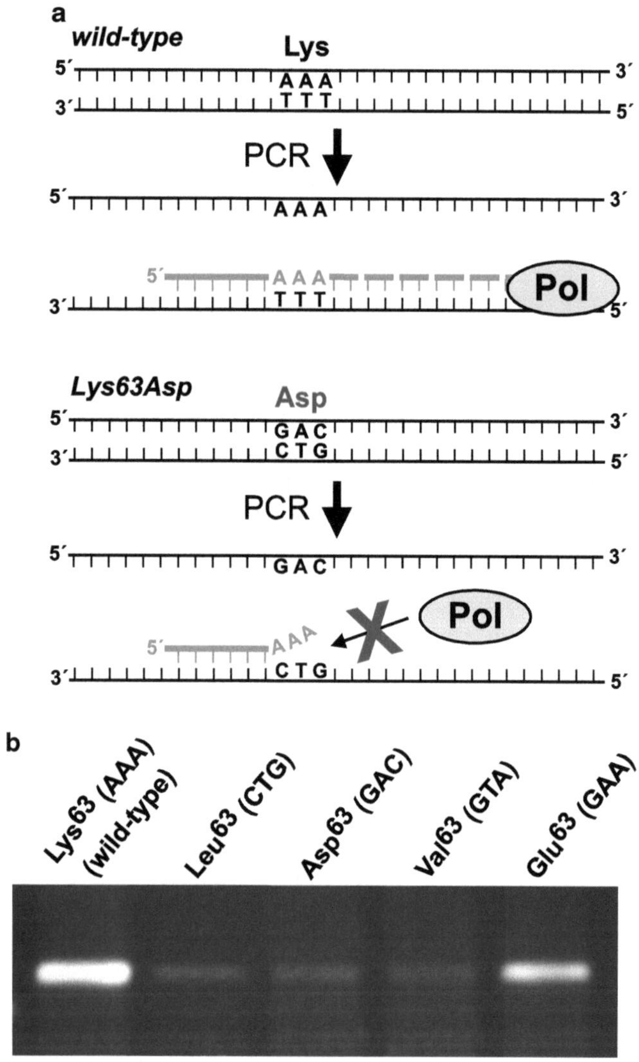

Fig. 1. Mismatch amplification mutation assay-polymerase chain reaction (MAMA-PCR). (**a**) The MAMA-PCR strategy. Short primers are designed whose 3′-ends anneal to the wild-type sequence of the target codon. PCR of the wild-type gene will result in efficient amplification. In contrast, Taq DNA polymerase will not extend the primer from the mutated codon due to the 3′ mismatched nucleotides. (**b**) MAMA-PCR screen of erythromycin-resistant *E. coli* mutants. Wild-type and erythromycin-resistant cells containing mutations in the *rplD* gene (encoding ribosomal protein L4) were subjected to whole-cell MAMA-PCR.

35 mL of ice-cold water. Mix gently and centrifuge at 6,000 × *g* for 7 min at 4°C to collect cells.

6. After centrifugation, carefully remove the supernatant with a sterile glass pipette. Water-washed cells are loose and it is difficult to decant the supernatant without disturbing the pellet.
7. Resuspend the cell pellet in 10 mL ice-cold water, swirl gently to resuspend cells, and centrifuge 6,000 × *g* for 7 min at 4°C to collect cells.

8. Remove water wash with a 10-mL sterile glass pipette, taking care not to disturb the pellet. The cell pellet will be very loose now and decanting the supernatant will result in cell loss! Add 1 mL of ice-cold sterile water to the pellet and swirl gently to resuspend cells.
9. Transfer the washed cells to a precooled 1.5 mL microfuge tube on ice and centrifuge at 14,000 × *g* for 1 min at 4°C to collect cells.
10. Carefully remove the supernatant with a micropipette and resuspend the cells in 200 μL of ice-cold sterile water. Avoid excessive pipetting, which may damage cells and decrease transformation efficiency. Store cells on ice until electroporation. Coordinate the procedure such that electrocompetent cells are ready just prior to electroporation. Extensive incubation reduces Red protein levels, resulting in decreased recombination efficiency. Alternatively, electrocompetent cells may be resuspended in 15% ice-cold glycerol and stored at −80°C for later use. Frozen electrocompetent cells have much lower transformation efficiency than freshly prepared cells (2).

3.2. Electroporation and Selection

1. Four separate transformations should be performed: (1) Red-induced cells plus oligo, (2) Red-induced cells without oligo, (3) uninduced cells plus oligo, and (4) uninduced cells without oligo. Note that electroporation of uninduced cells with DNA may yield transformants because oligonucleotides can recombine independent of Beta expression (4).
2. Aliquot 10 μL of oligonucleotide solution (0.5–2.5 μM) into a sterile microfuge tubes and place on ice. Aliquot 10 μL of sterile water to the other tubes as negative controls.
3. Add 50–100 μL of electrocompetent cells to the aliquoted DNA (or water) immediately before electroporation. Tap gently to mix and pipet into a prechilled, 1 mm electroporation cuvette. Ensure that the cell suspension is free of air bubbles, and remove all condensation from the cuvette electrode surface using a Kimwipe.
4. Pulse cells at 1.80 kV in an electroporator and immediately add 1.0 mL of sterile LB to the cuvette with a pipetman. Transfer the cell suspension to a sterile 1.5 mL microfuge tube. Expected time constants range from 4.5 to 5.5 ms for successful electroporations. If the sample arcs during the pulse (usually accompanied by an "Arc" error message on the pulser), discard the sample. Although it is possible to repulse after arcing, the number of viable cells decreases significantly after arcing. If successive arcs occur, it is likely that the cells have not been properly washed to remove salts. Electroporation cuvettes may be washed, sterilized, and reused several times (see Note 2).

5. Spread 100 µL of electroporated cell suspension onto an LB agar plate (no antibiotic) that has been overlayed with a nylon (or nitrocellulose) filter (see Note 3). Agar plates should be prewarmed and filters should be applied prior to plating the cell suspension. Incubate plates up to 3 h at 30°C (or 37°C) to allow recovery of electroporated cells prior to antibiotic selection. Immobilization of transformed cells allows for reasonable quantification of allele frequencies, because each antibiotic resistant colony represents an individual recombination event within a single cell (see Fig. 2 for an example). The recovery step may also be conducted in liquid media, but fast-growing mutants will multiply during this incubation and therefore will likely be over-represented on the antibiotic selection plate.

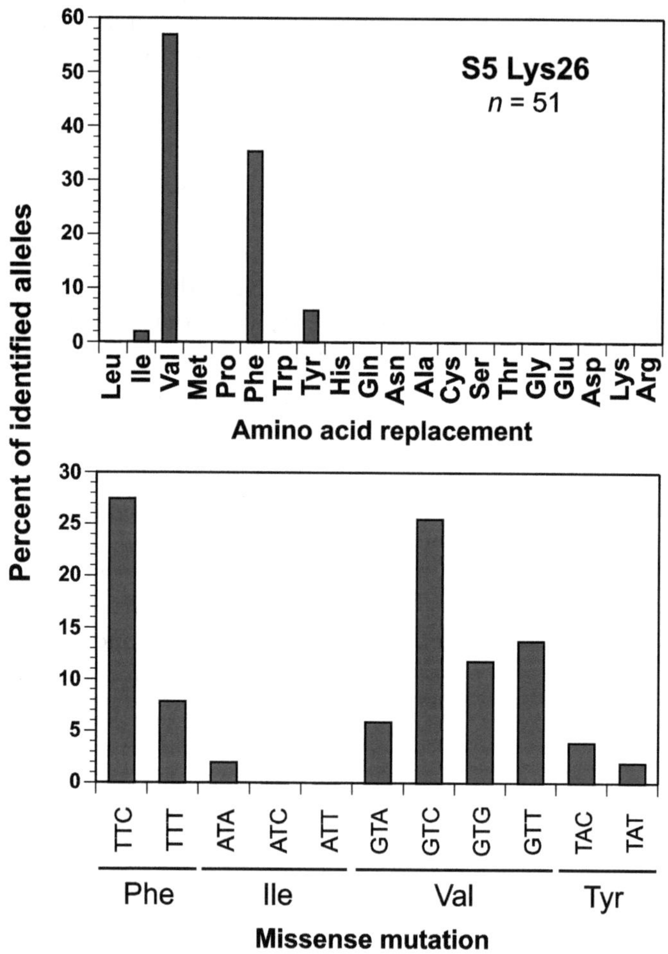

Fig. 2. Isolated spectinomycin resistance mutations from the rpsE-K26X oligonucleotide library. The *rpsE* gene was sequenced from 51 spectinomycin resistant mutants isolated from a single library recombineering experiment. All mutants contained changes of Lys26 to hydrophobic residues. Percentages of the identified Lys26 missense mutations are presented.

6. After recovery, use sterilized forceps to transfer the filters to LB agar plates containing 25 μg/mL spectinomycin (or other appropriate antibiotic/selection), and incubate for 24–48 h. Recovery for up to 72–96 h may be required to isolate slow growing mutants. After growth at 37°C, the selected mutants should have lost the Red protein expression plasmid and will be sensitive to ampicillin (or chloramphenicol) (see Note 4).

3.3. Isolation of Antibiotic Resistant Colonies and MAMA-PCR Screening

1. Pick single colonies and streak onto fresh selective agar plates to confirm antibiotic resistance. We use large petri plates (150×15 mm) with a 5×5 grid for this secondary selection step. Incubate the secondary selection plate overnight at 37°C.
2. The isolated mutants may then be screened for targeted mutations using mismatch amplification mutation assay-PCR (MAMA-PCR) on whole cells. Remove a small portion of a bacterial colony from the secondary selection plate and resuspend the cells in 20 μL of sterile water.
3. Set up PCR reactions as follows: 17.5 μL water, 2.5 μL of 10× PCR buffer, 2.5 μL of 2 mM dNTPs, 0.25 μL each of forward and reverse primer, 1–2 μL of mutant cell suspension, and 0.5 μL Taq DNA polymerase (see Note 5). Overlay with mineral oil.
4. The following MAMA-PCR cycling program works well for 25 μL reactions using a Perkin-Elmer 480 thermocycler:

   94°C for 2 min, 30 s – 1 cycle
   94°C for 1 min, 65°C 1 min, 72°C 1 min – 25 cycles
   72°C for 10 min – 1 cycle
5. Mix 10 μL of PCR product with 5 μL of gel loading buffer and load onto a 1% agarose gel prepared in 1× TAE buffer. Run gel at 100 V and stain with ethidium bromide. Always include a wild-type control for comparison to mutant PCR products. Typical results are as shown in Fig. 1.
6. Mutants that have passed the secondary selection and the MAMA-PCR screen are then sequenced to identify mutations (see Note 6). Table 1 shows the predicted ribosomal protein S5 variants encoded by the spectinomycin-resistant mutants isolated in this procedure. Note that for some randomized positions, complex mutations affecting adjacent codons will be isolated (see Note 7).
7. To confirm that the recombineered mutations are responsible for the selected phenotype, we transfer all mutations into the wild-type genetic background using recombineering or bacteriophage P1-mediated transduction. For a detailed protocol on bacteriophage P1-mediated transduction, see Chap. 10 of this volume.

Table 1
Predicted S5 proteins from spectinomycin-resistant *E. coli* mutants[a]

Val25	Lys26
V25F	**K26F**
V25P	**K26I**
V25W	**K26V**
V25D, ΔG27	**K26Y**
V25R, ΔG27	
Δ(S22-K23), **V25L**	
ΔV25	
S22A, Δ(K23-V25)	
K23T, T24N, ΔV25	
T24K, Δ(V25-K26)	

[a]Codons corresponding to S5 residues Val25 and Lys26 were randomized using separate oligonucleotide libraries. Transformed cells were selected on media containing 25 μg/mL spectinomycin. Boldfaced alleles indicate a missense mutation within the target codon. One letter amino acid code is used throughout

4. Notes

1. In principle, MMR defective strains can be used to obtain higher frequencies of oligonucleotide recombination, allowing for efficient site-directed mutagenesis directly onto the chromosome without functional selection (7). However, we have found that library oligonucleotide library recombineering in the Δ*mutS* MMR-deficient background can induce dozens of unintended silent mutations in the target gene. Therefore, we recommend the use of MMR proficient strains.
2. Used electroporation cuvettes should be washed extensively (>10 times) with water, then rinsed three times with 95% ethanol and stored in a closed sterile container. Depending on the manufacturer, electroporation cuvettes can be reused several times. However, cuvettes deteriorate with multiple uses and if arcing becomes frequent, then the cuvettes should be discarded. We have found that both autoclaving and bleach treatment significantly reduce the lifespan of electroporation cuvettes.
3. For many selections, nylon and nitrocellulose membranes perform equally well. However, nitrocellulose filters bind

hydrophobic antibiotics, such as macrolides, and can therefore interfere with some selections. It is critical that the membranes lie flat on the agar surface, with no underlying air bubbles. Use only sterile forceps to manipulate filters. Rinse forceps with 95% ethanol and flame between uses. If plating results in a confluent lawn of cells, serially dilute the transformed cells until well-isolated colonies are obtained.

4. After isolation of mutants, it may be necessary to remove the Red protein expression plasmid for downstream applications. Plasmids, pSIM5, pSIM6, and pKD46 all contain the *repA101* temperature-sensitive replication origin (17, 18), so growth at 37°C without antibiotic selection should result in plasmid loss. If the recombineering plasmid is still retained after growth at 37°C, streak the cells onto an antibiotic-free LB agar plate and incubate overnight at 42°C. Isolate individual colonies and screen for sensitivity to ampicillin (pSIM6, pKD46) or chloramphenicol (pSIM5) to identify plasmid-free mutants.
5. Mismatch screening primers are typically short (17–18 nt) with GC-content of 50–70% and T_m of approximately 50–55°C. The 3′-terminal three nucleotides of the screening primer should anneal to the wild-type sequence of the target codon. MAMA-PCR should be conducted with Taq DNA polymerase, or other thermostable DNA polymerases that lack 3′–5′ exonucleolytic activity. This activity will repair the mismatched nucleotide residues at the 3′ end of the MAMA-PCR primer and yield high levels of PCR product.
6. We typically amplify the target gene from isolated mutants using whole-cell PCR. The resulting PCR products are sequenced using an additional primer that is nested within the amplification primers. The use of nested sequencing primers greatly reduces the background signals that typically plague PCR product sequencing reads.
7. We suspect that untargeted mutations in adjacent codons are generated at low frequency in most randomization experiments. Though rare, these complex mutations will dominate the mutant pool if the more abundant simple missense mutations fail to confer antibiotic resistance. These unintended mutations significantly increase the complexity of the mutant pool, which may be informative in some instances.

Acknowledgment

This work was supported by grant R01 GM078634 from the National Institutes of Health.

References

1. Court D. L., Sawitzke J. A., and Thomason L. C. (2002) Genetic engineering using homologous recombination. *Annu Rev Genet* **36,** 361–388.
2. Thomason L., Court D. L., Bubunenko M., Costantino N., Wilson H., Datta S., et al. (2007) Recombineering: genetic engineering in bacteria using homologous recombination. *Curr Protoc Mol Biol* **Chapter 1,** Unit 1 16.
3. Baba T., Ara T., Hasegawa M., Takai Y., Okumura Y., Baba M., et al. (2006) Construction of *Escherichia coli* K-12 in-frame, single-gene knockout mutants: the Keio collection. *Mol Syst Biol* **2,** 2006 0008.
4. Swingle B., Markel E., Costantino N., Bubunenko M. G., Cartinhour S., and Court D. L. (2010) Oligonucleotide recombination in Gram-negative bacteria. *Mol Microbiol* **75,** 138–148.
5. van Kessel J. C., Marinelli L. J., and Hatfull G. F. (2008) Recombineering mycobacteria and their phages. *Nat Rev Microbiol* **6,** 851–857.
6. Ellis H. M., Yu D., DiTizio T., and Court D. L. (2001) High efficiency mutagenesis, repair, and engineering of chromosomal DNA using single-stranded oligonucleotides. *Proc Natl Acad Sci U S A* **98,** 6742–6746.
7. Costantino N., and Court D. L. (2003) Enhanced levels of lambda Red-mediated recombinants in mismatch repair mutants. *Proc Natl Acad Sci U S A* **100,** 15748–15753.
8. Diner E. J., and Hayes C. S. (2009) Recombineering reveals a diverse collection of ribosomal proteins L4 and L22 that confer resistance to macrolide antibiotics. *J Mol Biol* **386,** 300–315.
9. Holberger L. E., and Hayes C. S. (2009) Ribosomal protein S12 and aminoglycoside antibiotics modulate A-site mRNA cleavage and transfer-messenger RNA activity in *Escherichia coli. J Biol Chem* **284,** 32188–32200.
10. DeWilde M., and Wittmann-Liebold B. (1973) Localization of the amino-acid exchange in protein S5 from an *Escherichia coli* mutant resistant to spectinomycin. *Mol Gen Genet* **127,** 273–276.
11. Funatsu G., Nierhaus K., and Wittmann-Liebold B. (1972) Ribosomal proteins. XXII. Studies on the altered protein S5 from a spectinomycin-resistant mutant of *Escherichia coli. J Mol Biol* **64,** 201–209.
12. Funatsu G., Schiltz E., and Wittmann H. G. (1972) Ribosomal proteins. XXVII. Localization of the amino acid exchanges in protein S5 from two *Escherichia coli* mutants resistant to spectinomycin. *Mol Gen Genet* **114,** 106–111.
13. Itoh T. (1976) Amino acid replacement in the protein S5 from a spectinomycin resistant mutant of *Bacillus subtilis. Mol Gen Genet* **144,** 39–42.
14. Kehrenberg C., and Schwarz S. (2007) Mutations in 16S rRNA and ribosomal protein S5 associated with high-level spectinomycin resistance in *Pasteurella multocida. Antimicrob Agents Chemother* **51,** 2244–2246.
15. He X., Miao V., and Baltz R. H. (2005) Spectinomycin resistance in rpsE mutants is recessive in *Streptomyces roseosporus. J Antibiot (Tokyo)* **58,** 284–288.
16. Kirthi N., Roy-Chaudhuri B., Kelley T., and Culver G. M. (2006) A novel single amino acid change in small subunit ribosomal protein S5 has profound effects on translational fidelity. *RNA* **12,** 2080–2091.
17. Datsenko K. A., and Wanner B. L. (2000) One-step inactivation of chromosomal genes in *Escherichia coli* K-12 using PCR products. *Proc Natl Acad Sci USA* **97,** 6640–6645.
18. Datta S., Costantino N., and Court D. L. (2006) A set of recombineering plasmids for gram-negative bacteria. *Gene* **379,** 109–115.
19. Cha R. S., Zarbl H., Keohavong P., and Thilly W. G. (1992) Mismatch amplification mutation assay (MAMA): application to the c-H-ras gene. *PCR Methods Appl* **2,** 14–20.

Chapter 6

Microarray-Based Genetic Footprinting Strategy to Identify Strain Improvement Genes after Competitive Selection of Transposon Libraries

Alison K. Hottes and Saeed Tavazoie

Abstract

Successful strain engineering involves perturbing key nodes within the cellular network. How the network's connectivity affects the phenotype of interest and the ideal nodes to modulate, however, are frequently not readily apparent. To guide the generation of a list of candidate nodes for detailed investigation, designers often examine the behavior of a representative set of strains, such as a library of transposon insertion mutants, in the environment of interest. Here, we first present design principles for creating a maximally informative competitive selection. Then, we describe how to globally quantify the change in distribution of strains within a transposon library in response to a competitive selection by amplifying the DNA adjacent to the transposons and hybridizing it to a microarray. Finally, we detail strategies for analyzing the resulting hybridization data to identify genes and pathways that contribute both negatively and positively to fitness in the desired environment.

Key words: Genetic footprinting, *Escherichia coli*, Strain engineering, Transposon, Bacterial genetics, Microarray analysis, Statistics

1. Introduction

Strain engineering starts with an existing cellular network and determines how best to modify that network to optimize a phenotype of interest, such as production of a metabolite. Complicating the design process, however, is the biological reality that multiple cellular pathways affect many phenotypes of commercial and medical importance, such as ethanol tolerance and antibiotic susceptibility (1, 2). While mutations can be directed to regions of interest (3), exploring all possible cellular networks that are within even a few mutational steps of the original network is not currently feasible. Fortunately, although not all

James A. Williams (ed.), *Strain Engineering: Methods and Protocols*, Methods in Molecular Biology, vol. 765,
DOI 10.1007/978-1-61779-197-0_6, © Springer Science+Business Media, LLC 2011

mutations are additive, many are (1, 2). Thus, discovering single perturbations that influence a phenotype is a productive first step toward identifying combinations of mutations likely to further enhance a phenotype.

Transposon mutagenesis is a convenient way to generate a collection of strains each with a single mutation in a readily identifiable location (4–8). Transposon insertions can produce a wide range of phenotypes from null alleles caused by insertions in coding regions, to overexpression phenotypes resulting from insertions in intergenic regions that increase the expression of neighboring genes, to hypomorphs produced by insertions in the extreme 3′ end of genes. Furthermore, many commercial companies, such as Epicentre Biotechnologies, Finnzymes, and New England Biolabs offer transposons and transposases with desirable properties such as high transposition efficiency and low insertion site sequence bias.

Although some studies have tested large numbers of transposon insertion mutants individually (7), working with a library en masse is frequently more convenient and cost-effective (9). Subjecting a transposon library to a competitive selection enriches for strains with insertions that increase fitness and depletes the library of insertions that decrease fitness. Insertions that enhance fitness are obviously relevant to strain engineering. Since many insertions that decrease fitness are in genes essential to the behavior of interest, such genes are good candidates for targeted upregulation. Thus, strain engineering requires knowledge of both beneficial and deleterious insertion locations. Strongly beneficial insertion locations can often be identified by individually mapping the location of the transposon insertions in a number of cells isolated from a population after competitive selection. Individual colony methods, however, are not suitable for identifying insertion locations that decrease fitness or that increase fitness only moderately. More global methods, however, can characterize the full distribution of transposon insertion locations in a population before and after a selection and provide quantitative information about the contribution of each gene to a phenotype.

Here, we first discuss key considerations for designing an informative competitive transposon library selection. We then describe how to selectively amplify the DNA adjacent to transposons and hybridize it to a microarray to quantify the distribution of transposon insertion locations in a population. Finally, we address the main issues in data analysis: array normalization, identification of transposon insertion sites that cause fitness effects significant at a chosen false discovery rate (FDR), and discovery of pathways underlying the phenotype of interest.

The protocols presented were developed by Badarinarayana et al. (9) and Girgis et al. (10) using *Escherichia coli*, but should be readily adaptable to other organisms. A wide variety of related protocols are available (e.g., see refs. 11, 12).

2. Materials

2.1. Competitive Library Enrichment

1. Transposon insertion library, preferably frozen at –80°C in single-use aliquots. Figure 1 shows a typical transposon and the specific elements needed for this protocol.
2. Enrichment-specific materials.
3. LB + 30% glycerol: 0.5% yeast extract (w/v), 0.5% NaCl (w/v), 1% tryptone (w/v), and 30% glycerol (v/v). Autoclave to sterilize. Store at room temperature (see Note 1).
4. Dry ice.
5. Ethanol.

2.2. Genetic Footprinting

2.2.1. Restriction Digestion

1. Lysis buffer: Prepare just before use, per sample, combine 96 μl water, 12 μl 10× NEBuffer 2 [500 mM NaCl, 100 mM Tris–HCl, 100 mM $MgCl_2$, 10 mM dithiothreitol pH 7.9 (New England Biolabs, Ipswich, MA)], and 6 μl Triton X-100.
2. Alkaline phosphatase, 1 U/μl (Roche). Store at 4°C.
3. HinP1I, 10 U/μl (New England Biolabs or equivalent). Store at –20°C.
4. MspI, 20 U/μl (New England Biolabs or equivalent). Store at –20°C.

2.2.2. Y-Linker Ligation

1. 3 M sodium acetate pH 5.2: Use acetic acid for pH adjustment. Autoclave to sterilize; store at room temperature.
2. Ethanol chilled at –20°C.
3. 70% ethanol chilled at –20°C.

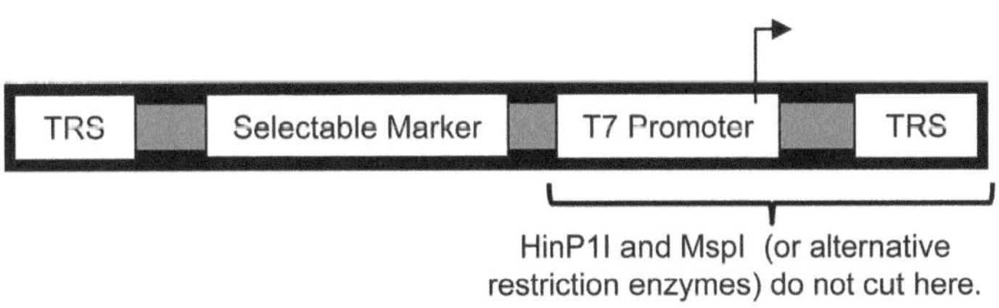

Fig. 1. Transposon structure and required components. Transposon ends contain transposase recognition sequences (TRS) that are recognized by the corresponding transposase. Transposons typically also contain a selectable marker that can facilitate selecting for strains that contain the transposon. The protocol presented here requires the presence of an outward-reading T7 promoter near one of the transposon's ends. Additionally, the protocol assumes that the HinPl1 and MspI restriction enzymes do not cut between the T7 promoter and the end of the transposon. Otherwise, alternative restriction enzymes must be substituted (see Note 7). The modified Tn5 transposon described in ref. (10) that meets these criteria and was used to develop the methods described herein is available upon request.

4. Y-linker (40 pmol/μl): Purchase the following HPLC-purified primers:

   5′ – ACTACGCACGCGACGAGACGTAGCGTC – 3′ (YCG5) and

   5′ – P-CGGACGCTACGTCCGTGTTGTCGGTCCTG – 3′ (YCG3).

   Note that YCG3 is phosphorylated on the 5′ end. Dissolve each in water at a concentration of 100 pmol/μl. In a PCR tube, combine 30 μl primer YCG5, 30 μl primer YCG3, 7.5 μl 10× annealing buffer [1 M NaCl, 100 mM Tris–HCl (pH 8.0), 10 mM EDTA (pH 8.0)], and 7.5 μl water. Using a thermocycler, heat the mixture at 94°C for 1 min and then drop the temperature in 2°C increments every 30 s until reaching 26°C. The reaction may be scaled up as needed. Y-linker should be frozen at –20°C in single use aliquots (e.g., 25 μl aliquots are ideal for processing samples in batches of eight).

5. T4 DNA ligase (400 U/μl) and 10× buffer [500 mM Tris–HCl, 100 mM $MgCl_2$, 10 mM ATP, 100 mM dithiothreitol, pH 7.5] (New England Biolabs or equivalent). Store at –20°C.

6. QIAquick PCR Purification Kit (Qiagen, Valencia CA)

2.2.3. Repair Nicks

1. 10× NEBuffer 2 [500 mM NaCl, 100 mM Tris–HCl, 100 mM $MgCl_2$, 10 mM dithiothreitol pH 7.9 (New England Biolabs)]. Store at –20°C.

2. dNTP mix: 2.5 mM each of dATP, dCTP, dGTP, and dTTP. Store at –20°C.

3. *E. coli* DNA polymerase I (10 U/μl) (New England Biolabs or equivalent). Store at –20°C.

2.2.4. Amplify Transposon-Adjacent DNA by PCR

1. Water.

2. dNTP mix: 2.5 mM each of dATP, dCTP, dGTP, and dTTP. Store at –20°C.

3. Primer Y-COMP (5′-ACTACGCACGCGACGAGACG-3′), 10 μM. This primer anneals to the complement of the single-stranded part of the Y-linker (see Fig. 2). Store at –20°C.

4. Primer T7-UPSTRM, 10 μM. This primer, in conjunction with primer Y-COMP should amplify the end of the transposon, including the T7 promoter (see Fig. 2). Store at –20°C.

5. Ex Taq polymerase and 10× Ex Taq buffer (Takara). Store at –20°C.

6. QIAquick PCR Purification Kit (Qiagen).

7. Nuclease-free water.

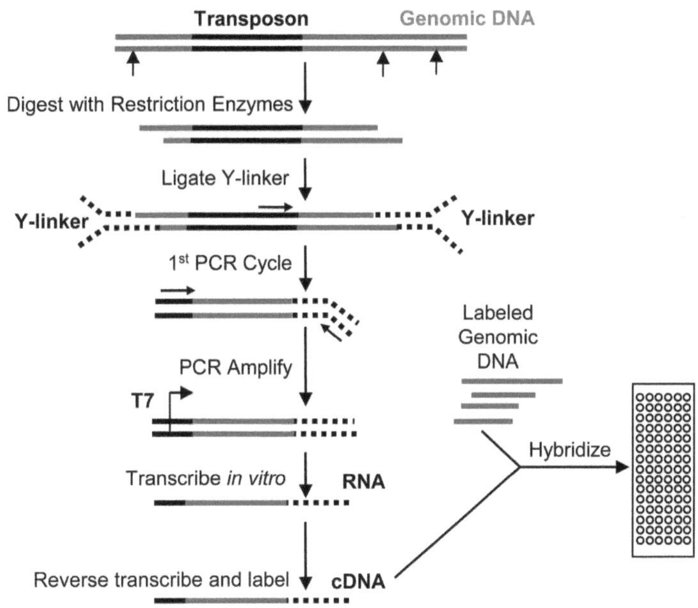

Fig. 2. Genetic footprinting protocol overview. First, genomic DNA from the transposon insertion library is digested with restriction enzymes; the DNA adjacent to a transposon insertion serves as the marker for the insertion site. Then, a Y-linker with an overhang compatible with the restriction digestion is ligated to the DNA. Next, PCR is used to amplify the ends of the transposons and the adjacent DNA. During the first PCR cycle, the primer from the transposon primes the synthesis of DNA complementary to one strand of the Y-linker. The second PCR primer then anneals to the newly synthesized DNA and participates in subsequent rounds of amplification. To reduce the nonlinearities introduced by PCR, the number of cycles is limited as much as possible. To obtain sufficient product for hybridization, the DNA adjacent to the transposon is further amplified by in vitro transcription using a T7 promoter located on the transposon. The resulting RNA is then typically converted into cDNA and labeled in a way suitable for the chosen microarray hybridization technology. Finally, a microarray is used to quantify the fraction of the library population with transposon insertions near each array probe (modified from ref. (10), which was published by Public Library of Science as an open-access article under a Creative Commons Attribution License).

2.2.5. Further Amplify Transposon-Adjacent DNA Using In Vitro Transcription

1. MEGAscript T7 Kit (Ambion Inc., Austin, TX).
2. RNeasy Mini Kit (Qiagen).

2.2.6. Microarray Hybridization

1. Genomic DNA from the transposon library's parental strain: DNA should be fragmented to an appropriate size and suitably labeled for hybridization using the chosen microarray platform (see Note 2).
2. Reagents needed to synthesize cDNA suitably labeled for the chosen microarray platform from RNA.
3. Reagents needed for a microarray hybridization.

3. Methods

3.1. Competitive Library Enrichment

1. Subject the transposon insertion library to the experimental conditions of interest (see Note 3 and Fig. 3).
2. Preserve samples of the population throughout the course of the experiment by mixing equal volumes of culture and LB + 30% glycerol, snap-freezing in dry ice and ethanol, and storing at –80°C. Archival samples allow for detailed studies of the progression of the selection and can also be searched for mutants with transposon insertions in sites of interest.
3. At times of interest, collect samples for genetic footprinting (see Note 4). For each sample, pellet ~10^7 cells by centrifugation, remove all supernatant possible using a pipette, and store the pellet at –80°C until needed.

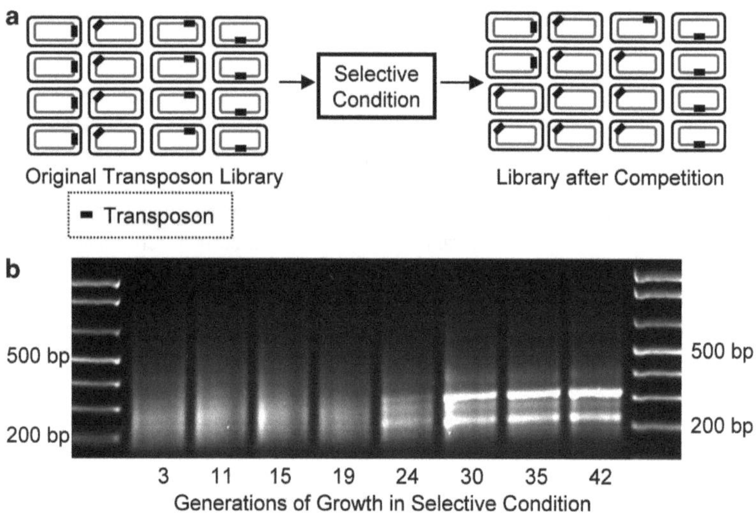

Fig. 3. Library diversity as a function of generations of competitive selection. (**a**) The original, high-diversity transposon library is subjected to a competitive selection that increases the abundance of strains with beneficial transposon insertions and reduces the abundance of strains with deleterious transposon insertions. Ideally, a selection should span enough generations to detectably magnify the abundance of strains with small fitness increases over the wild-type strain, but not so many generations that both strains of average and below-average fitness drop out of the population completely. (**b**) Samples of a transposon library propagated in defined media with aspartic acid as the sole carbon source for the indicated number of generations were subjected to genetic footprinting. The resulting PCR products (the output of Subheading 3.2.4) were then run on a 2% agarose gel. DNA band sizes are indicated in the far left and right lanes. The presence of discrete bands indicates that a clone reached high density in the population. The clone either contained a highly beneficial transposon insertion or, as happens more commonly, a beneficial spontaneous mutation that allowed the endogenous transposon insertion to hitchhike to prominence. In our experience, spontaneous mutations typically become problematic after about 20 generations.

3.2. Genetic Footprinting

See Fig. 2 for an overview of the procedure.

3.2.1. Restriction Digestion

1. Thaw the sample pellet briefly at room temperature and suspend it in 114 µl lysis buffer.
2. Transfer 48 µl of cells to each of two PCR tubes (see Note 5).
3. Incubate the tubes at 99°C for 40 s in a thermocycler to lyse the cells, and then cool to room temperature.
4. Add 1 µl alkaline phosphate to both tubes (see Note 6).
5. Add 1 µl HinP1I to one tube and 1 µl MspI to the other (see Note 7). Mix.
6. Incubate at 37°C for 3 h.
7. Heat at 65°C for 20 min to deactivate the restriction enzymes (see Note 8).

3.2.2. Ligate Y-Linker

1. Combine the two restriction digests.
2. Add 10 µl 3 M sodium acetate (pH 5.2) and transfer the mixture to a microfuge tube.
3. Add 0.3 ml of −20°C ethanol and mix.
4. Freeze at −20°C for at least 1 h.
5. Centrifuge at >13,000 × *g* for 10 min at 4°C (maximum RPM in microfuge).
6. Pour off the supernatant without disturbing the pellet.
7. Add 0.5 ml −20°C 70% ethanol.
8. Centrifuge at >13,000 × *g* for 10 min at 4°C.
9. Pour off the supernatant without disturbing the pellet.
10. Centrifuge the tube briefly to collect the remaining liquid in the bottom of the tube.
11. Pipet out the residual liquid.
12. Allow the pellet to dry to remove the remaining ethanol. This can either be done in a speed-vac for ~1 min or in a fume hood for ~30 min. Do not over-dry.
13. Resuspend the pellet in 23 µl water, 3 µl 10× T4 DNA ligase buffer, and 3 µl Y-linker. Keep on ice.
14. Add 1 µl T4 DNA ligase.
15. Place the sample in a floating microfuge tube rack in a container with 2 l of room temperature water. Place the container with water in a 4°C room overnight. Alternatively, the sample can be ligated at 16°C overnight.
16. Clean up the sample using a Qiaquick PCR purification kit according to the manufacturer's directions. In the last step, elute in 26 µl of water; approximately 24 µl will flow through.

3.2.3. Repair Nicks (see Note 9)

1. To the 24 μl sample, add 3 μl 10× NEBuffer 2, 2 μl dNTP mix, and 1 μl *E. coli* DNA polymerase I.
2. Incubate at 25°C for 2 h.
3. Inactive the enzyme by heating at 75°C for 20 min.

3.2.4. Amplify Transposon-Adjacent DNA by PCR

1. Combine the following in order: 25.8 μl water, 5 μl 10× Ex Taq buffer, 4 μl dNTP mix, 5 μl T7-UPSTRM primer, 5 μl Y-COMP primer, 5 μl of nick-repaired ligation product, and 0.2 μl Ex Taq polymerase.
2. Heat in a thermocycler at 94°C for 2 min. Then, cycle at 94°C for 30 s, 68°C for 30 s, and 72°C for 3 min 30 times (see Note 10). Finally, heat at 72°C for 10 min.
3. Clean up the sample using a Qiaquick PCR purification kit according to the manufacturer's directions. In the last step, elute in 30 μl nuclease-free water.
4. If desired, visualize the sample on a 2% agarose gel as in Fig. 3b.

3.2.5. Further Amplify Transposon-Adjacent DNA Using In Vitro Transcription

1. Combine the following components (from the MEGAscript T7 kit) in a PCR tube at room temperature: 2 μl each of ATP, CTP, GTP, and UTP solutions (8 μl total), 2 μl of 10× reaction buffer, 1 μg of PCR product from the reaction above, and enough nuclease-free water to bring the total volume to 18 μl (see Note 11).
2. Add 2 μl T7 enzyme mix (from kit).
3. Incubate for 4 h at 37°C.
4. Add 1 μl TURBO DNase (2 U/μl) from the MEGAscript T7 kit and incubate for 15 min at 37°C.
5. Purify the RNA using the RNeasy Mini Kit according to the manufacturer's directions. In the final step, elute in 40 μl RNase-free water.

3.2.6. Microarray Hybridization

1. Select a microarray platform (see Note 2).
2. For two-color, comparative platforms, prepare a labeled, genomic DNA reference (see Note 12). See Girgis et al. (10) or the array manufacturer's instructions.
3. Synthesize cDNA suitably labeled for the chosen microarray platform from the in vitro transcribed RNA.
4. Hybridize the sample to the chosen array.

3.3. Data Analysis

This section focuses on the analysis of samples either hybridized to single channel platforms (e.g., Affymetrix arrays) or hybridized to two-channel platforms (e.g., Agilent arrays) using genomic DNA as a common reference. Data sets from competitive selections, similar to expression data sets, are large and for reasons of expense typically contain few repetitions. The large number of genes per array necessitates an awareness of the number of false positives expected due to multiple hypothesis testing (see Note 13). The small number of repetitions favors the use, at least initially, of simple analysis techniques with few parameters to fit. Here, we describe basic analysis techniques that work well with most data sets; numerous alternative algorithms are described in the literature that may be helpful in special situations (see refs. 13–15 for a sampling of reviews).

3.3.1. Obtain Data Describing the Composition of the Transposon Library Prior to Competitive Selection

1. For comparative purposes, process and hybridize *at least* three samples of the original, unselected library. Five samples are commonly used (10). To make the null distribution as accurate as possible, each sample should be processed independently starting with the genetic footprinting step (Subheading 3.2).

3.3.2. Perform Suitable Within-Array Normalization (see Note 14)

1. Compensate for background and off-target hybridization as dictated by the technology.
2. Additionally, for two-channel arrays, scale the signals so that the contribution of each channel is equal (10). In other words, the sum of the signal from the first channel over all of the probes should be equal to the sum of the signal from the second channel over all the probes.
3. Combine data from all probes representing each gene as appropriate for the array.

3.3.3. Employ Between-Array Normalization to Correct for Signal Strength Variations Between Arrays

1. Identify the genes that are present on all of the unselected library hybridizations and all of the experimental samples of current interest. This step does not distinguish between experimental and reference samples; all of the arrays should be processed together.
2. For a one-channel technology, let $s_{i,j}$ be the signal from the ith gene on the jth array; for a two-channel technology, let $s_{i,j}$ be the ratio of the competitive enrichment signal and the genomic DNA signal for the ith gene on the jth array.
3. For each array, compute t_j, the total signal from array j for all genes present on all arrays. That is, find $t_j = \sum_{i=1}^{N} s_{i,j}$, where the index, i, runs over the N genes with

   signal present on all arrays.
4. For each array, j, replace $s_{i,j}$ with $s_{i,j}C/t_j$ where C is an arbitrary constant chosen to put the numbers on a convenient scale. Make the replacement for all genes, not just those with valid signals on all arrays.

3.3.4. Calculate z-scores

A z-score, $z_{i,j}$, should be calculated for each gene, i, and each hybridization, j, of the competitively selected library.

1. Let μ_i be the average of the normalized signal for gene i from the hybridizations of the unselected library.
2. Let σ_i be the standard deviation of the normalized signal for gene i from the hybridizations of the unselected library.
3. Define $z_{i,j} = (s_{i,j} - \mu_i)/\sigma_i$ where $s_{i,j}$ is the normalized signal calculated above (see Note 15). A positive z-score indicates that the fraction of strains with insertions in or near gene i increased during the selection; a negative z-score indicates that the fraction of strains with insertions in or near gene i decreased during the selection. The normalization by σ_i accounts for the expected variability of each gene.

3.3.5. Identify Genes that Changed Compared to the Unselected Library

1. Let z be the significance threshold.
2. Consider a gene i to have caused a significant effect in competitive selection j if $|z_{i,j}| > z$ where $z_{i,j}$ is the z-score calculated in Subheading 3.3.4 (see Note 16).

3.3.6. Estimate the False Discovery Rate

The FDR is the fraction of the set deemed significant using a particular z-score, z, that is expected to consist of false positives (16).

1. Let S be the number of significant genes from Subheading 3.3.5.
2. Use the hybridizations of the unselected library as a model for the null distribution. For each gene, randomly remove one of the measurements from the set of unselected library hybridizations and designate it as "signal." Then, calculate z-scores as in Subheading 3.3.4. Take care *not* to use the data designated as "signal" in calculating the per-gene means and standard deviations. See Fig. 4a.
3. Calculate *FP*, the expected number of false positives. FP is the number of samples in the null distribution with z-scores of greater magnitude than z, the significance threshold used in Subheading 3.3.5.
4. The FDR is FP/S. See Fig. 4b, c.

3.3.7. Combine Data from Multiple Competitive Selections, if Available

1. If three or more samples are available for each gene, use the median.
2. If only two samples are available, and both have z-scores of the same sign, use the one closest to zero; otherwise assign a z-score of zero (see Note 17).
3. If desired, reestimate the FDR by generating a null distribution that reflects how multiple samples were combined.

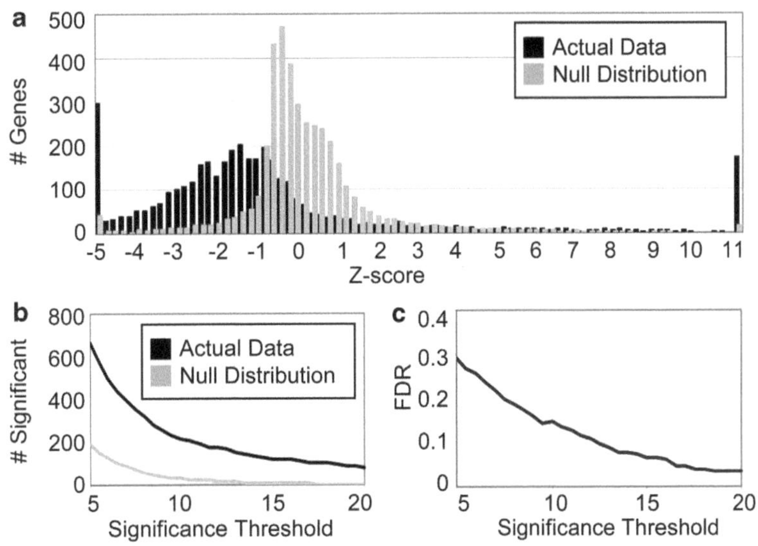

Fig. 4. Calculating the false discovery rate (FDR) as a function of the significance threshold. *Z*-scores relative to five hybridizations of the original, unselected library were calculated for data from a competitive selection to find *E. coli* mutants that remain motile in high salt concentrations. A null distribution was simulated by treating one of the five reference samples for each gene as data as described in Subheading 3.3.6. A global component equal to one-tenth of the average standard deviation was added to the standard deviation of each gene (see Note 15). (**a**) The histogram displays the *z*-scores for the real data and the null distribution. The real data has a larger spread and heavier tails than the null distribution indicating that the library contained some mutants of above- and below-average fitness. During the course of the selection, several strains became a substantial part of the population and reduced the prevalence of the average mutant. As a result, the mean *z*-score for the real data is lower than the mean *z*-score of the null distribution. (**b**) A gene was considered significant if the absolute value of its *z*-score was greater than the indicated threshold. (**c**) As the significance threshold decreases, both the estimated FDR and the number of true positives increase. The FDR will not necessarily increase monotonically as the number of true positive increases, but it usually does. All data were published in Girgis et al. (10).

3.3.8. Search for Pathways that Contributed to Fitness in the Competition

1. Pathway analysis looks for commonalities among the genes with similar *z*-scores. By examining the data set as a whole, *z*-scores that are individually too small to be considered significant can still contribute to the identification of large-scale patterns.
2. Many pathway analysis tools are available. In particular, the Tavazoie lab has developed iPAGE (17), which identifies pathways and gene ontology (GO) terms (18) that are enriched or depleted for each range of *z*-scores. See Fig. 5 for an example.

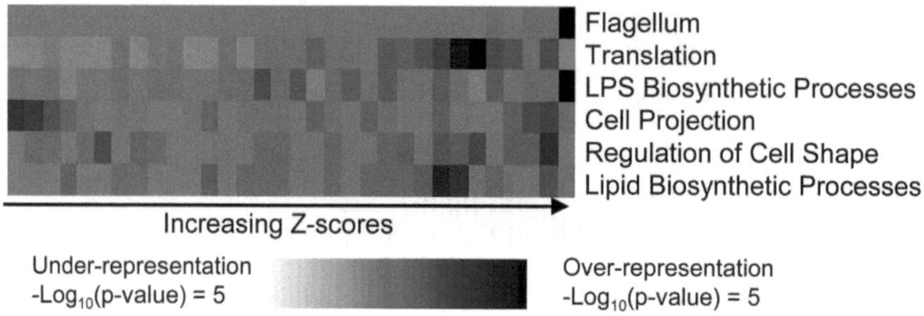

Fig. 5. Using iPAGE (17) to identify pathways involved in χ-phage susceptibility. *Z*-scores from a competitive selection to find *E. coli* mutants with reduced sensitivity to χ-phage were calculated relative to five hybridizations of the original, unselected library. A global component equal to one half of the average standard deviation was added to each gene's standard deviation (see Note 15). Data from two independent repetitions were combined by taking the value closest to zero when the repetitions had the same sign and using a value of zero otherwise (see Subheading 3.3.7). Columns, from left to right, correspond to equally populated bins of increasing *z*-scores; values of zero are present in the second through fifth columns from the right. The darker (lighter) the rectangle, the more the range of *z*-score was enriched (depleted) for the indicated functional category; no significant regions of depletion were identified in this data set. The results suggest that LPS or flagella defects increase χ-phage resistance while defects in cell projection processes (e.g., fimbrial-like proteins) increase susceptibility (10). iPAGE can detect functional enrichments in middle ranges of *z*-scores as well as in the most extreme ranges. For example, *z*-scores just below zero are enriched for genes with products involved in translation; members of the set, which consists mainly of genes encoding essential ribosomal proteins, were largely absent in the library both before and after the selection. Data came from Girgis et al. (10). *LPS* lipopolysaccharides.

4. Notes

1. Unless stated otherwise, solutions and media should be made with deionized water.
2. We have successfully used Affymetrix tiling arrays (unpublished), Agilent oligo arrays (unpublished), and in-house arrays containing a PCR product from each open reading frame (ORF) (1, 2, 10, 19). The size of the features on an array (i.e., 25 mers, 60 mers, or ~1 kb ORFs) determines the precision with which the technology will be able to resolve transposon insertion locations. The combination of the density of the features and the size of the transposon-adjacent DNA amplified, which is set by the restriction enzymes used in the protocol, determines which transposon insertion locations will contribute signal to the hybridization. Other considerations are a lab's familiarity with a particular platform and the availability of the needed infrastructure.
3. During competitive selections, the minimum population size should be kept large enough to avoid unwanted bottlenecking. Additionally, as determining the ideal length (generations) for an enrichment a priori is difficult, taking samples at multiple times is advisable.
4. If feasible, collect all samples at similar growth stages and conditions, such as stationary phase. Otherwise, cells from

the fastest growing cultures will have more DNA near the origin of replication, which will inflate the number of copies of insertions near the origin compared to the terminus of replication (20). If such growth rate differences are unavoidable, consult Vora et al. (21) for an example of a windowing approach that can be used to correct for the resulting chromosome position biases.

5. Samples are suspended in a slight excess of lysis buffer as the Triton X-100 causes bubbles that make it difficult to use the whole volume.
6. The inclusion of alkaline phosphatase, as suggested by Girgis et al. (10), prevents genomic DNA segments from ligating to each other instead of Y-linker.
7. Since transposon insertion sites too close to restriction enzyme cut sites do not yield identifiable DNA segments, two separate restriction digests are used. Ensure that the restriction enzymes do not cut between the T7 promoter and the end of the transposon. If the chosen restriction enzymes do not leave a 5′-CG overhang, then the Y-linker sequence will need to be adjusted.
8. Alkaline phosphatase cannot be heat-inactivated, and prolonged storage of the mixture at 4°C may result in DNA degradation. Either proceed immediately to the next step or store samples at −20°C.
9. DNA polymerase I repairs the nicks between the 5′-ends of the genomic DNA and the Y-linker, which exist because the genomic DNA was dephosphorylated. Unfortunately, the enzyme can be finicky, resulting in little or no PCR product in the next step. Omitting alkaline phosphatase from the restriction digest, similar to Badarinarayana et al. (9), obviates the need for DNA polymerase I repair and increases both the signal and the background.
10. The number of PCR cycles should be kept to a minimum to reduce nonlinear amplification biases.
11. Take standard precautions, such as using filter tips, to avoid introducing RNases into the sample.
12. Instead of comparing each sample to a common reference (genomic DNA), two samples can also be compared directly as was done in Goodarzi et al. (22). Typically, the use of a common reference facilitates the meta-analysis of data from a large number of competitions.
13. For simplicity, the analysis procedure discusses genes instead of probes. Repeating the analyses with the probes treated individually may provide insights into regions of genes, such as segments that code for protein domains, that affect fitness differentially. Additionally, many probes or probe sets represent intergenic regions, which can be treated similarly to genes.

14. A variety of commercial software performs all of the steps of Subheading 3.3.2. For example, the MAS5 algorithm (23–25) commonly used with Affymetrix arrays performs background corrections, combines all of the probes for each gene, and scales the final results so that sets of arrays will have similar scaling, which may reduce the need to perform between-array normalization (see Subheading 3.3.3).
15. The normalization procedure assumes that the abundance of most mutants in the population remains relatively constant throughout a selection. Some stringent selections that cause the abundance of all but the fittest mutants to decrease appreciably, however, can pose analysis problems if the level assigned to genes present in only negligible amounts shifts. A slight change in mean "signal" from absent genes coupled with the typically small standard deviation of the unselected library signal of essential genes (i.e., genes in which the cell cannot tolerate transposon insertions in the conditions used for library construction) can cause large *z*-scores to be associated with essential genes. The difficulty can be largely overcome by adding a small constant, which represents the global variability of the array, to all of the gene-specific standard deviations used in calculating *z*-scores (1). That is, each σ_i can be replaced with $\sigma_i + \sigma_{\text{global}}$, where σ_{global} is, for example, one-tenth to one half the average σ_i.
16. If only beneficial insertions are of interest, neglect the absolute value symbol and consider only positive *z*-scores.
17. The procedure described reduces the number of false positives at the possible expense of an increase in the number of false negatives. Averaging is avoided as the noise caused by spontaneous mutations that can cause some transposon insertions to hitchhike to prominence can result in extreme outliers that do not follow a Gaussian distribution. For similar reasons, all repetitions should be biologically independent and go through separate competitive enrichments.

Acknowledgments

We are grateful to Hany Girgis for developing and optimizing many of the techniques described here and to Hani Goodarzi for developing iPAGE. Work in the Tavazoie lab was supported by grants from NSF (CAREER), DARPA (BIOS), NIGMS (P50 GM071508), and the NIH Director's Pioneer Award (1DP1OD 003787-01).

References

1. Girgis H. S., Hottes A. K., and Tavazoie S. (2009) Genetic architecture of intrinsic antibiotic susceptibility. *PLoS One* 4, e5629.
2. Goodarzi H., Bennett B. D., Amini S., Reaves M. L., Hottes A. K., Rabinowitz J. D., and Tavazoie, S. (2010) Regulatory and metabolic rewiring during laboratory evolution of ethanol tolerance in *E. coli*. *Mol. Syst. Biol.* 6, 378.
3. Wang H. H., Isaacs F. J., Carr P. A., Sun Z. Z., Xu G., Forest C. R., and Church G. M. (2009) Programming cells by multiplex genome engineering and accelerated evolution. *Nature* 460, 894–898.
4. Akerley B. J., Rubin E. J., Novick V. L., Amaya K., Judson N., and Mekalanos J. J. (2002) A genome-scale analysis for identification of genes required for growth or survival of *Haemophilus influenzae*. *Proc. Natl. Acad. Sci. U. S. A.* 99, 966–971.
5. Dziva F., van Diemen P. M., Stevens M. P., Smith A. J., and Wallis T. S. (2004) Identification of *Escherichia coli* O157 : H7 genes influencing colonization of the bovine gastrointestinal tract using signature-tagged mutagenesis. *Microbiology* 150, 3631–3645.
6. Gonzalez M. D., Lichtensteiger C. A., and Vimr E. R. (2001) Adaptation of signature-tagged mutagenesis to *Escherichia coli* K1 and the infant-rat model of invasive disease. *FEMS Microbiol. Lett.* 198, 125–128.
7. Jacobs M. A., Alwood A., Thaipisuttikul I., Spencer D., Haugen E., Ernst S., Will O., Kaul R., Raymond C., Levy R., Chun-Rong L., Guenthner D., Bovee D., Olson M. V., and Manoil C. (2003) Comprehensive transposon mutant library of *Pseudomonas aeruginosa*. *Proc. Natl. Acad. Sci. U. S. A.* 100, 14339–14344.
8. Salama N. R., Shepherd B., and Falkow S. (2004) Global transposon mutagenesis and essential gene analysis of *Helicobacter pylori*. *J. Bacteriol.* 186, 7926–7935.
9. Badarinarayana V., Estep P. W., 3rd, Shendure J., Edwards J., Tavazoie S., Lam F., and Church G. M. (2001) Selection analyses of insertional mutants using subgenic-resolution arrays. *Nat. Biotechnol.* 19, 1060–1065.
10. Girgis H. S., Liu Y., Ryu W. S., and Tavazoie S. (2007) A comprehensive genetic characterization of bacterial motility. *PLoS Genet.* 3, 1644–1660.
11. Winterberg K. M., and Reznikoff W. S. (2007) Screening transposon mutant libraries using full-genome oligonucleotide microarrays. *Methods Enzymol.* 421, 110–125.
12. Baldwin D. N., and Salama N. R. (2007) Using genomic microarrays to study insertional/transposon mutant libraries. *Methods Enzymol.* 421, 90–110.
13. Do J. H., and Choi D. K. (2006) Normalization of microarray data: single-labeled and dual-labeled arrays. *Mol. Cells.* 22, 254–261.
14. Speed T., and Zhao H. (2009) Microarrays. *Stat. Methods Med. Res.* 18, 531–532.
15. Steinhoff C., and Vingron M. (2006) Normalization and quantification of differential expression in gene expression microarrays. *Brief Bioinform.* 7, 166–177.
16. Tusher V. G., Tibshirani R., and Chu G. (2001) Significance analysis of microarrays applied to the ionizing radiation response. *Proc. Natl. Acad. Sci. U. S. A.* 98, 5116–5121.
17. Goodarzi H., Elemento O., and Tavazoie S. (2009) Revealing global regulatory perturbations across human cancers. *Mol. Cell.* 36, 900–911.
18. Ashburner M., Ball C. A., Blake J. A., Botstein D., Butler H., Cherry J. M., Davis A. P., Dolinski K., Dwight S. S., Eppig J. T., Harris M. A., Hill D. P., Issel-Tarver L., Kasarskis A., Lewis S., Matese J. C., Richardson J. E., Ringwald M., Rubin G. M., and Sherlock G. (2000) Gene ontology: tool for the unification of biology. The Gene Ontology Consortium. *Nat. Genet.* 25, 25–29.
19. Amini S., Goodarzi H., and Tavazoie S. (2009) Genetic dissection of an exogenously induced biofilm in laboratory and clinical isolates of *E. coli*. *PLoS Pathog.* 5, e1000432.
20. Cooper S., and Helmstetter C. E. (1968) Chromosome replication and the division cycle of *Escherichia coli* B/r. *J. Mol. Biol.* 31, 519–540.
21. Vora T., Hottes A. K., and Tavazoie S. (2009) Protein occupancy landscape of a bacterial genome. *Mol. Cell.* 35, 247–253.
22. Goodarzi H., Hottes A. K., and Tavazoie S. (2009) Global discovery of adaptive mutations. *Nat. Methods* 6, 581–583.
23. Hubbell E., Liu W. M., and Mei R. (2002) Robust estimators for expression analysis. *Bioinformatics* 18, 1585–1592.
24. Liu W. M., Mei R., Di X., Ryder T. B., Hubbell E., Dee S., Webster T. A., Harrington C. A., Ho M. H., Baid J., and Smeekens S. P. (2002) Analysis of high density expression microarrays with signed-rank call algorithms. *Bioinformatics* 18, 1593–1599.
25. Affymetrix. (2002) Statistical Algorithms Description Document http://www.affymetrix.com/support/technical/whitepapers/sadd_whitepaper.pdf Accessed June 22, 2010

Chapter 7

Optimization of Synthetic Operons Using Libraries of Post-Transcriptional Regulatory Elements

Daniel E. Agnew and Brian F. Pfleger

Abstract

Constructing polycistronic operons is an advantageous strategy for coordinating the expression of multiple genes in a prokaryotic host. Unfortunately, a basic construct consisting of an inducible promoter and genes cloned in series does not generally lead to optimal results. Here, a combinatorial approach for tuning relative gene expression in operons is presented. The method constructs libraries of post-transcriptional regulatory elements that can be cloned into the noncoding sequence between genes. Libraries can be screened to identify sequences that optimize expression of metabolic pathways, multi-subunit proteins, or other situations where precise stoichiometric ratios of proteins are desired.

Key words: Synthetic biology, Promoter, Operon, Ribosome binding site, Intergenic sequence, Megaprimer PCR, Metabolic engineering, mRNA stability, Transcription termination

1. Introduction

A major use of biotechnology is the production of a metabolite or protein of interest – be it a pharmaceutical, biofuel, or another important compound – in a microbial host. The design of new production strains often requires the expression of a number of genes in concert (1, 2). To facilitate this approach, especially in prokaryotic hosts, genes can be grouped into a synthetic operon (3). Unfortunately, a basic construct consisting of an inducible promoter and genes cloned in series does not generally lead to optimal gene expression and metabolite production (4, 5). Worse, expression of poorly designed operons can lead to accumulation of undesired intermediate products, reduced growth rates, and in some cases cell death from metabolite toxicity (6). As understanding of transcription and translation increases, the selection of

James A. Williams (ed.), *Strain Engineering: Methods and Protocols*, Methods in Molecular Biology, vol. 765,
DOI 10.1007/978-1-61779-197-0_7, © Springer Science+Business Media, LLC 2011

appropriate regulatory sequences becomes a more complex process. Methods of altering transcription initiation, mRNA stability, transcript secondary structure, translation initiation, and translation elongation are known (5, 7–9). However, current methods provide few quantitative relationships between primary sequence and protein expression. In addition, the optimal level of expression for a particular protein is frequently unknown a priori. This combination makes rational operon design challenging. Here, a combinatorial method of operon design is presented wherein an optimal intergenic sequence is selected from a library of regulatory elements that control post-transcriptional processes.

Methods of altering gene expression exist for each step in the process of producing a protein from the corresponding DNA sequence. To control transcription, one can choose from an array of promoters that have been characterized in both native and heterologous hosts, see Table 2 in Jana and Deb (10). Methods for producing libraries of synthetic promoters using basic PCR techniques have also been used to tune transcription levels (11). Rates of translation initiation can be controlled by altering the sequence of the ribosome binding site (RBS) located 5′ of each gene (12). In recent work, an in silico model was developed for predicting RBS strength. Predictions were confirmed in vivo for engineered RBS with relative strengths spanning a 100,000-fold range (5). The key feature of this model is its ability to account for the sequence context of the RBS with respect to the surrounding nucleotides. Translation elongation rates can be altered by the codon usage and mRNA secondary structure of sequences located immediately 3′ of the start codon (e.g., ATG) (13). Altering the secondary structure of intergenic region mRNA to include hairpins or RNase sites has also been shown to be effective for tuning expression of synthetic operons by altering the rates of transcription termination, mRNA decay, translation initiation. (14). Unfortunately, the coupling of transcription, translation, and mRNA turnover in a prokaryotic cell complicate efforts to precisely engineer gene expression using directed approaches. Consequently, combinatorial approaches are attractive because they can be used to identify sequences which address each level simultaneously. Here, a method of generating libraries of post-transcriptional regulatory elements that can be incorporated into the intergenic sequence of synthetic operons is presented. Libraries can be screened to select the intergenic sequence which results in the optimal level of gene expression.

The method for synthesizing and inserting libraries of regulatory sequences is illustrated in Figs. 1 and 2, and outlined below (see Subheadings 3.1–3.3). Briefly, 100–400 bp, or larger, intergenic sequences are randomly assembled from two or more regions comprised of moderate length (<60 bp) oligonucleotides. Each region is designed to share overlapping sequence with

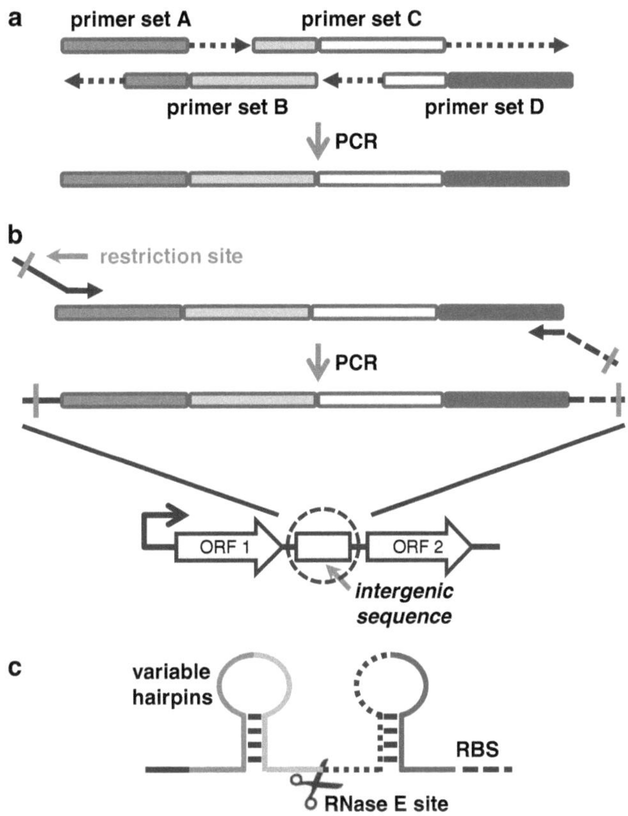

Fig. 1. Construction of a bicistronic operon with variable intergenic sequences. (**a**) Primer sets are designed with terminal homology to neighboring sets and variable internal sequences. Sets A and D may also contain unique restriction sites (vertical bars) at their 5′ ends, to facilitate cloning. Primer sets are combined in a single reaction to create a library of intergenic sequences. (**b**) Intergenic sequences can be further amplified for cloning between ORFs 1 and 2 in a previously constructed vector. (**c**) Example mRNA secondary structure of intergenic sequence. Regions are designed to contain features such as hairpins, RNase sites, protein binding sites, RBS, etc.

neighboring regions. The overlapping sequences permit base pairing of the 3′ terminus of an oligonucleotide from one region with the 5′ terminus of an oligonucleotide from a neighboring region. For each region, a diverse set of oligonucleotides containing regulatory sequences is designed. In a PCR tube containing a mixture of all oligonucleotides, complementary sequences base pair with one another and are extended by DNA polymerase. Through iterative rounds of elongation, a library of chimeric DNA molecules containing one oligonucleotide from each region can be constructed. The chimeric DNA sequences can encode a mixture of various regulatory elements, including hairpins of various size/strength, aptamer binding domains, specific RBS sequences, or other protein binding sites for regulatory proteins such as RNases. Primary libraries can be amplified using terminal

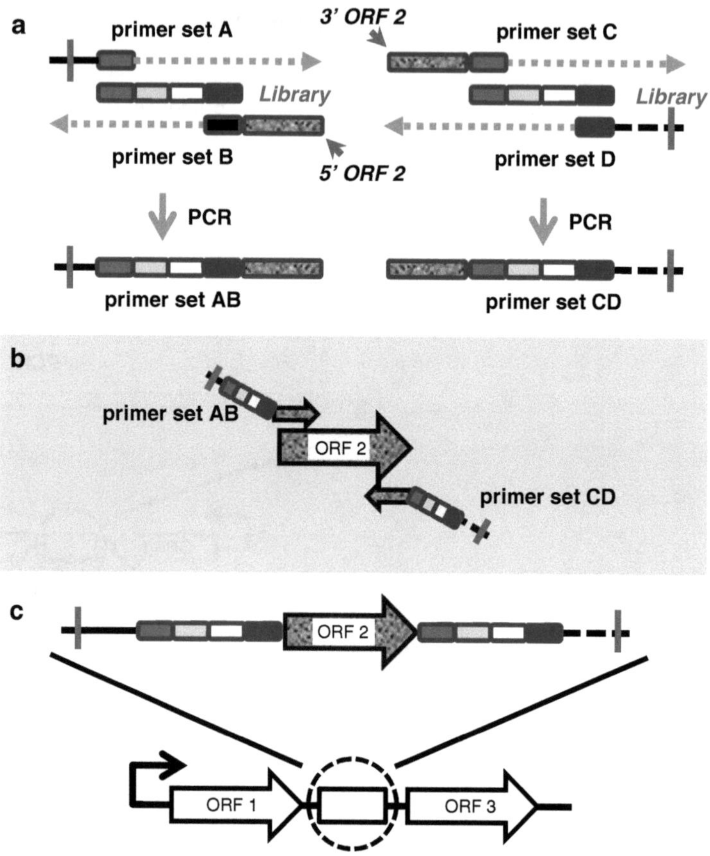

Fig. 2. Construction of a tricistronic operon with variable intergenic sequences using megaprimer PCR. (**a**) Primer sets are designed with terminal homology to neighboring sets and variable internal sequences. Sets A and D may also contain unique restriction sites at the 5′ and 3′ ends, respectively, to facilitate cloning. Primer sets A and B are combined in a separate reaction from C and D to create a library of megaprimers with terminal homology to ORF 2. (**b**) ORF 2 is amplified with megaprimers generated in Fig. 2a. (**c**) ORF 2 and surrounding intergenic sequences are inserted between ORFs 1 and 3 in a previously constructed vector. As in Fig. 1, primer sets are designed to contain features such as hairpins, RNase sites, protein binding sites, RBS, etc.

sequences that are conserved across all oligonucleotides in the first and last region. Amplified libraries provide sufficient DNA for cloning into expression vectors to yield dicistronic operons (Subheading 3.2) or megaprimers used to construct libraries of tricistronic (or larger) operons (Subheading 3.3).

2. Materials

2.1. Library Assembly, Amplification, and Purification

1. Nuclease-free water. Store at room temperature.
2. PCR amplification buffer (e.g., 10× *Taq* PCR buffer). Store at −20°C.

3. 40× dNTP stock solution: Equimolar mixture of dATP, dGTP, dCTP, and dTTP used at a stock concentration of 10 mM (2.5 mM each dNTP). Store at −20°C in 5–10 μL aliquots.
4. 50× PCR primers stock solution. The sequences of PCR primers depend on the characteristics of the intergenic sequences to be assembled (see Notes 1–3). In general, we order custom-synthesized primers shipped in a lyophilized state. Primers are resuspended to a concentration of 200 μM in nuclease-free water or 10 mM Tris–HCl (pH 7.5) and stored at −20°C.
5. DNA polymerase (e.g., *Taq* polymerase used at a stock concentration of 5 U/μL). Store at −20°C.
6. 100% DMSO stock solution (optional).
7. 50 mM $MgCl_2$ in nuclease-free water (optional).
8. 50 mM $MnSO_4$ in nuclease-free water (optional).
9. Gel extraction, nucleotide removal, and PCR cleanup kits (e.g., Qiagen, Promega). These are commercially available. Reagents should be stored and used according to the manufacturer's instructions.

2.2. Insertion of an Oligonucleotide Library into a Bicistronic Operon

1. Expression vector harboring a desired promoter, at least two desired open reading frames, a terminator, origin of replication, and resistance marker. In the process of cloning the expression vector, insert a DNA sequence to facilitate the cloning of the intergenic DNA library, e.g., unique restriction sites (see Notes 4–8). Expression vectors can be purified using standard protocols (15) or commercial purification kits (Qiagen, Promega, etc.). Useful plasmids and promoters can be found in Tables 1 and 2, respectively. Store plasmids at −20°C in TE buffer (10 mM Tris HCl, 1 mM EDTA, pH 8.0).
2. Commercial restriction enzymes and corresponding buffers (optional, see Notes 5 and 7). Store each according to the manufacturer's instructions.
3. Electrocompetent (>10^{10} transformants per μg DNA) *E. coli* strains such as DH10B. These can be purchased (Invitrogen, Promega, etc.) or prepared in house (15). Store aliquots at −80°C.
4. Rich medium for bacterial growth. For example, Luria Broth (LB) and LB agar plates for growth of *E. coli*: 10 g tryptone, 5 g yeast extract, 10 g NaCl, dissolved in milliQ water to 1 L final volume. Include 15 g agar for preparation of agar plates (15). Autoclave to sterilize.
5. Laboratory strains for expression such as K12 MG1655 (ATCC 700962), DH10B (Invitrogen), Top10 (Invitrogen), and BL21(DE3) (New England Biolabs) (16).

Table1
Properties of selected expression vectors

DNA construct	Origin of replication	Copy number	Classification	Examples	Source
pBluescript vectors	ColE1 (29)	300–500	High copy	pBlueScriptII (30)	(31)
pUC vectors	pMB1 (32)	500–700	High copy	pUC19 (33)	(34)
Gateway vectors	pMB1	500–700	High copy	pCR8/GW/TOPO	Invitrogen
pGEM vectors	pMB1	300–400	High copy		Promega
pBR322 and derivatives	pMB1	15–20	Low copy		(35)
pET vectors	pMB1	~40	Low copy	pET-28a(+)	Novagen
pBBR1 and derivatives	pBBR1 (36)	~10	Low copy	pBBR1MCS (37)	(37)
pACYC and derivatives	p15A (38)	10–12	Low copy	pACYC184	(39)
pSC101 and derivatives	pSC101	~5	Very low copy		(40)

Table 2
Properties of selected bacterial promoters

Promoter	Strength	Inducer	Regulator	Source
T7	Strong	None/thermal	None	(41)
T7:Lac	Strong	IPTG/allolactose	LacI/LacI^Q	(42)
Lac	Weak	IPTG/allolactose	LacI/LacI^Q	(33)
Tac	Strong	IPTG/allolactose	LacI/LacI^Q	(43)
Trc	Strong	IPTG/allolactose	LacI/LacI^Q	(44)
BAD	Moderate	Arabinose	AraC	(7)
λp_L, λp_R	Strong	None	λ repressor, *cI*	(45)
Tet	Moderate	Anhydrotetracycline	Tet^R	(46)
Pro	Strong	2-Methyl-citrate	PrpR	(47)

IPTG isopropyl β-D-1-thiogalactopyranoside

2.3. Megaprimer Library Cloning for Optimization of Tricistronic Operons

1. High fidelity DNA polymerase (e.g., Phusion, Thermo Scientific). Store at −20°C.
2. Additional materials listed in Subheadings 2.1 and 2.2.

3. Methods

3.1. Library Assembly, Amplification, and Purification

1. For review of standard PCR procedures, please see *Molecular Cloning* (15) or *PCR Protocols* (17).
2. Dilute oligonucleotide sets to 400 μM in water (or 10 mM Tris–HCl pH 7.5) such that the mixture is equimolar in each, unless a bias for a desired oligonucleotide is desired.
3. To assemble the library (Fig. 1b) make the following PCR master mix: 40 nmol of oligonucleotide mixture, 1× polymerase buffer, 250 μM dNTP mix, five units of polymerase and nuclease-free water to a final volume of 100 μL. Mix thoroughly by pipetting up and down. Do not vortex.
4. Run the following thermocycler protocol: 95°C for 2 min, cycle – 15 s at 95°C, 30 s at 72°C, and 20 + 5 s/cycle at 72°C – for 35 rounds, 72°C for 10 min.
5. Purify the resulting DNA mixture using a nucleotide cleanup kit or a DNA purification kit capable of binding small (<200 bp) DNA fragments. Run an agarose gel to validate assembly reactions. Store assembly at −20°C.
6. To amplify a library for cloning (Fig. 1b), make the following master mix: 1× PCR buffer, 250 μM dNTPs (optional 0–1 mM $MnSO_4$), 2 μL of assembly mixture, 600 nM amplification primers, five units of DNA polymerase and nuclease-free water to a final volume of 100 μL.
7. Run the following thermocycler protocol: 95°C for 2 min, cycle – 95°C for 20 s melting temperature + 3°C for 15 s, 72°C for 30 s – for 35 rounds, 72°C for 10 min.
8. Purify the resulting DNA mixture, as above, and run a gel to verify amplification. The amplified mixture can now be cloned into an operon or stored at −20°C.

3.2. Insertion of an Oligonucleotide Library into a Bicistronic Operon

1. For a review of standard cloning procedures, please see *Molecular Cloning* (15).
2. Construct an expression vector harboring a desired promoter, the two desired open reading frames (ORFs), a terminator, origin of replication, and resistance marker. In the process of cloning the expression vector, insert a DNA sequence to facilitate the cloning of the intergenic DNA library, e.g., unique restriction sites (see Note 5).

3. Use the desired cloning technique to insert the DNA library into the expression vector.

3.3. Megaprimer Library Cloning for Optimization of Tricistronic Operons

1. This protocol will use megaprimer PCR (17) to amplify the central open reading frame of a tricistronic operon (Fig. 2) using DNA libraries as primers. The resulting product will be cloned into an expression vector containing the first and third open reading frames of a tricistronic operon.
2. Construct an expression vector containing a desired promoter, the first and third open reading frames, a terminator, origin of replication, and resistance marker. In the process of cloning the expression vector, insert a DNA sequence to facilitate the cloning of the megaprimer PCR product, e.g., unique restriction sites (see Note 5).
3. Design primers to amplify DNA libraries and construct megaprimers. In the 5′ tail of primer A and primer D (Fig. 2), attach restriction sites for cloning the megaprimer PCR product into the expression vector. In the 5′ tail of primer B, insert the sequence of a primer for amplifying the 5′ end of the central open reading frame. In the 5′ tail of primer C, insert the sequence of a primer for amplifying the 3′ end of the central open reading frame.
4. To assemble the two library fragments (Fig. 2a), run two amplification reactions as above (see steps 3–4 in Subheading 3.1) using primer sets A, B and C, D.
5. Purify each megaprimer using a nucleotide removal kit and quantify the DNA using a spectrophotometer (e.g., NanoDrop) at A_{260}. Estimate the average size from a gel standard. Estimate the molar concentration of the primers using the following conversion factor: approximate molecular weight of double stranded DNA = (number of nucleotides × 607.4 g/mol) + 157.9 g/mol (18).
6. To amplify ORF2 for cloning (Fig. 2b) assemble the following master mix: 1× PCR buffer, 250 μM dNTPs, 100–300 nM megaprimer AB product, 100–300 nM megaprimer CD product, 0.1–0.5 μg of central ORF template, five units of high fidelity DNA polymerase, and nuclease-free water to a final volume of 100 μL.
7. Run the following thermocycler protocol: 95°C for 3 min, cycle – 95°C for 60 s melting temperature + 3°C for 2 min, 72°C for 1–2 min/kb – for 35 rounds, 72°C for 10 min.
8. Purify the resulting PCR fragments using a PCR cleanup kit. If necessary, amplify the products with a standard PCR using primers A, D.
9. Clone the library into the expression vector containing the first and third open reading frames.

4. Notes

1. Libraries can be constructed to include sequences that control transcription termination, mRNA turnover, and/or translation initiation. Pfleger et al. describes a basic design strategy where two hairpins were separated by an RNase E cleavage site (14). The RNase E site directs cleavage of the primary mRNA transcript and positions the two hairpins at the ends of the two secondary transcripts. Each hairpin's characteristics (size, ΔG_f^0, primary sequence) will have a different impact on the stability of the secondary transcripts. To construct a library of intergenic sequences, the base construct was broken into four sections with three overlapping sequences placed in the loops of the hairpins and at the RNase E site. Diversity was introduced into the library by including oligonucleotides that expanded and contracted the stems of each hairpin, introduced bulges, RNase sites, and sequences that would sequester a RBS. Alternative strategies could include termination sequences (e.g., poly U tracts), aptamers, or protein binding sites. Libraries can be constructed to include RBSs or use sites incorporated on to the expression vector. Smaller libraries that contain sequences targeting one post-transcriptional mechanism, e.g., libraries of ribosome binding (19) sites or intrinsic transcription terminators (20), have been used to alter gene expression and could be used to optimize expression from operons.
2. The Mfold web server (21) is the standard method of predicting mRNA secondary structure. In designing libraries, the intergenic sequence and 50 bases of flanking coding sequence were input to Mfold. The effect of secondary structure surrounding the RBS on translation rates has been correlated (22) and can be used to guide design of libraries. Alternatively, work by Voigt and coworkers has led to the development of a tool to predict RBS strength (5).
3. When designing primer sets for assembling regulatory sequence libraries, the following design criteria should be used. Overlapping regions should be 15–20 bases in length, possess moderate melting temperatures ($50°C < T_m < 72°C$), and be devoid of internal base pairing (within the overlap region). Primers for amplifying libraries, making megaprimers, and amplifying intergenic sequence–gene–intergenic sequence libraries should be at least 20 bases in length and follow standard primer design characteristics.
4. As an alternative to the one shot megaprimer PCR method described here, a set of two sequential megaprimer reactions can be run. In the first, a primer that amplifies the central

open reading frame in the forward direction can be used with the 3′ megaprimer to generate a gene-library amplicon. Once purified, this library can be used as a template for a second PCR using the 5′ megaprimer library and primer D from the megaprimer assembly reactions.

5. The number of background colonies (from the vector only) that result from cloning an intergenic sequence library can be reduced by adding a sacrificial DNA sequence in between the desired cloning sites. In Pfleger et al., a 1,000-bp fragment of the *lacZ* gene was cloned between the unique restriction sites used for cloning intergenic libraries (14). When completely digested, the desired expression vector generated fragments that were distinct from singly cut or uncut plasmid DNA. This facilitated isolation of the desired fragment via gel extraction and reduced unwanted background colonies.

6. This protocol describes the use of traditional restriction enzyme cloning to insert intergenic sequence libraries into expression vectors. The protocol could be adapted to incorporate modern cloning strategies. The synthetic biology research community advocates the use of standardized cloning protocols to facilitate collaborative research. By using restriction enzymes that generate compatible ends, ligation products can be generated that do not contain the original restriction sites at the fusion point. A BioBrick standard has been established at the Massachusetts Institute of Technology to take advantage of this strategy to perform serial cloning of standardized DNA fragments (23). If intergenic sequence libraries were assembled using a BioBrick strategy, the resulting fragments could be cloned into many expression systems in parallel.

7. Commercially available restriction enzymes or alternative cloning procedures, such as enzyme free cloning (24) or recombination based techniques (25), should be used to insert DNA libraries into desired operons. Enzyme-free cloning is a modern method of cloning where long complementary overhangs are generated on inserts and vectors by PCR or exonuclease activity. Base pairing between the long overhangs is sufficiently stable to eliminate the need for in vitro ligation reactions. Expression vectors containing standardized insertion sites have been constructed (26). Libraries could be designed to incorporate this method of cloning. Similarly, cloning methods using homologous recombination, both in yeast and in vitro (27, 28), have been developed. These methods could be used to construct larger (i.e., greater than three gene) operon libraries.

8. The protocol described above is capable of generating large libraries where the upper bound is likely determined by

transformation efficiency. In order to use this method to optimize expression from an operon, a method of screening good clones from bad must be established. In Pfleger et al., an auxotrophic mutant was used to prescreen libraries for functional operons (14). If high-throughput screening or selection methods are not available, the size of the library should be reduced and/or the library should be designed to target specific ranges of expression.

Acknowledgments

Work in the authors' lab was sponsored by the University of Wisconsin-Madison Graduate School. Daniel Agnew is the recipient of an NIH Biotechnology Training Program Graduate Fellowship (NIH 5 T32 GM08349).

References

1. Khosla C., and Keasling J. D. (2003) Timeline - Metabolic engineering for drug discovery and development. *Nature Reviews Drug Discovery* **2**, 1019–1025.
2. Martin V. J. J., Pitera D. J., Withers S. T., Newman J. D., and Keasling J. D. (2003) Engineering a mevalonate pathway in *Escherichia coli* for production of terpenoids. *Nat. Biotechnol.* **21**, 796–802.
3. Baga M., Goransson M., Normark S., and Uhlin B. E. (1988) Processed messenger-RNA with differential stability in the regulation of *E. coli* pilin gene expression. *Cell* **52**, 197–206.
4. Nishizaki T., Tsuge K., Itaya M., Doi N., and Yanagawa H. (2007) Metabolic engineering of carotenoid biosynthesis in *Escherichia coli* by ordered gene assembly in Bacillus subtilis. *Appl. Environ. Microbiol.* **73**, 1355–1361.
5. Salis H. M., Mirsky E. A., and Voigt C. A. (2009) Automated design of synthetic ribosome binding sites to control protein expression. *Nat. Biotechnol.* **27**, 946-U112.
6. Kizer L., Pitera D. J., Pfleger B. F., and Keasling J. D. (2008) Application of functional genomics to pathway optimization for increased isoprenoid production. *Appl. Environ. Microbiol.* **74**, 3229–3241.
7. Guzman L. M., Belin D., Carson M. J., and Beckwith J. (1995) Tight Regulation, Modulation, and High-Level Expression by Vectors Containing the Arabinose PBAD Promoter. *J. Bacteriol.* **177**, 4121–4130.
8. Carrier T. A., and Keasling J. D. (1999) Library of synthetic 5‘ secondary structures to manipulate mRNA stability in *Escherichia coli. Biotechnol. Prog.* **15**, 58–64.
9. Smolke C. D., Carrier T. A., and Keasling J. D. (2000) Coordinated, differential expression of two genes through directed mRNA cleavage and stabilization by secondary structures. *Appl. Environ. Microbiol.* **66**, 5399–5405.
10. Jana S., and Deb J. K. (2005) Strategies for efficient production of heterologous proteins in *Escherichia coli. Appl. Microbiol. Biotechnol.* **67**, 289–298.
11. Hammer K., Mijakovic I., and Jensen P. R. (2006) Synthetic promoter libraries - tuning of gene expression. *Trends Biotechnol.* **24**, 53–55.
12. Kozak M. (2005) Regulation of translation via mRNA structure in prokaryotes and eukaryotes. *Gene* **361**, 13–37.
13. Seo S. W., Yang J., and Jung G. Y. (2009) Quantitative Correlation Between mRNA Secondary Structure Around the Region Downstream of the Initiation Codon and Translational Efficiency in *Escherichia coli. Biotechnol. Bioeng.* **104**, 611–616.
14. Pfleger B. F., Pitera D. J., D Smolke C., and Keasling J. D. (2006) Combinatorial engineering of intergenic regions in operons tunes expression of multiple genes. *Nat. Biotechnol.* **24**, 1027–1032.
15. Sambrook J., Russell D. W. (2001) *Molecular cloning : a laboratory manual* Cold Spring Harbor Laboratory, Cold Spring Harbor, N.Y.
16. Studier F. W., Rosenberg A. H., Dunn J. J., and Dubendorff J. W. (1990) Use of T7

RNA-polymerase to direct expression of cloned genes. *Methods Enzymol.* **185**, 60–89.
17. Bartlett J. M. S., and Stirling D., (Eds.) (2003) *PCR Protocols*, Vol. **226**, second ed., Humana Press.
18. (2010) Ambion's Appendix - DNA and RNA Molecular Weights and Conversions, Applied Biosystems. http://www.ambion.com/techlib/append/na_mw_tables.html.
19. Meynial-Salles I., Cervin M. A., and Soucaille P. (2005) New tool for metabolic pathway engineering in *Escherichia coli*: One-step method to modulate expression of chromosomal genes. *Appl. Environ. Microbiol.* **71**, 2140–2144.
20. Graham J. E. (2004) Sequence-specific Rho-RNA interactions in transcription termination. *Nucleic Acids Res.* **32**, 3093–3100.
21. Zuker M. (2003) Mfold web server for nucleic acid folding and hybridization prediction. *Nucleic Acids Res.* **31**, 3406–3415.
22. Desmit M. H., and Vanduin J. (1994) Translational initiation on structured messengers: another role for the Shine-Dalgarno interaction. *J. Mol. Biol.* **235**, 173–184.
23. Canton B., Labno A., and Endy D. (2008) Refinement and standardization of synthetic biological parts and devices. *Nat Biotechnol.* **26**, 787–793.
24. Aslanidis C., and Dejong P. J. (1990) Ligation-independent cloning of PCR products (LIC-PCR). *Nucleic Acids Res.* **18**, 6069–6074.
25. Datsenko K. A., and Wanner B. L. (2000) One-step inactivation of chromosomal genes in *Escherichia coli* K-12 using PCR products. *Proc Natl Acad Sci U.S.A.* **97**, 6640–6645.
26. Stols L., Gu M. Y., Dieckman L., Raffen R., Collart F. R., and Donnelly M. I. (2002) A new vector for high-throughput, ligation-independent cloning encoding a tobacco etch virus protease cleavage site. *Protein Expression Purif.* **25**, 8–15.
27. Benders G. A., Noskov V. N., Denisova E. A., Lartigue C., Gibson D. G., Assad-Garcia N., Chuang R. Y., Carrera W., Moodie M., Algire M. A., Phan Q., Alperovich N., Vashee S., Merryman C., Venter J. C., Smith H. O., Glass J. I., and Hutchison C. A. (2010) Cloning whole bacterial genomes in yeast. *Nucleic Acids Res.* **38**, 2558–2569.
28. Gibson D. G., Young L., Chuang R. Y., Venter J. C., Hutchison C. A., and Smith H. O. (2009) Enzymatic assembly of DNA molecules up to several hundred kilobases. *Nat. Methods* **6**, 343-U341.
29. DeWitt W., and Helinski D. R. (1965) Characterization of colicinogenic factor El from a non-induced and a mitomycin C-induced Proteus strain. *J. Mol. Biol.* **13**, 692–703.
30. Altingmees M. A., and Short J. M. (1989) pBlueScript II: gene mapping vectors. *Nucleic Acids Res.* **17**, 9494–9494.
31. Short J. M., Fernandez J. M., Sorge J. A., and Huse W. D. (1988) Lambda ZAP: a bacteriophage lambda expression vector with in vivo excision properties. *Nucleic Acids Res.* **16**, 7583–7600.
32. Betlach M., Hershfield V., Chow L., Brown W., Goodman H. M., and Boyer H. W. (1976) Restriction endonuclease analysis of bacterial plasmid controlling EcoRI restriction and modification of DNA. *Fed Proc* 35, 2037–2043.
33. Yanischperron C., Vieira J., and Messing J. (1985) Improved M13 phage cloning vectors and host strains - nucleotide-sequences of the M13mp18 and pUC19 vectors. *Gene* 33, 103–119.
34. Vieira J., and Messing J. (1982) The pUC plasmids, an M13mp7-derived system for insertion mutagenesis and sequencing with synthetic universal primers. *Gene* **19**, 259–268.
35. Bolivar F., Rodriguez R. L., Greene P. J., Betlach M. C., Heyneker H. L., Boyer H. W., Crosa J. H., and Falkow S. (1977) Construction and characterization of new cloning vehicles. II. Multipurpose cloning system. *Gene* **2**, 95–113.
36. Antoine R., and Locht C. (1992) Isolation and molecular characterization of a novel broad-host-range plasmid from Bordetella bronchiseptica with sequence similarities to plasmids from gram-positive organisms. *Mol. Microbiol.* **6**, 1785–1799.
37. Kovach M. E., Phillips R. W., Elzer P. H., Roop R. M., and Peterson K. M. (1994) pBBR1MCS: a broad-host-range cloning vector. *BioTechniques* **16**, 800-&.
38. Cozzarelli N.R, Kelly R. B., and Kornberg A. (1968) A minute circular DNA from *Escherichia coli* 15. *Proc Natl Acad Sci U.S.A.* **60**, 992–999.
39. Chang A. C. Y., and Cohen S. N. (1978) Construction and characterization of amplifiable multicopy DNA cloning vehicles derived from p15A cryptic miniplasmid. *J. Bacteriol.* **134**, 1141–1156.
40. Cohen S. N., and Chang A. C. Y. (1973) Recircularization and autonomous replication of a sheared R-factor DNA segment in *Escherichia coli* transformants: (plasmid/transformation/antibiotic resistance/DNA). *Proc Natl Acad Sci U.S.A.* **70**, 1293–1297.
41. Tabor S., and Richardson C. C. (1985) A bacteriophage T7 RNA polymerase promoter

system for controlled exclusive expression of specific genes. *Proc Natl Acad Sci U.S.A.* **82**, 1074–1078.

42. Giordano T. J., Deuschle U., Bujard H., and McAllister W. T. (1989) Regulation of coliphage T3 and coliphage T7 RNA polymerases by the lac repressor-operator system. *Gene* **84**, 209–219.
43. Deboer H. A., Comstock L. J., and Vasser M. (1983) The tac promoter - a functional hybrid derived from the trp and lac promoters. *Proc Natl Acad Sci U.S.A.* **80**, 21–25.
44. Brosius J., Erfle M., and Storella J. (1985) Spacing of the -10 and -35 regions in the tac promoter: effect on its in vivo activity. *J. Biol. Chem.* **260**, 3539–3541.
45. Elvin C. M., Thompson P. R., Argall M. E., Hendry P., Stamford N. P. J., Lilley P. E., and Dixon N. E. (1990) Modified bacteriophage lambda promoter vectors for overproduction of proteins in *Escherichia coli. Gene* **87**, 123–126.
46. Skerra A. (1994) Use of the tetracycline promoter for the tightly regulated production of a murine antibody fragment in *Escherichia coli. Gene* **151**, 131–135.
47. Lee S. K., and Keasling J. D. (2005) A propionate-inducible expression system for enteric bacteria. *Appl. Environ. Microbiol.* **71**, 6856–6862.

Chapter 8

Marker-Free Chromosomal Expression of Foreign and Native Genes in *Escherichia coli*

Chung-Jen Chiang, Po Ting Chen, Shan-Yu Chen, and Yun-Peng Chao

Abstract

Genetic manipulation of *Escherichia coli* strains for desired traits is the most applied strain engineering approach in industrial applications. For chromosomal insertion of genes and controlled expression of genomic genes in *E. coli*, the replicon-free and markerless method is described based on a series of conditional-replication plasmids called phage-integration vectors. They mainly carry the multiple cloning site and the prophage attachment site, which are sandwiched by two FRT sites. With the aid of the phage integrase from conditional-replication helper plasmids, the passenger genes of either foreign or native type incorporated into the integration vectors can be specifically integrated into bacterial genome at the prophage attachment site. Finally, the inserted DNA containing replicon and/or selective markers in integrants can be eliminated by the act of the FLP recombinase provided from a conditional-replication helper plasmid.

Key words: Genomic engineering, Markerless, Lycopene, Flippase, Recombinase

1. Introduction

Escherichia coli strains are widely applied for industrial bioprocessing. It always requires the genetic manipulation of *E. coli* with a new or improved trait. This is usually done through enhanced expression of key endogenous and alien genes using plasmid-based expression vectors. However, manifold copies of plasmid-encoded genes seem unnecessary for pathway engineering of microbes (1). Additionally, the physiological stress in bacteria might arise due to the redundant copy of DNAs, consequently leading to segregational loss or internal rearrangements of plasmids (2). Indeed, antibiotics are frequently used as a selective pressure to ensure the stable inheritance of plasmids in *E. coli*. For some applications, antibiotics are prohibited for use to prevent fermentation products

James A. Williams (ed.), *Strain Engineering: Methods and Protocols*, Methods in Molecular Biology, vol. 765,
DOI 10.1007/978-1-61779-197-0_8,

from drug contamination. Moreover, the potential risk of plasmids spreading the antibiotic-resistant trait to other microbes in nature also exits (3).

Apparently, all the above-mentioned problems caused by using plasmids can be circumvented if chromosome engineering of *E. coli* is applied. A series of vectors called CRIM plasmids were developed to achieve genomic insertion of target genes. They mainly consist of a conditional-replication origin and a phage attachment (*attP*) site (4). With the aid of the phage integrase (Int) from a helper plasmid, these plasmids can be integrated into the bacterial attachment (*attB*) site. However, the complication of using CRIM plasmids still exists because the replication origin and the antibiotic-resistant marker carried in plasmids are retained in integrants' genomes after integration. Therefore, on the basis of the backbone of CRIM plasmids, a series of phage-integration vectors have been recently developed with incorporation of the FRT site-bracketed DNA consisting of multiple cloning site (MCS) and an *attP* locus for phage λ, HK022, Ø80, P21, and P22 (5). These integration vectors are able to facilitate the site-specific insertion of target genes into *E. coli* chromosome and to allow later elimination of replicons and selective markers by FLP flippase. With the phage-integration vectors, we were able to engineer the *E. coli* genome in a marker-free way to achieve high and stable production of lycopene.

1.1. Phage-Integration Vectors

Table 1 summarizes the detailed characteristics of strains and plasmids described herein. As shown in Fig. 1a, these phage-integration vectors contain MCS, *attP*, conditional-replication R6K origin, the *cat* gene (encoding aminoglycoside 3′-phosphotransferase that confers on chloramphenicol resistance), and the phage tL3 transcription terminator. These vectors can only be maintained in an *E. coli* strain harboring the *pir* gene (encoding the replication protein for the R6K origin), such as strain DH5α (*pir*), with the supplement of chloramphenicol. Mediated by the function of phage Int provided on a heterologous plasmid, the recombination of the *attP* in the plasmid with the *attB* in bacterial genome is readily achieved, leading to the integration of the plasmid DNA into bacterial genome at the attachment locus (Fig. 1b). In addition, MCS and *attP* are flanked by two FRT sites. This allows for the later removal of the R6K origin and *cat* gene with FLP after insertion (Fig. 1c). These integration vectors carry five different types of *attP* and permit insertion of genes into various *attB* in bacterial genome, including phage λ, HK022, Ø80, P21, and P22 (Table 1; see Note 1).

1.2. Incorporation of Target Genes into Phage-Integration Vectors

The target gene of either foreign or native nature is cloned into an expression vector carrying the promoter of interest. This will result in the controllable expression of the target gene. The artificial promoter could be *tac*, *araBAD*, T7 promoter, etc.

Table 1
Strains and plasmids useful for genomic engineering of *E. coli*

Strain	Relevant characteristics	Sources
DH5α	*deoR endA1 gyrA96 hsdR17 supE44 thi1 recA1lacZM15*	Lab collection
DH5α (pir)	As DH5α but contains *pir*	Lab collection
BL21(DE3)	F^- *dcm gal ompT hsdS*(r_B^- m_B^-) (DE3)	Novagen
Plasmid		
pHK-Cm	Integration vector with $attP_{HK}$ and cat^+	(5)
pPhi80-Cm	Integration vector with $attP_{Phi}$ and cat^+	(5)
pP21-Cm	Integration vector with $attP_{P21}$ and cat^+	(5)
pP22-Cm	Integration vector with $attP_{P22}$ and cat^+	(5)
pLamda-Cm	Integration vector with $attP_{\lambda}$ and cat^+	(5)
pCP20	Expressing FLP with bla^+	(11)
pINT-ts	Expressing phage λ Int with bla^+	(4)
pAH69	Expressing phage HK022 Int with bla^+	(4)
pAH121	Expressing phage P21 Int with bla^+	(4)
pAH123	Expressing phage Ø80 Int with bla^+	(4)
pAH130	Expressing phage P22 Int with bla^+	(4)

Abbreviations: *attP* site for phage HK022 ($attP_{HK}$), λ ($attP_{\lambda}$), Ø80 ($attP_{Phi}$), P21 ($attP_{P21}$), and P22 ($attP_{P22}$); *bla*, the gene encoding β-lactamase

We prefer T7 promoter, which is known for its strong strength and specifically requires T7 RNA polymerase for activation (6, 7). The latter feature permits the compartmentalization of cell growth and production phase. In addition, a library of promoter variants with a wide range of strength is also accessible (8). This appears to be attractive for use when the issue of tuning the expression of genomic genes is addressed (9, 10). The vectors are then integrated into the genome using a phage integrase expressing helper plasmid (Table 1). The selective marker and the replication origin are excised from the integrant using helper plasmid pCP20 expressing FLP -mediated FRT recombination (11).

1.3. Genomic Engineering of E. coli for Lycopene Production

The utility of phage-integration vectors was illustrated by genomic engineering of *E. coli* for lycopene production. This was achieved with recruitment of three heterologous genes containing *gps*, *crtI*, and *crtB* to extend the existing pathway in *E. coli*, thereby leading to the synthesis of lycopene.

To do so, the three genes were cloned in a cistronic way from *Archaeoglobus fulgidus* (12) and *Erwinia herbicola* Eho10 (13) by PCR. By incorporation into plasmid pET-20bI (14), the resulting

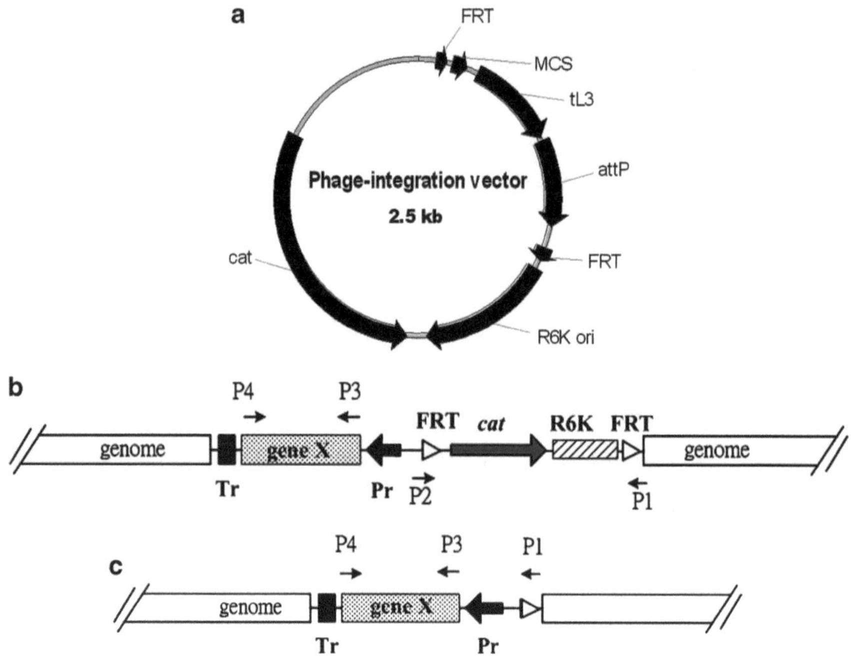

Fig. 1. Phage integration vector application. (**a**) The physical map of a typical phage-integration vector. The MCS consists of *Pst*I, *Sal*I, *Xba*I, *Bam*HI, *Sma*I, *Kpn*I, *Sac*I, and *Eco*RI. The target gene with a promoter and optionally a transcription terminator is incorporated into the MCS. Alternatively, the vector encoded tL3 terminator is utilized as the Tr. (**b**) Schematic illustration of insertion of a phage-integration vector into *E. coli* chromosome. The target gene X with a promoter (Pr) and a transcription terminator (Tr) incorporated into the MCS of a phage-integration vector is shown. Mediated by the phage Int provided in *trans* from a Int expression plasmid (Table 1), the phage-integration vector is inserted into *E. coli* genome as shown (see Note 11). (**c**) Schematic illustration of the inserted DNA after removal of the replication origin and marker by pCP20-encoded FLP recombinase-mediated recombination at the vector-encoded FRT sites.

gene cluster (*gps-crtI-crtB*) was placed under the control of the T7 promoter. Subsequently, the T7 promoter-driven gene cluster and the T7 transcription terminator was produced by PCR and subcloned into phage-integration vector pP21-Km (Table 1) to give plasmid pP21-Crt (5).

Following the protocol outlined in Subheading 3.2, plasmid pP21-Crt was then transformed into strain BL21(DE3) carrying helper plasmid pAH121. After selection for integrants exhibiting chloramphenicol resistance but ampicillin sensitivity (indicating the loss of the helper plasmid pAH121), colony PCR was carried out to verify the inserted DNA in integrants based on the protocol described in Subheading 3.3. As depicted in Fig. 2, the inserted *crt* gene cluster was identified with primers T3/gps and T4/crtB (5), which correspond to primers P3 and P4 in Fig. 1b. With primers T1-T2 (5), corresponding to primers P1 and P2 in Fig. 1b, the result confirmed the presence of the DNA containing the R6K-replication origin and *cat*. The inserted origin and *cat* were removed with the aid of FLP supplied by helper plasmid pCP20 as described in Subheading 3.4. Integrants were selected for exhibiting sensitivity to both chloramphenicol (loss of marker)

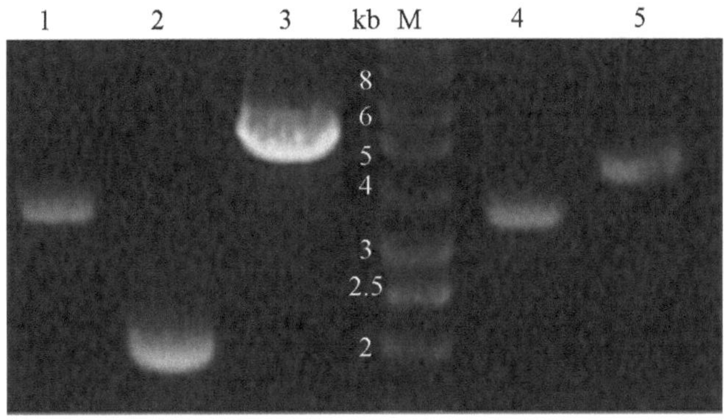

Fig. 2. Verification of the gene insertion and the DNA removal by agarose gel electrophoresis. Refer to text for more details. Keys: *lane 1*, verification of the inserted *crt* gene cluster with primers T3/gps-T4/crtB; *lane 2*, verification of the inserted DNA containing the replication origin and the marker with primers T1-T2; *lane 3*, verification of the inserted DNA containing the replication origin, the marker, and the *crt* gene cluster with primers T1-T4/crtB; *lane 4*, verification of the inserted *crt* gene cluster with primers T3/gps-T4/crtB after FLP-mediated deletion of the DNA; *lane 5*, the remaining plasmid DNA backbone containing the inserted *crt* gene cluster with primers T1-T4/crtB after FLP-mediated deletion of the DNA; *lane M*, the DNA marker.

and ampicillin (loss of pCP20). With colony PCR, the presence of the *crt* gene cluster without the replication origin and *cat* was identified with primers T1-T4/gps and T3/gps-T4/crtB (Fig. 2). The resulting strain was designated BL21-Crt, and it carries a genomic copy of the *crt* gene cluster under control of the T7 promoter.

With the phage-integration vector, we were also able to manipulate the expression level of the native *dxs* gene. The function of *dxs* is responsible for the first committed step leading to the synthesis of lycopene. Therefore, *dxs* fused to the T7 promoter was constructed and incorporated into phage-integration vector pHK-Cm to give plasmid pHK-dxs (5). Essentially following the protocols as outlined in Subheadings 3.2–3.4, strain BL21-Crt was engineered to produce strain BL21-CrtD1 free of the marker. This resulting strain carries an extragenomic copy of *dxs* that is driven by the T7 promoter. As diagramed in Fig. 1b and c, the insertion and deletion of DNA events were examined by colony PCR using primers T1-T2 (5) and T1-T4/dxs, respectively.

Consequently, shake-flask cultures of the constructed strain were carried out with LB medium plus 0.4% glucose. Upon induction with IPTG, lycopene production was measured along the time course (5). After fermentation for 36 h, 5 mg/L and 55 mg/L lycopene were obtained for strain BL21-Crt and BL21-CrtD1, respectively. This case study demonstrates the application of the marker-free chromosomal expression system described herein for engineering *E. coli* metabolic pathways.

2. Materials

1. *E. coli* strains: DH5α, DH5α(*pir*), and BL21(DE3) competent cells (see Note 2).
2. Helper plasmids: pINT-ts, pAH69, pAH121, pAH123, and pAH130 (Table 1; see Note 3).
3. FRT plasmid: pCP20 (Table 1).
4. Phage-integration vectors: pHK-Cm, pPhi80-Cm, pP21-Cm, pP22-Cm, and pLamda-Cm (Table 1; see Note 4).
5. Luria–Bertani (LB) medium: 10 g/L Bacto tryptone, 5 g/L Bacto yeast extract, 10 g/L NaCl. Sterilize by autoclaving.
6. LB agar: LB medium containing 1.5% agar.
7. LB agar + ampicillin: LB agar containing 30 μg/mL ampicillin (added from stock after cooling the autoclaved medium).
8. LB agar + chloramphenicol: LB agar containing 20 μg/mL chloramphenicol (added from stock after cooling the autoclaved medium).
9. Antibiotic reagents: 34 mg/mL chloramphenicol in ethanol and 50 mg/mL sodium ampicillin in water. Refrigerate.
10. Molecular cloning reagents: restriction enzymes, T4 DNA ligase, Pfu polymerase, and DreamTaq polymerase (Fermentas).
11. PCR DNA purification kit: NucleoSpin Extraction Kit (Clontech).
12. Plasmid purification kit: QIAprep Spin Miniprep Kit (Qiagen).
13. 10× Taq PCR buffer: 100 mM Tris–HCl (pH 8.3), 500 mM KCl, 15 mM $MgCl_2$.
14. 10× dNTP nucleotide mix: 2 mM each of dATP, dCTP, dGTP, and dTTP.
15. 0.1 M $MgCl_2$.
16. 0.1 M $CaCl_2$.
17. Loading buffer: 2.5 g/L bromophenol blue, 40% sucrose.
18. 5× TBE buffer: 54 g/L Tris–HCl, 27.5 g/L boric acid, 20 mL/L 0.5 M EDTA (pH 8.0).
19. DNA size marker.
20. 5 mg/mL Ethidium bromide (10,000×) stock solution.
21. Primer P1: CCTTCTGCGAAGTGATCTTCCGTC.
22. Primer P2: CTGCAGAATGAAGTTCCTATTCCGAAG.
23. Custom target gene-specific primers P3 and P4: see Fig. 1b.

3. Methods

Based on the phage-integration vectors, the general procedure for performing genomic insertion of target genes is described below.

3.1. Incorporation of Target Genes into Phage-Integration Vectors

1. The target gene of either foreign or native nature is amplified by PCR with a proofreading DNA polymerase (e.g., Pfu) and primers that incorporate a restriction site of choice to facilitate cloning into an expression vector containing the promoter of choice (i.e., to create a promoter-gene fusion construct).
2. The PCR amplified DNA is then purified with a commercial PCR cleanup kit (e.g., NucleoSpin Extraction Kit) and digested with the restriction enzymes to create the restriction sites that are originally incorporated into the primers.
3. The digested DNA is resolved by agarose gel electrophoresis and the target DNA fragment is purified with a commercial gel-purification kit (e.g., NucleoSpin Extraction Kit).
4. Ligate the purified DNA into an expression vector carrying the artificial promoter of interest. We prefer T7 promoter because a variety of T7 promoter-carrying plasmids are commercially available (see Note 5). This will result in the controllable expression of the target gene.
5. Transform the resulting plasmid into DH5α (or equivalent) and antibiotic-resistant transformants are selected after plating on selective medium.
6. From these transformants, the composite plasmid is isolated and examined for the correctness of the clone by restriction digestion or, preferably, DNA sequencing.
7. Amplify the target gene associated with the promoter and, if necessary, terminator (see Note 6) from the constructed plasmid by PCR with primers that incorporate a restriction site to facilitate cloning into the phage integration vector. We usually incorporate *Pst*I and *Sma*I into the promoter and terminator-specific primers, respectively. This allows directional cloning of the promoter-gene of interest cassette upstream of the vector-encoded tL3 terminator; however, alternative restriction sites are also present in the vectors (Fig. 1a).
8. The PCR-amplified DNA is treated in a similar way as described above and ligated into the linearized phage-integration vector. The resulting vector is then transformed into strain DH5α (*pir*) and transformants exhibiting resistance to chloramphenicol are selected. Clones are verified by restriction

digestion and sequencing. In the case where the T7 promoter is not used, simply amplify the DNA containing the fusion of the target gene with its functional promoter and splice the amplified DNA into the MCS of phage-integration vectors in the correct orientation. This will incorporate the phage tL3 terminator downstream of the target gene (Fig. 1a).

3.2. Genomic Insertion of Target Genes Using the Phage-Integration Vector

Genomic insertion of target genes is carried out as follows.

1. A host strain free of *pir*, such as BL21 (DE3), is first transformed with the helper plasmid expressing phage Int (Table 1) and is selected for resistance to ampicillin at 30°C after plating on LB + ampicillin plates (see Note 7).
2. Pick a single colony of the strain harboring the helper plasmid from the LB agar plate plus ampicillin and transfer into a capped flask containing 5 mL LB medium containing ampicillin.
3. Maintain the culture in a shaker at 30°C and 200 rpm for overnight.
4. Inoculate the overnight culture into a capped flask containing 10 mL fresh LB medium to obtain an initial cell density of 0.08 at OD_{550} (optical density at 550 nm wave length).
5. Maintain the seeding culture in a shaker at 30°C and 200 rpm until the cell density reaches around 0.3 at OD_{550}.
6. Expose the bacterial culture to 39°C for 30 min (see Note 8).
7. Transfer 3 mL bacterial culture to a sterilized capped tube and keep on ice for 10 min.
8. Spin the cells down by brief centrifugation at 5,000 rpm in a bench top centrifuge.
9. Remove the supernatant and add 2 mL cold 0.1 M $MgCl_2$.
10. Gently flip the tube to dissolve the cell pellets.
11. Repeat step 8 and add 1.5 mL cold 0.1 M $CaCl_2$ after removal of the supernatant.
12. Repeat step 10 and keep the dissolved culture on ice for 20 min.
13. Repeat step 8 and add 100 μL cold $CaCl_2$ (0.1 M) after removal the supernatant. Repeat step 10 and keep the competent cells on ice until use.
14. Add 50–100 μg of the phage-integration vector to the competent cells.
15. Gently mix and keep on ice for 30 min.
16. Transfer the tube to the water bath at 42°C for 2 min.
17. Add 2 mL fresh LB into the tube and keep in the water bath at 39°C for 2 h.

18. Centrifuge the bacterial culture at 3,000 rpm (2,000 × *g*) in a microcentrifuge for 2 min.
19. Remove the supernatant and pipette out the remaining cells.
20. Spread on LB agar plates supplemented with chloramphenicol at 39°C for overnight or longer.
21. Cell colonies appearing on plates are picked and patched onto LB agar plates containing ampicillin and chloramphenicol, respectively. Consequently, the integrants are picked for exhibiting sensitivity to ampicillin (indicating loss of the helper plasmid) and resistance to chloramphenicol (indicating stable integration).

3.3. Verification of Inserted DNA by Colony PCR

As indicted in Fig. 1b, further examination is required to verify true integrants using colony PCR with primers P1 and P2 (see Note 9). Additionally, a second PCR should be performed using primers P3 and P4 (see Note 10). The colony PCR is carried out as outlined below:

1. The PCR master mix is prepared and kept on ice until use. Per reaction set up a 20-μL PCR master mix containing 2 μL 10× Taq PCR buffer, 2 μL 10× dNTP mix, 2 μM each primer, and 2 U DreamTaq polymerase.
2. Pick a number of well-grown colonies on chloramphenicol-containing plates with a sterilized toothpick and streak on the side of a PCR tube containing 20 μL PCR master mix.
3. Cycle each reaction as follows: 95°C 5 min, then 20 cycles of 95°C 1 min, 55°C 1 min, 72°C 1 min/kb.
4. Remove 6 to 1 μL of loading buffer and run on a 0.8% agarose gel in TBE buffer alongside a DNA size marker.
5. Stain the gel with ethidium bromide and analyze by using an Image analyzer (e.g., AlphaImger EP, Apha Innotech. USA).

3.4. Elimination of Inserted Replication Origin and Selection Marker

To eliminate the inserted DNA containing the selective marker and the replication origin, the integrant is transformed with temperature-sensitive helper plasmid pCP20. This FLP expression plasmid is resistant to ampicillin at 30°C (11). The procedure essentially follows the protocol outlined in Subheading 3.2 except that the thermal challenge in step 6 (that induces FLP and eliminates the plasmid) is conducted by shifting 30–42°C for 30 min. The resulting integrants are then spread on nonselective LB medium agar at 39°C for overnight. Cell colonies appearing on plates are again picked and patched onto LB agar plates containing ampicillin and chloramphenicol, respectively. Consequently, the integrants are picked for exhibiting sensitivity to both ampicillin (indicating loss of the helper plasmid) and chloramphenicol (indicating removal of the replication origin and marker).

The DNA deletion event is then verified by colony PCR as outlined in Subheading 3.3 except using primer combinations P1-P4 and P3-P4 (Fig. 1c), respectively.

4. Notes

1. To perform the site-specific genomic insertion of genes, the phage-integration vector is used in an association with its helper plasmid expressing phage Int. Therefore, integration vector pHK-Km, pPhi80-Km, pP21-Km, pP22-Km, and pLambda-Km are paired with the use of helper plasmid pAH69, pAH123, pAH121, pAH130, and pINT-ts, respectively.
2. For more information on *E. coli* strains, contact Dr. Yun-Peng Chao, Department of Chemical Engineering, Feng Chia University, 100 Wenhwa Road, Taichung, Taiwan 40724; e-mail: ypchao@fcu.edu.tw.
3. These helper plasmids can be obtained from The *E. Coli* Genetic Stock Center (CGSC) at Yale University.
4. For more information on phage-integration vectors, contact Dr. Yun-Peng Chao, Department of Chemical Engineering, Feng Chia University, 100 Wenhwa Road, Taichung, Taiwan 40724; e-mail: ypchao@fcu.edu.tw.
5. For obtaining pET-serious vectors, consult Merck Chemicals Co.
6. This is particularly important to include the T7 terminator when the T7 promoter is used.
7. Beware to keep the strain harboring the Int-expressing helper plasmid at low temperature (below 30°C) all the time. These helper plasmids carry a thermosensitive nature of the replication protein and are easily lost at high temperatures (4).
8. These helper plasmids carry the phage Int under control of the heat-inducible λP_R promoter. Phage Int is forced to produce upon temperature upshift, while the helper plasmid can be simultaneously cured from the host strain due to the thermosensitive nature of the replication protein (4).
9. The inserted DNA of 2 kb containing the replication origin and the marker can be amplified with primers P1 and P2 (see Fig. 1b).
10. Referring to Fig. 1b, primers P3 and P4 are designed to have the sequence complementary to the 5′- and 3′-terminus of the inserted foreign gene, respectively. For a native gene

insertion, primer P3 is preferably designed to complement the 5′-terminus of the artificial promoter.

11. Note that primer T4/dxs (5), corresponding to primer P4 in Fig. 1b, contains a sequence complementary to the 3′-terminus of *dxs*.

Acknowledgments

This work was supported by National Science Council of Taiwan (NSC 98-2622-E-035-011-CC1 and NSC-98-2221-E-035-029-MY3), China Medical School (CMU97-282), and Ministry of Economic Affairs (99-EC-17-A-10-SI-156).

References

1. Jones K. L., Kim S. W., and Keasling J. D. (2000) Low-copy plasmids can perform as well as or better than high-copy plasmids for metabolic engineering of bacteria. *Metab. Eng.* **2**, 328–38.
2. Peredelchuk M. Y. and Bennett G. N. (1997) A method for construction of *E. coli* strains with multiple DNA insertions in the chromosome. Gene **187**, 231–238.
3. Julian A., Hanak J. and Cranenburgh R. M. (2001) Antibiotic-free plasmid selection and maintenance in bacteria, In *Recombinant Protein Production with Prokaryotic and Eukaryotic Cells: A Comparative View on Host Physiology* (Merten, O.-W., Mattanovich, D., Lang, C., Larsson, G., Neubauer, P., Porro, D., Postma, P., Teixeira de Mattos, J., and Cole, J. A. ed.). Kluwer Academic, Dordrecht, Netherlands, pp 121–134.
4. Haldimann A. and Wanner B. L. (2001) Conditional-replication, integration, excision, and retrieval plasmid-host systems for gene structure-function studies of bacteria. *J. Bacteriol.* **183**, 6384–6393.
5. Chiang C. J., Cheng P. T. and Chao Y. P. (2008) Replicon-free and markerless methods for genomic insertion of DNAs in phage attachment sites and controlled expression of chromosomal genes in *Escherichia coli*. *Biotechnol. Bioeng.* **101**, 985–995.
6. Tabor S. and Richardson C. C. (1985) A bacteriophage T7 RNA polymerase/promoter system for controlled exclusive expression of specific genes. *Proc. Natl. Acad. Sci. USA.* **82**, 1074–1078.
7. Studier F. W. and Moffatt B. A. (1986) Use of bacteriophage T7 RNA polymerase to direct selective high-level expression of cloned genes. *J. Mol. Biol.* **189**, 113–130.
8. Imburgio D., Rong M., Ma K. and McAllister W. T. (2000) Studies of promoter recognition and start site selection by T7 RNA polymerase using a comprehensive collection of promoter variants. *Biochem.* **39**, 10419–10430.
9. Alper H., Fischer C., Nevoigt E. And Stephanopoulos G. (2005) Tuning genetic control through promoter engineering. *Proc. Natl. Acad. Sci. USA.* **102**, 12678–12683.
10. Meynial-Salles I., Cervin M. A. and Soucaille P. (2005) New tool for metabolic pathway engineering in *Escherichia coli*: one-step method to modulate expression of chromosomal genes. *Appl. Environ. Microbiol.* **71**, 2140–2144.
11. Datsenko K. A. and Wanner B. L. (2000) One-step inactivation of chromosomal genes in *Escherichia coli* K-12 using PCR products. *Proc. Natl. Acad. Sci. USA.* **97**, 6640–6645.
12. Wang C. W., Oh M. K. and Liao J. C. (1999) Engineered isoprenoid pathway enhances astaxanthin production in *Escherichia coli*. *Biotechnol Bioeng.* **62**, 235–241.
13. Perry K. L., Simonitch T. A., Harrison-Lavoie K. J. and Liu S. T. (1986) Cloning and regulation of *Erwinia herbicola* pigment genes. *J. Bacteriol.* **168**, 607–612.
14. Wang Z. W., Law W. S. and Chao Y. P. (2004) Improvement of the thermoregulated T7 expression system by using the heat-sensitive *lacI*. *Biotechnol Prog.* **20**, 1352–1358.

Chapter 9

Array-Based Synthetic Genetic Screens to Map Bacterial Pathways and Functional Networks in *Escherichia coli*

Mohan Babu, Alla Gagarinova, Jack Greenblatt, and Andrew Emili

Abstract

Cellular processes are carried out through a series of molecular interactions. Various experimental approaches can be used to investigate these functional relationships on a large-scale. Recently, the power of investigating biological systems from the perspective of genetic (gene–gene or epistatic) interactions has been evidenced by the ability to elucidate novel functional relationships. Examples of functionally related genes include genes that buffer each other's function or impinge on the same biological process. Genetic interactions have traditionally been investigated in bacteria by combining pairs of mutations (e.g., gene deletions) and assessing deviation of the phenotype of each double mutant from an expected neutral (or no interaction) phenotype. Fitness is a particularly convenient phenotype to measure: when the double mutant grows faster or slower than expected, the two mutated genes are said to show alleviating or aggravating interactions, respectively. The most commonly used neutral model assumes that the fitness of the double mutant is equal to the product of individual single mutant fitness. A striking genetic interaction is exemplified by the loss of two nonessential genes that buffer each other in performing an essential biological function: deleting only one of these genes produces no detectable fitness defect; however, loss of both genes simultaneously results in systems failure, leading to synthetic sickness or lethality. Systematic large-scale genetic interaction screens have been used to generate functional maps for model eukaryotic organisms, such as yeast, to describe the functional organization of gene products into pathways and protein complexes within a cell. They also reveal the modular arrangement and cross talk of pathways and complexes within broader functional neighborhoods (Dixon et al., Annu Rev Genet 43:601–625, 2009).

Here, we present a high-throughput quantitative *Escherichia coli* Synthetic Genetic Array (eSGA) screening procedure, which we developed to systematically infer genetic interactions by scoring growth defects among large numbers of double mutants in a classic Gram-negative bacterium. The eSGA method exploits the rapid colony growth, ease of genetic manipulation, and natural efficient genetic exchange *via* conjugation of laboratory *E. coli* strains. Replica pinning is used to grow and mate arrayed sets of single gene mutant strains and to select double mutants en masse. Strain fitness, which is used as the eSGA readout, is quantified by the digital imaging of the plates and subsequent measuring and comparing single and double mutant colony sizes.

While eSGA can be used to screen select mutants to probe the functions of individual genes, using eSGA more broadly to collect genetic interaction data for many combinations of genes can help reconstruct a functional interaction network to reveal novel links and components of biological pathways as well as unexpected connections between pathways. A variety of bacterial systems can be investigated,

James A. Williams (ed.), *Strain Engineering: Methods and Protocols*, Methods in Molecular Biology, vol. 765,
DOI 10.1007/978-1-61779-197-0_9, © Springer Science+Business Media, LLC 2011

wherein the genes impinge on a essential biological process (e.g., cell wall assembly, ribosome biogenesis, chromosome replication) that are of interest from the perspective of drug development (Babu et al., Mol Biosyst 12:1439–1455, 2009). We also show how genetic interactions generated by high-throughput eSGA screens can be validated by manual small-scale genetic crosses and by genetic complementation and gene rescue experiments.

Key words: *Escherichia coli*, Conjugation, Double mutant, Hypomorphs, Epistasis, Genetic interaction, Network, Synthetic lethality or sickness, Aggravating, Alleviating, Suppression

1. Introduction

While physical interactions and computational inferences can indicate which bacterial proteins are associated into complexes or are evolutionarily linked (1, 2), these inferences do not necessarily reveal the nature of the functional relationship at a pathway level (3–5). With the exception of metabolic pathways (6), genetic dependencies and pathway architecture in bacteria have been traditionally explored in a stepwise manner by isolating second-site genetic modifier mutations in classic suppressor/enhancer screens. For example, the loss of functional alleles in two genes may cause "synthetic lethality" or "synthetic sickness" even if neither mutation alone significantly reduces cell viability. Such aggravating (or negative) genetic interactions usually result when the products of two genes either jointly control an essential process or encode subunits of an essential protein complex or function within redundant or convergent pathways such that one can functionally compensate for, or buffer, defects in the other. Conversely, alleviating (or positive) genetic interactions (e.g., suppression) typically arise between genes operating at different steps within the same pathway (7–10).

Despite wide conservation, only ~300 protein coding genes are essential for *Escherichia coli* viability under standard laboratory growth conditions (11). The fact that the other ~3,900 genes are dispensable for viability presumably reflects, in part, the adaptability and robustness of microbial processes to environmental perturbations (12, 13). Obvious backup controls that "buffer" system failures include DNA repair systems, protein chaperones, and duplicated genes (i.e., paralogs) (14, 15). However, alternate types of functional dependencies and genetic redundancy likely exert more pervasive stabilizing effects on the molecular networks and phenotypes of cells (16, 17). Indeed, the organization of gene products into redundant parallel pathways and buffered functional modules can explain the dispensability of most genes (18).

1.1. eSGA and GIANT

The development of a high-throughput Synthetic Genetic Array (SGA) method for detecting genetic dependencies by systematically assaying the fitness of digenic combinations by mating and meiotic assortment of unlinked nonessential gene deletions allowed for the genome-scale mapping of genetic interaction networks in *Saccharomyces cerevisiae* (19–22). A related screening strategy, termed Epistatic MiniArray Profiling (23), or E-MAP, has likewise been used to generate comprehensive maps of alleviating and aggravating interactions by making quantitative measurements of fitness of strains deleted for pairs of genes representing a particular functional neighborhood of interest. For example, comprehensive genetic interaction profiling has revealed the biological roles of individual genes, the extent of pleiotropy and specialization among the subunits of protein complexes, and the extensive nature of pathway cross talk and redundancy among components of the chromatin remodeling machinery (9), protein secretion apparatus (23), and RNA processing systems (24).

Although *E. coli* does not have a sexual life cycle in the same manner as *S. cerevisiae*, like many other bacterial species, it is naturally capable of conjugation and genetic exchange. This ability for genetic transfer and the efficiency of the ensuing homologous recombination are exploited for making double mutants in two analogous but independently developed methods for investigating *E. coli* genetic interactions in GIANT-coli (for *G*enetic *I*nteraction *AN*alysis *T*echnology for *E. coli*) and eSGA (for *E.* coli *s*ynthetic *g*enetic *a*rrays)(25, 26). Although the readout in both approaches, digenic mutant fitness, is determined by systematically generating and measuring the colony sizes of double mutants, we focus on the eSGA procedure in this Chapter. In addition to scoring deletion mutants, analogous to what has been done in *S. cerevisiae*, hypomorphic alleles (e.g., temperature-sensitive conditional alleles with reduced gene function (27) or mRNA perturbation (DAmP) alleles (28)) can likewise be used to assess the genetic interaction patterns of essential bacterial genes. Genes in the same pathway typically display alleviating interactions with each other and highly correlated patterns of aggravating interactions with genes in other overlapping pathways (4, 19). Hierarchical clustering of the patterns of genetic interactions measured for various genes associated with a biological system of interest can, therefore, reveal the arrangement of genes into pathways (23).

1.2. Conjugation

In bacterial conjugation, DNA is unidirectionally transferred from a male (F^+ or Hfr) strain to a female (F^-) strain. Male *E. coli* donor strains harbor a self-transmissible low-copy plasmid (Fertility, or F factor) (29), which encodes functions necessary to promote its own transfer into a naïve (female) F^- bacterial cell via conjugation (29).

Transfer of the F factor is catalyzed upon nicking of the double-stranded DNA at the origin of transfer, *oriT*, followed by rolling-circle replication and transfer of single-stranded DNA through a sex pilus into the recipient F^- cell (30). Occasionally, the F factor integrates into the host chromosome (31), giving rise to High frequency of recombination (Hfr) male strains. An Hfr donor transfers host DNA, concomitantly with the F factor, to a conjugated F^- recipient strain. The transfer starts with loci adjacent to *oriT* and continues in a unidirectional manner until mating is physically interrupted or the transfer is complete. Once in the recipient, the donor DNA is integrated into the corresponding genomic locus by homologous recombination (14). To achieve high-throughput, arrayed sets of single gene deletion, mutant recipient (F^-) and donor (Hfr) strains are grown on solid media, are conjugated by replica pinning, and the double mutants are selected.

Our high-throughput, quantitative eSGA assay is used to make large numbers of double mutants in parallel, as ordered colonies on plates, to investigate genetic interactions on genomic scale (25). In eSGA, first, a male donor Hfr strain is constructed with a single marked "query" gene mutation. The donor is constructed by targeted introduction of a specific gene mutation, representing either a full deletion of the entire open reading frame (in the case of nonessential genes) or a point mutation (for essential genes), using a cassette marked with chloramphenicol-resistance (Cm^R). Temperature-inducible expression of an integrated phage lambda Red-mediated homologous recombination system is used to dramatically improve the efficiency of homologous recombination (25, 32) (Fig. 1). Conjugation is then performed by pinning the donor onto an arrayed set of viable kanamycin-resistance marked (Kan^R) single gene deletion *E. coli* K-12 F^- recipient strains (11). A genome-wide single-gene deletion recipient collection representing 3,968 nonessential genes of *E. coli* is publicly available (11). Since essential genes cannot be deleted, we typically perturb the 3′ UTR to destabilize transcript abundance to generate hypomorphic alleles (25). After genomic DNA transfer and homologous recombination into the recipient host, the resulting double mutants are selected by replica pinning onto solid medium containing both antibiotics. The screening process can be speed up using a robotic pinning device, but manual strain manipulation is readily performed with a handheld device. Growth fitness, the primary eSGA readout, is recorded by imaging the plates and measuring the respective colony sizes (25). A workflow diagram for the entire process is shown in Fig. 2.

As with yeast, the ability to perform eSGA screens easily and systematically allows one to examine the genetic interaction profiles of both individual genes and the components of entire systems in *E. coli* in an unbiased, comprehensive manner (3, 25), permitting the investigation of bacterial gene function at a higher

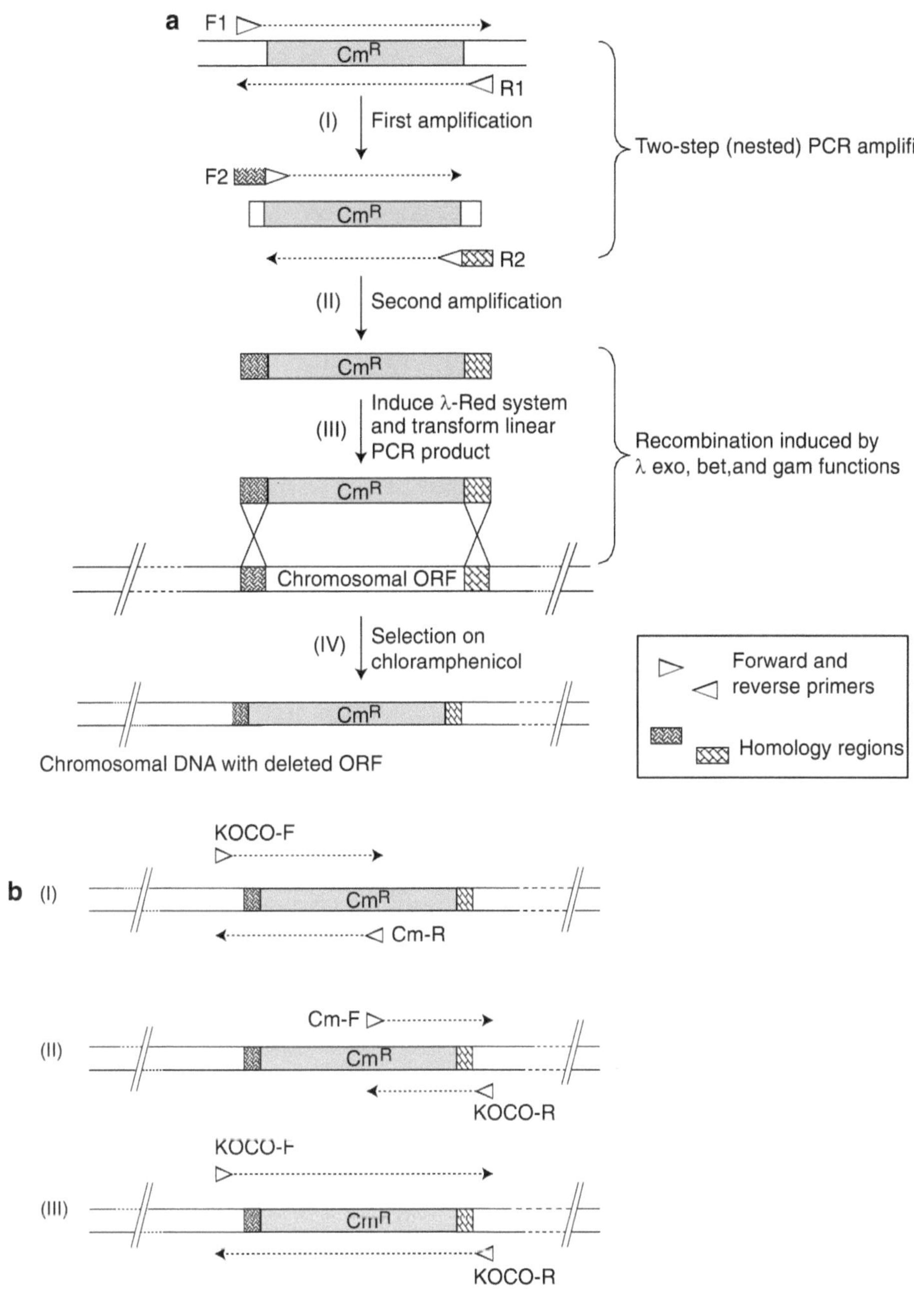

Fig. 1. Donor construction and confirmation. Panel A: Construction of eSGA donor mutant strains by deletion of *E. coli* chromosomal ORF in Hfr Cavalli. In the first step, the chloramphenicol resistance (Cm^R) cassette, with short adjacent regions, is amplified from the pKD3 plasmid using primers F1 and R1. In the second step, the Cm^R region of the plasmid is amplified and 45-nt homology regions, for site-specific recombination, are added using primers F2 and R2. In the third step, the product of the second amplification is transformed into the Hfr Cavalli strain after the λ-Red system derepression to specifically replace the target ORF with the Cm^R. In the fourth step, the mutants having the Cm^R are selected on Chloramphenicol. Panel B: The gene deletions in Hfr Cavalli are confirmed by separate PCRs with three primer sets. The first primer set consists of a 20-nt flanking primer, located 200 bp upstream of the targeted region (KOCO-F), and reverse (Cm-R) primer complementary to the Cm^R cassette sequence (shown in *top* panel). The second set includes a forward (Cm-F) primer, annealing to the Cm^R cassette sequence, and a reverse flanking confirmation primer (KOCO-R), which should be designed to anneal 200 bp downstream of the 3′ end of the deleted gene (shown in *middle* panel). The third PCR includes KOCO-F and KOCO-R primers (shown in *lower* panel). See Subheading 3 for details.

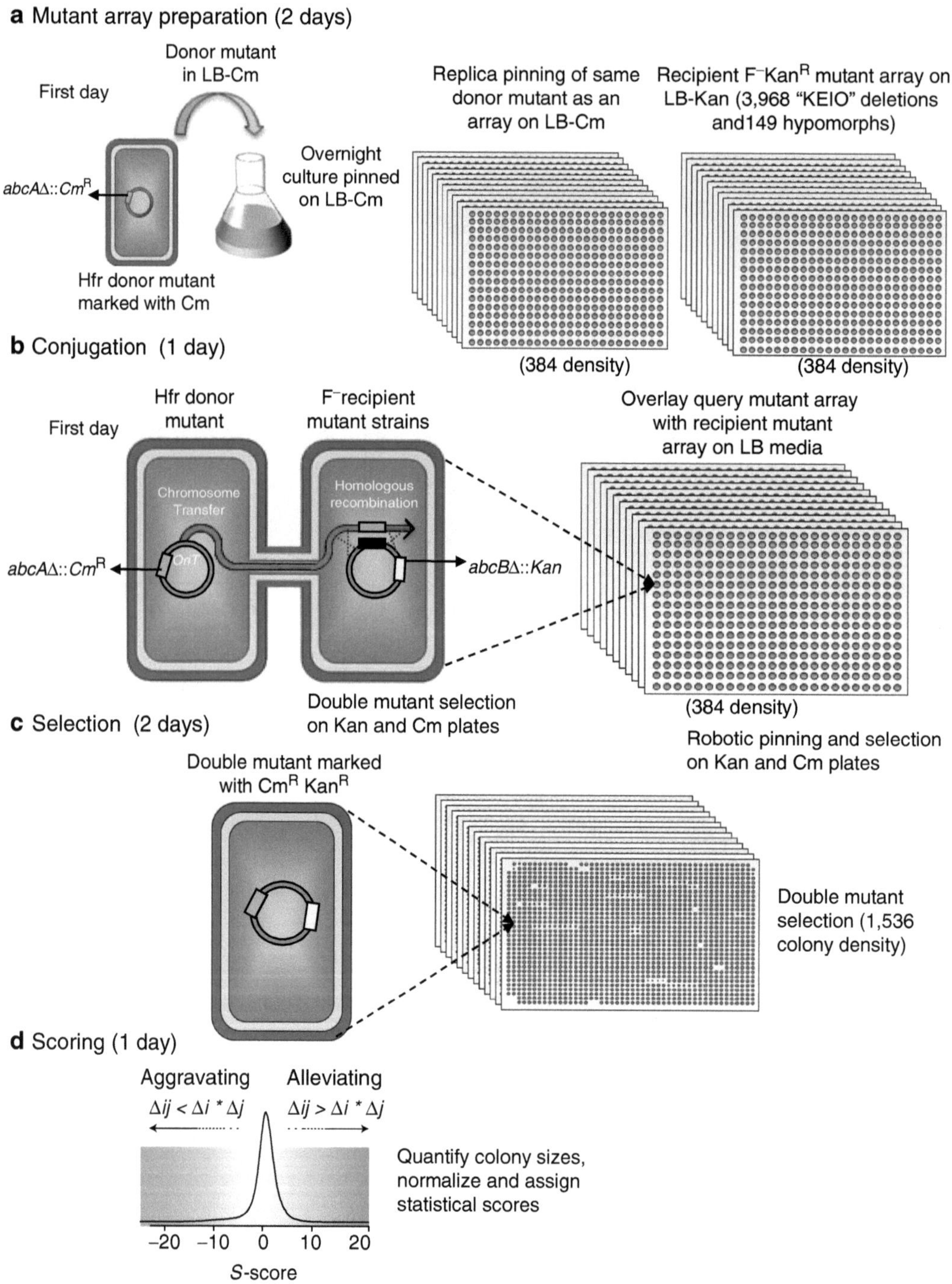

Fig. 2. Systematic double-mutant construction in *E. coli* using eSGA. Schematic summary of key eSGA steps: (**a**) Overnight Hfr query mutant strain (marked with Cm^R), grown overnight in LB-Cm, is pinned onto LB-Cm plates in 384 format. Simultaneously, the recipient F- mutant array strain (marked with Kan^R) is pinned onto LB-Kan plates. (**b**) After overnight growth the Hfr query strain colonies and recipient array colonies are pinned over each other on LB plates. Conjugation ensues, with DNA transfer initiating at a specific origin of transfer, *oriT*, and proceeding via a rolling circle mechanism of replication. The donor chromosome undergoes homologous recombination with the recipient chromosome (marked as "X in broken lines"). (**c**) The resulting colonies are pinned onto plates containing both Kan and Cm for selection of double mutants. (**d**) The double-mutant plates are then imaged and colony sizes are scored to identify aggravating (synthetic lethal and synthetic sick) and alleviating (buffering) interactions.

order, pathway level. Functional relationships are usually identified by the nonmultiplicative growth fitness of the double mutants. For example, combinations of mutations in enzymes participating in two alternate pathways (e.g., Isc, Suf) that generate iron–sulfur cluster prosthetic groups widely used in multiple bacterial processes (33) are synthetic lethal by eSGA. In such cases, genetic interactions between two alleles can be measured based on how the phenotype of an organism lacking both alleles (double mutant) differs from that expected when the phenotypes of single mutations are combined (13). Based on this proposition, the genetic interactions ε_{IJ} can be defined between mutations I and J in terms of any quantitative phenotype P as the difference between an observed ($P_{IJ,\ observed}$) and an expected ($P_{IJ,\ expected}$) phenotype of the double mutants if no interaction exists between the two mutations:

$$\varepsilon_{IJ} = P_{IJ}^{observed} - P_{IJ}^{expected}$$

This mathematical equation depends on our ability to compute $P_{IJ}^{expected}$ as a function based on the combined effect of two individual mutations with phenotypes $P_{I,\ observed}$ and $P_{J,\ observed}$. For example, if loss of allele I ($P_{I,\ observed}$) results in a growth rate 0.8 times the wild-type growth rate, whereas loss of allele J ($P_{J,\ observed}$) results in a growth rate of 0.9, then the expected ($P_{IJ}^{expected}$) growth rate of the double mutant (lacking alleles I and J) would be 0.72 times that of the wild-type. This neutral model assumes that two genes do not normally impact each other, and in fact, experimental observations support the intuitive idea that genetic interactions are rare (21, 23). In cases when deletion of two genes causes a more deadly effect than the fitness reduction expected from the combined loss of individual genes, the two genes are said to have negative or aggravating interaction (e.g., synthetic sickness or lethality). Such interactions often identify proteins that function in distinct but parallel pathways in a given process (4). Alternatively, when double mutant has a better than expected fitness, the two genes are said to have positive or alleviating interaction (e.g., suppression).

To date, we have performed over 160 genome-wide eSGA screens using donors bearing mutations in diverse query genes, including hypomorphic alleles of several essential *E. coli* genes. Since linkage may bias the interaction score for the pairs of genes within 30 kbp from each other, we eliminate the 30 kbp regions on either side of the query gene from analysis (25). This means that unrelated genes are not said to interact with the query gene simply because linkage prevented formation of the double mutant, producing an apparent synthetic lethal phenotype. On the contrary, this also means that relationships between genes with related function – such as often functionally related genes within the same operon – cannot be investigated (25). Nevertheless, our screens have revealed hundreds of novel genetic interactions suggestive of pathway relationships (34). Typically, the number of genetic interactions per screen, at an average of 30 synthetic lethal interactions per bacterial query gene, is roughly similar to the number of interactions found in yeast (22). Although genetic interactions reveal many interesting connections suggestive of novel mechanistic links, analyzing the network for concurrence with protein–protein interactions (PPI), membrane protein expression levels, coexpression, and functional connections predicted by genomic-context methods – such as the conservation of gene order (operons); gene fusions; operon recombination frequencies derived from intergenic distances of predicted operons across genomes and phylogenetic profiling – can be especially informative. For example, like the buffered pathways that are linked by genetic interactions, genes encoding subunits of protein complexes tend to buffer each other and will also be connected by genetic interactions. Previous E-MAP studies in yeast have shown that genes exhibiting alleviating genetic interactions are more likely to encode proteins that are physically associated (9, 35), and examination of our recent envelope data in *E. coli* revealed a similar overall tendency toward alleviating interactions between genes encoding protein subunits of the same bacterial complex (data not shown). In addition, comparative genomic procedures can be performed in conjunction with genetic interaction data to investigate the evolutionary significance of the putative functional relationships detected in *E. coli* and in other proteobacterial species and Prokaryotic taxa. Alterations to the components of the pathways and functional modules in virulent *E. coli* strains and other pathogens can illuminate divergent functional adaptations. Analogous comparisons of budding and fission yeast (~400 million years divergence) have shown that although there is a highly significant conservation of synthetic lethal genetic interactions, substantial rewiring of functional modules occurs between distantly related eukaryotes (35). The interaction data generated for *E. coli* may be similarly used to gain insight into the pathway architecture of other microbes for which functional annotations are largely lacking. Collectively, our results have confirmed the place of genetic interaction screens using eSGA, or conceptually similar alternate methods like GIANT-coli (26) and next-gen TnSeq (36), to illuminate novel functional interactions in *E. coli* often missed by other experimental or computational approaches. Since many of the *E. coli* genes are widely conserved across microbes (1), and because antibiotic sensitivity is often enhanced in combination with certain gene mutations (i.e., underlining the field of chemical genomics) (37), any functional relationships illuminated by eSGA may be exploited for designing innovative combination drug therapies.

In this chapter, we describe the key steps required to perform a single whole-genome eSGA screen and subsequent data analysis procedures that allow identifying the quantitative defects in double-mutant fitness to reveal the functional dependencies and pathway redundancy. These steps include generating query gene donor strains, performing conjugation using high-density colony plates, and processing digital images for statistical scoring. We also briefly describe how genetic interaction scores can be used in formulating a testable biological hypothesis. Overall, these protocols are readily implemented in a lab with experience in basic microbiological techniques, can be scaled up through the use of commercial robotic platforms, and make use of *E. coli* lab strains that are either publicly accessible (11) or readily generated in-house (1, 25).

2. Materials

2.1. Media, Stock Solutions, and Reagents

1. Luria–Bertani (LB) medium: Solid medium is prepared by dissolving 25 g of LB powder and 20 g of agar in 1,000 mL of distilled water. Liquid LB medium is prepared without agar.
2. SOC medium (Invitrogen).
3. Kanamycin antibiotic stock: Dissolve 50 mg/mL of kanamycin (Kan) in double-distilled water and filter-sterilize using a 0.22-μm millipore filter. The filter-sterilized Kan stock solution is stored in single use aliquots at −20°C.
4. Chloramphenicol antibiotic stock: Dissolve 34 mg/mL of chloramphenicol (Cm) in 95% ethanol and filter-sterilize using a 0.22-μm millipore filter. The filter-sterilized Cm stock solution is stored in single use aliquots at −20°C.
5. Ampicillin antibiotic stock: Dissolve 100 mg/mL of ampicillin (Amp) in double distilled water and filter-sterilize using a 0.22-μm millipore filter. The filter-sterilized Amp stock solution is stored in single use aliquots at −20°C.
6. LB medium + Cm plates: Prepare solid LB medium and prior to pouring add Cm to 34 μg/mL final concentration. Pour in plates for pinning (see Item 5 in Subheading 2.3).
7. LB medium + Kan plates: Prepare solid LB medium and prior to pouring add Kan to 50 μg/mL final concentration. Pour in plates for pinning (see Item 5 in Subheading 2.3).
8. LB medium + Cm/Kan plates: Prepare solid LB medium and prior to pouring add Cm to 34 μg/mL and Kan to 50 μg/mL final concentration. Pour in plates for pinning (see Item 5 in Subheading 2.3).

9. 10% ice-cold sterile glycerol and ice-cold sterile distilled water for preparing competent cells.
10. 70% sterile glycerol stock for long-term strain storage.
11. Desalted custom primers: KOCO-F and KOCO-C (20-nt primers 200 bp away from the deletion site) designed as described further in Subheading 3.1.4. These can be purchased from a commercial supplier. Resuspend to 50 μM in 10 mM Tris–HCl pH 8.
12. Desalted custom primers: F2 and R2 (20-nt constant regions based on pKD3 sequence and 45-nt custom homology regions) designed as described further in Subheading 3.1.1. These can be purchased from a commercial supplier. Resuspend to 50 μM in 10 mM Tris–HCl pH 8.
13. Desalted pKD3-based constant primers. F1: 5′-AGATTG CAGCATTACACGTCTT-3′; R1: 5′-GGCTGACATGGG AATTAGC-3′; Cm-R: 5′-TTATACGCAAGGCGACA AGG-3′; Cm-F: 5′- GATCTTCCGTCACAGGTAGG-3′. These can be purchased from a commercial supplier. Resuspend to 50 μM in 10 mM Tris–HCl pH 8.
14. Taq DNA polymerase (Fermentas) and 10× buffer for PCR amplifying the Cm template to make the gene tagging or replacement cassettes.
15. dNTP mix: 10 mM each (Fermentas).
16. Genomic DNA isolation kit (Promega).
17. Plasmid DNA and PCR purification kits (Qiagen).

2.2. Equipment

1. Thermal cycler for standard PCR (BioRad iCycler). The amplification conditions may need to be optimized if a different thermal cycler is used.
2. Equipment and supplies for standard agarose gel electrophoresis: gel box, power supply, buffer TBE, DNA ladder, as well as ethidium bromide and UV transilluminator for visualizing the DNA.
3. Electroporator (BioRad MicroPulser).
4. A centrifuge for pelleting cultures. We typically use a Beckman Coulter TJ-25 centrifuge with TS-5.1-500 rotor 42°C water bath shaker, 0°C ice-slurry shaker, 32°C shaker, 32°C plate incubator.
5. The 96-pin and 384-floating pin handheld pinning device (V & P Scientific, Inc.; VP 409 or VP 438FP3 work well for manual copying 96 and 384 density plates. VP 384FP1 works well for pinning 384 density colonies to 1,536 format) for the manual pinning of single and double mutants. Colony copiers from V&P Scientific (for source and target plates), to guide

pins to consistent positions on all plates. The colony copier to be used will depend on the chosen handheld device. (Optional: To automate and speed up the pinning process, we use the RoToR-HDA benchtop robot (Singer Instruments), fitted with 96 or 384-pin density pads for the replica-pinning procedures).

6. Digital camera with close-up (macro) imaging capabilities – for plate imaging (10 megapixels minimum resolution).
7. To allow for automatic colony size quantization, reduce variation in lighting between the plates, and to make imaging faster, use a camera stand (Kaiser) and a Canon digital camera with remote shoot and close-up (macro) imaging capabilities (10 megapixels minimum resolution). Affix the camera to the Kaiser, 50 cm above the surface of the stand. Make an adapter for the plates that, once attached to the Kaiser stand, will hold the plates in place and provide the background to allow automated plate and colony detection. For this, first, prepare the cardboard placeholder: trace a plate that you will be using on a ≥0.3-cm thick cardboard, leaving *ca.* 3–4 cm of cardboard on each side of the plate (trim the cardboard, if necessary). Cut the cardboard along the trace lines and discard the center piece. Second, make a narrow white edge that will aid in automatic location of the plate borders. For this, obtain a piece of thin white plastic; cut it to match the outer measurements of the cardboard. Attach the plastic to the cardboard with tape along the outer edges. Cut out the center of the plastic such that a ~0.3 cm edge is visible from the center opening of the cardboard. Third, to provide a background to the plates, obtain a piece of black velvet larger than the central opening in the plastic. Tape the velvet, around its outside edge, to the plastic (on the opposite side of the cardboard) so that the velvet does not have any wrinkles. Now, the narrow white plastic edge and the velvet pile (the reverse side of the velvet will not provide sufficiently even background) should be visible through the central opening in the cardboard. Place the entire plate adaptor in the middle of the Kaiser stand, with velvet side down and the cardboard side up, so that the plate's borders are exactly vertical and horizontal in the photograph. Affix (using black electrical tape) the adaptor to the stand – completely and evenly cover the cardboard with tape, leaving the white plastic edges and the velvet free of tape. Once the adaptor assembly is complete and attached to the stand, place light boxes (two parallel Testrite 16 × 24 units; Freestyle Photographic supplies) on the sides of the Kaiser stand, 15–25 cm from the adaptor – to provide light during imaging. Cover the light boxes with white or blue bench liners to reduce the light intensity, if necessary. Cover the lower

half of the light boxes with 1–2 (additional) layers of bench liners to distribute the light primarily upwards. This makes the light distribution around the whole imaged plate more even, which aids in later plate quantizations. Lastly, shield the set up from external light – to reduce day-to-day and time-of-day variation in images – by covering the set up with black light-impermeable or nonreflective material. For taking a plate image, place the plate in the center opening of the cardboard and keep the set up shielded from outside light.

2.3. Pinning System, Plates and Accessories for Working with Cultures

1. Sterile 50-mL polypropylene tubes and 1.5-mL microcentrifuge tubes for preparing competent cells and transformation. These need to be chilled on ice for at least 10 min prior to use.
2. 250-mL flasks for growing cell cultures.
3. Sterile 0.2-cm electroporation cuvettes. These need to be chilled on ice for at least 10 min prior to use.
4. 15-mL sterile culture tubes for recovery of double mutants following a transformation.
5. Rectangular plates (Nunc) for manual pinning (or optional: rectangular plates (Singer Instruments) – for all robotic replica-pinning procedures).
6. 384-well microtitre plates (Nunc) for constructing a stock copy culture of the F^- recipient single gene deletion mutants.
7. Distilled water, 70% ethanol (v/v), 95% ethanol, and cheesecloth for cleaning the reusable handheld pinning device. Three suitable rectangular liquid containers – to hold the liquids. These should be large enough to fit all pins of the handheld pinning device (for example, large enough solid tip box lids). Fold the cheesecloth and place it in the first container, cover it with distilled water for *ca.* 0.5 cm. This first water station will be used to remove the cells from the pins. Pour 70% ethanol in the second container. During this step, the floating pins are sterilized after washing them in water (the height of the ethanol in this container should be about 0.7 cm – slightly higher than the liquid level in the first container). In the third container, place 95% ethanol (*ca.* 0.9 cm height). After a 70% ethanol wash, briefly dip the pins in the 95% ethanol and flame them to sterilize. The pins of the handheld pinning device are washed successively in water, 70% ethanol, 95% ethanol and are flamed prior to each pinning step. (Optional: V&P Scientific offers reservoirs that can be used instead of containers described above. V&P Scientific also has Pin Cleaning Paint Pad with 4 mm nylon bristles, which can be used in place of cheesecloth).

8. Optional: cleaning accessories for robot plastic pinning pads: short- and long-pin plastic pads can be reused up to three times. Sterilize the pads, following each use, by soaking them in 10% bleach overnight, rinsing with distilled water, washing in 70% ethanol, and air-drying the pads in a flow hood under UV. Store the sterile pads in sealed sterile plastic bags until use.

2.4. Bacterial Strains and Plasmids

1. Hfr Cavalli strain with an integrated temperature inducible λ-Red system (JL238) (25) is used as the parental strain for the construction of Hfr query mutant donor strains.
2. pKD3 (38) as PCR template for chloramphenicol resistance marker.
3. The Keio *E. coli* F^- recipient nonessential single gene deletion mutant collection (11) can be obtained from the National BioResource Project (NBRP) of Japan: (http://www.shigen.nig.ac.jp/ecoli/strain/top/top.jsp).
4. Potentially hypomorphic *E. coli* F^- recipient SPA-tagged essential gene strains (39) can be obtained from open biosystems: (https://www.openbiosystems.com/GeneExpression/NonMammalian/Bacteria/EcoliTaggedORFs/).

3. Methods

3.1. Construction of Query Deletion Mutants in an Hfr Cavalli Donor Strain

Donor strains are made by using homologous recombination to replace the query gene with a "Cm" resistance marker in an engineered *E. coli* Hfr Cavalli strain, which bears an integrated temperature-inducible λ-Red recombination system to improve integration efficiency (25). In making the donors, the following steps are performed: first, the gene mutagenesis (e.g., deletion) cassette is amplified from pKD3 plasmid (Subheading 3.1.1). Second, competent cells are prepared from the Hfr Cavalli strain with the λ-Red system (Subheading 3.1.2). Third, the cassette is transformed into the competent cells and mutants are selected (Subheading 3.1.3). Fourth, the gene mutation is confirmed by PCR of individual clones (Subheading 3.1.4) and validated mutants are stored for future use (Subheading 3.1.5).

3.1.1. Preparation and Generation of the Linear DNA Mutagenesis Cassette

The linear DNA cassette to replace (i.e., delete) a target open reading frame with a "Cm" resistance marker is synthesized by a two-step (nested) PCR amplification from the pKD3 plasmid (38). Nested amplification (Fig. 1) is employed to minimize the chance of transforming and recovering an intact plasmid when making the donors.

1. In the first two rounds of amplifications, ~45 ng of pKD3 is used as a template in PCR with forward F1 and reverse R1 primers to produce a 1,070 bp product. PCR is carried out in 50 μL reactions containing 1× reaction buffer with $(NH_4)_2SO_4$, 4 mM $MgCl_2$, 200 μM dNTPs, 2 μM of each primer, 1.25 U *Taq* DNA polymerase (Fermentas) per 50 μL. The following cycling conditions are used: 3 min denaturation at 95°C is followed by 30 cycles of 1 min denaturation at 95°C, 1 min annealing at 56°C, and 2 min extension at 72°C. Subsequently, a 10 min final extension at 72°C and a final hold at 4°C are performed.
2. Perform agarose gel electrophoresis to confirm that the correct fragment was obtained.
3. PCR purify the obtained fragment following standard manufacturer protocol (Qiagen) and dilute the purified product (5 ng/μL) in sterile distilled water.
4. The purified linear product is stored in −20°C for use as a template for all subsequent second-step amplifications.
5. Set up a second (nested) PCR with gene-specific (e.g., knockout) primers, F2 and R2, that have 45-nt of homology to the gene of interest at the 5′-ends (to allow for homologous recombination), immediately upstream and downstream of the target, and 20 nt at the 3′-ends to prime the synthesis of the "Cm" marker. The size of the produced fragment is 1,123 bp. The reactions are set up with ~2.5 ng of template PCR (from Step 4 above) per 50 μL reaction. The remaining conditions are as described in Step 1 in Subheading 3.1.1.
6. Confirm using agarose gel electrophoresis that the expected product is obtained.
7. PCR purify the obtained fragment (Qiagen) following standard manufacturer protocol, eluting into 30 μL sterile distilled water. Dilute the product to ~50 ng/μL. The purified product can be stored in −20°C for long-term storage prior to use in transformation (Subheading 3.1.3).

3.1.2. Preparation of Competent Cells for Donor Mutant Construction

1. Inoculate a single Hfr Cavalli colony in 5 mL of LB medium with 2.5 μL of 100 μg/mL ampicillin. Incubate the strain overnight at 32°C with shaking at 220 rpm.
2. Inoculate 1 mL of the saturated overnight culture into 70 mL of fresh LB medium with 35 μL of 100 mg/mL ampicillin in a 250-mL flask.
3. Incubate the culture at 32°C with shaking until an optical density (OD) of ~0.5–0.6 is obtained (~2 h) (see Note 1).
4. When an OD reaches ~0.5–0.6, transfer the culture to a water bath for heat induction of the λ-Red recombination system at 42°C for 15 min with shaking at 160 rpm (see Note 2).

5. Transfer the culture to a chilled ice-slurry water bath for 10–20 min at 160 rpm to stop the induction. Make sure to keep the cells cold from this point until after transformation (Step 2 in Subheading 3.1.3).
6. While keeping the culture on ice, divide it equally into two prechilled 50-mL polypropylene tubes (~35 mL culture per tube) and centrifuge at 4,400 × *g* for 6 min at 4°C.
7. Decant the supernatant; resuspend the cell pellets in ~10 mL of ice-cold sterile distilled water. To ensure cell survival, do not vortex the cells past this stage! Gentle pipetting may be used if necessary to dislodge the cells. Add ice-cold sterile distilled water up to 40 mL. The cells should resuspend easily by gentle inversion once they are dislodged. Perform centrifugation under the same conditions as mentioned in Step 6.
8. Repeat Step 7 using 10% glycerol instead of water.
9. Decant the supernatant, resuspend each cell pellet in 20 mL ice-cold sterile 10% glycerol and centrifuge again as described in Step 6.
10. Decant the supernatant and resuspend the cell pellet in 500 μL of ice-cold 10% glycerol.
11. Aliquot 50 μL of the cell suspension into individual prechilled 1.5-mL microcentrifuge tubes and proceed with transformation (see Note 3).

3.1.3. Electroporation and Donor Mutant Selection

1. Add 100 ng of purified gene deletion cassette in 2 μL of ice-cold water (from Step 7 in Subheading 3.1.1) to the prepared competent cells (from Step 11 in Subheading 3.1.2). Flick the tube – do not pipette-mix. Allow suspension to sit on ice for 5 min.
2. Transfer the suspension of ice-cold competent cells and DNA to a prechilled electroporation cuvette. Electroporate the cell mixture using 2.5 kV, 25 μF, 200 Ω setting (applies for 0.2-cm cuvettes) and immediately add 1 mL of room temperature SOC medium.
3. Transfer the electroporated cells in SOC medium with a sterile Pasteur pipette into a 15-mL culture tube and incubate at 32°C for 1 h with orbital shaking at 220 rpm.
4. After incubation, centrifuge the cells in a Beckman at 4,400 × *g* for 5 min (at room temperature).
5. Remove approximately 850 μL of the supernatant and resuspend the cell pellet in the remaining liquid.
6. Spread the cells on LB plates containing 34 μg/mL Cm.
7. Incubate the plates at 32°C overnight (see Note 4).
8. Pick and streak out 2–3 individual transformants on LB-Cm plates for mutant confirmation. Also, streak the same transformants on LB-Kan to make sure that the strains are not Kan resistant.

3.1.4. PCR Confirmation of Successful Gene Deletion

1. Grow overnight the individual knockout strains in liquid LB, complemented with 34 μg/mL Cm, and isolate the genomic DNA following the manufacturer's instructions (Promega) (see Note 5).
2. The DNA is amplified in three separate 25 μL reactions, with three different sets of knockout confirmation primers. All reactions are performed under conditions described in Step 1 of part Subheading 3.1.1, but with ~150 ng of genomic DNA template. After the PCR products are synthesized, the products are run out on an agarose gel to confirm that the correct fragments were obtained.
3. The first primer set consists of a 20-nt flanking primer, located 200 bp upstream of the targeted region (KOCO-F), and Cm-R primer, which is complementary to the "Cm" cassette sequence. This amplification is expected to produce a 445-nt amplicon (Fig. 1).
4. The second set includes a forward primer (Cm-F), annealing to the "Cm" cassette sequence, and a reverse flanking confirmation primer (KOCO-R), which should be designed to anneal 200 bp downstream of the 3 end of the deleted gene. This amplification reaction is expected to produce a 309-nt amplicon (Fig. 1).
5. The third PCR contains KOCO-F and KOCO-R primers. This reaction is required to verify that the selected strain is not a merodiploid, with one gene locus having been replaced by the cassette and another duplicated but otherwise wild-type gene copy still present. This amplification is expected to produce a 1.433 kb product (Fig. 1).

3.1.5. Storage of Confirmed Query Deletion Donor Mutant Strains Prior to Screening

1. Strains can be stored at 4°C for up to a month.
2. Overnight cultures of successfully confirmed recombinant donor strain clones for each query mutant are placed in individually labeled cryovials, supplemented with 15% glycerol (by volume), vortexed for 20 s, frozen on dry ice and transferred to −80°C for long-term storage.

3.2. Arraying an E. coli F⁻ Recipient Strain Collection for Genome-Wide eSGA Screens

1. The entire Keio *E. coli* single gene deletion mutant collection (3,968 strains; F^- BW25113) (11) is replicated by pinning to twenty-four 384-well microplates containing 80 μL of liquid Luria–Bertani (LB) medium supplemented with 50 μg/mL kanamycin per well. Each strain is pinned into one well, without replicates.
2. To make room for border control strain, which will aid in within and between plate normalization as well as in colony quantization's, remove inoculated media from the outermost wells of each plate (from Step 1 above) and transfer to new plates, again leaving the outermost wells empty.

3. Similarly, make negative control spots by removing two strains from the inner wells of each plate to a new plate. The strains should be removed from a different location in each plate. Thus, these negative control spots are expected to be empty in the recipient and double-mutant plates, ensuring that there were no processing errors when numbering the plates or in plate orientation during imaging or pinning.
4. Fill the empty wells, where cultures have been removed in Steps 2 and 3 above, with LB containing 34 μg/mL of chloramphenicol.
5. Inoculate Keio deletion strain JW5028 (11) into a 500-mL flask with 150 mL LB, containing 50 μg/mL of kanamycin. In our whole-genome eSGA experiments, this particular deletion mutant showed the smallest number of interactions and is, thus, used as a border control. Since the same recipient strain is used in all the border wells and all recipient plates, it is expected to grow equally well across all plates of the same screen. Thus, measurement of the border control strain growth should help with between plate normalization. Further, since the left and the right as well as the top and the bottom border sports should grow and be detected by the imaging software equally well, comparing the border colony sizes provides a good control for efficient automatic quantization and is useful for within plate normalization.
6. Grow the arrayed strains as well as the JW5028 culture overnight at 32°C with 190 rpm orbital shaking to (OD) of ~0.4–0.6 at 600 nm.
7. After the overnight growth, remove completely the LB-Cm from all border wells and replace it with 80 μL of JW5028 overnight culture.
8. Supplement each well in the recipient plates with 15% glycerol by volume. Gently pipette-mix the liquids, without spilling, to uniformly distribute the glycerol, and store the plates at −80°C for long-term storage.

3.3. Construction of *E. coli* Double Mutants Using an Arrayed Strain Mating Procedure

Once a query donor mutant strain is available and the recipients have been arrayed, genome-wide eSGA screens can be performed (see Note 6). The construction and screening process for generating and evaluating large sets of double mutants is divided into four steps: (1) mutant array preparation, (2) conjugation, (3) selection of double mutants on "Cm" and "Kan" –containing plates, and (4) plate imaging and quantification of individual colony sizes. Construction of double mutants in high-density gridded arrays (Fig. 2) involves replica pinning and incubations over 6 days. We suggest performing at least three independent biological replicate experiments. One experiment, without replicates, is described below:

1. First day: The Hfr donor strain, bearing the query gene replacement (marked with Cm, from Subheading 3.1), is grown overnight at 32°C in rich LB liquid medium (with 34 μg/mL of Cm).
2. Second day: The thawed collection of ordered recipient mutants is pinned in a 384-format onto solid LB plates supplemented with 50 μg/mL Kan. Simultaneously, the donor query strain is pinned in 384 densities onto the same number of LB plates supplemented with 34 μg/mL of Cm.
3. Third day: Conjugation plates are made by (1) pinning the donor from the 384-spot overnight donor plates onto solid LB plate and (2) pinning the arrayed recipients, also grown overnight in 384-spot density, over the freshly pinned donor (see Note 7).
4. Fourth day: Conjugants are subjected to first round of selection: each 384-density conjugation plate is pinned onto one solid LB plate containing Cm and Kan (see Note 8).
5. Fifth day: Double-mutant colonies from the first double drug selection plate are repinned onto a second double drug selection plate in 1,536-spot format, such that each first selection colony is represented on the second selection plate by four colonies (i.e., array each 384 density plate by pinning four times onto one plate to make one 1,536 density plate. Thus, the new 1,536 density plate will have four colonies from any one colony on the 384 density source plate). In each step of the above process, the plates are usually incubated for 16–36 h at 32°C (see Note 9). The double mutants generated by the plate-based assay can be randomly checked by PCR, as described in Fig. 3, to confirm the correctness of the mutations.
6. Sixth day: The second selection double-mutant plates are photographed to record the growth of all colonies for quantitatively assessing mutant fitness (Subheading 3.4) and analyzing the interactions between gene pairs (Subheading 3.5).

3.4. Data Processing and Score Generation

3.4.1. Quantitative Image Analysis and Plate Colony Size Normalization

1. The size of each colony is measured by processing plates in batch mode or individually using specialized colony imaging software (19, 25). The output is a raw colony size measurement for each of the colonies present on the plate.
2. The raw colony measurements can be normalized using multiple normalization and filtering steps to correct for systematic biases in colony growth and measurement within and between plates such as plate edge effects, interplate variation effects, uneven image lighting, artifacts due to physical curvature of the agar surface, competition effects for neighboring colonies and possible pinning defects, as well as differences in growth time (25). These systematic artifacts are independent

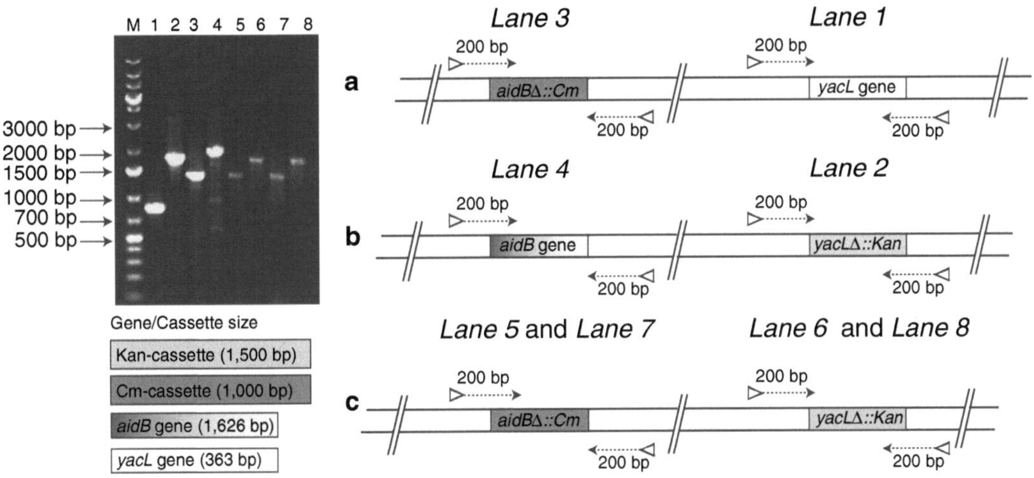

Fig. 3. PCR confirmation of deletions at two unrelated loci, combined through conjugation. Following 24 h conjugation of Δ*aidB*-Cm donor deletion strain (**a**) and Δ*yacL*-Kan recipient deletion strain (**b**), the selected double mutants (**c**) are confirmed by PCR. The sizes of the Cm and Kan cassettes, as well as of *aidB* and *yacL* loci, in base pairs, are shown in parentheses. Amplifications with flanking primers, KOCO-F and KOCO-R (as shown in Fig. 1), from strains with the target gene replaced by Cm^R or Kan^R cassette are expected to produce 1,400 and 1,900 bp products, respectively. On the contrary, amplifications from the isolates containing a wild-type copy of the target gene result in a product equal to the size of the gene plus 400 bp. *Lanes from left to right*: M. DNA marker; *Lane 1*. Δ*aidB*-Cm donor deletion strain amplified with *yacL* knockout confirmation primers; *Lane 2*. Δ*yacL*-Kan recipient deletion strain amplified with *yacL* knockout confirmation primers; *Lane 3*. Δ*aidB*-Cm query deletion strain amplified with *aidB* knockout confirmation primers; *Lane 4*. Δ*yacL*-Kan recipient deletion strain amplified with *aidB* knockout confirmation primers; *Lanes 5* and *7*; and *6* and *8* show amplifications from two independently constructed double mutants (Δ*aidB*-Cm ⁎ Δ*yacL*-Kan), amplified with *aidB* and *yacL* knockout confirmation primers, respectively. The molecular weights, in basepair, are shown beside arrows to the left of the gel. The *lanes 1–4* serve as controls and show that donor and recipient are mutants for *aidB* and *yacL* only, respectively, each bearing a wild-type copy of the second locus. *Lanes 5 to 8* show the same PCR amplifications using conjugant genomic DNA and confirm that PCR products corresponding to both mutant loci are present in both isolates.

of the growth properties of *E. coli* deletion mutants used in the study and should be corrected, as they give rise to spurious growth fitness estimates.

3. A correction can be applied for the larger trend in colony sizes typically observed in the outermost colony rows and columns (due to lower competition for nutrients than found in the center of the plate). Thus, for each edge row or column, the colony sizes can be scaled such that the median size in that row or column is equal to the median size of the colonies in the center of the plate (see Note 10).
4. Further, normalization can be applied to account for the differences in the growth phenotype of the donor mutations. Since most mutations in the array have little or no growth defect, and most of the mutation combinations will not interact (4, 40), we typically normalized the colony sizes according to the peak of the histogram of colony sizes on a given plate (see Note 11).

5. Finally, the results are analyzed statistically to take into account the reproducibility and the deviation (standard variance) from the median sizes of the replicate colony measurements.

3.4.2. Generation of Genetic Interaction Scores

1. The normalized median colony growth sizes are used to generate a genetic interaction score (*S*) for each gene pair as follows:

$$S-\text{score}=\frac{\mu_{Exp}-\mu_{Cont}}{\sqrt{\frac{S_{\text{var}}}{n_{Exp}}+\frac{S_{\text{var}}}{n_{Cont}}}}$$

Where, $S_{\text{var}}=(\text{var}_{\text{Exp}}\times(n_{\text{Exp}}-1)+\text{var}_{\text{Cont}}\times(n_{\text{Cont}}-1))/(n_{\text{Exp}}+n_{\text{Cont}}-2)$; var_{Exp} = the maximum variance of the normalized colony sizes for the double mutant; var_{Cont} = median of the variances in the normalized double-mutant colony sizes from the reference set; n_{Exp} = number of measurements of double-mutant colony sizes; n_{Cont} = the median number of experimental replicates over all the experiments; μ_{Exp} = median normalized colony sizes of the double mutants; and μ_{Cont} = median of normalized colony sizes for all double mutants arising from the single donor mutant strain. The *S*-score reflects both the statistical confidence of a putative digenic interaction as well as the biological strength of the interaction. Strong positive *S*-scores indicate alleviating (e.g., suppression) effects (i.e., growth rate of the double mutant is better than the product of the growth rates of the two respective single mutants), suggesting that the interacting genes participate in the same pathway (25), while significant negative *S*-scores reflect synthetic sickness or lethality (i.e., growth rate of the double mutant is worse than the product of the growth rates of the two single mutants), which is often suggestive of membership in parallel redundant pathways (25).

2. Closely linked genes should exhibit negative *S*-scores because recombination between these genes happens with greatly lower frequencies, which results in a failure to form double mutants. Based on our analysis in the previous study (25), the linkage suppression is evident and inversely proportional to the genetic distance, but is generally undetectable 30 kbp from the donor query gene loci. Although functionally linked genes are often located in the same operon on the *E. coli* chromosome, it is recommended to remove from further assessment all recipient genes located within 30 kbp of a query locus. Nonetheless, linkage suppression can be used to confirm the correctness of a particular query mutation (i.e., linked loci should exhibit a detectable linkage effect).

3. After taking into account of all the above steps, a final dataset is created for functional analysis with a single *S*-score generated for each pair of tested genes (see Note 12).

3.4.3. Assessing the Genetic Interaction Scores

1. The interaction *S*-scores from the genome-wide screens can be plotted to examine the normality of the *S*-score distribution. If the *S*-scores approximate a normal distribution, then the next step would be to see what significant fraction of *S*-scores is observed in the two tails of the distribution of the experimental dataset compared to chance alone (see Note 13).
2. The genetic interaction data can be randomized to evaluate whether the *S*-scores calculated for pairs of genes in the genome-wide screens reflect true genetic interactions or are due to residual position effects of the strains on the plate. For example, one can scramble the dataset's row and column coordinates using the original raw colony sizes of each gene extracted from the image extraction software. The *S*-scores are then recalculated.
3. *S*-scores deviating significantly from the mean represent candidates for functional associations. One can then evaluate the extent to which available knowledge – about functionally related gene pairs – is reflected at various *S*-score thresholds (see Note 14).
4. Candidate interacting gene pairs can be considered to be functionally bridged if they share a physical interaction or another known (annotated) functional relationship (25). With this assumption, one can generate a positive prediction rate (TP/(TP+FP)) for the genetic interaction data, where true positives (TP) can be calculated for functional associations predicted by eSGA that have links from other experimental studies, while false positives (FP) can be defined as eSGA hits without any supporting evidence. Using this strategy, one can determine not only the optimal threshold score but also the rate of positive prediction with increasing |*S*-score| value.
5. Since genetic interaction (*S*) scores, deviating significantly from the mean, represent candidates for functional associations (37), one can also make use of pathway level annotations to examine the genetic interaction network at a more global mechanistic level (23, 41). A wealth of curated information is contained in public functional annotation databases such as KEGG (42), EcoCyc (43) and MultiFun (44), and other sources of functional associations (1), to determine if putative genetic interactors have related functions, which should reflect the overall accuracy of the filtered dataset. Specifically, one can evaluate the extent to which genes belonging to a particular functional category (i.e., annotation term) show statistical

enrichment for genetic interactions at various *S*-score thresholds after correcting for multiple-hypothesis testing. To accomplish this, one can calculate enrichment between each pair of functional processes by performing a hypergeometric distribution analysis (45) with correction for multiple hypothesis testing (46). After accounting for the false discovery rate, the scored data should show a marked corresponding increase in enrichment for both aggravating and alleviating gene pairs (*p*-value 0.05) between select biological processes with increasing *S*-score values. This alternative strategy can also serve to define a confidence threshold for selecting the putative genetic interactions for further analysis.

6. The reliability of high-confidence genetic interactions can be confirmed by reanalyzing individual donor–recipient digenic mutant combinations using custom miniarray conjugation assays (25). For example, the specificity of the interaction of *pdxB* with *iscS* in the iron-sulfur cluster biosynthesis pathway from Butland et al. (25) can be explained by looking at the curated functional annotations. The gene *pdxB* is involved in the biosynthesis of pyridoxal 5-phosphate (PLP), which is a cofactor for IscS cysteine desulfurases, suggesting a plausible link. The reliability of the putative interaction was confirmed by manually conjugating the $iscS\Delta::Cm^R$ query deletion mutant strain to the $pdxB\Delta::Kan^R$ recipient deletion mutant strain, and then by reciprocally conjugating $pdxB\Delta::Cm^R$ as query deletion strain to $iscS\Delta::Kan^R$ as the recipient deletion mutant strain. As shown in Fig. 4a, the *iscS* mutant displayed synthetic lethality with *pdxB*, and a similar aggravating interaction was observed when $pdxB\Delta::Cm^R$ was used as donor, implying *iscS* and *pdxB* are functionally related. Such small-scale validation experiments are useful for verifying putative genetic interactions generated from the high-throughput eSGA screens.
7. One can also perform rescue experiments with one or both mutant alleles by gene complementation using an inducible expression plasmid from the public *E. coli* ORFeome collection (47).

3.5. Discerning Pathway Level Relationships

1. Genes in the same pathway are expected to display closely correlated patterns of genetic interactions with genes in parallel, functionally redundant pathways.
2. Groupings reflecting such biological relationships can be visualized by two-dimensional hierarchical clustering of the *S* scores to suggest the possible membership of novel genes in known pathways and biological processes. A representative example is shown in Fig. 4b, where the components of the functionally redundant Isc and Suf pathways form distinct clusters that are linked together by extensive aggravating interactions.

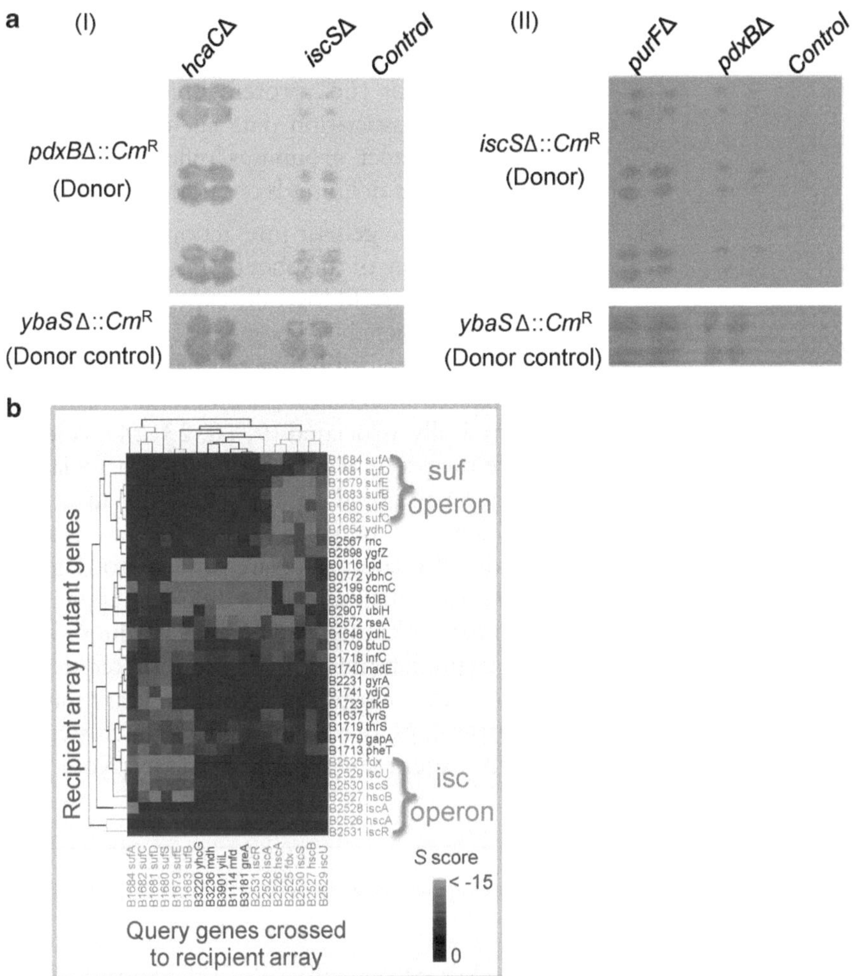

Fig. 4. Validation of genetic interactions using custom miniarrays and clustering of genetic interaction profiles. Panel A: Confirmation of genetic interactions of *iscS* and *pdxB* strains using custom miniarrays. The eSGA miniarray confirmations, shown as plate images, were performed by crossing a *pdxB* query mutant strain (marked with Cm^R) with an *iscS* recipient deletion strain (marked with Kan^R) and vice versa. Both donor–recipient combinations, *pdxB* and *iscS* (I) as well as *IscS* and *pdxB* (II), displayed an SSL relationship. The *hcaC*Δ::*Kan*^{R} and *purF*Δ::*Kan*^{R} recipient mutants were used as positive recipient controls (no SSL interactions expected) and "Control" represents no recipient control (i.e., no recipient mutant is included in the array during the screening process). The *ybaS*Δ was used as a positive donor control strain, as no SSL relationship is expected with the indicated recipient mutants. Panel B: A subset of the high confidence genetic interaction network highlighting the known Isc (*gray* print) and Suf (*gray* print) pathway genes with similar patterns of genetic interactions. Both pathways function in the same process and their patterns of genetic interactions cluster together. *Columns*: recipient array genes. *Gray* represents aggravating (negative *S*-score) interactions and *black* represents the absence of genetic interactions. The array strains with the supposed *iscR* and *hscA* mutations were defective. The *ydhD* gene displays interactions with the Isc pathway. The network clustering analysis shown was performed on data from Butland et al. (25).

3. A hypergeometric distribution function (45) can be applied to evaluate pathway cross talk (i.e., determine the significance of interactions between combinations of pathways). Likewise, nonthreshold based enrichment algorithms such as Gene Set Enrichment Analysis (48) may allow for a more sensitive

detection of trends and modularity in the genetic interaction networks. Integration of genetic interaction networks with physical associations (i.e., protein–protein interactions) and other functional association data such as genomic contexts can potentially detect groupings reflective of higher order functional modules defining biological systems.

4. The similarity of the genetic interaction profiles should represent the congruency of the phenotypes of the two mutations. For instance, one would logically expect both measures to be indicative of whether the two genes act in the same pathway. In fact, gene pairs exhibiting alleviating interactions with highly correlated interaction profiles tend to encode proteins that are physically associated (9, 10, 23, 24). Additionally, in cases where the proteins do not physically associate, products of such gene pairs tend to act coherently in a biochemical pathway. Such cases are particularly informative, as these very close functional relationships are difficult to detect by other methods. Cluster analysis can be therefore applied to genetic interaction networks to group genes according to profile similarity and to predict functions of unannotated genes (21–23, 25, 35, 49). Since clustering algorithms vary, putative functional relationships determined through clustering require independent experimental verification (see Note 15).

4. Notes

1. At an OD between 0.5 and 0.6, E. *coli* culture is at an exponential growth phase with cells dividing at a maximal rate, allowing for the preparation of highly competent cells.
2. At this temperature, the temperature-sensitive λ cI-repression is removed, which leads to the expression of λ *exo*, *bet*, and *gam* genes that in turn make the cells more recombinogenic. A maximum level of recombination is reached for induction times between 7.5 and 17.5 min, with reduction in efficiencies occurring for times longer than 17.5 min (32).
3. The competent cells, resuspended in 10% glycerol, can be stored for up to 6 months by first freezing the cells on dry ice and then transferring them to −80°C for storage. Prior to subsequent transformation (Subheading 3.1.3), the cells are thawed on ice for 10–15 min.
4. Include positive and negative transformation controls. For a negative control, we suggest a mock electroporation of competent cells without any PCR product to test for the presence of contaminating strains. As a positive control, use a PCR product (or a plasmid) that is known to generate transformants.

5. Colony PCR can also be performed to more rapidly confirm the replacement of the target gene with the Cm cassette. This method does not require the purification of genomic DNA; however, care has to be taken to pick a very small volume of cells (*ca.* 0.5 mm^2) using a plastic pipette tip. Furthermore, a 5 min initial denaturation step is used in this case.
6. The construction and screening process for generating and evaluating large sets of double mutants can be applied for epistatic miniarray profiling studies as well (23, 34). In this case, the donor query strain is crossed against an array of recipient genes selected to answer a specific biological question.
7. In the standard eSGA screening process, the donor and recipient strains are incubated for conjugation at an approximate cell ratio of ~1:1. We observed that the selection of conjugants using only one antibiotic (kanamycin or chloramphenicol), followed by selection using both antibiotics, results in highly variable double mutant colonies. This additional selective outgrowth step is deemed to be advantageous for two reasons. First, it has been previously reported that having >1 colony measurement greatly increases the accuracy of the fitness estimates allowing for the natural measure of variation as the standard deviation. Second, it has been noted during the development process that direct double antibiotic selection on the initial conjugation plate leads to highly variable colony size distributions, which severely impairs the accuracy of strain fitness estimates.
8. For standard genome-wide eSGA screens, each individual Keio gene deletion mutant (11) is arrayed in a row of four replicates: two copies of each of the two biological replicates ("Isolate 1 and 2", which are independently confirmed deletion strains for the same gene). Further, the entire collection is screened twice on separate plates to establish score reproducibility and statistical significance. Therefore, in total, each combination of mutations is assayed using a total of eight replicate colony measurements.
9. We suggest allowing conjugation to take place over a 24 h period because only a few viable conjugants were obtained for short mating periods (<12 h) (25). Monitor the growth of colonies on plates to determine appropriate incubation time for the other steps. The time may vary from screen to screen, since different mutants may have different growth rates.
10. There may be cases in which one is unable to detect or quantify genetic interactions because one of the mutations causes an *E. coli* strain to have an unusual mutant colony morphological phenotype (e.g., spreading or mucoid) that precludes colony transfer or conjugation, which in turn

masks or exaggerates growth defects. Measurements for such colonies are removed from further analysis.

11. Systematic variability can also arise due to batch-to-batch effect, where the size of a colony estimated for one set of screens completed at approximately the same time using the same media preparation differs from the values estimated for another set of screens completed at a different time. If such batch-to-batch variability is evident, it is preferable to compute the expected colony sizes independently for each batch.
12. In cases where two *S*-scores are produced from two independent biological experiments, we suggest calculating and substituting in either a median or an average score.
13. A randomized dataset can be created for each query mutant screen where the row and column coordinates and the raw colony sizes of each recipient mutant are shuffled. A randomized dataset to control for score distributions obtained for each query mutant screen is permuted 1,000 times. *S*-scores are then calculated for each query mutant screen from this permuted randomized dataset. The difference in tails between the experimental and the randomized datasets can be then used to assess the statistical significance of aggravating or alleviating genetic interactions.
14. Predicted functional associations between *E. coli* genes can be obtained from public databases such as STRING (50) or eNET (1).
15. Since *E. coli* pathways have evolved adaptive mechanisms to withstand environmental perturbations and since certain functional relationships are likely manifested only under particular physiological conditions (3), one can subject the array of viable double mutants to the additional stress of growth on minimal glucose medium or at high temperature (i.e., at 42°C). This strategy is likely to reveal condition-dependent genetic interactions. For example, Gross and colleagues have shown condition-dependent synthetic genetic interactions involving the outer membrane-associated lipoprotein *pal* in a media-dependent manner (e.g., *pal-yfgL*, *pal-ompA*, *pal-cpxR*) (26). The use of rich LB medium for the conjugation step typically results in more rapid and consistent growth of recipient mutants from the KEIO collection. Cells grown on minimal media require much longer incubation times (a total of approximately 2 weeks to complete a full genome screen in minimal medium versus 6 days on rich medium). Further, screens, where conjugation and selection are performed on minimal media, display far more variability in colony sizes and conjugation success compared to rich media screens. Thus, we incubate all plates at 32°C during the conjugation and selection steps. For experiments that require other conditions

to study genetic interactions, conjugation and selection can be performed in standard conditions and then the double mutants can be replicate-pinned and grown in the condition of interest (e.g., on different media or at a different temperature). Thus, a new set of condition-specific genetic interactions can be generated.

Acknowledgments

We thank members of the Emili and Jack Greenblatt laboratories for assistance and advice. AG is a recipient of Vanier Canada Graduate Scholarship. This work was supported by a Canadian Institute of Health Research (CIHR) grant (MOB-82852) to JG and AE.

References

1. Hu P., Janga S. C., Babu M., Diaz-Mejia J. J., Butland G. (2009) Global functional atlas of *Escherichia coli* encompassing previously uncharacterized proteins. *PLoS Biol* **7**, e1000096.
2. Pereira-Leal J. B., Levy E. D., Teichmann S. A. (2006) The origins and evolution of functional modules: lessons from protein complexes. *Philos Trans R Soc Lond B Biol Sci* **361**, 507–517.
3. Babu M., Musso G., Diaz-Mejia J. J., Butland G., Greenblatt J. F., Emili A. (2009) Systems-level approaches for identifying and analyzing genetic interaction networks in *Escherichia coli* and extensions to other prokaryotes. *Mol Biosyst* **12**, 1439–1455.
4. Beltrao P., Cagney G., Krogan N. J. (2010) Quantitative genetic interactions reveal biological modularity. *Cell* **141**, 739–745.
5. Beyer A., Bandyopadhyay S., Ideker T. (2007) Integrating physical and genetic maps: from genomes to interaction networks. *Nat Rev Genet* **8**, 699–710.
6. Motter A. E., Gulbahce N., Almaas E., Barabasi A. L. (2008) Predicting synthetic rescues in metabolic networks. *Mol Syst Biol* **4**, 168.
7. Bandyopadhyay S., Kelley R., Krogan N. J., Ideker T. (2008) Functional maps of protein complexes from quantitative genetic interaction data. *PLoS Comput Biol* **4**, e1000065.
8. Boone C., Bussey H., Andrews B. J. (2007) Exploring genetic interactions and networks with yeast. *Nat Rev Genet* **8**, 437–449.
9. Collins S. R., Miller K. M., Maas N. L., Roguev A., Fillingham J., Chu C. S., Schuldiner M., Gebbia M., Recht J., Shales M., Ding H., Xu H., Han J., Ingvarsdottir K., Cheng B., Andrews B., Boone C., Berger S. L., Hieter P., Zhang Z., Brown G. W., Ingles C. J., Emili A., Allis C. D., Toczyski D. P., Weissman J. S., Greenblatt J. F., Krogan N. J. (2007) Functional dissection of protein complexes involved in yeast chromosome biology using a genetic interaction map. *Nature* **446**, 806–810.
10. Fiedler D., Braberg H., Mehta M., Chechik G., Cagney G., Mukherjee P., Silva A. C., Shales M., Collins S. R., van Wageningen S., Kemmeren P., Holstege F. C., Weissman J. S., Keogh M. C., Koller D., Shokat K. M., Krogan N. J. (2009) Functional organization of the *S. cerevisiae* phosphorylation network. *Cell* **136**, 952–963.
11. Baba T., Ara T., Hasegawa M., Takai Y., Okumura Y., Baba M., Datsenko K. A., Tomita M., Wanner B. L., Mori H. (2006) Construction of *Escherichia coli* K-12 in-frame, single-gene knockout mutants: the Keio collection. *Mol Syst Biol* **2**, 2006.0008.
12. Hartwell L. H., Hopfield J. J., Leibler S., Murray A. W. (1999) From molecular to modular cell biology. *Nature* **402**, C47-52.
13. Mani R., St Onge R. P., Hartman J. L., Giaever G., Roth F. P. (2008) Defining genetic interaction. *Proc Natl Acad Sci USA* **105**, 3461–3466.

14. Kuzminov A. (1999) Recombinational repair of DNA damage in *Escherichia coli* and bacteriophage lambda. *Microbiol Mol Biol Rev* **63**, 751–813.
15. Seoighe C., Wolfe K. H. (1998) Extent of genomic rearrangement after genome duplication in yeast. *Proc Natl Acad Sci USA* **95**, 4447–4452.
16. Dixon S. J., Costanzo M., Baryshnikova A., Andrews B., Boone C. (2009) Systematic mapping of genetic interaction networks. *Annu Rev Genet* **43**, 601–625.
17. Dixon S. J., Fedyshyn Y., Koh J. L., Prasad T. S., Chahwan C., Chua G., Toufighi K., Baryshnikova A., Hayles J., Hoe K. L., Kim D. U., Park H. O., Myers C. L., Pandey A., Durocher D., Andrews B. J., Boone C. (2008) Significant conservation of synthetic lethal genetic interaction networks between distantly related eukaryotes. *Proc Natl Acad Sci USA* **105**, 16653–16658.
18. Cork J. M., Purugganan M. D. (2004) The evolution of molecular genetic pathways and networks. *Bioessays* **26**, 479–484.
19. Costanzo M., Baryshnikova A., Bellay J., Kim Y., Spear E. D., Sevier C. S., Ding H., Koh J. L., Toufighi K., Mostafavi S., Prinz J., St Onge R. P., VanderSluis B., Makhnevych T., Vizeacoumar F. J., Alizadeh S., Bahr S., Brost R. L., Chen Y., Cokol M., Deshpande R., Li Z., Lin Z. Y., Liang W., Marback M., Paw J., San Luis B. J., Shuteriqi E., Tong A. H., van Dyk N., Wallace I. M., Whitney J. A., Weirauch M. T., Zhong G., Zhu H., Houry W. A., Brudno M., Ragibizadeh S., Papp B., Pal C., Roth F. P., Giaever G., Nislow C., Troyanskaya O. G., Bussey H., Bader G. D., Gingras A. C., Morris Q. D., Kim P. M., Kaiser C. A., Myers C. L., Andrews B. J., Boone C. (2010) The genetic landscape of a cell. *Science* **327**, 425–431.
20. Pan X., Yuan D. S., Xiang D., Wang X., Sookhai-Mahadeo S., Bader J. S., Hieter P., Spencer F., Boeke J. D. (2004) A robust toolkit for functional profiling of the yeast genome. *Mol Cell* **16**, 487–496.
21. Tong A. H., Evangelista M., Parsons A. B., Xu H., Bader G. D., Page N., Robinson M., Raghibizadeh S., Hogue C. W., Bussey H., Andrews B., Tyers M., Boone C. (2001) Systematic genetic analysis with ordered arrays of yeast deletion mutants. *Science* **294**, 2364–2368.
22. Tong A. H., Lesage G., Bader G. D., Ding H., Xu H., Xin X., Young J., Berriz G. F., Brost R. L., Chang M., Chen Y., Cheng X., Chua G., Friesen H., Goldberg D. S., Haynes J., Humphries C., He G., Hussein S., Ke L., Krogan N., Li Z., Levinson J. N., Lu H., Menard P., Munyana C., Parsons A. B., Ryan O., Tonikian R., Roberts T., Sdicu A. M., Shapiro J., Sheikh B., Suter B., Wong S. L., Zhang L. V., Zhu H., Burd C. G., Munro S., Sander C., Rine J., Greenblatt J., Peter M., Bretscher A., Bell G., Roth F. P., Brown G. W., Andrews B., Bussey H., Boone C. (2004) Global mapping of the yeast genetic interaction network. *Science* **303**, 808–813.
23. Schuldiner M., Collins S. R., Thompson N. J., Denic V., Bhamidipati A., Punna T., Ihmels J., Andrews B., Boone C., Greenblatt J. F., Weissman J. S., Krogan N. J. (2005) Exploration of the function and organization of the yeast early secretory pathway through an epistatic miniarray profile. *Cell* **123**, 507–519.
24. Wilmes G. M., Bergkessel M., Bandyopadhyay S., Shales M., Braberg H., Cagney G., Collins S. R., Whitworth G. B., Kress T. L., Weissman J. S., Ideker T., Guthrie C., Krogan N. J. (2008) A genetic interaction map of RNA-processing factors reveals links between Sem1/Dss1-containing complexes and mRNA export and splicing. *Mol Cell* **32**, 735–746.
25. Butland G., Babu M., Díaz-Mejía J. J., Bohdana F., Phanse S., Gold B., Yang W., Li J., Gagarinova A. G., Pogoutse O., Mori H., Wanner B. L., Lo H., Wasniewski J., Christopolous C., Ali M., Venn P., Safavi-Naini A., Sourour N., Caron S., Choi J. Y., Laigle L., Nazarians-Armavil A., Deshpande A., Joe S., Datsenko K. A., Yamamoto N., Andrews B. J., Boone C., Ding H., Sheikh B., Moreno-Hagelseib G., Greenblatt J. F., Emili A. (2008) eSGA: *E. coli* synthetic genetic array analysis. *Nat Methods* **5**, 789–795.
26. Typas A., Nichols R. J., Siegele D. A., Shales M., Collins S. R., Lim B., Braberg H., Yamamoto N., Takeuchi R., Wanner B. L., Mori H., Weissman J. S., Krogan N. J., Gross C. A. (2008) High-throughput, quantitative analyses of genetic interactions in *E. coli. Nat Methods* **5**, 781–787.
27. Davierwala A. P., Haynes J., Li Z., Brost R. L., Robinson M. D., Yu L., Mnaimneh S., Ding H., Zhu H., Chen Y., Cheng X., Brown G. W., Boone C., Andrews B. J., Hughes T. R. (2005) The synthetic genetic interaction spectrum of essential genes. *Nat Genet* **37**, 1147–1152.
28. Breslow D. K., Cameron D. M., Collins S. R., Schuldiner M., Stewart-Ornstein J., Newman H. W., Braun S., Madhani H. D., Krogan N. J., Weissman J. S. (2008) A comprehensive strategy enabling high-resolution functional analysis of the yeast genome. *Nat Methods* **5**, 711–718.

29. Firth N., Ippen-Ihler K. Skurray R.A. (1996) Structure and function of the F-factor and mechanism of conjugation. *Escherichia coli and Salmonella: Cellular and Molecular Biology, Vol. 1 (Neidhardt, F.C., Ed.), ASM Press, Washington, DC,* 2377–2401.
30. Ippen-Ihler K. A., Minkley E. G., Jr. (1986) The conjugation system of F, the fertility factor of *Escherichia coli. Annu Rev Genet* **20**, 593–624.
31. Chumley F. G., Menzel R., Roth J. R. (1979) Hfr Formation Directed by Tn10. *Genetics* **91**, 639–655.
32. Yu D., Ellis H. M., Lee E. C., Jenkins N. A., Copeland N. G., Court D. L. (2000) An efficient recombination system for chromosome engineering in *Escherichia coli. Proc Natl Acad Sci USA* **97**, 5978–5983.
33. Johnson D. C., Dean D. R., Smith A. D., Johnson M. K. (2005) Structure, function, and formation of biological iron-sulfur clusters. *Annu Rev Biochem* **74**, 247–281.
34. Babu M., Díaz-Mejía J. J., Phanse S., Vlasblom J., Gagarinova A., Graham C., Ding H., Hu P., Yousif F., Nazarians-Armavil A., Pogoutse O., Ali M., Peer A., Margalit H., Wodak S. J., Moreno-Hagelsieb G., Greenblatt J. F., Emili A. (2010) Genetic interaction profiling reveals the global functional organization of cell-envelope pathways in *Escherichia coli. Cell (Submitted).*
35. Roguev A., Bandyopadhyay S., Zofall M., Zhang K., Fischer T., Collins S. R., Qu H., Shales M., Park H. O., Hayles J., Hoe K. L., Kim D. U., Ideker T., Grewal S. I., Weissman J. S., Krogan N. J. (2008) Conservation and rewiring of functional modules revealed by an epistasis map in fission yeast. *Science* **322**, 405–510.
36. van Opijnen T., Bodi K. L., Camilli A. (2009) Tn seq: high-throughput parallel sequencing for fitness and genetic interaction studies in microorganisms. *Nat Methods* **6**, 767–772.
37. St Onge R. P., Mani R., Oh J., Proctor M., Fung E., Davis R. W., Nislow C., Roth F. P., Giaever G. (2007) Systematic pathway analysis using high-resolution fitness profiling of combinatorial gene deletions. *Nat Genet* **39**, 199–206.
38. Datsenko K. A., Wanner B. L. (2000) One-step inactivation of chromosomal genes in Escherichia coli K-12 using PCR products. *Proc Natl Acad Sci USA* **97**, 6640–6645.
39. Butland G., Peregrín-Alvarez J. M., Li J., Yang W., Yang X., Canadien V., Starostine A., Richards D., Beattie B., Krogan N., Davey M., Parkinson J., Greenblatt J., Emili A. (2005) Interaction network containing conserved and essential protein complexes in *Escherichia coli. Nature* **433**, 531–537.
40. Collins S. R., Schuldiner M., Krogan N. J., Weissman J. S. (2006) A strategy for extracting and analyzing large-scale quantitative epistatic interaction data. *Genome Biol* **7**, R63.
41. Ma X., Tarone A. M., Li W. (2008) Mapping genetically compensatory pathways from synthetic lethal interactions in yeast. *PLoS One* **3**, e1922.
42. Kanehisa M. (2002) The KEGG database. *Novartis Found Symp* **247**, 91–101; discussion 101–3, 119–28, 244–252.
43. Keseler I. M., Bonavides-Martinez C., Collado-Vides J., Gama-Castro S., Gunsalus R. P., Johnson D. A., Krummenacker M., Nolan L. M., Paley S., Paulsen I. T., Peralta-Gil M., Santos-Zavaleta A., Shearer A. G., Karp P. D. (2009) EcoCyc: a comprehensive view of *Escherichia coli* biology. *Nucleic Acids Res* **37**, D464-470.
44. Serres M. H., Riley M. (2000) MultiFun, a multifunctional classification scheme for *Escherichia coli* K-12 gene products. *Microb Comp Genomics* **5**, 205–222.
45. Le Meur N., Gentleman R. (2008) Modeling synthetic lethality. *Genome Biol* **9**, R135.
46. Benjamini Y., Yekutieli D. (2001) The control of the false discovery rate in multiple testing under dependency. *Annals of Statistics* **29**, 1165–1188.
47. Saka K., Tadenuma M., Nakade S., Tanaka N., Sugawara H., Nishikawa K., Ichiyoshi N., Kitagawa M., Mori H., Ogasawara N., Nishimura A. (2005) A complete set of *Escherichia coli* open reading frames in mobile plasmids facilitating genetic studies. *DNA Res* **12**, 63–68.
48. Subramanian A., Tamayo P., Mootha V. K., Mukherjee S., Ebert B. L., Gillette M. A., Paulovich A., Pomeroy S. L., Golub T. R., Lander E. S., Mesirov J. P. (2005) Gene set enrichment analysis: a knowledge-based approach for interpreting genome-wide expression profiles. *Proc Natl Acad Sci USA* **102**, 15545–15550.
49. Wong S. L., Zhang L. V., Tong A. H., Li Z., Goldberg D. S., King O. D., Lesage G., Vidal M., Andrews B., Bussey H., Boone C., Roth F. P. (2004) Combining biological networks to predict genetic interactions. *Proc Natl Acad Sci USA* **101**, 15682–15687.
50. von Mering C., Jensen L. J., Kuhn M., Chaffron S., Doerks T., Kruger B., Snel B., Bork P. (2007) STRING 7--recent developments in the integration and prediction of protein interactions. *Nucleic Acids Res* **35**, D358–362.

Chapter 10

Assembling New *Escherichia coli* Strains by Transduction Using Phage P1

Sean D. Moore

Abstract

A protocol is described that allows the transfer of genetic material from one *Escherichia coli* strain to another using bacteriophage P1. P1 transduction can be used to construct new bacterial strains containing multiple alleles, to restore a locus to wild type, to move specific genetic markers from one strain to another, to relocate different mutant genes to a common genetic background, and to evaluate second-site suppression of a mutant allele. Because of these abilities, P1 transduction remains a staple in the arsenal of genetic tools that have kept *E. coli* at the forefront of model bacterial systems. The protocol incorporates some updated steps and discusses general principles of bacteriophage handling and the infection process.

Key words: Bacteriophage, Phage, P1, Transduction, Transduce, Titer, Plaque, Cross-streak, *E. coli*

1. Introduction

There are few instances in science where a simple technology persists for decades while retaining its original utility. Nearly 60 years ago, the transduction of genetic markers by bacteriophage was first described for *Salmonella* (1). Soon after, the same process was described in *Escherichia coli* using the bacteriophage P1, and this phage is still used as a workhorse in a wide variety of genetics experiments and strain constructions (2). During a P1 infection, a small number of abnormal virus-like particles are produced that contain segments of DNA that are not derived from the replicated phage genome, but rather from other intracellular DNAs such as the host chromosome or plasmids (3–5). These particles retain the ability to subsequently inject the packaged, nonphage DNA into a recipient bacterium and enable the recombination of

James A. Williams (ed.), *Strain Engineering: Methods and Protocols*, Methods in Molecular Biology, vol. 765, DOI 10.1007/978-1-61779-197-0_10, © Springer Science+Business Media, LLC 2011

this DNA with the recipient chromosome. Aberrant DNA packaging by P1 can occur anywhere along the chromosome, so any region of the host DNA can be moved from one strain to another if there is a selection for a gene contained within it (so-called "generalized" transduction). Transduction is carried out by first infecting a donor host with P1 and allowing a lysate to form that contains transducing particles. After sterilization, the lysate is used to infect a recipient strain and recombinants are then selected by plating on a suitable medium. Typically, P1 transduction is implemented using protocols that control the infection cycle with the use of lytic mutant of P1 (P1*vir*) and calcium availability to regulate infectivity (6). The P1 capsid can accommodate ~110 Kb of DNA, and transduction is usually, but not always, mediated by a single DNA fragment (4). Therefore, with a library of suitably separated selectable markers, the entire *E. coli* genome can be represented as transducible fragments in fewer than 100 different phage lysates.

The protocol for P1 transduction has remained practically unchanged since its inception. The protocol in this chapter is a refined protocol used in our lab that contains additional commentary and some slight modifications that have improved reliability in our lab, especially with researchers performing transduction for the first time. Most likely, the testing (Subheading 3.1) and troubleshooting (Subheading 3.5 and 3.6) steps will not be required unless an attempted transduction fails.

2. Materials

2.1. Pretesting Donor and Recipient E. coli Strains

1. Donor strain: Any *E. coli* strain capable of allowing the replication of phage P1. The simplest way to determine the suitability of the donor host is to prepare an infected culture and observe lysis. Alternately, a plaque assay or streak test can be used to monitor replication ability.
2. Recipient strain: Any *E. coli* with an active homologous recombination system and that is capable of being infected by P1. Although the recipient host need not be able to fully replicate P1, as with the donor strain, the ability to complete a replication cycle is a fair measure of competence for infection.
3. Antibiotic stock: The choice depends on the marker being transduced. A 100 mg/mL ampicillin stock solution is prepared by solubilizing the sodium salt of ampicillin at 200 mg/mL in 50 mM Tris–HCl, pH 8.0 and diluting with ethanol to 100 mg/mL. A 25 mg/mL kanamycin stock solution is prepared in water. A 5 mg/mL tetracycline stock solution is prepared in 50% ethanol. A 15 mg/mL chloramphenicol

stock solution is prepared in ethanol. Antibiotics not prepared in alcohol should be filter-sterilized. Acid- or base-labile antibiotics should be prepared in a suitable buffer (e.g., 25 mM MOPS, HEPES, or Tris). Store at −20°C up to 1 year.

4. LB plates and broth: 5 g NaCl, 10 g tryptone, 5 g yeast extract per liter. The pH may be adjusted to 7–7.5 prior to autoclaving by the addition of either NaOH or KOH. LB is used for the more common drug-resistance marker transductions. Plates receive 12–15 g/L agar. Store at room temperature up to 3 months.
5. Defined medium plates: Used for transduction of nutrient-restrictive markers (e.g., amino-acid prototrophy). Any formulation that does not permit growth of the recipient host but that does allow growth of the donor. Use of a rich, defined medium allows for rapid growth and selection. Our current preferred formulation, per liter: 5 g sodium chloride (NaCl, ~86 mM), 10 g tryptone (unless selecting for amino acid anabolism), 12 g agar, 0.5 g ammonium sulfate ($(NH_4)_2SO_4$, ~3.8 mM). 0.3 g dipotassium phosphate (K_2HPO_4, ~1.7 mM final), 1 g sodium bicarbonate ($NaHCO_3$, ~12 mM), 2.5 mM $MgCl_2$, 1 mL of vitamin mixture (reagent 4), and a suitable sugar source to 0.2%.
6. Vitamin mixture for the defined medium: It can be prepared as described by Neidhardt (7) or more easily by dissolving a high-potency multivitamin containing in 10 mL of water, letting the insoluble matter settle out, and adding 1 mL of the solution to one liter of the medium. Our pill formulation supplies vitamins A, C, D, E, K1, B-1 (thiamine), B-2, B-6, and B-12, as well as niacin, folate, biotin, calcium, iron, iodine, zinc, selenium, copper, manganese, chromium, molybdenum, potassium, boron, nickel, tin, vanadium, lutein, and lycopene. If the solution is to be prepared in advance, the vitamin should be dissolved in a nonmetabolizable buffer containing a dissociable metal chelator, a reducing agent, and filter-sterilized (e.g., 25 mM MOPS-K, 10 mM tricine, 30 mM 2-mercaptoethanol, pH 7.4). Always test the medium's ability to support growth of the recipient with the addition of the limiting nutrient to ensure that you will be selecting for transfer of the relevant marker.

2.2. Donor Lysate Preparation

1. Bacteriophage P1*vir*: a mutant P1 that has lost the ability to form lysogens (always enters the lytic cycle). Several P1*vir* mutants have been developed, and we use the phage from a stock maintained at the Coli Genetic Stock Center (CGSC #12133, Department of Molecular, Cellular and Developmental Biology, Yale University, New Haven, CT). The origin of this phage was traced to the laboratory of Jun-Ichi Tomizawa

(personal communication from John Wertz, Yale University) and is described in reference (5). The phage should only need to be acquired once because lysates are stable for years at 4°C and regenerated for each transduction experiment.

2. An overnight (stationary) culture of donor bacteria grown with selection for the marker to be transduced.
3. 1 M $CaCl_2$: Either autoclaved or filter-sterilized. Store at room temperature up to 3 months.
4. 2.5 M $MgCl_2$: Either autoclaved or filter-sterilized. $MgSO_4$ is a suitable substitute. Magnesium is required for the stable condensation of DNA inside the phage head. These magnesium salt crystals are very hygroscopic and stocks should be prepared from the entire contents of newly opened containers and stored as liquids. Store at room temperature up to 5 years.
5. 20% glucose: Either autoclaved or filter-sterilized. Store at room temperature up to 3 months.
6. P1-LB: LB supplemented, per mL, with 5 μL of $MgCl_2$ stock (~12.5 mM), 5 μL of $CaCl_2$ stock (~5 mM), and 5 μL of 20% glucose stock (~0.1%). Optionally, add 2-mercaptoethanol to ~14 mM (1/1,000 dilution of the pure reagent) to aid in lysis. P1-LB can be prepared each time it is needed or in a larger batch and refiltered to ensure sterility. Store at room temperature for up to 3 months.
7. Chloroform stabilized with 1% ethanol or 0.1% isoamyl alcohol. Chloroform is flammable and should be used with adequate ventilation. Store at room temperature for up to 3 months.
8. LB plates and broth. See Subheading 2.1 item 4. Used for the more common drug-resistance marker transductions.
9. Defined medium plates. See Subheading 2.1 item 5.

2.3. Transduction: Phage Infection, Recovery, and Selection

1. Fresh stationary culture of the recipient strain. An "overnight" culture works fine.
2. 1 M sodium citrate, pH 5.5. Prepare 1 M citric acid, raise pH with NaOH to 5.5, filter-sterilize or autoclave. Store at room temperature up to 1 year.
3. LB-citrate: LB broth supplemented with 1/10th volume 1 M Na-citrate, pH 5.5 (100 mM final).
4. Phosphate-buffered saline (PBS; optional): 150 mM NaCl, 50 mM NaH_2PO_4, pH adjusted to ~7.4 with KOH and filter-sterilized. Not needed for drug-resistance transductions. Store at room temperature.

2.4. Purification of the Transductant and Confirmation

1. Selective plates with 1–5 mM sodium citrate: Generally, these are made by spreading the antibiotic and citrate on the surface of a plate in the morning as needed. Per plate

(estimated 25 mL media per plate if 50 plates per liter), prepare a mixture containing 25 μL of 1 M sodium citrate stock, enough selective agent from stock solutions for 25 mL [e.g., ampicillin is used at 50–150 μg/mL (12.5–37.5 μL of stock), kanamycin at 15–30 μg/mL (15–30 μL of stock), tetracycline at 10–20 μg/mL (50–100 μL of stock), and chloramphenicol at 15–30 μg/mL (25–50 μL of stock)], and sterile water to a final volume of 100 μL. Spread the mixture evenly and allow to soak in completely. Let the drug diffuse for several hours before use.

2. PCR primers that can either diagnostically detect the transduced marker location or amplify the marker region for sequencing.
3. Colony-resuspension buffer: 25 mM NaCl, 1 mM $MgCl_2$, 25 mM HEPES, pH 7.4. Filter-sterilize. This is an optional solution, water will generally suffice for resuspending the colony, but the bacteria will die if they stand for extended periods of time in water.

2.5. Troubleshooting: Plaque Formation, Cross-streaking, and Titering

1. P1 top agar: P1-LB medium supplemented with 7 g/L (0.7%) agar and 1 mM dithiothreitol (alternately, 14 mM 2-mercaptoethanol). Adjust pH to 7.5–8.0 with KOH. Low pH restricts disulfide isomerization, which is required for efficient cell lysis (8). The top agar can be allowed to solidify and stored up to 3 months at room temperature. The agar is remelted in a microwave when needed. For pouring, the agar should be melted, sterilely dispensed into 3–4 mL aliquots in prewarmed tubes in a 43–47°C block. The agar should be cooled to this temperature completely before addition of bacteria. Agar tubes can be prepared up to a day in advance.
2. P1-streak plate: LB plate with 10 mM $MgCl_2$ and 5 mM $CaCl_2$. This can be prepared by spreading a mixture of the two compounds (100 μL of 2.5 M $MgCl_2$ and 125 μL of 1 M $CaCl_2$ stocks) on an LB plate (assuming 25 mL volume of the plate) and letting the plate rest for several hours.
3. Sterile flat toothpicks, inoculation loop, or thin wooden sticks.

3. Methods

3.1. Pretesting Donor and Recipient E. coli Strains

1. Streak the strain to be the donor on a selective medium to obtain isolated colonies. While this step may seem trivial, it not only provides a clean source of the donor strain but also verifies if the donor has the genes necessary for growth under selective conditions.

2. Streak the recipient strain on two plates: one with and one without the selection compound. In general, antibiotics are used at concentrations lower than those used for maintaining multicopy plasmids or for selection with a very high cell count (e.g., electroporation). For example, during an electroporation transformation experiment with a multicopy plasmid that confers kanamycin resistance, many viable cells will be plated and they will absorb the free drug even if they are not resistant, thus lowering the effective concentration on the plate. In this case, 50 μg/mL may be required to prevent background growth. With a phage transduction, fewer viable cells are plated and a lower dose of 10–15 μg/mL will allow robust colony development and selection. Ensure that the recipient forms colonies under nonselective conditions and does not form colonies under selective conditions.
3. Several *E. coli* genome sequences are now publicly available. Use the genome sequences of strains closely related to your donor and recipient to determine the location of your marker to be transduced (if known) and take note of other important mutations that are within ~100 Kb of either side of the marker. The closer they are to your intended gene, the more likely they will also be transduced from the donor and replace the copy in the recipient (see Note 1).
4. Ensure that the restriction/modification systems of the recipient and donor are compatible and that both strains are *recA*+ (see Notes 2 and 3).

3.2. Donor Lysate Preparation

1. Grow a small (~1 mL) overnight culture with selection for the marker in LB. For each lysate to be made, prepare 1 mL of P1-LB. Use the overnight donor culture to inoculate a fresh 1 mL culture LACKING SELECTION at a 1/100 dilution in a clear culture tube (see Note 4). Grow the bacteria with aeration at a suitable temperature (keeping any cold-sensitive or temperature-sensitive phenotypes in mind). This culture will become the phage lysate used for the transduction infection and the presence of antibiotics will inhibit the necessary growth of the recipient and the transduction may not work at all.
2. When the culture reaches early log phase (light, but noticeable turbidity, $\sim 2 \times 10^8$ cells per mL), add 40 μL of a preexisting phage stock with a confirmed transduction activity (generally 10^9–10^{10} plaque-forming units per mL, see Note 5). Mix the tubes immediately after adding the phage to allow for an even dispersion and return the tubes to the incubated shaker. While not critical, as a precaution we make an effort to avoid using a phage stock that contains the same selectable marker as the donor strain. This avoids the remote possibility

of inadvertently transducing the wrong marker. Make sure to not pipette any chloroform into the culture. When performing transductions for the first few times, include the following controls: a mock culture (no donor cells that receive 40 μL of phage) and a mock infection (a culture that does not receive phage).

3. Allow the infection to proceed until there is evidence of lysis (1–3 h). In some cases, the culture will turn completely clear, in others there will be noticeable aggregated debris that is visible when the tube is held with backlighting. If the culture keeps growing without evidence of lysis, either too little phage has been used or the culture has been infected too late or the host is not able to support P1 replication. If the culture stops growing and maintains the turbidity at the time of infection, too much phage has been used (see Subheading 2.5 and Note 5).
4. Add enough chloroform to the tube to ensure saturation (50–100 μL). Mix the sample well and let stand for 5–10 min. Transfer the samples to a microfuge tube and centrifuge to pellet debris and any unlysed cells (5 min at maximum *g* in a microfuge is more than enough).
5. Transfer the cleared supernatant to a clean microfuge tube, add 50 μL of chloroform to the tube and close it. Label with the donor strain, the relevant marker, and the date. Store the lysate at 4°C. Over time, the chloroform will evaporate from the tube; the lysate will still be sterile.

3.3. Transduction: Phage Infection, Recovery, and Selection

1. Prepare the recipient strain by growing a 1 mL overnight culture in LB. Transfer the culture to a microfuge tube and harvest the cells at ~3,000 × *g* for ~3 min. Aspirate the growth medium and resuspend the cells in P1-LB. Some protocols call for a resuspension at the same volume as the amount harvested, but a more-concentrated culture will generally perform better (see Note 5). Therefore, resuspend the cells in a 200–500 μL volume; 100 μL is needed for each transduction. It helps to have an extra aliquot available in case of a mistake when adding phage.
2. For each transduction, transfer a 100 μL aliquot of phage stock to a sterile microfuge tube, leave the cap open, and place the tube in a 37°C incubator for 15–30 min. This step allows the chloroform in the phage stock to evaporate to less-than-saturated. It is not essential, but reduces the chance of killing all of the recipient cells before transduction.
3. Add 100 μL of cells to the 100 μL phage aliquot and rapidly mix with the pipettor to allow an evenly distributed infection. Close the tube and briefly vortex. Transfer the tubes to a 37°C

incubator (unless the recipient is temperature-sensitive). A convenient shaking method is to drop all of the tubes into a flask and shake them en masse in an air incubator. The infection has now started: the phage particles adhere to the cells, inject their DNA, and because of the *vir* mutation, enter the lytic cycle.

4. 30 min after starting the infection, remove the tubes from the incubator, briefly spin the samples to collect the culture away from the cap, open each in a rack and add 200 μL of 1 M sodium citrate, pH 5.5. On top of this, add 1 mL of fresh LB (without calcium), close the tubes, and return them to the shaker. This step chelates the free calcium that is required for an efficient P1 infection. New infections will now be inhibited and only previously infected cells will generate new phage (see Note 5).
5. Incubate the infected cells long enough to allow recombination and expression of the transduced marker to selectable levels, generally 1 h at 37°C or 2 h at 30°C. Bear in mind that if your selection requires the depletion of an existing gene product in the recipient cell, a longer incubation may be required to allow survival under selective conditions. During this time, the majority of cells infected with viable P1 phage (nontransducing) will lyse, releasing many progeny phage. The citrate prevents their killing of the cells infected with transducing particles. You will have generated another high-titer phage stock mixed with living bacteria that either were not infected or received nonphage DNA.
6. Transfer the tubes to a microfuge and collect intact cells by centrifuging at 5,000 × *g* for 5 min. During the spin, prepare LB-citrate broth; 100 μL per sample is required.
7. Remove the supernatant using a 1 mL pipettor taking care not to disturb the cell/debris pellet and discard the medium. Use of an aspirator here can cause the accidental removal a large portion of the cell pellet because the lysed cell genomic DNA forms a viscous mesh that is integrated with the pelleted material.
8. Optional: If selection will be carried out for a nutrient, it is advisable to resuspend the cells in 1 mL of PBS supplemented with 5 mM sodium citrate, pH 5.5 and reharvest them to remove residual nutrients of the LB.
9. Add 100 of LB-citrate (or PBS with citrate) to each tube, cap, and vortex to resuspend the cells. Depending on the dynamics of the infection, the cells may readily redisperse or remain clumped. Make an effort to disperse the cells fully if you need to count the transductant colonies. Presence of a few clumps will cluster the transductants on the plate, which is not usually a concern.

10. Transfer the entire contents of the tube to a selective plate and distribute the sample with a spreader. The mixture can sometimes be viscous from remaining genomic DNA and slow to absorb. Incubate the plates to allow colony growth. Generally, anywhere from 10 to 500 colonies should form depending on the location of the transduced marker, the health of the cells, and the proportions of phage to bacteria during the infection.

3.4. Purification of the Transductant and Confirmation

1. The plate containing the transductant colonies is rife with free P1 phage and potentially genomic DNA fragments from the donor. Therefore, directly screening the transduction marker using PCR can cause false-positives. The resulting colonies should be secondarily restreaked for colony isolation on a selective plate containing 1–5 mM citrate. Generally, four colonies of each transduction plate are transferred with a toothpick to a quadrant of a new plate and a wooden stick or inoculation loop used to streak the cells for isolation. Allow the cells to form colonies overnight. Skipping this step can result in P1 contamination of the transductant, which may only become evident in subsequent outgrowth experiments.
2. The next day, pipette 5 μL of colony resuspension buffer into sterile 0.5 mL snap-cap tubes, one for each restreaked strain.
3. Using a sterile pipette tip, acquire cells from the center of a colony from each of the restreaked quadrants and mix the cells into the buffer.
4. Use 1 μL of this cell suspension as a template in a PCR reaction designed to identify the transduced marker. It is better to have primers that will yield a diagnostic PCR product for both the parental and transduced gene locus. High-fidelity enzymes do not generally perform well when screening intact bacteria; Taq polymerase is recommended. If the PCR product is to be sequenced, a larger reaction (25–50 μL) will provide enough material for subsequent gel purification.
5. Add 100 μL of LB to the remainder of each resuspension and store the cells at 4°C.
6. Use half of the cell suspension to grow a liquid culture under selection for the new marker from a PCR-positive strain. Prepare freezer stocks of the strain (see Note 6).

3.5. Troubleshooting: Plaque Formation and Titering

1. Plaque formation on the donor and recipient hosts requires multiple rounds on infection and lysis by the phage and is a clear indicator of the ability of the phage to replicate. Additionally, the titer of the virus can be determined to optimize the infection protocol so that bacteria are infected with no more than one phage on average (see Note 7). Prepare a mid-log phage culture of each strain, ice the culture and

harvest the bacteria at 4°C. Resuspend the cells at 1/10th the harvested culture volume in ice-cold LB. These preparations should last for several days at 4°C.

2. Distribute 3–4 mLs of molten P1 top agar into sterile glass tubes with caps in a prewarmed 45°C block. ALLOW TO COOL to 45°C. Hot agar will kill a percentage of the bacteria and can reduce the apparent phage titer substantially.
3. Add 100 μL of the log-phase bacterial stock to a tube, vortex with care taken not to splash the agar out of the tube, and pour the contents of the tube onto an LB plate. Immediately tilt the plate to create an even distribution of the agar. Let cool and solidify at room temperature. The surface should be smooth and shiny.
4. Prepare serial dilutions of a phage stock in P1-LB. A reasonable dilution series can be made by pipetting 990 μL of P1-LB into each of two tubes and 90 μL of broth into each of three (5 total). Add 10 μL of the phage stock to the first tube (10^{-2} dilution), mix well, and transfer 10 μL to the next tube (4). Transferring 10 μL serially to each of the remaining tubes with 90 μL of broth will generate 10^{-5}, 10^{-6}, and 10^{-7} dilutions.
5. Invert the plates containing the bacteria-laden top agar. Mark a row of four dots across the plate spaced approximately 2 cm apart and at least 1 cm from the plate edge. If titering several phage, several rows can be made. At most 4–5 rows of four dots each should be made per plate. Label above the columns to indicate the dilution to be spotted (e.g., "10^{-4}", "10^{-5}" ... or just "4", "5" ...). Label each row with a brief name for the phage stock being titered. There is no need to spot the 10^{-2} dilution.
6. Place the plates upright and transfer 10 μL aliquots of each phage sample into the marked dot corresponding to the phage and dilution. To form an even plaque distribution, partially form a small drop from the tip of the pipettor and carefully lower the drop on the surface. Once contact is made, smoothly deliver the remaining liquid into the drop. It will spread to ~0.75 cm. Be careful not to splash sample when ejecting the rest of the liquid. Repeat the spotting for each phage dilution. Do not disturb the plate until the drops have completely soaked in.
7. Incubate the plate overnight at an appropriate temperature.
8. Count the plaques the next morning. The spot technique has a variability of ~20% for any given spot, which is usually accurate enough to establish a usable estimate of the phage count. If more accurate titers are needed, multiple spots can be prepared for a given dilution and the numbers averaged. Plaque quality and apparent titer can be heavily influenced by the quality of the bacterial lawn (see Note 8).

9. Back-calculate the phage titer per mL. For example, 5 plaques on the 10^{-7} spot equates to 5×10^9 plaque-forming units per mL.
10. Adjust either the amount of bacteria or phage such that the number of phage per bacterium in the transduction step is between 0.1 and 1. Even at a multiplicity of infection of one, a substantial number of bacteria will receive two phages and be killed (the chance that the two phages are both transducing particles is fleetingly small).

3.6. Troubleshooting: Cross-streaking

1. A simple and rapid method to determine the susceptibility of the bacteria to be infected and support replication is a cross-streak. This method does not allow virus titer to be determined but is a fast way to check a strain to ensure that it can be infected. Mark a straight line across the back of a P1 streak plate. Prepare a phage line by pipetting 25–50 μL of a phage stock to the edge surface of a plate held at an angle. Using the nonpipetting hand (holding the plate), gently control the tilt angle of the plate as the phage is delivered to let a narrow drip line form in a straight line across the surface of the plate following the marked line. As the drop reaches the opposite end, reverse the tilt angle and evenly distribute the phage in the drip line. Let the phage soak into the plate.
2. Invert the plate and label starting points for the cross streak perpendicular to the phage line. These can be spaced ~1 cm apart if multiple strains are being tested. Place the plate upright again. Include a strain that is known to be susceptible to P1.
3. Using a sterile, flat toothpick, gently touch the top of a colony of test bacteria and, with the toothpick held at a 45° angle, gently drag the toothpick from the labeled portion of the plate through the phage line and continue almost to the other side of the plate. Barely any pressure needs to be applied, and just a slight scratch should appear.
4. Incubate the plate overnight.
5. Read the streaks. A fully susceptible strain that supports robust phage replication will not grow past the phage line. A partially susceptible strain will form a spotted streak past the line as will bacteria that become lysogens (this should not happen with P1*vir*). A reduction in the width or quality of the bacterial streak as it crosses the phage line may indicate that the bacteria can be infected, but that it does not support replication well. Using a phage stock with a low titer can also cause colonies to form past the phage line. Bacteria that show no evidence of phage infection are immune.

4. Notes

1. For the construction of a new strain of *E. coli* by transduction, the investigator needs to be cognizant of other relevant genetic markers that are in the vicinity of the marker to be transduced. If there is a need to transduce a gene into a strain that has another important marker within ~90 Kb of the transduced gene, then some method of either selecting or screening for a strain with both markers is required. Because the P1 transducing particle will inject a randomly packaged ~110 Kb fragment of DNA from the donor strain and the transduced marker that is being selected can be positioned anywhere along the strand, there will be a variety of recombination events that can give rise to a recombinant colony in the population that is selected (Fig. 1).

   While this process may appear to be an inconvenience, in reality this feature makes the transduction procedure a very powerful tool. Phenotypic variance in the resulting transductants will reveal that a mutant gene near the transduced marker is potentially affecting a pathway related to the transduced gene.

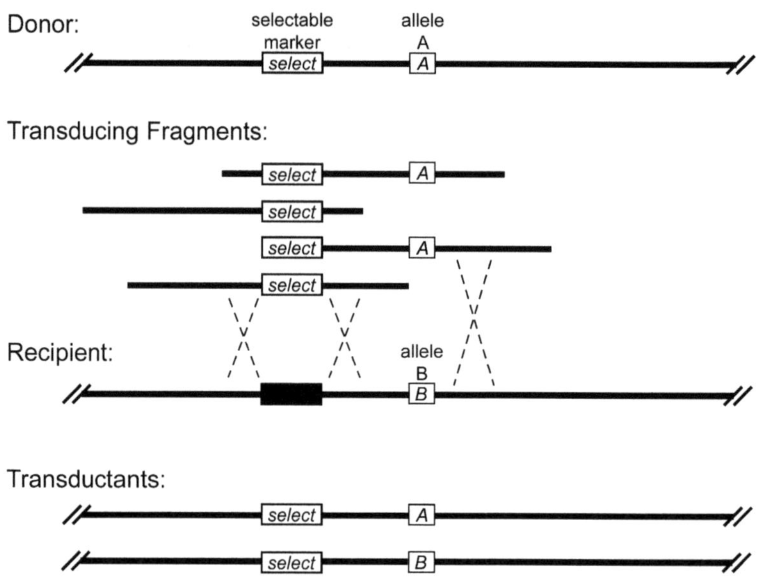

Fig. 1. Schematic illustrating a variable genetic outcome from a single transduction experiment. A selectable gene (*select*) in the donor chromosome (*black line*) will be packaged into the transducing particle population with a wide variety of flanking genetic material. Upon recombination with the recipient chromosome, the resulting strain may have either allele A (*A*) or allele B (*B*), as well as the selected marker. Thus, in a single transduction experiment each resulting colony arises from a double crossover event, and each will have a random amount of donor DNA replacing the recipient DNA in the vicinity of the selected marker.

2. Be aware of the restriction systems present in the donor and recipient. If the donor DNA is not modified in accord with the recipient restriction system, the injected DNA will be cleaved, significantly reducing the transduction efficiency. It may be necessary to first move the marker into a suitable *restriction*−, *modification*+ strain before moving it into the final recipient.
3. P1 requires RecA for efficient replication and RecA is required for the homologous recombination needed to incorporate the marker in the recipient chromosome (9, 10). Unfortunately, there are many situations where the intended recipient cell line is *recA*⁻, for example host strains used for the stable maintenance of a plasmid with homology to the chromosome. We circumvent this limitation using either a donor strain containing a transducible *recA* deletion or a support plasmid supplying RecA for P1 replication (BW26547, CGSC #7652) [11]. After transducing desired markers into a recipient, the resulting strain is subsequently transduced to *recA::kan*. The cellular dose of RecA suffices for the recombination event before the gene disruption.
4. Improper reagent preparation is another common cause of failed transductions. Double-check that the strains were grown at permissive temperatures, that the citrate is adjusted to pH 5.5 (and not straight citric acid), that there is no preservative in a stock solution (e.g., azide or EDTA), and that the plates have the correct amount of antibiotic (tenfold errors are common with students). Also, another all-too-common mistake is to include an antibiotic in the culture when preparing the donor lysate. The antibiotic will stop the growth of the recipient when mixed with the donor phage and prevent recovery and expression of the transduced marker.
5. The titer of the phage relative to the concentration of bacteria is an important factor both in generating a high-titer transducing lysate and in the efficiency of transduction. A reasonable stock of P1 may have ~5×10^9 plaque-forming units (pfu) per mL. Adding 40 μL of such a stock to a milliliter of an early log phase culture of donor strain (~2×10^8 per mL) as outlined in the protocol to generate a transducing lysate will place the multiplicity of infection (MOI) right around 1 phage per bacterium. Phage distributions on the bacteria are governed by a Poisson distribution (12), so with an MOI of 1, ~37% of the bacteria remain uninfected, ~37% of the bacteria get a single infection, and the rest of the bacteria receive more than one phage. After ~45 min of infection with P1*vir*, the infected cells will lyse, liberating ~100 phage per infected cell (13). These liberated phage spread to infect any remaining uninfected cells and, ~45 min later, all cells in the

tube should have been infected and most will have lysed. Remaining intact cells were most likely infected with too many phages to remain viable.

Counterintuitively, adding the same volume of a lower titer phage stock (older perhaps) can yield a higher titer in the final lysate. This occurs when a small percentage of cells gets productively infected while the uninfected cells continue to divide and become hosts for a massive lysis event. With an MOI of 0.1, about 90% of the cells remain uninfected, after 45 min, the uninfected cell count may be ~5×10^8 per mL, which will then become completely infected by the ~2×10^9 phage liberated from the first infection cycle. Each of the cells from this secondary infection then liberates 100 phage particles, driving the virus titer well over 10^{10} per mL.

During the transduction step, 100 μL of phage is added to 100 μL of recipient cells. Here, having too high of an MOI will multiply-infect a large fraction of the cells and prevent transduction because the cells will most likely have received an infectious particle regardless of whether a transducing particle coinfected. For example, a 5×10^9 per mL phage stock mixed with a stationary culture of *E. coli* at ~1×10^9 cells per mL will yield an MOI of 5, which reduces the number of singly infected cells to less than 5%. An easy precaution to avoid this situation is to concentrate the recipient cells by resuspending them in less volume than the harvested overnight culture. Doing so will not reduced the number of transducing particles, but will increase the chance that they infect a cell by themselves. Typically, rather than titer each phage stock, we concentrate the overnight recipient cells to 3–5 times their original volume. Doing so may leave a larger proportion uninfected, but will increase the number of transduction events.

6. P1 or other contamination in the transductant stock will cause subsequent out-growths to either lyse or yield highly variable results. Stocks can be "cleaned" by reisolating single colonies on a citrate plate.
7. For the reasons outlined in Note 5, adding the phage too early or too late during lysate preparation can dramatically reduce the lysate titer. Additionally, working with donor strains that grow slowly or that do not fully support P1 replication will also affect the resulting titer. In these cases, when transduction efficiencies are very low, titering the phage stock and adjusting the ratios of phage to cells can usually allow transduction. Target an MOI of 1 for the lysate preparation and an MOI of 0.1–0.5 for the transduction step.
8. If the plaques are too small, the amount of bacteria added to the top agar can be reduced to lengthen the time it takes for the bacteria to reach saturation. Once the bacteria in the top

layer have grown to a stationary lawn, the P1 plaques will not develop much further because the bacteria are not actively growing. If the lawn appears grainy or speckled, either too few bacteria were added or they had died before the agar was poured. Do not let the bacteria/top-agar solution sit for extended periods of time before the layer is poured.

References

1. Zinder N.D., and Lederberg J. (1952) Genetic exchange in *Salmonella*. *J Bacteriol.* **64**, 679–699.
2. Lennox E.S. (1955) Transduction of linked genetic characters of the host by bacteriophage P1. *Virology* **1**, 190–206.
3. Adams J.N., and Luria S.E. (1958) Transduction by bacteriophage P1: Abnormal phage function of the transducing particles. *Proc Natl Acad Sci U.S.A.* **44**, 590–594.
4. Coren J.S., Pierce J.C., and Sternberg N. (1995) Headful packaging revisited: the packaging of more than one DNA molecule into a bacteriophage P1 head. *J Mol Biol* **249**, 176–184.
5. Ikeda H., and Tomizawa J.I. (1965) Transducing fragments in generalized transduction by phage P1. I. Molecular origin of the fragments. *J Mol Biol* **14**, 85–109.
6. Bertani G. (1951) Studies on lysogenesis. I. The mode of phage liberation by lysogenic *Escherichia coli*. *J Bacteriol* **62**, 293–300.
7. Neidhardt F.C., Bloch P.L., and Smith D.F. (1974) Culture medium for enterobacteria. *J Bacteriol* **119**, 736–747.
8. Xu M., Arulandu A., Struck D.K., Swanson S., Sacchettini J.C., and Young R. (2005) Disulfide isomerization after membrane release of its SAR domain activates P1 lysozyme. *Science* **307**, 113–117.
9. Lusetti S.L., and Cox M.M. (2002) The bacterial RecA protein and the recombinational DNA repair of stalled replication forks. *Annu Rev Biochem* **71**, 71–100.
10. Zabrovitz S., Segev N., and Cohen G. (1977) Growth of bacteriophage P1 in recombination-deficient hosts of *Escherichia coli*. *Virology* **80**, 233–248.
11. Datsenko K.A., and Wanner B.L. (2000) One-step inactivation of chromosomal genes in Escherichia coli K-12 using PCR products. *Proc Natl Acad Sci U.S.A.* **97**, 6640–6645.
12. Ellis E.L., and Delbrück M. (1939) The growth of bacteriophage. *J Gen Physiol* **22**, 365–384.
13. Walker J.T., and Walker D.H. (1980) Mutations in coliphage P1 affecting host cell lysis. *J Virol* **35**, 519–530.

Part II

Saccharomyces cerevisiae

Chapter 11

Yeast Bioinformatics and Strain Engineering Resources

Audrey L. Atkin

Abstract

Saccharomyces cerevisiae, commonly known as baker's or budding yeast, is an attractive organism for design-based engineering because it is an industrially important organism with a well-annotated genome sequence and an extensive collection of resources for molecular analyses. This chapter describes the utility of Saccharomyces Genome Database for analysis of *S. cerevisiae* genes and identification of homologs, strategies for integration and analysis of gene expression data, and the genetic resources available for doing experiments using *S. cerevisiae*.

Key words: Yeast bioinformatics tools, Yeast genome resources, Yeast homologs, Computational prediction of promoter structure, Primary mRNA sequence, Yeast genetic resources

1. Introduction

The yeast *Saccharomyces cerevisiae* is an attractive organism for design-based strain engineering. This organism, commonly known as baker's or budding yeast, is an industrially important organism. For example, it is used in the food industry for brewing, winemaking, and production of bread and in the bioenergy industry for production of ethanol. As a consequence, there are well-established large-scale production methods utilizing this organism. In addition, it is also a model eukaryotic genetic organism. Its genome was the first eukaryotic genome completely sequenced and assembled. The yeast genome is also well annotated. The genome sequence is available from the Saccharomyces Genome Database (SGD) (1). SGD also provides information about the genes and their biological functions, and resources and tools for exploring this data (2). In addition, there is substantial knowledge of yeast gene expression from experimental studies. For example, many yeast transcription factors and their binding

James A. Williams (ed.), *Strain Engineering: Methods and Protocols*, Methods in Molecular Biology, vol. 765,
DOI 10.1007/978-1-61779-197-0_11, © Springer Science+Business Media, LLC 2011

sites are known. There is also information about mRNA sequence from a number of genome-wide studies. Integration of information from SGD with data from gene expression studies is a powerful tool for modeling gene expression.

The bioinformatics resources for *S. cerevisiae* are complemented by simple and reliable genetic methods for manipulation of its genome (3–5). There is an extensive collection of genetic resources available for experiments using yeast. These resources include strain collections, plasmids, and antibodies. Together, this combination of tools, resources, and knowledge makes *S. cerevisiae* a powerful resource for rationale-based strain engineering.

This chapter describes how to search for and analyze *S. cerevisiae* genes using the Saccharomyces Genome Database (SGD), how to identify homologs, tools and strategies for integration of data on gene expression, and the genetic resources available for doing experiments using *S. cerevisiae*.

2. Materials

Broadband Internet connection and Web browser software such as Safari 3.0, Firefox 3.0, Internet Explorer 7, or higher ones.

3. Methods

3.1. Search and Analysis of S. cerevisiae Genes Using the Saccharomyces Genome Database

The Saccharomyces Genome Database (SGD; http://www.yeastgenome.org/) is a public resource with a collection of data and tools for genetic and proteomic analyses of *S. cerevisiae*. This database provides access to the sequences and annotations for *S. cerevisiae*. Annotated *S. cerevisiae* genes can be found using the known standard name, an alias, or systematic name for particular genes to search the database. Alternatively, the database can be queried with keywords such as function, mutant phenotype, and interactions. Yeast homologs for proteins or genes of interest can be found by BLAST searches of *S. cerevisiae* sequence datasets. This database is maintained and updated by SGD curators. The SGD also maintains the *S. cerevisiae* Gene Name Registry, a complete list of all gene names used in *S. cerevisiae*. The data stored in SGD can be searched or downloaded. SGD contains links to other yeast information sources, such as the Community wiki page and the Biosci Yeast Archive. The Community wiki has community-edited information about *S. cerevisiae* genes, gene products, and resources, and the Biosci Yeast Archive is a public newsgroup that is intended to promote communication between researchers worldwide.

There are currently 7,031 *S. cerevisiae* annotated genes in the SGD database. Basic information, a gene summary with references, and resources for each annotated gene or chromosomal feature (e.g., centromere) are organized on locus summary pages (Fig. 1). Basic information provided include (1) standard name, systemic name, alias, feature type and description, and name description, (2) gene ontology annotations, (3) pathways, (4) phenotypes of strains with mutations in the gene, (5) interactions, (6) sequence information, and (7) posttranslational modifications. The gene ontology annotations are controlled vocabulary terms for describing gene product characteristics. Much of the information summarized in the basic information section is hyperlinked to detailed information with references. More detailed information relevant to each locus can also be accessed using the tabs at the top of the page. Resources include the following: (1) tools for retrieval of chromosomal and genetic maps in the region of the locus, (2) access to relevant literature, (3) tools for retrieval and analysis of the gene and protein sequences, (4) protein information and structure, (5) localization and phenotype resources, (6) interacting genes or proteins, (7) tools for comparison of the protein sequence to proteins of other budding yeast species, *C. elegans*, and mammals, and (8) links to functional analyses. The functional analyses link to gene expression data from individual genome-wide studies. Alternatively, the Expression Connection Summary enables the simultaneous retrieval of *S. cerevisiae* gene expression data for a given locus from a large number of published experiments.

The following exercise is an example of the database capabilities for finding information about annotated genes. In this exercise, we use SGD to find the locus summary page for *PGA1*. Next, we use the resources and basic information on the PGA1 locus summary page to identify the genes encoding proteins that have the same molecular function as Pga1p, and find the protein that is in the same complex as Pga1p. Finally, we link to the locus page for this second protein and from there find homologs in other model organisms.

3.1.1. Basic Search for Information on the S. cerevisiae Gene PGA1

1. Access the SGD home page at http://www.yeastgenome.org/.
2. Enter PGA1 (see Note 1) into the basic search text box at the top left of the page (see Note 2). Click the submit button to link to the PGA1 locus summary page. The resulting "PGA1/YNL158W Summary" page is shown in Fig. 1.

3.1.2. Find Gene Products with the Same Molecular Function as PGA1

1. Begin on the "PGA1/YNL158W Summary" page and scroll down to the section labeled "Molecular Function."
2. Click on the link labeled "mannosyltransferase activity" to view the genes encoding proteins with the same activity.

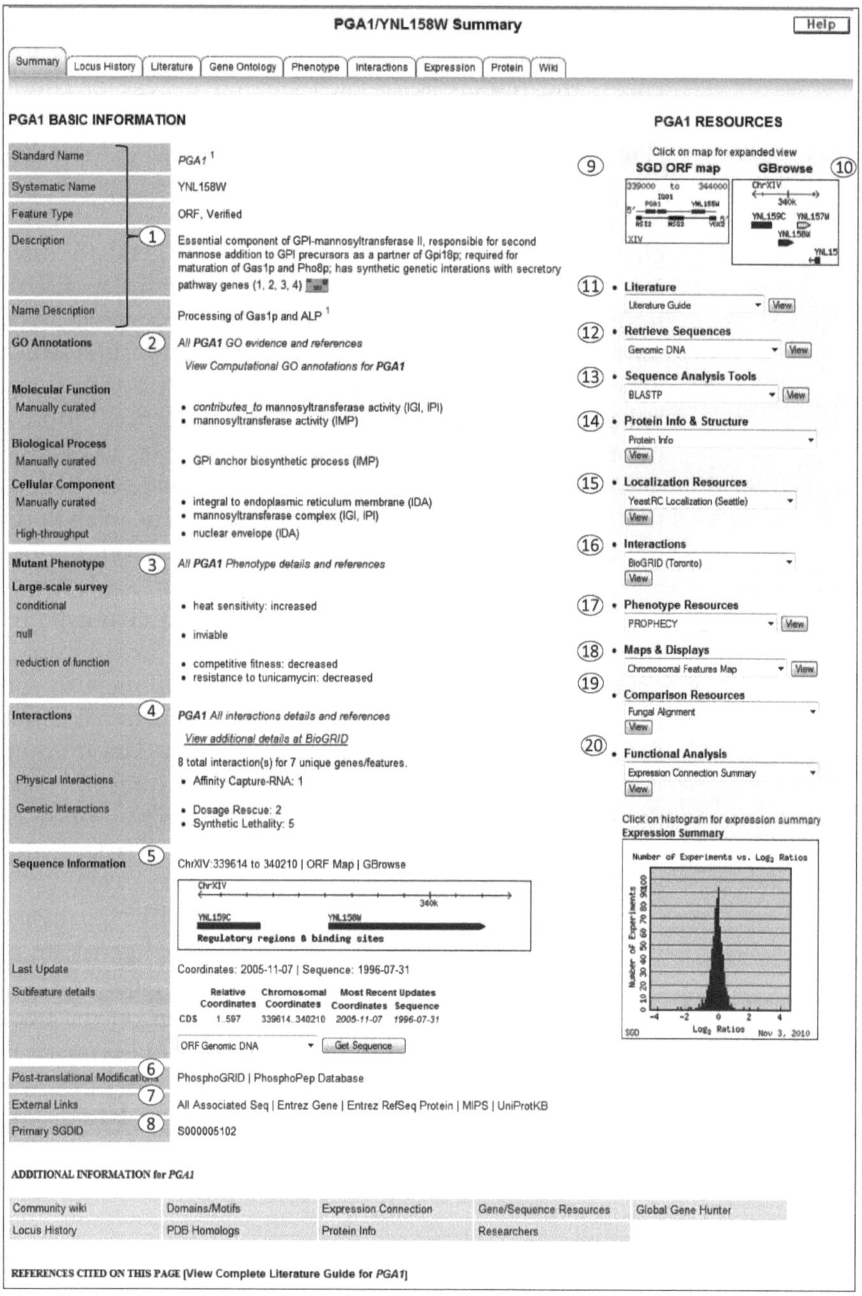

Fig. 1. SGD's Summary page for PGA1. The left-hand column of all Summary pages lists the standard name, the systemic name, an alias (missing if there are no aliases as in this case), feature type and description, and name description (1). This column also has the GO Annotations (2), Mutant Phenotype (3), Interactions (4) Sequence Information, Posttranslational Modifications (6), External Links (7), and Primary SGDID (8). This bar serves as a navigation tool for information about the locus. The right column on each Summary page provides links to resources for the locus. The resources include tools for retrieval of chromosome (9) and genetic maps (10) in the region of the locus, access to literature (11), tools for sequence retrieval (12) and analysis (13), protein information and structure (14), localization (15), interactions (16), phenotypes (17), maps and displays (18), tools for comparison of sequences to other organisms (19), and links to functional analyses (20).

3. At the top of the page, find the definition for mannosyltransferase activity.
4. Scroll down to the table of "Genes Annotated with this Term." Examine the loci that have been annotated with this term, the annotation method, reference, and who is credited with assigning the term.
5. Scroll back to the top of the page and click on the view ontology "GO tree view" icon. This will open a page in a new window with the network of terms related to "mannosyltransferase activity."
6. Return to the "PGA1/YNL158W Summary" page and scroll down to the section labeled "Cellular Component."
7. Click on the link labeled "mannosyltransferase complex" to view the other gene that encodes the protein that functions with Pga1.
8. Scroll down to the table of "Genes Annotated with this Term". Examine the loci that have been annotated with this term, the annotation method, reference, and who is credited with assigning the term.

3.1.3. Search for Homologs of GPI18, Which Encodes a Mannosyltransferase Complex Protein, in Other Model Organisms

1. Beginning on the "GO term: mannosyltransferase complex" page at the table of "Genes Annotated with this Term" click on "GPI18/YBR004C" to link to the "GPI18/YBR004C Summary" page.
2. Scroll down to the "Comparison Resources" pull-down menu in the resources section on the right side of the page.
3. Click on the small arrow to the right of the pull-down menu.
4. Select "Ortholog Search (P-POD)" (see Note 3) and click the button labeled "view" to see an alignment between Gpi18 protein from *S. cerevisiae* and other *Saccharomyces* species.
5. Click on the box under "Ortholog Identification (OrthoMCL 2.0b6)" to view a phylogenetic tree and summary table of the homologs in model organisms.
6. Click on "Jalview Alignment" to see a color coded multiple sequence alignment of the proteins.
7. Return to the summary and click on a link in the "Protein (Synonyms)" column in the "A. thaliana" row of the summary table to link to detailed information about that protein.

3.2. Identification of Homologs in S. cerevisiae

You may also start with a gene of interest from another organism and search for *S. cerevisiae* homologs using a BLAST search strategy. The WU-BLAST2 tool is available though SGD. WU-BLAST2 stands for Washington University Basic Local Alignment Search Tool Version 2.0 (6). The emphasis of this tool is to find regions of protein or nucleotide sequence similarity quickly and reliably with minimum loss of sensitivity.

1. A query sequence can be used to search for *S. cerevisiae* homologs by either uploading a local TEXT file, or putting a query sequence into the textbox on the BLAST Web page (http://www.yeastgenome.org/cgi-bin/blast-sgd.pl). The query sequence can be either a nucleotide or protein sequence, in a FASTA, GCG, and RAW sequence format.
2. Choose the appropriate BLAST program or search mode to run. Five search modes in the WU-BLAST suite are available: blastp, blastn, blastx, tblastn, and tblastx. For example, choose blastp to compare an amino acid query sequence against the *S. cerevisiae* protein sequence database or blastn to compare a nucleotide query sequence against the *S. cerevisiae* nucleotide sequence database.
3. Choose one or more sequence datasets to search.
4. Select Run WU-BLAST (see Note 4).

Once a *S. cerevisiae* gene of interest has been identified, SGD has a set of essential sequence analysis tools for the design of experiments using the gene. With these sequence analysis tools, you can determine the encoded protein sequence, design primers, make restriction fragment maps, and determine restriction fragment sizes.

3.3. Gene Expression

Gene expression involves transcription and RNA processing to produce a mature mRNA that is transported to the cytoplasm where it is translated, stored, or degraded. Each one of these steps can be regulated to ensure the correct amount of gene product is synthesized only when it is needed. Expression of genes involved in the same or related processes is often coordinately regulated. This coordinate regulation can be at any individual step, or combination of the steps in gene expression. For example, transcription is regulated by the combination of transcription factors that bind to each promoter. For this reason, it is often important to determine what transcription factors regulate a gene. Functional genomics studies have resulted in identification of the sequences that many transcription factors bind. Some of these have been mapped onto the yeast genome and are available from SGD. For others, the information is available to make computational predictions. Together, reasonable models of the functional transcription factor binding sites in promoter regions can now be constructed.

Gene expression is also regulated posttranscriptionally. Posttranscriptional regulation in yeast is simpler than other model eukaryotes because *S. cerevisiae* does not have a functional RNA interference pathway. As a consequence, the majority of the posttranscriptional regulation is likely carried out by RNA binding proteins. Many of known RNA binding sites are located within 3′-UTRs. Some proteins bind sites within the open reading frame, and still others tend to bind sites in the 5′-UTR. Thus, to study posttranscriptional regulation, it is important to determine the primary sequence of mRNAs. Recent analyses of the yeast transcriptome have provided the information needed to predict primary mRNA structures of many yeast mRNAs.

We use a combination of available information and computational tools to predict promoter structure and primary mRNA structure for *HIS4* as an example of strategies for integration and analysis of yeast gene expression data.

3.3.1. Computational Prediction of S. cerevisiae Promoter Structure

Promoters consist of two parts: a core that binds the basal transcription apparatus, and transcription factor binding sites. These sites bind transcription factors that activate or repress transcription either directly by affecting RNA polymerase function or by modifying chromatin. Most genes are regulated by multiple transcription factors. Many transcription factors are known in yeast, and for many of these, the binding sites have been discovered. Yeast promoters are unique for model eukaryotes because they are compact and located exclusively upstream of transcription start sites. The transcription factor binding sites in yeast promoters typically are positioned within the ~600 bp upstream of the ATG and rarely beyond 1,000 bp. This compact organization of yeast promoters and available information about transcription binding sites makes computational prediction of promoter structure possible.

In this exercise, we use a three-step process to identify known functional transcription factor binding sites within the promoter region for the *HIS4* gene. First, we find the known transcription factor binding sites that have been annotated in SGD. Second, we use *Saccharomyces cerevisiae* Promoter Database (SCPD; http://rulai.cshl.edu/SCPD/) to search for putative binding sites of characterized transcription factors in the *HIS4* promoter region. Third, we use a comparative genetic approach to determine which potential binding sites are conserved in closely related yeast species.

Find known transcription factor binding sites that have been annotated in SGD.

1. Begin on the GBrowes map for HIS4/YCL030C.
2. Make sure "All on" is selected for the Regulatory regions and binding sites option.
3. Get the chromosomal coordinates for known transcription factor binding sites located upstream of the *HIS4* ORF by

mousing over the orange diamonds and the ORF by mousing over the red arrow for HIS4/YCL030C.

Identify potential binding sites for known transcription factors.

1. Retrieve the promoter region sequence for *HIS4* by going to The Promoter Database of *Saccharomyces cerevisiae* (SCPD; http://rulai.cshl.edu/SCPD/) (see Note 5). Click on the "Retrieve promoter sequences" link. Put HIS4 in the Gene names or ORFs box and reset the Upstream from Start codon ATG to −1,000 and the Downstream from start codon ATG to +3. Click the "submit" button. Copy the sequence for the *HIS4* promoter region.
2. Return to the SCPD home page and click on the "Search for predefined putative regulatory elements using matrix and consensus."
3. Paste the *HIS4* promoter region sequence into the box and click the "submit" button. You will retrieve a tab-delimited dataset of the potential transcription factor binding sites in the *HIS4* promoter region.

Comparative genetic analysis to identify conserved potential binding sites.

1. Align the sequence of the *HIS4* promoter region from *S. cerevisiae* with homolog from closely related yeast strains by going to the Comparison Resources on the "HIS4/YCL030C Summary" page. Select Fungal Alignment from the drop-down menu and click on the "View" button.
2. Select two or more sequences for alignment from the left box and then select "Upstream sequence" from the Pick a sequence type box. Click the "Align" button.
3. Determine which putative transcription factor binding sites are conserved by mapping the putative sites retrieved from SGD and SCPD onto the aligned sequences. Binding sites that are conserved are more likely to be functional (7).

Support for predicted promoter sequences can be obtained experimentally by testing for the potential interaction by chromatin immunoprecipitation to confirm that the predicted transcription factor(s) really bind the promoter region, and by testing for changes in transcript levels in transcription factor mutants. Additional support can come from coexpression of genes that are predicted to be regulated by the same transcription factor(s). Find genes that are potentially coregulated by going to the summary pages of the potential transcription factors and scroll down to the Regulatory Role section. Click on the "Predicted Binding Site Locations" to find other sites that the transcription regulator binds. The expression pattern for the potentially coregulated genes can be found by using the Expression Connection tool under Functional Analysis on the Summary page. The Expression Connection tool provides access to many published genome-wide expression studies. These studies can be queried to identify genes whose expression is coordinated with a query gene or ORF.

3.3.2. Strategy to Predict Primary Sequences for S. cerevisiae mRNAs

Primary mRNA sequence is defined here as the entire sequence of the mature mRNA encompassing the 5′-UTR, ORF, and 3′-UTR. Primary mRNA sequences can be predicted by combining information from genome-wide analysis of mRNA length, global identification of transcription start sites, open reading frame (ORF) lengths, and prediction of 3′-end processing sites (8) (Fig. 2). In this exercise, we use this information to predict the primary mRNA sequence of *HIS4* mRNA.

1. Begin on the "HIS4/YCL030C Summary."
2. Scroll down the page to the Sequence information. Determine the ORF length from the relative coordinates, and note the chromosomal coordinates (see Note 6).
3. Scroll to the top of the page and click on the GBrowes map.
4. Reset the scroll/zoom to "show 5 Kb" to zoom in on the *HIS4* chromosomal region (see Note 7).
5. Scroll down the page to Regulatory regions and binding sites and click the "All on" box. Then, click the "All on" box for the Transcript information. Click "update image."
6. Get the chromosomal coordinates for known transcription start sites mapped by Muira et al. (9) and Zhang and Dietrich (10) by mousing over the horizontal orange arrows and vertical orange arrows, respectively. Determine the 5′-UTR lengths

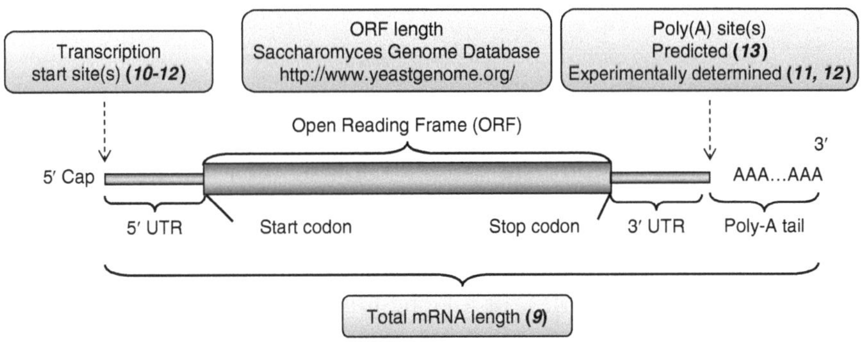

Fig. 2. Resources for prediction of *S. cerevisiae* mRNA sequence (adapted from (8)). *S. cerevisiae* mRNA lengths were determined genome-wide by Hurowitz and Brown (11). mRNA open reading frame (ORF) lengths are annotated in SGD. The location(s) of transcription start sites have been determined by three genome-wide studies (9, 10, 13). The transcription start sites mapped by Zhang and Dietrich (10) and Muira et al. (9) are now also available from SGD. Polyadenylation sites were mapped by Muira et al. (9) and Nagalakshmi et al. (13) or can be predicted using the mRNA 3′-processing site predictor, a tool generated by Graber et al. ((14); http:/harlequin.jax.org/polyA/).

by the absolute value of the chromosomal coordinate for the A of the start codon minus the chromosomal coordinate of the transcription start site (see Notes 8 and 9).

7. Get the chromosomal coordinates for the known poly(A) sites mapped by Muira et al. (9) by mousing over the dark blue arrows distal to the *HIS4* ORF to find the arrows that are annotated with "3′-end: poly(A)" (see Notes 10, 11 and 12).
8. Find the *HIS4* mRNA length in the Web supplement for Hurowitz and Brown (11). Go to http://microarray-pubs.stanford.edu/vnorth/, click on "Data" and then scroll down and click on "Transcript lengths." Use the systematic name to search the tab-delimited data for the virtual Northern measured transcript length.
9. Assess your predictions by comparing the sum of the lengths for the 5′ UTR, ORF, and predicted 3′ UTR to the virtual Northern transcript length.

Predicted mRNA structures can be experimentally validated using the 5′ and 3′ RACE Systems for Rapid Amplification of cDNA Ends (Invitrogen Corporation, Carlsbad, CA).

3.4. Genetic Resources for Experiments Using Yeast

There is a wealth of resources available for yeast genetic experiments. Many of these resources were developed for genome-wide analyses and include collections of yeast mutants, plasmids, yeast strains with epitope and fluorescent tagged alleles, and antibodies. A summary of these resources is described in Table 1. Most of these resources are available for individual genes of interest or as a collection.

4. Notes

1. The basic search option works with the standard name, the systematic name, or an alias. If you need to investigate genes or proteins without knowing their names, you can search for a class of similarly named genes using the wildcard character (e.g., a search for UPF* brings up UPF1, UPF2, or UPF3). You can also search with the name of a protein or protein complex, or a Gene Ontology term. This alternative search strategy will bring up a list of the gene summary pages where this text occurs. Each gene name in the hit list resulting from the search is hyperlinked to the corresponding gene summary page.
2. Gene summary pages are also accessible by using the full search form.

Table 1
Commercially available genetic resources for experiments using yeast

Resource	Description	Supplier(s)
Yeast mutant strains		
Yeast Knockout strains	A collection of over 6,000 gene disruption mutants developed by the Saccharomyces Genome Deletion project. The Community Posting Page provided by the Saccharomyces Genome Deletion Project enables users of the mutant collection to share information about the collection	1. American Type Culture Collection (ATCC), Manassas, VA 2. Open Biosystems Products, Huntsville, AL 3. Invitrogen. Corporation, Carlsbad, CA 4. European *Saccharomyces cerevisiae* archive for functional analysis (EUROSCARF, http://web.uni-frankfurt.de/fb15/mikro/euroscarf/)
Yeast Tet-Promoter collections	A collection of 800 essential yeast genes for which their native promoter has been replaced with a Tet-titratable promoter. Expression of the gene can be switched off by the addition of doxycycline to the growth medium (15)	Open Biosystems Products, Huntsville, AL
Yeast Insertional Mutant strains	A collection of 3,600 transposon insertion mutants	Open Biosystems Products, Huntsville, AL
Plasmids		
VectorDB	Annotations and sequence information for many vectors commonly used in molecular biology	http://genome-www.stanford.edu/vectordb/
Yeast Genomic Tiling Collection	A collection of 1,588 plasmid clones containing *S. cerevisiae* genome fragments in an *E. coli*-yeast 2-micron *LEU2* shuttle vector. This collection is a virtually complete overlapping clone collection of the entire *S. cerevisiae* genome (16)	Open Biosystems Products, Huntsville, AL
Yeast ORF Collection	The Yeast ORF Collection enables robust protein expression and purification for over 4900 *S. cerevisiae* genes. Each plasmid construct can be expressed in yeast or *E. coli* (17)	Open Biosystems Products, Huntsville, AL
Systems for discovery of novel protein–protein or protein–DNA interactions		
Two-hybrid systems	A system that utilizes protein–protein interactions to reconstitute transcription activator activity. Allows for selection and verification of novel protein–protein interactions	Clontech Laboratories Inc., Mountain View, CA

(continued)

Table 1 (continued)

Resource	Description	Supplier(s)
One-hybrid systems	A system to screen for novel DNA-binding proteins	Clontech Laboratories Inc., Mountain View, CA
Yeast strains with Epitope or fluorescent tagged alleles		
Molecular Bar-coded yeast ORFs (MoBY)	A collection of over 4,900 *Saccharomyces cerevisiae* (S288C) Open Reading Frames (ORFs) expressed from their endogenous promoter and tagged with a unique molecular bar code. The MoBY ORF Collection enables simultaneous measurement of all transformants in a single pooled culture using a bar code microarray readout (18)	Open Biosystems Products, Huntsville, AL
TAP-tagged collection	A collection of yeast strains each expressing a single C-terminal TAP-tagged protein from its endogenous promoter (19)	Open Biosystems Products, Huntsville, AL
HA-tagged collection	The HA-Tagged Yeast Collection contains over 2,400 yeast mutagenized strains each producing a single protein with an inserted triple hemagglutinin epitope tag (3× HA) (20)	Open Biosystems Products, Huntsville, AL
GST-tagged collection	A collection of more than 5,000 yeast strains that each overexpresses a different yeast open reading frame (ORF) when induced with galactose. ORFs are plasmid-encoded, tagged at the N-terminus with GST, and expressed under control of the *GAL1/10* promoter (21)	Open Biosystems Products, Huntsville, AL
Antibodies		
	Polyclonal antibodies directed against proteins from *S. cerevisiae*. Intended for Western blotting, immunoprecipitation, immunofluorescence, or ELISA	Santa Cruz Biotechnology, Inc. Santa Cruz, CA

3. P-POD is an acronym for The Princeton Protein Orthology Database (http://ppod.princeton.edu/, (12)). This database is designed to allow users to find and visualize the phylogenetic relationships among predicted orthologs to a query gene from 12 model organisms. For P-POD Version 3 (July 2009), families of predicted orthologs were determined using either OrthoMCL or MultiParanoid, and larger families using Jaccard Clustering. Multiple sequence alignments and evolutionary trees for each family were also produced using MAFFT and PhyML. Reconciliation and orthology analysis of these trees was carried out using Notung.
4. Most of the options to run the program effectively are presented with defaults. For many searches, the defaults are appropriate. However, it may be necessary to customize the options. For example, if the input sequence is less than 30 characters, the default cutoff score value should be changed to something less than 100 to avoid missing matches. A complete description of BLAST options and parameters is available from the BLAST documentation at NCBI (http://www.ncbi.nlm.nih.gov/blast/blast_help.shtml).
5. Putative binding sites for known transcription factors can also be found using the transcription factor database (TRANSFAC; http://www.biobase-international.com/index.php?id=transfac). Full access to this database requires a subscription. Academic and nonprofit organizations have free access to reduced functionality versions of the TRANSFAC products and tools. With a paid subscription, users have access to data and tools not available in the free versions.
6. Some genes have introns, and this is indicated in the sequence information on the gene summary page for the relevant genes.
7. The choices to Scroll/Zoom are preset. The best choice depends on the size of the ORF. Select the preset setting that is just larger than the ORF + 1 Kb size.
8. This page will also show any known uORFs. The chromosomal coordinates for uORFs can be retrieved by mousing over the diamond.
9. Additional mapped 5′-UTR coordinates are available in Table S4 in the supporting online material of Nagalakshmi et al. (13).
10. Some genes have multiple 3′ end processing sites.
11. Additional poly (A) sites have been mapped by Nagalakshmi et al. (13), and the coordinates are available in Table S4 in the supporting online material.
12. The most likely Poly(A) sites for yeast genes have been predicted by Graber et al. ((14); http://harlequin.jax.org/polyA/). To access this data, click on the "most likely 3′-UTR for all predicted yeast genes" link and search the tab-delimited data with the systematic name(s) of interest.

The number to the right of the systemic name is the position of the most likely polyA cleavage site in the 500 nt following the stop codon. This is followed by the DSM/HMM score for that site, which is a measure of the relative likelihood of the site being used as the cleavage site. Note that about 3% of the genes that have UTRs extending beyond the arbitrary cutoff of 500 nt that was used for this analysis.

References

1. Cherry J. M., Ball C., Weng S., Juvik G., Schmidt R., Adler C., Dunn B., Dwight S., Riles L., Mortimer R. K., Botstein D. (1997) Genetic and physical maps of *Saccharomyces cerevisiae*. *Nature* **387**, 67–73.
2. Issel-Tarver L., Christie K. R., Dolinski K., Andrada R., Balakrishnan R., Ball C. A., Binkley G., Dong S., Dwight S. S., Fisk D. G., Harris M., Schroeder M., Sethuraman A., Tse K., Weng S., Botstein D., Cherry J. M. (2002) Saccharomyces Genome Database. *Methods Enzymol.* **350**, 329–346.
3. Guthrie C., Fink G. R., (Eds.) (2002) *Guide to Yeast Genetics and Molecular and Cell Biology, Part B*, Vol. **350**, Academic Press, San Diego.
4. Guthrie C., Fink G. R., (Eds.) (2002) *Guide to Yeast Genetics and Molecular and Cell Biology, Part C*, Vol. **351**, Academic Press, San Diego.
5. Lundblad V., Struhl K. (2008) Yeast, In *Current Protocols in Molecular Biology* (Ausubel, F. M., Brent, R., Kingston, R. E., Moore, D. D., Geidman, J. G., Smith, J. A., and Struhl, K., Eds.), pp 13.11-13.17.18, John Wiley and Sons, Inc.
6. Lopez R., Silventoinen V., Robinson S., Kibria A., and Gish W. (2003) WU-Blast2 server at the European Bioinformatics Institute. *Nucleic Acids Res.* **31**, 3795–3798.
7. Kellis M., Patterson N., Endrizzi M., Birren B., Lander E. S. (2003) Sequencing and comparison of yeast species to identify genes and regulatory elements. *Nature* **423**, 241–254.
8. Kebaara B. W., Atkin A. L. (2009) Long 3'-UTRs target wild-type mRNAs for nonsense-mediated mRNA decay in *Saccharomyces cerevisiae. Nucleic Acids Res.* **37**, 2771–2778.
9. Miura F., Kawaguchi N., Sese J., Toyoda A., Hattori M., Morishita S., Ito T. (2006) A large-scale full-length cDNA analysis to explore the budding yeast transcriptome. *Proc. Natl. Acad. Sci. U.S.A.* **103**, 17846–17851.
10. Zhang Z., Dietrich F. S. (2005) Mapping of transcription start sites in *Saccharomyces cerevisiae* using 5' SAGE. *Nucleic Acids Res.* **33**, 2838–2851.
11. Hurowitz E. H., Brown P. O. (2003) Genome-wide analysis of mRNA lengths in *Saccharomyces cerevisiae. Genome Biol.* **5**, R2.
12. Heinicke S., Livstone M. S., Lu C., Oughtred R., Kang F., Angiuoli S. V., White O., Botstein D., Dolinski K. (2007) The Princeton Protein Orthology Database (P-POD): a comparative genomics analysis tool for biologists. *PLoS One* **2**, e766.
13. Nagalakshmi U., Wang Z., Waern K., Shou C., Raha D., Gerstein M., Snyder M. (2008) The transcriptional landscape of the yeast genome defined by RNA sequencing. *Science* **320**, 1344–1349.
14. Graber J. H., McAllister G. D., Smith T. F. (2002) Probabilistic prediction of *Saccharomyces cerevisiae* mRNA 3'-processing sites. *Nucleic Acids Res* **30**, 1851–1858.
15. Mnaimneh S., Davierwala A. P., Haynes J., Moffat J., Peng W. T., Zhang W., Yang X., Pootoolal J., Chua G., Lopez A., Trochesset M., Morse D., Krogan N. J., Hiley S. L., Li Z., Morris Q., Grigull J., Mitsakakis N., Roberts C. J., Greenblatt J. F., Boone C., Kaiser C. A., Andrews B. J., Hughes T. R. (2004) Exploration of essential gene functions via titratable promoter alleles. *Cell* **118**, 31–44.
16. Jones G. M., Stalker J., Humphray S., West A., Cox T., Rogers J., Dunham I., Prelich G. (2008) A systematic library for comprehensive overexpression screens in *Saccharomyces cerevisiae. Nat. Methods* **5**, 239–241.
17. Gelperin D. M., White M. A., Wilkinson M. L., Kon Y., Kung L. A., Wise K. J., Lopez-Hoyo N., Jiang L., Piccirillo S., Yu H., Gerstein M., Dumont M. E., Phizicky E. M., Snyder M., Grayhack E. J. (2005) Biochemical

and genetic analysis of the yeast proteome with a movable ORF collection. *Genes Dev.* **19**, 2816–2826.

18. Ho C. H., Magtanong L., Barker S. L., Gresham D., Nishimura S., Natarajan P., Koh J. L., Porter J., Gray C. A., Andersen R. J., Giaever G., Nislow C., Andrews B., Botstein D., Graham T. R., Yoshida M., Boone C. (2009) A molecular barcoded yeast ORF library enables mode-of-action analysis of bioactive compounds. *Nat. Biotechnol* **27**, 369–377.
19. Ghaemmaghami S., Huh W. K., Bower K., Howson R. W., Belle A., Dephoure N., O'Shea E. K., Weissman J. S. (2003) Global analysis of protein expression in yeast. *Nature* **425**, 737–741.
20. Kumar A., Agarwal S., Heyman J. A., Matson S., Heidtman M., Piccirillo S., Umansky L., Drawid A., Jansen R., Liu Y., Cheung K. H., Miller P., Gerstein M., Roeder G. S., Snyder M. (2002) Subcellular localization of the yeast proteome. *Genes Dev.* **16**, 707–719.
21. Zhu H., Bilgin M., Bangham R., Hall D., Casamayor A., Bertone P., Lan N., Jansen R., Bidlingmaier S., Houfek T., Mitchell T., Miller P., Dean R. A., Gerstein M., Snyder M. (2001) Global analysis of protein activities using proteome chips. *Science* **293**, 2101–2105.

Chapter 12

Delete and Repeat: A Comprehensive Toolkit for Sequential Gene Knockout in the Budding Yeast *Saccharomyces cerevisiae*

Johannes H. Hegemann and Sven Boris Heick

Abstract

Gene inactivation is an essential step in the molecular dissection of gene function. In the yeast *Saccharomyces cerevisiae*, many tools for gene disruption are available. Gene disruption cassettes comprising completely heterologous marker genes flanked by short DNA segments homologous to the regions to the left and right of the gene to be deleted mediate highly efficient one-step gene disruption events. Routinely, in more than 50% of analyzed clones, the marker cassette is integrated in the targeted location. The inclusion of *loxP* sites flanking the disruption marker gene allows sequence-specific Cre recombinase-mediated marker rescue so that the marker can be reused to disrupt another gene. Here, we describe a comprehensive toolbox for multiple gene disruptions comprising a set of seven heterologous marker genes including four dominant resistance markers for gene disruption, plus a set of Cre expression plasmids carrying eight different selection markers, four of them dominant.

Key words: Single gene disruption, Multiple gene disruptions, Sequence-specific recombination, Heterologous marker genes, Dominant marker genes, *loxP* site, Cre recombinase

1. Introduction

The creation of loss-of-function mutations may be regarded as the fundamental operation in experimental genetics. One of the most powerful techniques for reliably eliminating gene function involves the insertion of an extraneous DNA fragment into the coding sequence, which most often disrupts the reading frame but may also replace it entirely. In the budding yeast *Saccharomyces cerevisiae*, gene disruption can be accomplished in a targeted fashion using a one-step approach. This is based on the fact that linear DNA fragments carrying a selectable marker gene with homology regions to a yeast gene placed at either end integrate with high

James A. Williams (ed.), *Strain Engineering: Methods and Protocols*, Methods in Molecular Biology, vol. 765,
DOI 10.1007/978-1-61779-197-0_12, © Springer Science+Business Media, LLC 2011

efficiency at the corresponding chromosomal locus by homologous recombination (1, 2). The flanking homology regions can be as short as 40–50 bp, making it possible to generate the gene disruption cassettes by PCR (polymerase chain reaction), thus avoiding the need for time-consuming cloning steps. The steps in a PCR-mediated one-step gene disruption experiment are shown schematically in Fig. 1. The selectable marker genes employed on the disruption cassette consist of completely heterologous DNA to restrict homologous recombination to the short flanking segments at the ends of the disruption cassette. The first heterologous, dominant marker used for disruption purposes was the *kan*^{r} gene encoding resistance to Geneticin (G418). This has since been used to generate a collection of more than 6000 disruption strains each carrying a defined deletion in a particular yeast gene. This yeast-gene-knockout (YKO) collection is the primary source of single-gene loss-of-function mutants in *S. cerevisiae* (see Note 1) (3, 4). Meanwhile, the list of heterologous disruption marker genes has significantly broadened (Fig. 2a). A detailed description of currently available disruption cassettes and their use can be found elsewhere (2, 5). Besides regular gene disruption experiments this set of heterologous marker cassettes can also be used in convenient one-step marker switch (see Note 2). Many cellular functions are maintained by the functionally redundant products of several related genes. Thus, the analysis of such functions requires simultaneous disruption of more than one gene. A particularly striking example in yeast is the hexose transporter family. Here, concurrent knockout of at least 20 genes was necessary to completely inhibit growth on hexoses (6). Multiple gene disruption can be accomplished in two ways: (1) genes can be deleted sequentially using different gene disruption cassettes carrying distinct selectable markers; (2) the disruption cassette can be removed from the genome by mitotic or recombinase-mediated recombination, allowing the same disruption marker to be reused to disrupt the next gene of interest (overview in 2, 5). Recyclable disruption cassettes are normally preferred, as they provide the greatest flexibility for later manipulations of the resultant strain. In the following we will focus on the use of a series of seven completely heterologous *loxP*-flanked disruption cassettes, all of which can be efficiently removed a *loxP*-mediated site specific recombination by the Cre recombinase (7, 8) (see Note 3). The Cre protein can be expressed from eight different expression plasmids each carrying a different selection marker gene for yeast transformation to allow for greatest flexibility when using laboratory or industrial yeast strains (Fig. 3). In the meantime, similar or even identical loxP-flanked disruption cassettes have been created by others, all of which are compatible with the gene disruption cassettes discussed here (9, 10). Other removable disruption cassettes rely on the action of the Flp recombinase or depend on a mitotic recombination event and have been summarized elsewhere (2). In recent

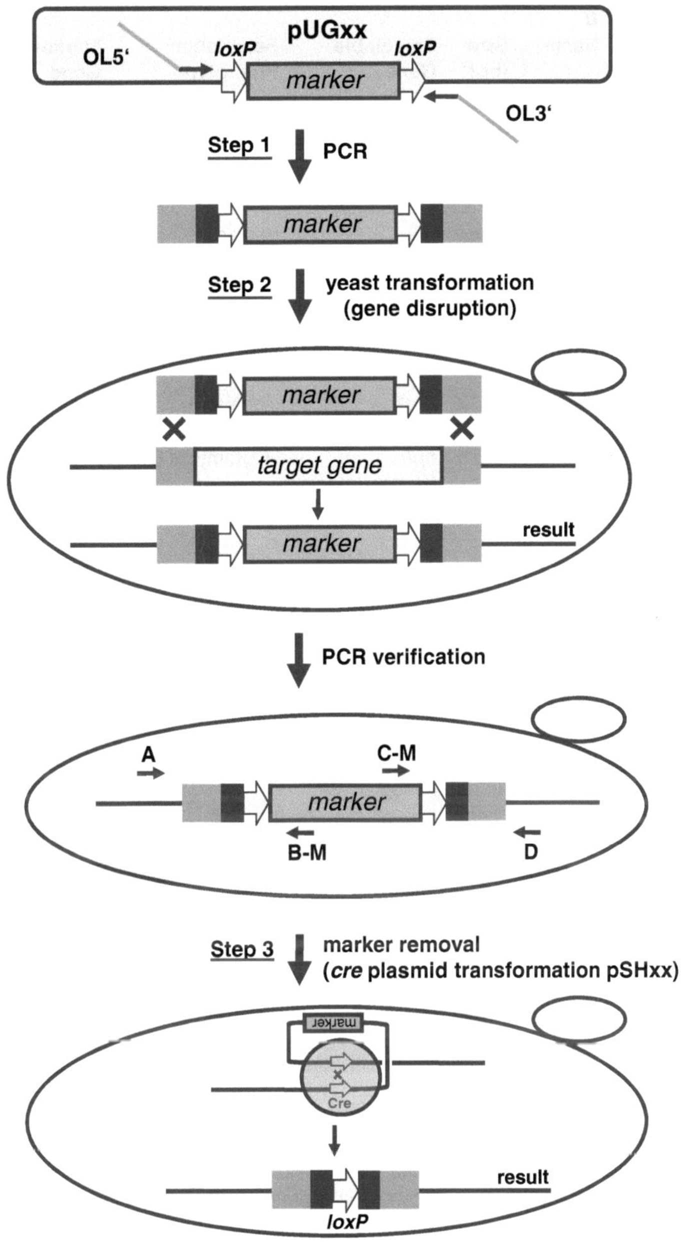

Fig. 1. General outline of the one-step gene disruption approach. A collection of seven marker plasmids (pUG series) carries the various selectable disruption marker genes, each flanked by *loxP* sites that allow their subsequent removal from the genome. In Step 1, the disruption cassette is generated by PCR, using oligonucleotides that carry at their 3′ ends sequences homologous to sequences left and right of the disruption marker gene and at their 5′ ends sequences homologous to sequences that flank the target gene. After yeast transformation (Step 2), the disruption cassette integrates via homologous recombination into the genome, replacing the target gene. PCR verification identifies yeast transformants harboring correctly integrated disruption cassettes. If required, marker rescue is initiated by transforming a Cre expression vector into the disruptant strain. Induction of Cre expression results in a strain in which the target gene has been replaced by a single *loxP* site (Step 3).

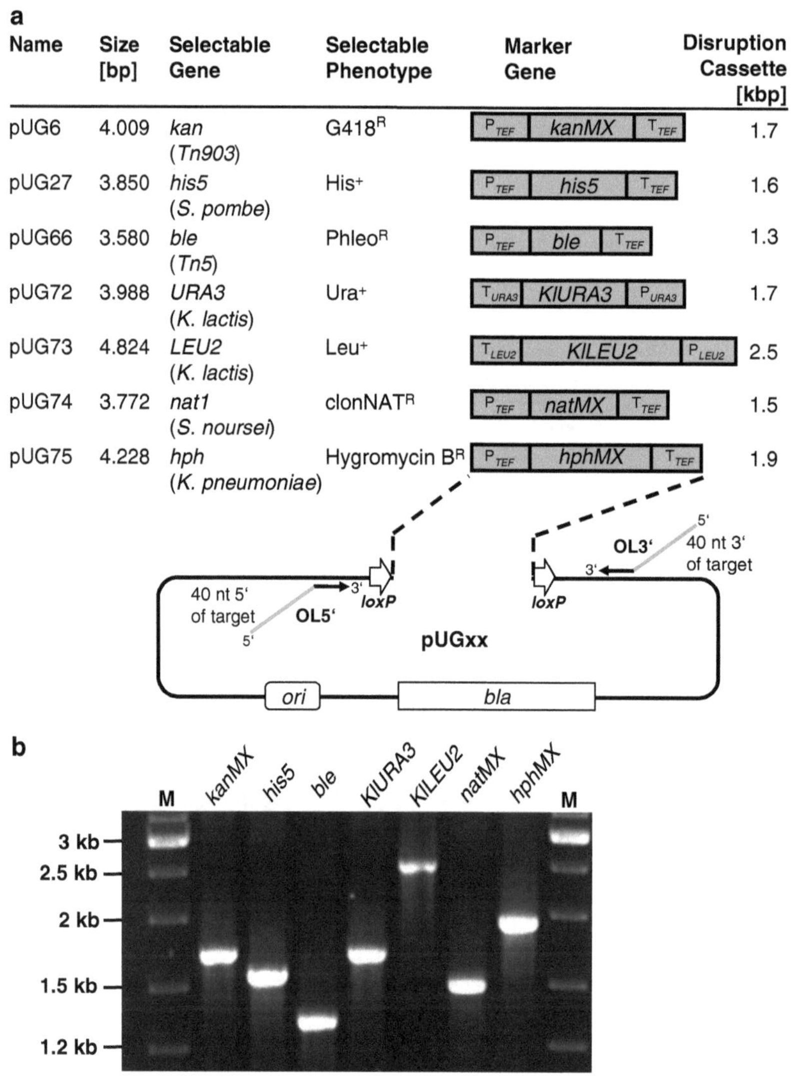

Fig. 2. The collection of seven pUG plasmids carrying *loxP*-flanked marker gene disruption cassettes. (**a**) The plasmids serve as templates for the generation of the individual disruption cassettes. All disruption marker genes originate from organisms other than *S. cerevisiae*, and, thus, will not recombine with the *S. cerevisiae* genome. Expression of the marker genes is controlled by the *TEF2 Ashbya gossypii* promoter (P_{TEF}) and terminator (T_{TEF}), while the two *Kluyveromyces lactis* genes are expressed from their own regulatory sequences (P and T respectively). Since these disruption marker genes exhibit no homology to the yeast genome, recombination between the yeast genome and internal parts of the disruption cassettes (which would result in chromosomal misintegration) is minimal, thus maximizing the frequency of correct integration. All seven disruption cassettes can be generated by PCR using the same primer pair, OL5′ and OL3′. The two primers comprise 19 or 22 3′ nucleotides complementary to sequences in the pUG plasmids flanking the disruption cassettes and 40 5′ nucleotides complementary to sequences upstream or downstream of the genomic target sequence to be deleted. (**b**) The correct size and the quality of the amplified disruption cassettes can be verified on a 0.7% agarose gel. The complete plasmid sequences can be found in GenBank under the following accession numbers: pUG6: AF298793; pUG27: AF298790; pUG66: AF298794; pUG72: AF298788; pUG73: AF298792 (7, 8), pUG74: HQ401268; pUG75: HQ401269 (Heick and Hegemann, unpublished). (*bla* confers resistance against ampicillin in *E. coli*, *ori* origin of replication in *E. coli*, *bp* base pairs, *kbp* kilo base pairs, *nt* nucleotides, *M* GeneRuler™ DNA Ladder Mix (ready-to-use, Fermentas), *P* promoter, *T* terminator, *TEF* translation elongation factor).

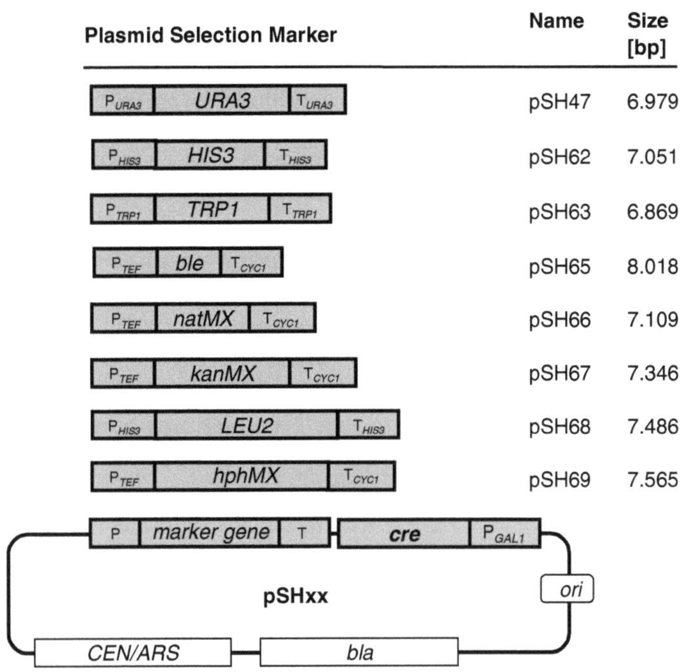

Fig. 3. The collection of eight Cre-expressing pSH plasmids. Cre expression is regulated by the galactose-inducible *GAL1* promoter. Transfer of yeast cells transformed with these plasmids to galactose medium results in expression of Cre, followed by Cre-induced recombination of the *loxP* sites flanking the disruption marker gene, leaving behind a single *loxP* site at the former site of integration of the disruption cassette. Eight different plasmid selection markers maximize the use of the Cre system. The size of the plasmids is given in bp. The complete plasmid sequences can be found at GenBank under the following accession numbers: pSH47: AF298782; pSH62: AF298785; pSH63: AF298789; pSH65: AF298780 (7, 8); pSH66: HQ412576; pSH67: HQ412577; pSH68: HQ401270; pSH69: HQ412578; (Heick and Hegemann, unpublished). (*bla* confers resistance against Ampicillin in *E. coli*, *ori* origin of replication in *E. coli*, *bp* base pairs, *P* promoter, *T* terminator, *TEF* translation elongation factor, *CYC1* cytochrome C).

years, there has been an upsurge in the use of Cre/lox-mediated gene disruption systems in *S. cerevisiae* (for example (11)). Moreover, this gene deletion tool has been successfully adapted to a variety of medically and industrially relevant yeasts, including *Candida*, *Cryptococcus*, *Hansenula*, *Kluyveromyces*, *Neurospora*, *Schizosaccharomyces* and *Yarrowia* (12–22).

2. Materials

2.1. Generation of Disruption Cassette

The pUG plasmids bear gene disruption cassettes comprising seven completely heterologous marker genes (*kan*, *his5*, *ble*, *URA3*, *LEU2*, *nat* and *hph*), each flanked by *loxP* sites (7, 8, *Heick and Hegemann, unpublished*) (Fig. 2a). Three disruption marker genes (*his5*, *URA3* and *LEU2*) complement the common auxotrophic markers *his3*, *ura3*, and *leu2* in *S. cerevisiae* laboratory strains, while four disruption cassettes harbor genes conferring resistance to the drugs Geneticin/G418 (*kan*), Phleomycin (*ble*), clonNAT (*nat*), or Hygromycin B (*hph*) and, thus, can be used as dominant markers to disrupt genes in virtually any yeast strain (prototrophic industrial, wild-type or laboratory strains) (see Note 3).

2.1.1. Primer Design

All seven disruption cassettes can be generated by PCR using the same oligonucleotides: (1) OL5′ (5′ CAGCTGAAGCTTCGTACGC 3′; hybridizes upstream of the P_{TEF} respectively of T_{URA3} and T_{LEU2} elements) and (2) OL3′ (5′ GCATAGGCCACTAGTGGATCTG 3′; downstream of T_{TEF}, respectively, of P_{URA3} and P_{LEU2} elements) and the pUGxx plasmids as template (Fig. 2a). The sequences flanking the target gene in the genome are added to the 5′ ends of these sequences as 40-nucleotide stretches that are homologous to sequences upstream of the ATG start codon and downstream of the stop codon, respectively (Fig. 1). The 40 base pairs of flanking sequence on each side are sufficient to ensure correct integration of the disruption cassette in approximately 80% of cases (see Note 4). The primers used to generate the disruption cassettes need to be of full length; otherwise, the chance of undesirable nonhomologous recombination increases (see Note 5). Care must also be taken that neighboring open reading frames (ORF) are not encroached upon by the gene disruption event. Every deletion should begin at least 500 bp upstream of the next start codon and end about 200 bp downstream of the next stop codon.

Many yeast genes and even chromosomal regions are duplicated in the genome. In these cases, it is necessary to verify that the 40-bp flanking sequences used for recombination are not repeated elsewhere in the genome. Moreover, many yeast genes are flanked by simple DNA sequences (e.g., poly(A/T) stretches downstream of a gene). Gene disruption cassettes carrying such segments in their targeting sequences will yield fewer transformants. In these cases, one should either choose a different 40-bp homology sequence or create a longer flanking homology sequence by adding a unique sequence to either end.

2.1.2. Preparative PCR to generate Disruption Cassette

1. Taq polymerase (available from various commercial sources; alternatively the enzyme can be purified from a recombinant *Escherichia coli* (*E. coli*) clone (23).
2. 10× PCR buffer: 750 mM Tris–HCl, pH 9.0, 200 mM $(NH_4)_2SO_4$, 0.1% (w/v) Tween 20. Store at −20°C.
3. 4 mM dNTPs.
4. 25 mM $MgCl_2$.
5. Targeting primers (OL5′ and OL3′-derived) (50 pM/μL).
6. Sterile ddH_2O.

All chemicals should be of highest quality.

2.2. Yeast Transformation

Yeast transformation is carried out using the method described in (24).

1. Carrier DNA (2 mg/mL): High-molecular-weight DNA (deoxyribonucleic acid sodium salt from salmon testes; Catalogue No. D1626, Sigma-Aldrich, Taufkirchen, Germany) is dissolved in sterile ddH_2O at 2 mg/mL. The DNA is dispersed into the solution by drawing it up and down repeatedly in a 10-mL pipette. The solution is then covered and mixed vigorously on a magnetic stirrer overnight in the cold room. Aliquots of about 1 mL are stored at −20°C. Before use, the DNA is denatured by boiling at 100°C for 10 min and then chilled on ice.
2. 1 M lithium acetate stock solution (LiAc), pH 8.4–8.9. The solution is prepared in ddH_2O, autoclaved, and stored at room temperature.
3. Polyethylene glycol (PEG 50% w/v): The PEG solution (MW 3350; P3640, Sigma) is made up to 50% (w/v) with ddH_2O and autoclaved. Directly create 2-mL aliquots after autoclaving and store them at −20°C (critical step). Avoid repeated thawing and freezing (use three times at most).
4. YPD medium: Mix 10 g yeast extract (e.g., 212750, BD, Heidelberg, Germany), 20 g peptone (e.g., 211677, Heidelberg, Germany), 13.5 g agar (for plates) (e.g., 214530, BD, Heidelberg, Germany), 2 mL adenine stock solution (2 mg/mL) in ddH_2O, 4 mL tryptophan stock solution (5 mg/mL) in ddH_2O, 20 g dextrose. Bring to 1 L with ddH_2O. Autoclave. (for details of yeast media, see (25)).
5. YPD + Geneticin: The active concentration of Geneticin (G418) may vary from lot to lot (500–800 μg/mg, w/w). It is crucial that a final *active* concentration of 200 μg/mL is used (G418 plates can be tested by plating single cells of a G418-sensitive strain: no visible microcolonies should form). Add 200 mg of active Geneticin (e.g., #345810, Calbiochem, Merck KGaA, Darmstadt, Germany) dissolved in 1 mL of sterile ddH_2O to 1 L of warm (~60°C) YPD medium.
6. YPD + Phleomycin: Add 10 μg/mL Phleomycin (Phleo) (#ant-ph-1, InvivoGen, San Diego, USA) to warm (~60°C) medium.
7. YPD + clonNAT: Add 100 μg/mL clonNAT (Nourseothricin-dihydrogen sulfate, #5001000, Werner BioAgents, Jena, Germany) to warm (~60°C) medium.
8. YPD + Hygromycin B: Add 300 μg/mL Hygromycin B (attention: highly toxic, e.g., #H3274, Sigma-Aldrich, Germany and worldwide) to warm (~60°C) medium.
9. SC-medium – His, Ura, or Leu: Mix 20 g dextrose, 20 g agar (for plates), 1.7 g yeast nitrogen base (YNB) w/o amino acids and ammonium sulfate, 5 g ammonium sulfate, 2 g dropout mix. The dropout powder mix consists of the constituents as

Table 1
Dropout powder mix for synthetic complete medium

Chemical	Amount (g)	Chemical	Amount (g)
Adenine	0.5	Leucine	10.0
Alanine	2.0	Lysine	2.0
Arginine	2.0	Methionine	2.0
Asparagine	2.0	*p*-Aminobenzoic acid	0.2
Aspartic acid	2.0	Phenylalanine	2.0
Cysteine	2.0	Proline	2.0
Glutamine	2.0	Serine	2.0
Glutamic acid	2.0	Threonine	2.0
Glycine	2.0	Tryptophan	2.0
Histidine	2.0	Tyrosine	2.0
Inositol	2.0	Uracil	2.0
Isoleucine	2.0	Valine	2.0

listed in Table 1, with the exception of the auxotrophic requirements supplied by the genes provided by the disruption plasmid (i.e., missing either His or Ura or Leu). The mixture needs to be vigorously agitated in a bottle containing sterile glass beads (Ø ~ 5 mm) for at least 15 min (shake for longer than you think necessary!). All chemicals should be of highest quality. Dissolve in 1 L of ddH_2O and adjust the pH to ~6.5 with 1 M NaOH (for details of yeast media, see (25)).

2.3. Verification of Correct Clone/Gene Disruption by PCR

2.3.1. Primer Design

To check that transformants have integrated the disruption cassette correctly, diagnostic PCR analyses are performed (Fig. 4a). The PCR primers A to D flanking the disrupted gene should be chosen such that the PCR products generated (PCR products of primers A, B, C, D and disruption cassette-specific primers B-M and C-M, as shown in Fig. 4a–b) are 500–1,000 bp long. Therefore, oligonucleotide A should bind about 300 bp upstream of the integration cassette in the genome while oligonucleotide D should be located about 300 bp downstream of the disruption cassette. The oligonucleotides B and C used to amplify the junctions extending from the endogenous gene into the adjacent genomic regions should bind within the target gene about 300 bp away from the start and stop codon. The primers should have melting temperatures of 63–67°C. The disruption cassette-specific primers B-M and C-M are listed in Fig. 4d.

2.3.2. PCR Verification

The required reagents are the same as those listed above (see Subheading 2.1.2).

a disrupted gene

b WT gene

c disrupted gene without marker gene

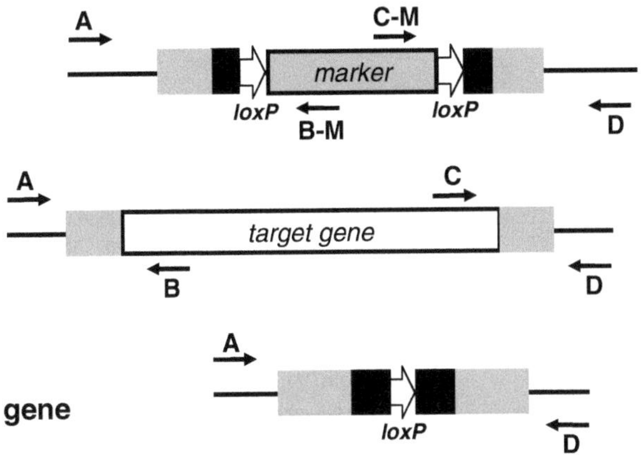

d verification primers for disruption cassettes (5'→3')

marker gene	B-M	C-M
kanMX, *his5*, *ble*, *natMX*, *hphMX*	GGATGTATGGGCTAAATG	CCTCGACATCATCTGCCC
KlURA3	CTAATAGCCACCTGCATTGG	CAGACCGATCTTCTACCC
KlLEU2	AGTTATCCTTGGATTTGG	ATCTCATGGATGATATCC

e PCR verification of the *ho* gene disruption

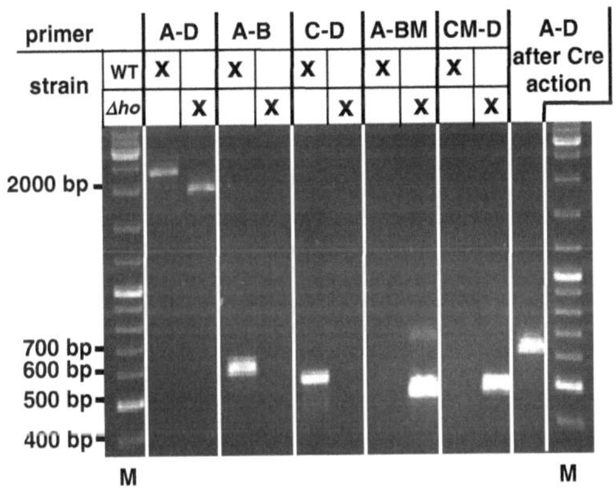

strain	primer combination	expected size in bp
WT	A-D	2.281
Δho		1.896
WT	A-B	570
Δho		-
WT	C-D	531
Δho		-
WT	A-BM	473
Δho		-
WT	CM-D	483
Δho		-
after Cre action	A-D	626

Fig. 4. PCR-based verification of a gene disruption in a haploid yeast strain. (**a**) Correct integration of the disruption cassette into the target gene can be efficiently diagnosed using a combination of target gene-specific (A and D) and disruption marker-specific (B-M and C-M) primers. PCR products of the expected size will be obtained only if the disruption cassette integrated successfully. (**b**) Presence of the wt target gene in a haploid strain (owing to a unsuccessful disruption) or in a diploid yeast strain (because of the second nontargeted wt allele) can be confirmed using a combination of the target gene-specific primers A, B, C, and D. (**c**) Cre-mediated removal of the disruption marker can be verified by PCR using primers A and D. (**d**) DNA sequences of the universal disruption cassette-specific primers B-M and C-M. (**e**) Example of a successful gene disruption experiment in a haploid yeast strain. The disruption cassette was generated by PCR using plasmid pUG74 as template and *HO*-specific oligonucleotides OL5′ and OL3′ (for sequences: see Note 6) and transformed into yeast strain CENPK2-1C (for genotype see Note 7). Colony-purified yeast transformants were checked for correct integration of the disruption cassette by PCR using target gene-specific primers A–D (for sequences: see Note 6) and the *nat*-specific B-M and C-M primers (see (**d**)). The size of the expected PCR products is given right-hand side. The successful gene disruption in a haploid yeast strain is indicated by the absence of PCR products for the primer combinations A-B and C-D (*wt* nontransformed wild type yeast strain is shown as control) and the presence of PCR products of the expected size for the primer combinations A-D, A-BM and CM-D (*Δho* yeast strain carrying the correctly disrupted *HO* gene). After transformation of a Cre expression plasmid in the disruptant yeast strain and induction of Cre expression, the disruption marker gene is excised by homologous recombination. Subsequently, the diagnostic PCR using primers A and D yields a correspondingly shorter PCR product (*see lane* "*A–D* after Cre action"). *M* GeneRuler™ DNA Ladder Mix, ready-to-use, Fermentas.

2.4. Marker Rescue/ Repeated Gene Disruption

The pSH plasmid series carries the *cre* gene under the control of the galactose-inducible *GAL1* promoter and is equipped with eight different selection marker genes to allow transformation into a variety of different auxotrophic or prototrophic laboratory or industrial yeast strains (7, 8, *Heick and Hegemann, unpublished*) (Fig. 3).

1. YPG medium: This is the same as YPD (see Item 4 in Subheading 2.2) except that 2% galactose, instead of glucose, is used as the carbon source.

3. Methods

Efficient insertion of a disruption cassette into the desired target gene requires that a linear DNA fragment is used for yeast transformation. The sequences that are homologous to sequences flanking the gene to be deleted must lie at the ends of the transforming DNA fragment, which also carries a marker gene that provides a selectable phenotype (usually prototrophy or resistance to drugs) (Fig. 1). Different selectable disruption marker genes are available, which encode resistance to drugs or prototrophy for amino acids or nucleotide bases. These entirely heterologous marker genes are all flanked by two *loxP* sites, which permit Cre-mediated recombination and so enable efficient marker rescue (7, 8) (Figs. 1 and 2a) (see Note 3). The disruption cassette is generated via PCR using oligonucleotides, whose 3′-terminal 19–22 nucleotides are homologous to sequences flanking the disruption marker gene on a plasmid, while their 5′ 40-nucleotides are homologous to sequences left and right of the gene to be deleted (Fig. 1, Step 1). The linear disruption cassette is then transformed into yeast cells using a highly efficient transformation protocol (Fig. 1, Step 2). The disruption cassette integrates into the genome by homologous recombination, precisely replacing the target gene. To confirm correct integration of the cassette into the genome, yeast transformants are analyzed by PCR using combinations of the corresponding target gene-specific and disruption cassette-specific primers (Fig. 1, PCR verification). PCR products of the expected size will be obtained only if the disruption cassette has integrated in the predicted manner. If one intends subsequently to remove the cassette from the genome, a Cre expression plasmid is transformed into the disruptant strain. Induction of Cre expression induces a *loxP*-mediated recombination event resulting in loss of the marker gene, leaving behind a single *loxP* site at the site of the deleted target gene (Fig. 1, product of Step 3). Finally, the Cre expression plasmid can be easily removed from the marker gene-less disruption strain by growth in nonselective liquid medium followed by plating onto nonselective plates. Colonies that have lost the Cre plasmid are identified by replica-plating onto selective plates.

3.1. Generation of Disruption Cassettes

The disruption cassettes are generated by preparative PCR. Any of the plasmids described in Fig. 2a can be used as a template.

1. The list of all ingredients required for the PCR reaction is given in Table 2a, while the PCR conditions are listed in Table 2b.

Table 2
PCR requirements for the disruption cassette and deletion verification

(A) PCR setup

		Aliquots needed for	
		Disruption cassette[a]	Verification of deletion[b]
Chemical	Stock	Amount (μL)	Amount (μL)
Primer 1	50 pmol/μL	2	0.5
Primer 2	50 pmol/μL	2	0.5
dNTPs	4 mM	5	1.25
$MgCl_2$	25 mM	6	1.5
Buffer	10×	10	2.5
Template	50 ng/μL	1	–
Taq polymerase	0.5 U/μL	1	0.5
Yeast cells	–	No	Yes
Sterile ddH_2O	–	73 μl	18.25
		100 μl	25 μl

(B) PCR Conditions

	Disruption cassette		Verification of deletion	
Step	Time	Temperature	Time	Temperature
Initial step	5 min	95°C	5 min	94°C
Denaturation	40 s	94°C	1 min 30 s	94°C
Annealing	1 min	58°C	2 min	c
Extension	2 min	68°C	c	72°C
Final extension	15 min	68°C	7 min	72°C
Cycles		25		35

[a]primers OL3′ and OL5′ as outlined in Fig. 2
[b]Primer pairs as outlined in Fig. 4
[c]Depending on the oligonucleotides you have designed for the verification and the expected product length, you have to adjust the annealing temperature and the elongation time

2. Each PCR should yield ~500 ng of product. For each transformation, one needs to combine the products of two reactions (in total ~1,000 ng).
3. Precipitate the PCR product and resuspend in 34 μL sterile ddH_2O (see Note 8).

3.2. Yeast Transformation (see ref. (24); Note 9)

1. Inoculate a yeast strain into 5 mL of YPD medium and incubate overnight on a rotary shaker at 30°C.
2. Determine the titer of the yeast culture by counting cells. Count budded cells as one cell. Some strains form clumps of cells, and these should be dispersed by vigorous vortexing before counting.
3. Transfer 2.5×10^8 cells to 50 mL of fresh YPD-medium to give 5×10^6 cells/mL.
4. Incubate the flask on a shaker at 30°C.

   It is important to allow the cells to complete at least two divisions. This will take 3–5 h. The transformation efficiency (transformants per μg plasmid per 10^8 cells) remains constant for 3–4 cell divisions.
5. When the cell titer has reached at least 2×10^7 cells/mL, harvest the cells by centrifugation at $1,600 \times g$ for 5 min, wash them in 25 mL of sterile ddH_2O, and resuspend them in 1 mL 0.1 M LiAc. Transfer the cell suspension to a 1.5-mL microfuge tube, centrifuge for 10 s at top speed ($10,000–13,000 \times g$) at room temperature, and discard the supernatant.
6. Boil the carrier DNA as described above (see Subheading 2.2).
7. Resuspend sufficient cells in 0.5 mL of 0.1 M LiAc (made fresh by tenfold dilution of 1 M LiAc stock) to yield a density of 2×10^9 cells/mL.
8. For each transformation reaction, pipette 50-μL samples into 1.5-mL microfuge tubes, centrifuge at top speed for 10 s and remove the supernatant.
9. Add the following components sequentially:

   240 μL PEG

   36 μL 1 M LiAc

   50 μL boiled carrier DNA

   34 μL DNA plus water (500–1,000 ng of the disruption cassette)

   Total: 360 μL
10. Vortex each tube vigorously until the cell pellet has been completely dispersed.
11. Incubate the cells for 30 min at 30°C.

12. Incubate the cells for 30–40 min at 42°C (the optimal time may vary for different strains).
13. Centrifuge at top speed for 10 s and remove the supernatant with a micropipette.
14. In case of selection for a prototrophy, resuspend the pellet in 200 µL of sterile ddH_2O and spread onto two selective plates, 100 µL per plate.
15. In case of selection for a resistance resuspend the cells in 1 mL of YPD and incubate for at least 1 h on a rotator at 30°C.
16. Centrifuge at top speed for 10 s and remove the supernatant.
17. Resuspend the pellet in 200 µL of sterile ddH_2O and spread onto two selective plates.
18. Incubate plates 3–5 days at 30°C. Expect between 10 and 100 transformants per plate.
19. G418 and Hygromycin B plates must be replica-plated onto fresh G418 or Hygromycin B plates after 24–36 h.

3.3. Verification of Correct Clone/Gene Disruption by PCR

To confirm that the disruption cassette has integrated correctly and replaced the intended target gene, prepare different PCRs as outlined in Fig. 4a, b. The primer combinations A/B-M and C-M/D will generate a specific PCR product only if the deletion cassette has integrated at the correct location (Fig. 4a).

In about 8% of the gene disruption events a gene deletion is accompanied by a duplication of the gene (duplication of the entire chromosome or of a particular chromosomal region). Therefore, one should confirm the absence of the target gene by PCR using primer combinations A/B and C/D (Fig. 4b). A PCR with oligonucleotides A and D amplifying the entire locus serves as a further check to ensure correct disruption. Here, care should be exercised in cases where the A/D PCR fragments obtained from the disrupted allele and from the wt allele are of similar size. Depending on the size of the DNA fragment you need to amplify, the incubation conditions for the A/D PCR may need to be modified. On average, between 50 and 80% of the transformants will be correct by PCR criteria.

In Fig. 4e, an example of a successful gene disruption experiment is presented. A *HO*-specific *natMX* disruption cassette was transformed into the haploid yeast strain CEN.PK2-1C and transformants were checked by verification PCR.

1. Colony-purify the yeast transformants on selective plates (use wild-type strain as negative control) and then on an YPD plate. For the PCR, always use freshly grown cells (no more than 2 days old).
2. To obtain cells for the PCR reaction, lightly touch the surface of a yeast colony with a yellow pipette tip so that you can just

barely see the cells on the end. Resuspend these cells in the PCR mix (see Note 10). Addition of too many cells or agarose contaminants will inhibit the PCR!

3. The PCR ingredients are listed in Table 2a, while the PCR conditions are summarized in Table 2b.

3.3.1. Important: Occurrence of Collateral Mutations

Every yeast transformation is mutagenic, i.e., randomly generates mutations in the genome. In gene disruption experiments, 5–10% of transformants will carry a second-site (or collateral) mutation resulting in a growth phenotype (2). To avoid this problem, one should work with diploid strains homozygous for the disruption. The diploid strain should be generated by crossing two independently obtained haploid disruption strains of opposite mating types, thus ensuring that most collateral mutations (which are recessive) are complemented. If one needs to work with haploid disruption strains, it is best to backcross the originally generated haploid disruption strain several times to the corresponding wild-type strain.

3.4. Marker Rescue/ Repeated Gene Disruption

To disrupt a second gene in a yeast strain, either one can use a disruption cassette with a different genetic marker or the original disruption marker gene can be removed from the genome so that the same marker can be used again. In the case of *loxP*-flanked disruption cassettes, one of the eight Cre expression plasmids has to be introduced into the strain. Induction of Cre expression by growing transformants in galactose-containing medium is followed by identification of yeast cells that have lost the disruption marker. Loss of the marker can be easily verified by (1) checking for growth on the appropriate medium and (2) by appropriate PCRs as outlined in Fig. 4c using primer pair A/D. Subsequently, the Cre plasmid is removed from this yeast strain, which is now ready for a second disruption experiment.

1. Transform with a suitable Cre expression plasmid (Fig. 3) as described in Subheading 3.2.
2. Select for transformants on selective medium (for pSH67 and pSH69 transformants need to be replica-plated after 24 h). Colony-purify single transformants.
3. Incubate single colonies in 5 mL of YPG medium overnight.
4. Plate 100–200 cells onto YPD plates and incubate them for 1 day at 30°C.
5. Replica-plate onto two plates: (1) selective for the marker on the disruption cassette and (2) on YPD. Alternatively, about 12 colonies can be streaked onto a selective and an YPD plate. Cells that fail to grow on the selective medium have lost the disruption cassette. Pick cells from the corresponding colonies/streak on the YPD plate. More than 50% of the colonies will have lost the disruption marker.

6. To verify marker loss, perform the appropriate PCR reactions as shown schematically in Fig. 4c (see Subheading 3.3).
7. To remove the Cre expression plasmid from a marker-minus yeast strain, incubate the cells in 5 mL of YPD medium overnight. The next morning use 200 μL of the cells to inoculate 5 mL of fresh YPD medium. In the evening, transfer 50 μL of these cells to 5 mL of fresh YPD medium. Always incubate the cells at 30°C on a rotator.
8. Plate 100–200 cells onto YPD plates and incubate for 1 day at 30°C.
9. Replica-plate onto two plates: (1) selective for the Cre-expressing plasmid and (2) on YPD. Alternatively, about 12 colonies can be streaked out onto a selective and an YPD plate. The cells that cannot grow on the selective medium have lost the *cre* plasmid (between 5 and 50% of the colonies should be positive). Corresponding colonies on the YPD plates can be streaked out on fresh YPD plates (see Note 11).
10. Finally, test again for loss of disruption cassette marker gene and Cre plasmid marker by streaking cells onto selective plates.

4. Notes

1. The entire YKO collection or single gene disruption strains thereof can be obtained from the following companies:

   EUROSCARF Institute of Molecular Biosciences, Johann Wolfgang Goethe-University, Max-von-Laue Strasse 9; Building N250, D-60438 Frankfurt, Germany, Fax: +49-69-79829527. e-mail: euroscarf@em.uni-frankfurt.de. http://web.uni-frankfurt.de/fb15/mikro/euroscarf/index.html

   American Type Culture Collection (ATCC), P.O. Box 1549 Manassas, Virginia 20108, USA. Phone: 703-365-2700. e-mail: news@atcc.org. http://www.biospace.com/company_profile.aspx?CompanyID=69904

   Invitrogen GmbH, Frankfurter Straße 129B, 64293 Darmstadt, Germany. e-mail: euroinfo@invitrogen.com. http://www.invitrogen.com

   Invitrogen Corporation, 1600 Faraday Avenue, Carlsbad, CA 92008, USA. Phone: 1.800.955.6288. Fax: 716.774.3157. e-mail: custom.services@invitrogen.com. http://clones.invitrogen.com/cloneinfo.php?clone=yeast

   Open Biosystems Products, 601 Genome Way, Huntsville, AL 35806, USA. Phone: (888) 412–2225, Fax: (256)

704–4849. e-mail: info@openbiosystems.com. http://www.openbiosystems.com/GeneExpression/Yeast/YKO/

2. The entire set of heterologous disruption cassettes can be used in one-step marker switch experiments to exchange markers within a strain. Since all cloned disruption marker genes are surrounded by identical DNA sequences an existing disruption cassette in the genome can be easily replaced with a different cassette simply by using PCR primers complementary to the flanking regions. Using the universal short oligonucleotides 5′ CAGCTGAAGCTTCGTACGC 3′ and 5′ GCATAGGCCACTAGTGGATCTG 3′, which hybridize upstream and downstream of the *loxP* sequences respectively in all pUGxx vectors (Fig. 2a), disruption cassettes can be generated harboring upstream 74 bp and downstream 66 bp of homology, respectively, to the other cassettes. Transformation of these cassettes into yeast will result in an efficient one-step marker exchange.
3. The cloned disruption cassettes (pUGxx plasmid series) and the various Cre expression plasmids (pSHxx plasmid series) are available from EUROSCARF (Frankfurt, Germany) (see Note 1 for complete address) or from our lab. Companies should contact Johannes H. Hegemann.
4. In rare cases, it may prove difficult to obtain correct transformants using the usual 40 bp of flanking homology, probably because homologous recombination is impeded (e.g., by a particular chromatin structure). Usually, extension of the homology regions to 90–100 bp solves this problem.
5. Make sure that oligonucleotides used to create the disruption cassette are of full length. The use of 5′-truncated oligonucleotides will reduce the efficiency of homologous recombination. To check the quality of oligonucleotides, one can load 2 μL of a 50 pmol/μL solution onto a 3–4% agarose gel. Comparison with control oligonucleotides of defined length gives a rough quality check.
6. Oligonucleotides used for the disruption of the *HO* gene and for its verification (5′→3′)[a]

OL5′	TATCCTCATAAGCAGCAATCAATTCTATC TATACTTTAAAcagctgaagcttcgtacgc
OL3′	ACTTTTATTACATACAACTTTTTAAACTA ATATACACATTgcataggccactagtggatctg
A	CCACGAAAAGTTCACCATAAC
B	TATTTGGTGGCATTTCTACC
C	TGGAGTGGTAAAAATCGAGT
D	AGTATCACAATTAAAATATTTG

[a]Lower case letters indicate nucleotides homologous to sequences to the left and right of the cloned disruption cassettes (see Fig. 2a).

7. Genotype of haploid yeast CEN.PK2-1C strain (26) used for disruption of the *HO* gene: *MAT*a *leu2-3,112 ura3-52 trp1-289 his3-Δ1MAL2-8C SUC2*
8. It is not necessary to separate the PCR product from the template plasmid DNA, as none of the pUG plasmids can replicate in yeast cells. If other cloned disruption cassettes are used as templates, this issue should be checked. If the plasmid used as template in the PCR to generate the disruption cassette is able to replicate autonomously in yeast cells (because it contains an ARS sequence), obviously many yeast transformants will carry the plasmid rather than the disruption cassette.
9. Details of the yeast transformation protocol can be found at http://home.cc.umanitoba.ca/~gietz/
10. Alternatively, you can boil about 5 μL of yeast cells in 50 μL 0.02 M NaOH for 15 min at 100°C and add 1 μL of this solution to the PCR mix.
11. The *GAL1* promoter expressing the Cre recombinase is already weakly active in glucose-containing media. If you are in a hurry, you can also incubate the cells for 2 days in YPD medium and then streak out and replica-plate onto selective and YPD plates. About 1–5% of the colonies will have lost the disruption marker.

References

1. Rothstein R. (1991) Targeting, disruption, replacement, and allele rescue: integrative DNA transformation in yeast. *Methods Enzymol.* **194**, 281–301.
2. Johnston M., Riles L. and Hegemann J. H. (2002) Gene disruption. *Methods Enzymol.* **350**, 290–315.
3. Winzeler E. A., Shoemaker D. D., Astromoff A., Liang H., Anderson K., Andre B., Bangham R., Benito R., Boeke J. D., Bussey H. *et al.* (1999) Functional characterization of the *S. cerevisiae* genome by gene deletion and parallel analysis. *Science* **285**, 901–906.
4. Giaever G., Chu A. M., Ni L., Connelly C., Riles L., Veronneau S., Dow S., Lucau-Danila A., Anderson K., Andre B. *et al.* (2002) Functional profiling of the *Saccharomyces cerevisiae* genome. *Nature* **418**, 387–391.
5. Hegemann J.H., Gueldener U., Koehler G.J., (2006) Gene disruption in the budding yeast *Saccharomyces cerevisiae*. In: Xiao W (ed) Yeast Protocols, 2nd edition, Humana Press, New Jersey
6. Wieczorke R., Krampe S., Weierstall T., Freidel K., Hollenberg C. P. and Boles E. (1999) Concurrent knock-out of at least 20 transporter genes is required to block uptake of hexoses in *Saccharomyces cerevisiae*. *FEBS Lett.* **464**, 123–128.
7. Güldener U., Heck S., Fiedler T., Beinhauer J. D. and Hegemann J. H. (1996) A new efficient gene disruption cassette for repeated use in budding yeast. *Nucleic Acids Res.* **24**, 2519–2524.
8. Gueldener U., Heinisch J., Koehler G. J., Voss D. and Hegemann J. H. (2002) A second set of *loxP* marker cassettes for Cre-mediated multiple gene knockouts in budding yeast. *Nucleic Acids Res.* **30**, e23.
9. Delneri D., Tomlin G.C., Wixon J.L., Hutter A., Sefton M., Louis E.J., Oliver S.G. (2000) Exploring redundancy in the yeast genome: an improved strategy for use of the cre-loxP system. *Gene.* **252**, 127–135.
10. Carter Z., Delneri D. (2010) New generation of loxP-mutated deletion cassettes for the

genetic manipulation of yeast natural isolates. *Yeast.* **27**, 765–775.

11. Fang F., Salmon K., Shen M.W., Aeling K.A., Ito E., Irwin B., Tran U.P., Hatfield G.W., Da Silva N.A., Sandmeyer S. (2010) A vector set for systematic metabolic engineering in *Saccharomyces cerevisiae. Yeast.* Sep 10 Epub.
12. Fickers P., Le Dall M.T., Gaillardin C., Thonart P., Nicaud J.M. (2003) New disruption cassettes for rapid gene disruption and marker rescue in the yeast *Yarrowia lipolytica. J Microbiol Methods.* **55**, 727–737.
13. Ribeiro O., Gombert A.K., Teixeira J.A., Domingues L.J. (2007) Application of the Cre-loxP system for multiple gene disruption in the yeast *Kluyveromyces marxianus. J Biotechnol.* **131**, 20–26.
14. Erler A., Maresca M., Fu J., Stewart A.F. (2006) Recombineering reagents for improved inducible expression and selection marker re-use in *Schizosaccharomyces pombe. Yeast.* **23**, 813–823.
15. Colot H.V., Park G., Turner G.E., Ringelberg C., Crew C.M., Litvinkova L., Weiss R.L., Borkovich K.A., Dunlap J.C. (2006) A high-throughput gene knockout procedure for *Neurospora* reveals functions for multiple transcription factors. *Proc Natl Acad Sci USA.* **103**, 10352–10357.
16. Dennison P.M., Ramsdale M., Manson C.L., Brown A.J. (2005) Gene disruption in *Candida albicans* using a synthetic, codon-optimised Cre-loxP system. *Fungal Genet Biol.* **42**, 737–748.
17. Qian W., Song H., Liu Y., Zhang C., Niu Z., Wang H., Qiu B.J. (2009) Improved gene disruption method and Cre-loxP mutant system for multiple gene disruptions in *Hansenula polymorpha. Microbiol Methods.* **79**, 253–259.
18. Iwaki T., Takegawa K. (2004) A set of loxP marker cassettes for Cre-mediated multiple gene disruption in *Schizosaccharomyces pombe. Biotechnol Biochem.* **68**, 545–550.
19. Steensma H.Y., Ter Linde J.J. (2001) Plasmids with the Cre-recombinase and the dominant nat marker, suitable for use in prototrophic strains of *Saccharomyces cerevisiae* and *Kluyveromyces lactis. Yeast.* **18**, 469–472.
20. Ikushima S., Fujii T., Kobayashi O. (2009) Efficient gene disruption in the high-ploidy yeast *Candida utilis* using the Cre-loxP system. *Biosci Biotechnol Biochem.* **73**, 879–884.
21. Patel R.D., Lodge J.K., Baker L.G. (2010) Going green in *Cryptococcus neoformans*: the recycling of a selectable drug marker.*Fungal Genet Biol.* **47**, 191–198.
22. Heinisch J.J., Buchwald U., Gottschlich A., Heppeler N., Rodicio R. (2010) A tool kit for molecular genetics of *Kluyveromyces lactis* comprising a congenic strain series and a set of versatile vectors.*FEMS Yeast Res.* **10**, 333–342
23. Pluthero F. G. (1993) Rapid purification of high-activity Taq DNA polymerase. *Nucleic Acids Res.* **21**, 4850–4851.
24. Gietz R. D. Woods R. A. (2002) Transformation of yeast by lithium acetate/single-stranded carrier DNA/polyethylene glycol method. *Methods Enzymol.* **350**, 87–96.
25. Sherman F. (2002) Getting started with yeast. *Methods Enzymol.* **350**, 3–41.
26. Entian K.D., Schuster T., Hegemann J.H., Becher D., Feldmann H., Güldener U., Götz R., Hansen M., Hollenberg C.P., Jansen G., Kramer W., Klein S., Kötter P., Kricke J., Launhardt H., Mannhaupt G., Maierl A., Meyer P., Mewes W., Munder T., Niedenthal R.K., Ramezani Rad M., Röhmer A., Römer A., Hinnen A., et al (1999) Functional analysis of 150 deletion mutants in *Saccharomyces cerevisiae* by a systematic approach. *Mol Gen Genet.* **262**, 683–702.

Chapter 13

Genome-Wide Transposon Mutagenesis in *Saccharomyces cerevisiae* and *Candida albicans*

Tao Xu, Nikë Bharucha, and Anuj Kumar

Abstract

Transposon mutagenesis is an effective method for generating large sets of random mutations in target DNA, with applicability toward numerous types of genetic screens in prokaryotes, single-celled eukaryotes, and metazoans alike. Relative to methods of random mutagenesis by chemical/UV treatment, transposon insertions can be easily identified in mutants with phenotypes of interest. The construction of transposon insertion mutants is also less labor-intensive on a genome-wide scale than methods for targeted gene replacement, although transposon insertions are not precisely targeted to a specific residue, and thus coverage of the target DNA can be problematic. The collective advantages of transposon mutagenesis have been well demonstrated in studies of the budding yeast *Saccharomyces cerevisiae* and the related pathogenic yeast *Candida albicans*, as transposon mutagenesis has been used extensively for phenotypic screens in both yeasts. Consequently, we present here protocols for the generation and utilization of transposon-insertion DNA libraries in *S. cerevisiae* and *C. albicans*. Specifically, we present methods for the large-scale introduction of transposon insertion alleles in a desired strain of *S. cerevisiae*. Methods are also presented for transposon mutagenesis of *C. albicans*, encompassing both the construction of the plasmid-based transposon-mutagenized DNA library and its introduction into a desired strain of *Candida*. In total, these methods provide the necessary information to implement transposon mutagenesis in yeast, enabling the construction of large sets of identifiable gene disruption mutations, with particular utility for phenotypic screening in nonstandard genetic backgrounds.

Key words: Transposon, Gene disruption, Insertional mutant, Genomics, Screen, Yeast, *Candida albicans*, *Saccharomyces cerevisiae*

1. Introduction

Transposons have long been utilized in the laboratory as a tool for the generation of random chromosomal mutations in a broad array of prokaryotes and eukaryotes (1–10). For these purposes, the transposable elements are routinely modified such that the

James A. Williams (ed.), *Strain Engineering: Methods and Protocols*, Methods in Molecular Biology, vol. 765,
DOI 10.1007/978-1-61779-197-0_13,

central genes within the transposon are replaced with appropriate markers and functional tags; the ends of the transposon are maintained to direct transposition. In this way, transposons can be used to generate gene disruptions, identify exons/introns, and construct misexpression alleles (11–13). As compared with other methods of mutagenesis, transposon-based approaches hold numerous advantages. Transposon mutagenesis is an excellent means to generate large numbers of mutations without an excess of labor. Also, the transposon insertions can be easily identified in mutants of interest by inverse PCR and other approaches, eliminating much of the difficulty associated with identifying causative mutations resulting from EMS mutagenesis or UV irradiation. Transposon mutagenesis, however, is not without its drawbacks. In particular, transposition is not strictly random, making it difficult to achieve saturating coverage of target DNA by transposon-based approaches (14). Also, as an obvious byproduct of transposition, the approach is not applicable for the generation of precisely directed mutations, as compared with PCR-based approaches for targeted start codon–stop codon gene replacement (15). On balance, however, transposon mutagenesis is a viable and effective method for many types of large-scale studies, particularly when directed approaches may be cost-prohibitive.

Transposon mutagenesis has been employed very successfully in a range of model organisms; here, we focus on the budding yeast *Saccharomyces cerevisiae* and the pathogenic yeast *Candida albicans* – two lower eukaryotes that have been studied extensively using transposon-based approaches. In these related yeasts, transposon mutagenesis is typically carried out with bacterial transposons (16–21), although the yeast transposable element Ty1 has also been used (22). Bacterial transposons exhibit less insertional bias than Ty1, enabling greater coverage of the target genome. Using bacterial transposons, mutagenesis of a plasmid-based yeast genomic DNA library is typically performed in vitro or in vivo in *Escherichia coli* (4). The resulting insertion alleles are subsequently shuttled into yeast, and the mutagenized DNA integrates into the genome by homologous recombination. As presented in Ross-Macdonald et al. (12) and Kumar et al. (13), Michael Snyder's group pioneered the large-scale application of this approach to generate extensive transposon insertion libraries of *S. cerevisiae* genomic DNA. These libraries were constructed using a specially modified bacterial transposon, such that a single chromosomal transposon insertion could yield a gene disruption allele, reporter-fusion construct, or epitope-tagged allele. Transposon insertion alleles of interest were sequenced, and in total, these plasmid-based collections encompass over 35,000 defined insertion alleles affecting approximately 60% of all annotated yeast genes. Plasmid-based insertion libraries can easily be prepared and introduced into a strain of interest; this is particularly useful for the analysis of gene function in nonstandard strains of *S. cerevisiae* (i.e., any strain for which targeted gene deletion collections are unavailable).

In *Candida albicans*, transposon mutagenesis has served as a powerful method for the generation of large sets of gene disruption mutants, as targeted PCR-based approaches (23) can be challenging to implement in *C. albicans* because of the decreased efficiency of DNA integration by homologous recombination relative to the efficiency observed in *S. cerevisiae*. Shuttle mutagenesis has been applied in *C. albicans* using modified bacterial transposons, as described in several significant studies (19–21). As *C. albicans* is diploid, many forward genetic screens for loss-of-function phenotypes employ heterozygous diploid insertions to screen for haploinsufficiency; this approach has proven effective in identifying genes that contribute to a given cell process and/or function within a signaling pathway. Although growing collections of targeted gene deletion constructs are now available in *C. albicans* (23–25), transposon mutagenesis remains an effective and informative means to probe gene function on a large scale. As also evident in *S. cerevisiae*, this is particularly true for loss-of-function screens in *C. albicans* that need to be performed in nonstandard or complex genetic backgrounds, in which targeted deletion mutants may be unavailable.

In this chapter, we present protocols for the implementation of transposon-based mutagenesis in *S. cerevisiae* and *C. albicans*. For studies in baker's yeast, protocols are provided detailing (1) the large-scale introduction of transposon-based insertion alleles in a desired strain of yeast; (2) the screening of transposon-mutagenized yeast strains for in-frame insertions, and (3) the subsequent identification of insertion sites in mutant strains of interest by inverse PCR. The availability of transposon insertion libraries in *S. cerevisiae* makes it unlikely that researchers will need to remutagenize yeast genomic DNA; consequently, the above protocols begin with available transposon insertion libraries as a starting point and indicate methods by which the library DNA can be introduced into a desired yeast strain. For the analysis of *C. albicans*, however, researchers may indeed need to mutagenize genomic DNA for a given screen, and thus, we provide here methods describing the following: (1) the construction of an appropriate bacterial transposon for mutagenesis, (2) its application toward the in vitro mutagenesis of a plasmid-based *C. albicans* genomic DNA library, (3) the introduction of transposon insertion alleles into *C. albicans*, and (4) a sample phenotypic screen. Collectively, these protocols outline the necessary steps in applying transposon mutagenesis for genetic screens in yeast, with particular utility for the rapid generation of large sets of identifiable loss-of-function and misexpression alleles in strains for which targeted deletion mutants may be unavailable (26).

2. Materials

2.1. Insertional Mutagenesis of S. cerevisiae

1. LB medium, sterile: 10 g tryptone, 5 g yeast extract, 5 g NaCl, 1 ml 1N NaOH, add water to 1 l. Autoclave.
2. 25 mg/ml Kanamycin: Store at 4°C for up to 3 months.
3. 5 mg/ml Tetracycline: In 50% ethanol. Store at –20°C for up to 1 year.
4. TE buffer, sterile: 10 mM Tris–Cl, pH 8.0, 1 mM EDTA, pH 8.0. Autoclave.
5. LB agar plates containing tetracycline and kanamycin: Prepare LB medium containing 15 g/l agar. After autoclaving cool to 45–50°C and add tetracycline to 3 mg/l and kanamcyin to 40 mg/l final concentration and pour plates.
6. *Not* I and *Alu* I restriction endonucleases (New England Biolabs, Ipswich, MA or equivalent).
7. One-Step buffer: 0.2 M lithium acetate, 40% (wt/vol) polyethylene glycol (PEG 4000), 100 mM 2-mercaptoethanol.
8. DNA: Sonicated salmon sperm DNA, 2 mg/ml.
9. Synthetic Complete (SC) dropout medium, sterile: Per liter, 1.3 g dropout powder, 1.7 g Yeast Nitrogen Base without Amino Acids/Ammonium Sulfate, 5 g Ammonium Sulfate, 20 g dextrose. Autoclave.
10. YPAD medium, sterile: 1% yeast extract, 2% peptone, 2% dextrose, 80 mg/l adenine. Autoclave.
11. 3-MM filter paper, cut to 4-in diameter discs (Whatman Inc., Clifton, NJ).
12. Glass petri dishes, 9-cm and 15-cm.
13. Chloroform.
14. 0.7 M Potassium phosphate, pH 7.0: Prepare 0.7 M solution of K_2HPO_4 and adjust pH to 7 with HCl.
15. 20 mg/ml X-gal: In 100% *N,N*-dimethylformamide.
16. X-gal plates, sterile: Per liter, 1.7 g Yeast Nitrogen Base without Amino Acids/Ammonium Sulfate, 5 g Ammonium Sulfate, 20 g dextrose, 20 g agar, 0.8 g dropout powder, NaOH pellet. Add water to 900 ml and autoclave. Cool to 45–50°C then add 100 ml 0.7 M potassium phosphate, pH 7.0, and 2 ml X-gal solution.
17. SC-Ura medium, sterile: Per liter, 1.3 g SC-Ura dropout powder, 1.7 g Yeast Nitrogen Base without Amino Acids/Ammonium Sulfate, 5 g Ammonium Sulfate, 20 g dextrose. Autoclave.
18. Clinical tabletop centrifuge.
19. 45°C water bath.

20. 96-well plate reader (optional).
21. Thermal cycler.
22. Blunt-end restriction enzyme and buffer (e.g., *Alu*I).
23. Oligo ABP1: GAAGGAGAGGACGCTGTCTGTCGAAGG TAAGGAACGGACGA-GAGAAGGGAGAG.
24. Oligo ABP2: GACTCTCCCTTCTCGAATCGTAACCGT TCGTACGAGAATCGCTGTCCTCTCCTTC.
25. Universal Vectorette (UV) Oligo: CGAATCGTAACCG TTCGTACGAGAATCGCT.
26. Primer PRSQZ: CGACGGGATCCCCCTTAACG.
27. Annealing buffer for inverse/vectorette PCR: 10 mM Tris–HCl, pH 8.0, 10 mM $MgCl_2$, 50 mM NaCl.
28. 10 mM rATP.
29. Primer M13(−47): CGCCAGGGTTTTCCCAGTCACGAC.
30. Taq DNA polymerase.
31. 80 mM dNTP mix: 20 mM with respect to each dNTP.

2.2. Insertional Mutagenesis of *C. albicans*

1. pGPS3 transposon donor plasmid (New England Biolabs, Ipswich, MA), customized.
2. Genomic DNA library for mutagenesis.
3. 10× Tn7 Mutagenesis Buffer: 250 mM Tris–HCl (pH 8.0), 20 mM ATP, 20 mM DTT (Dithiothreitol).
4. 300 mM magnesium acetate.
5. 3 M sodium acetate.
6. 1 M lithium acetate, sterile, (autoclaved).
7. TnsABC* Transposase: 7 μg/ml TnsA*, 10 μg/ml TnsB*, 20 μg/ml TnsC* in buffer containing 25 mM Tris–HCl (pH 7.9), 500 mM NaCl, 2 mM $MgCl_2$, 1 mM ATP, 0.5 mM DTT, 0.8 mM EDTA and 50% Glycerol (obtained from Nancy Craig's lab, Johns Hopkins University).
8. Restriction endonucleases: *Spe* I, PI-*Sce* I (VDE I), *Pvu* II with supplied buffers (New England Biolabs, Ipswich, MA or alternative).
9. ElectroMAX™ Stbl4™ cells (Invitrogen, Carlsbad, CA) or any library-efficient competent cells.
10. 95–100% ethanol; 70% ethanol.
11. 10× TE: 100 mM Tris–Cl, pH 8.0, 10 mM EDTA, pH 8.0.
12. TE: 10 mM Tris–Cl, pH 8.0, 1 mM EDTA, pH 8.0.
13. 15% glycerol, sterile (autoclaved).
14. Maxiprep plasmid isolation kit (Qiagen, Valencia, CA).
15. 10 mg/ml sonicated salmon sperm DNA; 10 mg/ml yeast tRNA.

16. 1 M LiAc.
17. TE-LiAC mix, sterile: 1 volume 10XTE, 1 volume 1 M LiAc, 8 volumes water.
18. 50% PEG (Polyethylene glycol, MW 3350): Filter-sterilized.
19. PEG-LiAC-TE mix: 8 volumes 50%PEG, 1 volume 10XTE, 1 volume 1 M LiAc.
20. 25 mg/ml kanamycin: Store at 4°C for up to 3 months.
21. 50 mg/ml ampicillin: Store at 4°C for up to 3 months.
22. LB plates with 40 mg/l kanamycin and 50 mg/l ampicillin.
23. YPD + uridine medium, sterile: 10 g yeast extract, 20 g bacto peptone, 20 g dextrose, 80 mg uridine in 1 l water. Autoclave.
24. YPD + uridine plates, sterile: YPD + uridine medium containing 15 g agar in 1 l water. Autoclave.
25. Spider medium, sterile: 10 g nutrient broth, 10 g mannitol, 2 g K_2PO_4, 13.5 g agar in 1 l water, pH 7.2 after autoclaving.
26. RNA extraction kit (Qiagen, Valencia, CA).
27. Ribopure Yeast kit (Ambion, Austin TX).
28. 0.1 M DTT (Dithiothreitol).
29. 100 mM dNTP mix: 25 mM with respect to each dNTP.
30. M-MLV reverse transcriptase with supplied First-Strand buffer (5×) (Invitrogen Corp., Carlsbad, CA).
31. Ribonuclease Inhibitor (40 U/μl).
32. Standard lab equipment: tabletop centrifuge, microcentrifuge, 45°C water bath, 25°C and 30°C shaking incubator, 30°C incubator, 65 and 75°C heat block, agarose gel equipment.
33. Oligonucleotide sequences:

    3′ RACE adapter: 5′ GCGAGCACAGAATTAATACGACT CACTATAGGT$_{12}$ 3′ (Ambion Inc., Austin TX)

    3′ RACE outer primer: 5′ GCGAGCACAGAATTAATACGA CT 3′

    URA3 Forward Primer: 5′ GACCTATAGTGAGAGAGCAG 3′
34. Phenol–chloroform (1:1): Equilibrated with 0.1 M Tris–HCl (pH 7.6).

3. Methods

3.1. Transposon Mutagenesis of S. cerevisiae

Protocols for the construction of plasmid-based transposon insertion libraries of *S. cerevisiae* genomic DNA have been described previously (4, 16), and insertional libraries (12, 13) are available upon request from the authors free of charge. The library consists of yeast genomic DNA derived from a strain lacking both its mitochondrial genome [ρ^-] as well as 2-micron DNA [cir^0]. Ten pools of this genomic DNA library were individually mutagenized using a modified tetracycline resistant Tn7-derived transposon (13, 27) (Fig. 1a). The transposon supports the generation of gene disruption alleles, reporter fusions, and epitope-tagged alleles as described more fully in Ross-Macdonald et al. (16) and Kumar et al. (13). Thus, each plasmid in the library contains an insert of yeast genomic DNA and a single Tn7 transposon insertion; since Tn7 exhibits transposition immunity (28), plasmids with multiple insertions will be rare. The kanamycin-resistant vector backbone of this library is small to minimize the possibility that a transposon insertion occurs in the plasmid rather than in the genomic DNA insert. Each mutagenized pool contains five genome equivalents of DNA, encompassing in excess of 300,000 independent insertions.

To utilize the transposon insertion library for functional analysis, the insertion alleles must be introduced into a desired strain of yeast by DNA transformation. Subsequently, mutagenized strains of interest can be selected, and the site of transposon insertion in a strain can be identified by inverse PCR. Protocols for these procedures are provided below. An overview of these steps is provided in Fig. 1b.

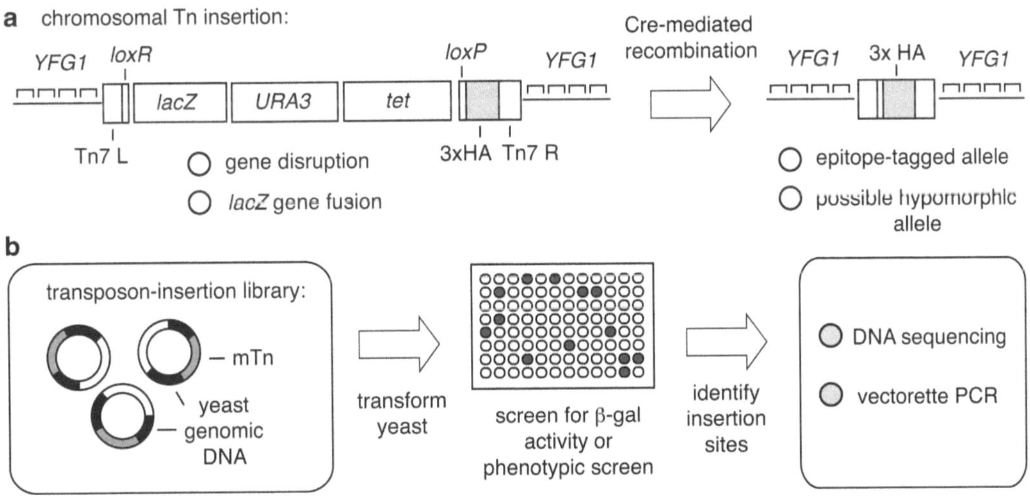

Fig. 1. Overview of the steps involved in applying transposon insertion libraries for mutagenesis in *Saccharomyces cerevisiae*. (**a**) Diagrammatic representation of the bacterial transposon used in the insertional library; the Tn7 transposon is shown here. An insertion in a sample gene (*YFG1*) is shown; brackets represent codons in the hypothetical gene sequence, and an in-frame insertion is represented here. The 3× HA sequence encodes three copies of the hemagglutinin (HA) epitope. Cre-mediated recombination at the transposon-encoded *lox* sites results in a 99-codon insertion element appropriate for epitope-tagging and the possible generation of hypomorphic alleles. (**b**) To utilize the transposon insertion library for mutagenesis, the library is amplified in *E. coli* and introduced into yeast by DNA transformation. Yeast transformants are screened for a desired phenotype, and insertion sites in mutants of interest are identified by DNA sequencing or inverse/vectorette PCR.

3.1.1. Amplifying Library DNA in E. coli for Transformation of Yeast

1. Introduce a suitable amount of mutagenized library DNA into any tetracycline- and kanamycin-sensitive *E. coli* strain by standard transformation procedures. Select transformants on LB plates supplemented with tetracycline and kanamycin (see Note 1).
2. Elute transformant colonies as follows: place 6 ml of LB medium onto each plate and scrape cells into a homogeneous suspension. Dilute an aliquot of this eluate into LB medium supplemented with tetracycline (3 μg/ml) and kanamycin (40 μg/ml) to yield a culture of nearly saturated cell density. Incubate at 37°C with aeration for 2–3 h.
3. Isolate plasmid DNA by any standard miniprep or large-scale protocol.

3.1.2. Transforming Yeast with the Mutagenized Library DNA

1. Digest a small aliquot of plasmid DNA (e.g., 1 μg) with *Not* I. Subsequently, analyze a portion of the reaction mixture by agarose gel electrophoresis to ensure release of transposon-mutagenized yeast DNA from the plasmid vector (see Note 2). Store the remaining reaction mixture for later use in step 4 below.
2. Grow a 10-ml culture of any desired *ura3* yeast strain to mid-log phase (a density of 10^7 cells/ml) maintaining appropriate selection if applicable; if no plasmids are present in the strain, then use YPD medium. To screen for disruption phenotypes, a haploid strain is often used; from previous experience, we estimate that 10% of transposon insertions in essential genes are viable. For the eventual analysis of hypomorphic mutants (e.g., for the study of essential genes), choose a *ura3 leu2* strain (see Note 3).
3. Pellet cells in a clinical tabletop centrifuge at 1,100 × *g* for 5 min. Wash once with 5 volumes of One-Step Buffer.
4. Resuspend cells in 1 ml One-Step Buffer supplemented with 1 ml of 2 mg/ml denatured salmon sperm DNA. Add 100-μl aliquots from this suspension to 0.1–1 μg *Not* I-digested plasmid DNA from step 1 (see Note 4). Vortex, and incubate at 45°C for 30 min.
5. Pellet cells and subsequently suspend in 400 μl SC-Ura medium. Spread 200-μl aliquots onto SC-Ura plates and incubate at 30°C for 3–4 days. Up to 1,000 transformants may be recovered per μg of transforming DNA (see Note 5).

3.1.3. Screening Yeast Transformants for β-Galactosidase Activity

1. The transposon used to generate this library contains a *lacZ*-reporter gene trap, such that if the transposon lands in-frame with surrounding gene coding sequence, a β-galactosidase protein chimera will be produced. To maximize detection of *lacZ* fusions expressed at low levels, patch transformant colonies onto YPD plates (supplemented with 80 μg/ml adenine if using an *ade2* host strain) at a density of up to 100 colonies per plate.

2. Place a sterile disc of Whatman 3-MM filter paper onto a plate of SC-Ura medium; repeat for as many plates as needed. Replicate transformant cells onto filter-covered plates using a multichannel pipettor or hand-pinning tool and incubate overnight at 30°C. Alternative growth conditions (e.g., growth on sporulation medium) may be substituted as desired.
3. Following overnight growth, lift filters from plates and place in the lid of a 9-cm glass petri dish. Place this lid inside a closed 15-cm petri dish containing chloroform to permeabilize the cells. Incubate for 10–30 min.
4. Place filters with colony side up onto fresh X-gal plates (5-bromo-4-chloro-3-indolyl-βD-galactopyranoside). Incubate inverted at 30°C for up to 3 days. After several days of growth, β-galactosidase levels can be reliably estimated from the observed intensity of blue staining (see Note 6).
5. Transformants containing in-frame *lacZ* -fusions may be screened for mutant phenotypes simply by incubating these mutants under desired growth conditions (e.g., in the presence and absence of a particular drug of interest).

3.1.4. Identifying Transposon Insertion Sites by Inverse PCR

Several methods are available for the identification of transposon insertion sites in mutants of interest, including direct sequencing of mutants and inverse or vectorette PCR-based approaches (18, 29, 30) (Fig. 2). In inverse PCR, genomic DNA is digested with a blunt-end restriction endonuclease possessing a 4–6 base pair recognition sequence. Digested DNA fragments are ligated to a

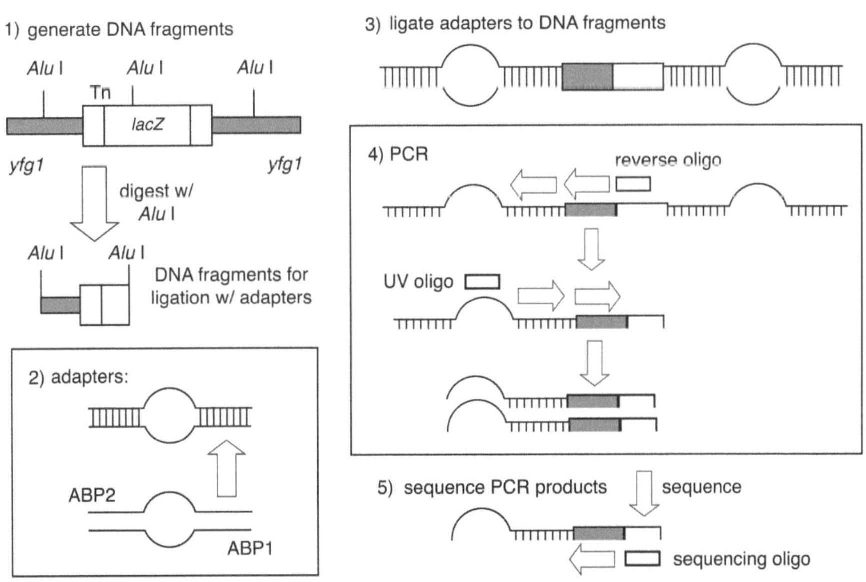

Fig. 2. Inverse or vectorette PCR for the identification of insertion sites in *S. cerevisiae* mutants of interest. The major steps are indicated. *Alu* I is indicated as an example restriction enzyme for this application; primer sequences are included in the Materials list.

pair of annealed primers containing a nonhomologous central region; these primer pairs form "anchor bubbles" flanking each genomic fragment. PCR is then performed using a primer complementary to the transposon and a primer identical to sequence within the anchor bubble. During the initial round of amplification, only the transposon primer can bind its template; however, during subsequent cycles, the anchor bubble primer can anneal to the extended transposon primer, resulting in selective amplification of DNA sequence adjacent to the point of transposon insertion. Protocols for this procedure are as follows.

1. Prepare genomic DNA by any standard protocol (see Note 7). Digest 5 μg of yeast genomic DNA with a blunt-end restriction endonuclease (such as *Alu* I) in a total volume of 20 μl. After overnight digestion, the enzyme is heat-inactivated by incubating 20 min at 65°C.
2. Primers ABP1 and ABP2 are annealed to each other to form the adaptor anchors by mixing 1 pmole of each primer in 200 μl of annealing buffer. The primer mixture is heated for 5 min at 95°C and allowed to slowly cool to 37°C.
3. The adaptors are ligated to the DNA fragments by adding 1 μl of the annealed primers, 0.25 μl of 10 mM ATP, 3 μl of 10× restriction buffer used in the digest (Buffers 1,2,3, and 4 from New England BioLabs are appropriate for use), and 24.25 μl H_2O to the 20 μl restriction digest mixture from Step 1. The ligation reaction is incubated overnight at 16°C.
4. A standard 100 μl PCR reaction is set up using 5 μl from the ligation mixture, 2.5 μl each of primers UV and M13(−47) at 20 μM, 5 U of Taq polymerase and 1 μl of 80 mM dNTP mix in a final volume of 100 μl. The PCR program consist of one cycle of 2 min at 92°C, followed by 35 cycles of 20 s at 92°C, 30 s at 67°C and 45–180 s at 72°C with a final extension of 90 s at 72°C. A 50 μl volume reaction can be used if preferred.
5. Analyze PCR products by gel electrophoresis. Each PCR product is gel-purified using standard protocols into a final volume of 30 μl TE. 10 μl of the purified product is sufficient for one sequencing reaction with primer PRSQZ (see Note 8).

3.2. Transposon Mutagenesis of *C. albicans*

Transposon-mediated gene disruption is one of the few widely available methods applicable for large-scale mutagenesis in the important human pathogen, *Candida albicans.* Various groups have used transposon mutagenesis to study gene function in this organism over the past few years by creating both heterozygous and homozygous mutants (19, 20, 31). In this text, we describe details of transposon mutagenesis of a *C.albicans* genomic DNA library to create heterozygous mutants for the study of a developmental program leading to hyphal growth. The first part of this section illustrates the use of the bacterial transposon Tn7 (32) in an in vitro mutagenesis reaction to create insertion mutants in a *Candida albicans* genomic DNA library. In this reaction, three transposase proteins act together to facilitate transposition – TnsA, TnsB, and TnsC* (33, 34). TnsB binds to the Tn7 sequences in the transprimer; TnsC* binds to the target DNA, and TnsA binds to the TnsB–DNA complex. The three proteins then enable the insertion of the transprimer into the target DNA molecules. Transposons insert randomly into target DNA; by virtue of transposition immunity, only one transposon insertion occurs within a single DNA molecule, so double insertions into a DNA fragment in the genomic library are unlikely (28).

For mutagenesis, the bacterial Tn7 transposon transprimer region was modified to include the *Candida albicans URA3*-based cassette, *URA3-dpl200* (35, 36) (see Note 9), which enables gene replacement by homologous recombination and counter-selection by 5 FOA. This *URA3* cassette was inserted adjacent to the kanamycin-resistance marker in the pGPS3 plasmid carrying Tn7. The following subsections describe protocols for introducing a genomic DNA library mutagenized with this Tn7-based transposon into the desired *C. albicans* strain as well as a sample phenotypic screen. The steps are summarized in Fig. 3.

3.2.1. Introducing the Transposon into a C. albicans Genomic DNA Library

1. 80 ng of a genomic DNA library derived from the *Candida albicans* strain WO-1 pEMBLY23 (NIH AIDS Research and Reference Reagent Program (37, 38)) is combined with 20 ng of the customized donor plasmid in a total reaction volume of 20 μl containing 1× TN7 mutagenesis buffer and 1 μl of the transposase TnsABC*. The mixture is incubated at 37°C for 10 min.
2. 1 μl of magnesium acetate (300 mM stock concentration) is then added, and the mixture is further incubated for 1 h at 37°C, followed by heat inactivation for 10 min at 75°C.
3. Digest with the restriction enzyme PI-*Sce* I for 3 h at 37°C to destroy any unreacted donor plasmid that might remain in the mix.
4. The mixture is then subjected to phenol extraction as follows. Add 100 μl phenol-chloroform mix and subject to centrifugation for 5 min at 12,750×*g* in microfuge. The mixture will separate into two layers. Remove the upper layer carefully and transfer to a clean microcentrifuge tube containing 250 μl 100% Ethanol, 10 μl NaAc (3 M stock concentration) and 0.5 μl tRNA (10 mg/ml stock concentration). Keep at –80°C for a minimum of 30 min (or –20°C for 1 h). Spin at 12,750×*g* for 30 min at 4°C and subsequently add 50 μl 70–80% ethanol. Spin for10 min at 12,750×*g* and resuspend in 30 μl 1× TE.

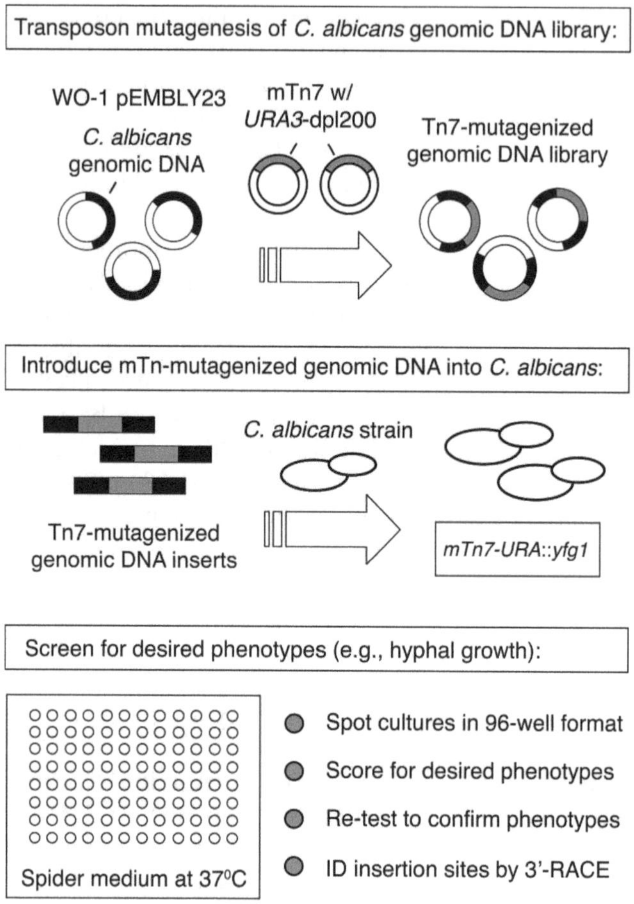

Fig. 3. Overview of steps involved in the transposon mutagenesis of *Candida albicans*. This diagram indicates use of the WO-1 genomic DNA library as described in the Methods text and the *URA3-dpl200* cassette incorporated in a modified Tn7 transposon. Specific media suggestions are included for the analysis of hyphal phenotypes in disruption mutants; alternative growth conditions can be used based on the chosen screen.

5. The mixture is then diluted tenfold, and 10 μl of the dilution is transformed by electroporation into ElectroMAX Stbl4 *E. coli* cells (Invitrogen). Transformants are plated on LB + ampicillin + kanamycin plates and incubated at 30°C for two days.
6. Multiple mutagenesis reactions are performed to allow maximum coverage (see Note 10). Cells from each mutagenesis reaction are harvested and stored in 15% glycerol. Plasmids are recovered using high-efficiency alkaline lysis (Maxiprep kit by Qiagen, Valencia, CA) for subsequent yeast transformation.

3.2.2. Creating C. albicans Mutant Strains from the Transposon Insertion Library

1. Digest 6 μg plasmid DNA (recovered from the mutagenesis reactions above) with *Pvu* II to release the genomic DNA fragments (see Note 11). Analyze a small fraction of this digest on an agarose gel to ensure that the digestion is

complete. Subject the rest of the digest to phenol extraction and elute in 25 μl TE buffer (see Note 12).

2. Grow a 5 ml culture of the desired *C. albicans* strain in appropriate medium such as YPD + uridine overnight (39). Add 100–500 μl of this overnight culture to 50 ml YPD + uridine to bring the culture to an OD_{600} of 0.1–0.2; incubate at 30°C with shaking for approximately 5 h until the culture reaches mid-log phase (OD_{600} of approximately 1).
3. Pellet the cells at 3,000 rpm (approximately 1,400 × *g*) for 5 min. Wash with 30 ml sterile water. Resuspend the cell pellet in 500 μl TE-LiAc.
4. Add 10 μl of salmon sperm DNA (10 mg/ml stock concentration) to two sterile microcentrifuge tubes. Add the entire volume of phenol-extracted DNA digest (25 μl) to one of the tubes and mix gently with a pipette tip. Add the same volume of sterile water or elution buffer to the other tube and mix; this will serve as a negative control. Add 100 μl of the resuspended cells to the microcentrifuge tubes and mix gently. Incubate the tubes at room temperature for 30 min.
5. Add 700 μl PEG-LiAC-TE mix to each tube and mix by inversion. Incubate at room temperature in a shaking water bath overnight.
6. Heat shock the cells by incubating in a 44°C water bath for 20 min (see Note 13) (40).
7. Pellet the cells at 3,000 rpm (1,400 × *g*) for 3 min. Add 150 μl sterile water to resuspend the pellet and plate on selective medium (SC-Ura in this case). Incubate the plates at 30°C for 2–3 days (see Note 14).
8. Freeze transformants in 15% glycerol in 96-well plates.

3.2.3. A Sample Phenotypic Screen

The transposon-insertion mutant strains may be used for any phenotypic screen of interest. Here, we describe a protocol to screen the mutants for hyphal growth phenotypes. The hyphal phenotype of the parental strain used for the transformation is compared to that of the mutants, and any alterations in hyphal growth are scored as positive. The screen is carried out in a 96-well format as described here.

1. Dispense 600 μl selective medium (SC-Ura in this case) in 96-well culture plates and inoculate with a small fraction of the pure colony (obtained as described above in Subheading 3.2.3) in the individual wells.
2. Allow strains to grow for approximately 24 h in a 30°C shaking incubator.
3. Using a hand-pinning tool (or multichannel pipettor), dispense a small amount (1–2 μl) onto the desired plates to be

used for the phenotypic screen (e.g., plates containing Spider medium for the analysis of hyphal growth phenotypes) (41).

4. Incubate the plates at 37°C for 5 days.
5. Colonies with altered hyphal growth relative to the starting strain are scored as positive. These strains should be retested to confirm the phenotype.

3.2.4. Identification of Transposon Insertions of Interest

This section describes the use of 3′ RACE to identify transposon insertion sites in mutants of interest. The procedure begins with extraction of cellular RNA from the strain using a standard extraction protocol and its subsequent conversion by reverse-transcription into cDNA. The cDNA is used as a template for amplification in a PCR reaction with one primer complementary to an adapter sequence and another complementary to the *URA3* selective marker. The details of the protocol are described here.

1. Inoculate 50 μl of the frozen stock of the desired strain in 4 ml SC-Ura and incubate overnight in a 30°C shaking incubator.
2. Extract total cellular RNA from the entire overnight culture (Ribopure Yeast kit).
3. Add 1–5 μg of extracted RNA to a nuclease-free microcentrifuge tube together with 1 μl 100 mM dNTP mix, 1 μl 3′ RACE adapter (500 μg/ml stock concentration) and sterile water for a total of 12 μl. Incubate at 65°C for 5 min and then keep on ice for 2 min.
4. Add 4 μl 5× First-Strand buffer, 2 μl DTT (0.1 M stock concentration) and 1 μl ribonuclease inhibitor (40 U/μl stock concentration). Mix gently and incubate at 37°C for 2 min.
5. Add 1 μl M-MLV reverse transcriptase (200 U) and mix gently. Incubate for 50 min at 37°C. Heat-inactivate the reaction for 15 min at 70°C.
6. Use 2–5 μl of the cDNA obtained in a standard PCR amplification reaction with primers complementary to the 3′ RACE adapter (3′ RACE outer primer) and the 3′ end of the *URA3* gene (*URA3* Forward Primer). The PCR product should comprise a fragment of the *URA3* marker along with the Tn7-insertion site in the genome. This PCR product can be sequenced directly using primers complementary to the *URA3* gene (see Note 15).

4. Notes

1. Approximately 10,000 transformants should be obtained per pool following overnight growth at 37°C. Electroporation may be useful if insufficient numbers of transformants are obtained by chemical transformation.

2. Upon electrophoresis, a distinct 2.1-kb band (corresponding to the vector) and broad 8-kb band should be visible: the broad 8-kb band consists of 2–3-kb inserts of yeast genomic DNA carrying the 6-kb transposon mutagenized construct.
3. Essential genes can be screened for haploinsufficiency by introducing the insertion library into a diploid strain of yeast. The Tn7 transposon used for mutagenesis carries internal *lox* sites such that the majority of the transposon can be removed by Cre-*lox* recombination. Following Cre-mediated excision of transposon sequence, residual sequence encoding an epitope tag remains. In many cases, this small residual epitope insertion sequence can result in a conditional or hypomorphic allele, of particular utility for the analysis of essential genes. For this purpose, choose a *ura3 leu2* strain, as the pGAL-Cre plasmid (for Cre-*lox* recombination), which carries the *LEU2* marker, will need to be introduced into the background strain.
4. Use a small quantity of transforming DNA to minimize generation of transformants containing more than one insertion.
5. To ensure 95% coverage of the genome (without regards to in-frame reporter activity), screen 30,000–50,000 colonies. To identify in-frame insertions within at least 95% of all yeast genes, screen approximately 180,000–200,000 transformants for β-gal activity.
6. We typically observe β-gal activity in 12–16% of transformants.
7. Care should be taken to obtain high-quality DNA, as this is critical to successful PCR amplification.
8. The PCR protocol provided here should yield approximately 200–400 ng of product, constituting sufficient template for at least 2–3 DNA sequencing reactions.
9. In place of the *URA3-dpl200* cassette, other markers may be used as well (e.g., the nourseothricin-resistance marker), particularly to avoid any position-specific effects from *URA3* insertion. The customized donor plasmid with the *URA3-dpl200* cassette is available upon request.
10. From past experience, we recommend carrying out a total of nine or more independent reactions to improve coverage of the target genomic DNA library.
11. A larger amount (2–6 μg) of the plasmid library DNA is used for each transformation reaction to ensure sufficient number of transformants per reaction.
12. This step is essential to purify and concentrate the DNA in a smaller volume for transformation.

13. The heat shock step may be performed at 42°C for 1 h instead of 44°C for 20 min. The optimal heat shock length should be determined empirically if initial results are inadequate.
14. In order to obtain pure colonies of transformants, it is advisable to restreak transformants onto fresh plates before any further analysis is carried out.
15. Alternatively, the PCR product can be cloned into a TA vector using standard cloning procedures. The vector DNA is then extracted using standard alkaline lysis; the DNA is sequenced, and BLASTN analysis can be used to identify the disrupted gene.

Acknowledgments

Research in the Kumar laboratory was supported by grant RSG-06-179-01-MBC from the American Cancer Society and National Institutes of Health grant 1R21A1084539-01.

References

1. Hensel M., Shea J. E., Gleeson C., Jones M. D., Dalton E., and Holden D. W. (1995) Simultaneous identification of bacterial virulence genes by negative selection. *Science* 269, 400–403.
2. Way J. C., Davis M. A., Morisato D., Roberts D. E., and Kleckner N. (1984) New Tn10 derivatives for transposon mutagenesis and for construction of lacZ operon fusions by transposition. *Gene* 32, 369–379.
3. Jacobs M. A., Alwood A., Thaipisuttikul I., Spencer D., Haugen E., Ernst S., Will O., Kaul R., Raymond C., Levy R., Chun-Rong L., Guenthner D., Bovee D., Olson M. V., and Manoil C. (2003) Comprehensive transposon mutant library of Pseudomonas aeruginosa. *Proc. Natl. Acad. Sci. U.S.A.* 100, 14339–14344.
4. Hoekstra M. F., Burbee D., Singer J., Mull E., Chiao E., and Heffron F. (1991) A Tn3 derivative that can be used to make short in-frame insertions within genes, *Proc. Natl. Acad. Sci. U.S.A.* 88, 5457–5461.
5. Smith V., Botstein D., and Brown P. O. (1995) Genetic footprinting: a genomic strategy for determining a gene's function given its sequence. *Proc. Natl. Acad. Sci. U.S.A.* 92, 6479–6483.
6. Devine S., and Boeke J. (1994) Efficient integration of artificial transposons into plasmid targets in vitro: a useful tool for DNA mapping, sequencing, and genetic analysis. *Nucleic Acids Res.* 22, 3765–3772.
7. Long D., Martin M., Sundberg E., Swinburne J., Puangsomlee P., and Coupland G. (1993) The maize transposable element system Ac/Ds as a mutagen in Arabidopsis: identification of an albino mutation induced by Ds insertion. *Proc. Natl. Acad. Sci. U.S.A.* 90, 10370–10374.
8. Karess R. E., and Rubin G. M. (1984) Analysis of P transposable element functions in Drosophila. *Cell* 38, 135–146.
9. Spradling A. C., Stern D. M., Kiss I., Roote J., Laverty T., and Rubin G. M. (1995) Gene disruptions using *P* transposable elements: An integral component of the Drosophila genome project. *Proc. Natl. Acad. Sci. U.S.A.* 92, 10824–10830.
10. Ivics Z., Hackett P. B., Plasterk R. H., and Izsvak Z. (1997) Molecular reconstruction of Sleeping Beauty, a Tc1-like transposon from fish, and its transposition in human cells. *Cell* 91, 501–510.
11. Burns N., Grimwade B., Ross-Macdonald P. B., Choi E.-Y., Finberg K., Roeder G. S., and Snyder M. (1994) Large-scale characterization of gene expression, protein localization and gene disruption in *Saccharomyces cerevisiae*. *Genes Dev.* 8, 1087–1105.

12. Ross-Macdonald P., Coelho P. S., Roemer T., Agarwal S., Kumar A., Jansen R., Cheung K. H., Sheehan A., Symoniatis D., Umansky L., Heidtman M., Nelson F. K., Iwasaki H., Hager K., Gerstein M., Miller P., Roeder G. S., and Snyder M. (1999) Large-scale analysis of the yeast genome by transposon tagging and gene disruption. *Nature* 402, 413–418.
13. Kumar A., Seringhaus M., Biery M., Sarnovsky R. J., Umansky L., Piccirillo S., Heidtman M., Cheung K.-H., Dobry C. J., Gerstein M., Craig N., and Snyder M. (2004) Large-Scale Mutagenesis of the Yeast Genome Using a Tn7-Derived Multipurpose Transposon. *Genome Res.* 14, 1975–1986.
14. Seringhaus M., Kumar A., Hartigan J., Snyder M., and Gerstein M. (2006) Genomic analysis of insertion behavior and target specificity of mini-Tn7 and Tn3 transposons in Saccharomyces cerevisiae. *Nucleic Acids Res.* 34, e57.
15. Winzeler E. A., Shoemaker D. D., Astromoff A., Liang H., Anderson K., Andre B., Bangham R., Benito R., Boeke J. D., Bussey H., Chu A. M., Connelly C., Davis K., Dietrich F., Dow S. W., Bakkoury M. E., Foury F., Friend S. H., Gentalen E., Giaever G., Hegemann J. H., Laub T. J. M., Liao H., Liebundguth N., Lockhart D. J., Lucau-Danila A., Lussier M., M'Rabet N., Menard P., Mittmann M., Pai C., Rebischung C., Revuelta J. L., Riles L., Roberts C. J., Ross-MacDonald P., Scherens B., Snyder M., Sookhai-Mahadeo S., Storms R. K., Véronneau S., Voet M., Volckaert G., Ward T. R., Wysocki R., Yen G. S., Yu K., Zimmermann K., Philippsen P., Johnston M., and Davis R. W. (1999) Fuctional characterization of the *S. cerevisiae* genome by gene deletion and parallel analysis. *Science* 285, 901–906.
16. Ross-Macdonald P., Sheehan A., Roeder G. S., and Snyder M. (1997) A multipurpose transposon system for analyzing protein production, localization, and function in *Saccharomyces cerevisiae*. *Proc. Natl. Acad. Sci. U.S.A.* 94, 190–195.
17. Kumar A., DesEtages S., Coelho P., Roeder G., and Snyder M. (2000) High-throughput methods for the large-scale analysis of gene function by transposon tagging. *Methods Enzymol.* 328, 550–574.
18. Kumar A., Vidan S., and Snyder M. (2002) Insertional mutagenesis: transposon-insertion libraries as mutagens in yeast. *Methods Enzymol.* 350, 219–229.
19. Davis D. A., Bruno V. M., Loza L., Filler S. G., and Mitchell A. P. (2002) Candida albicans Mds3p, a conserved regulator of pH responses and virulence identified through insertional mutagenesis. *Genetics* 162, 1573–1581.
20. Uhl M. A., Biery M., Craig N., and Johnson A. D. (2003) Haploinsufficiency-based large-scale forward genetic analysis of filamentous growth in the diploid human fungal pathogen C.albicans. *EMBO J.* 22, 2668–2678.
21. Nobile C. J., and Mitchell A. P. (2009) Large-scale gene disruption using the UAU1 cassette. *Methods Mol. Biol.* 499, 175–194.
22. Smith V., Chou K. N., Lashkari D., Botstein D., and Brown P. O. (1996) Functional analysis of the genes of yeast chromosome V by genetic footprinting. *Science 1996* 274, 2069–2074.
23. Blankenship J. R., Fanning S., Hamaker J. J., and Mitchell A. P. An extensive circuitry for cell wall regulation in Candida albicans. *PLoS Pathog.* 6, e1000752.
24. Homann O. R., Dea J., Noble S. M., and Johnson A. D. (2009) A phenotypic profile of the Candida albicans regulatory network. *PLoS Genet.* 5, e1000783.
25. Noble S. M., French S., Kohn L. A., Chen V., and Johnson A. D. Systematic screens of a Candida albicans homozygous deletion library decouple morphogenetic switching and pathogenicity. *Nat. Genet.* 42, 590–598.
26. Jin R., Dobry C. J., McCown P. J., and Kumar A. (2008) Large-scale analysis of yeast filamentous growth by systematic gene disruption and overexpression. *Mol. Biol. Cell* 19, 284–296.
27. Kumar A. (2008) Multipurpose transposon insertion libraries for large-scale analysis of gene function in yeast. *Methods Mol. Biol.* 416, 117–129.
28. Stellwagen A. E., and Craig N. L. (1997) Avoiding self: two Tn7-encoded proteins mediate target immunity in Tn7 transposition. *EMBO J.* 16, 6823–6834.
29. Ochman H., Gerber A. S., and Hartl D. L. (1988) Genetic applications of an inverse polymerase chain reaction. *Genetics* 120, 621–623.
30. Riley J., Butler R., Ogilvie D., Finniear R., Jenner D., Powell S., Anand R., Smith J. C., and Markham A. F. (1990) A novel, rapid method for the isolation of terminal sequences from yeast artificial chromosome (YAC) clones. *Nucleic Acids Res.* 18, 2887–2890.
31. Richard M. L., Nobile C. J., Bruno V. M., and Mitchell A. P. (2005) Candida albicans biofilm-defective mutants. *Eukaryot. Cell* 4, 1493–1502.
32. Craig N. L. (1991) Tn7: a target site-specific transposon. *Mol. Microbiol.* 5, 2569–2573.

33. Biery M., Stewart F., Stellwagen A., Raleigh E., and Craig N. (2000) A simple in vitro Tn7-based transposition system with low target site selectivity for genome and gene analysis. *Nucleic Acids Res.* 28, 1067–1077.
34. Stellwagen A., and Craig N. (1997) Gain-of-Function Mutations in TnsC, an ATP-Dependent Transposition Protein That Activates the Bacterial Transposon Tn7. *Genetics* 145, 573–585.
35. Wilson R. B., Davis D., Enloe B. M., and Mitchell A. P. (2000) A recyclable Candida albicans URA3 cassette for PCR product-directed gene disruptions. *Yeast* 16, 65–70.
36. Boeke J. D., Trueheart J., Natsoulis G., and Fink G. R. (1987) 5-Fluoroorotic acid as a selective agent in yeast molecular genetics. *Methods Enzymol.* 154, 164–175.
37. Slutsky B., Staebell M., Anderson J., Risen L., Pfaller M., and Soll D. R. (1987) "White-opaque transition": a second high-frequency switching system in Candida albicans. *J. Bacteriol.* 169, 189–197.
38. Baldari C., and Cesareni G. (1985) Plasmids pEMBLY: new single-stranded shuttle vectors for the recovery and analysis of yeast DNA sequences. *Gene* 35, 27–32.
39. Bensen E. S., Clemente-Blanco A., Finley K. R., Correa-Bordes J., and Berman J. (2005) The mitotic cyclins Clb2p and Clb4p affect morphogenesis in Candida albicans. *Mol. Biol. Cell* 16, 3387–3400.
40. Walther A., and Wendland J. (2003) An improved transformation protocol for the human fungal pathogen Candida albicans. *Curr. Genet.* 42, 339–343.
41. McNemar M. D., and Fonzi W. A. (2002) Conserved serine/threonine kinase encoded by CBK1 regulates expression of several hypha-associated transcripts and genes encoding cell wall proteins in Candida albicans. *J. Bacteriol.* 184, 2058–2061.

Chapter 14

Signature-tagged Mutagenesis to Characterize Genes Through Competitive Selection of Bar-coded Genome Libraries

Julia Oh and Corey Nislow

Abstract

The availability of collections of genome-wide deletion mutants greatly accelerates systematic analyses of gene function. However, each of the thousands of genes that comprise a genome must be phenotyped individually unless they can be assayed in parallel and subsequently deconvolved. To this end, unique molecular identifiers have been developed for a variety of microbes. Specifically, the addition of DNA "tags," or "bar codes," to each mutant allows all mutants in a collection to be pooled and phenotyped in parallel, greatly increasing experimental throughput. In this chapter, we provide an overview of current methodologies used to create such tagged mutant collections and outline how they can be applied to understand gene function, gene-gene interactions, and drug-gene interactions. Finally, we present a methodology that uses universal TagModules, capable of bar coding a wide range of microorganisms, and demonstrate its reduction to practice by creating tagged mutant collections in the pathogenic yeast *Candida albicans*.

Key words: Bar coding, Signature-tagged mutagenesis, Transposon mutagenesis, TagModules, Deletion collection

1. Introduction

1.1. Systematic Genome Annotation Using Parallel Phenotypic Assays

Next-generation sequencing methods have enabled the study of metagenomics, permitting systematic genomic studies of non-model microorganisms. However, characterization of gene function has significantly lagged the acceleration of genome decoding. To characterize and functionally annotate the wealth of genome sequence information now available, methodologies to rapidly and systematically determine gene function in a wide range of microorganisms are required. Indeed, functional annotation of

James A. Williams (ed.), *Strain Engineering: Methods and Protocols*, Methods in Molecular Biology, vol. 765,
DOI 10.1007/978-1-61779-197-0_14,

the majority of genes in a genome requires phenotypic characterization in a large range of treatment conditions or environments, as gene disruption or deletion phenotypes are often condition-specific (1). Considering the thousands of genes present in even the smallest microbial genome, combined with the potentially hundreds of experimental treatments required for detailed functional annotation, there is a clear need for methodologies to rapidly create knockouts of gene function and to multiplex experiments with such mutants.

In microorganisms, an effective solution to multiplex phenotypic assays has been accomplished by pooling mutants and growing them in a competitive environment, selecting against those strains that lack the gene product necessary for survival in that condition (2, 3). This approach requires a method to track the abundance of each individual mutant during the competitive growth. By introducing unique DNA tags, or bar codes, into each mutant, the tags can be used as a proxy to measure strain abundance. Following tag amplification from a mixed population and detection of the tags (via hybridization to a microarray containing the tag complements, or by tag counting using high-throughput sequencing), one can quantify strain abundance, which in turn reflects upon a gene's importance for growth in a particular condition.

1.2. Signature-Tagged Mutagenesis to Generate Tagged Mutants

The principle of introducing unique DNA tags to individual strains has been termed signature-tagged mutagenesis (STM). The first application of STM was in the random mutagenesis of the human pathogen *Salmonella typhimurium*, using small pools of DNA tags that were ligated to transposons (3). Following multiplexed growth in a murine model, tags were radiolabeled via PCR and quantified by hybridization to complementary probes. Following that landmark publication, a number of subsequent studies have used a variety of tagging, mutagenesis, and detection methods. Tags can range from sequences flanking transposon insertion sites (4, 5) to PCR products of alternative sizes generated by different sets of primers (6) or synthetic oligonucleotides (2, 3). These tags can be detected via PCR analysis (7), hybridization to tag complement microarrays, (2, 4, 8, 9), or high-throughput sequencing (5, 10). There are also numerous ways to generate the mutants that carry the tags. One useful method is to employ transposon mutagenesis either in vivo or in vitro. Direct in vivo mutagenesis has been most successfully employed in bacteria; transposons can be directly introduced into the organism of interest via transformation or conjugation (3). In shuttle and in vitro mutagenesis, a heterologous genomic library is mutagenized, and genomic fragments containing transposon insertions are then transformed into the target organism, disrupting gene function via homologous recombination (11, 12). This latter approach has been effective in not only bacteria but also yeast, such as *Saccharomyces cerevisiae*,

Candida albicans, and *Candida glabrata* (13–16). Finally, targeted gene disruption approaches typically rely on homologous recombination to knock out gene function; for example, upstream and downstream sequences that are part of a deletion cassette are recombined into specific genome locations (2, 17).

1.3. The Value of Archived, Tagged, Mutant Collections

The targeted deletion approach has been applied perhaps most widely to the budding yeast *S. cerevisiae.* This yeast has long been a model system for eukaryotes (as *Escherichia coli* has been used to model various prokaryotes), and studies in yeast have provided insight into a number of fundamental cellular processes, for example cell cycle progression (18) and gene expression regulation (19). Following the publication of the yeast genome (20) – the first eukaryotic genome to be sequenced – an international consortium made targeted knockouts of all ~6,000 open reading frames (ORFs) in the genome, including with each deletion a pair of 20 base DNA tags. This large-scale project, coordinated among many laboratories, has been an extraordinary resource for understanding gene function. Various forms of the tagged yeast knockout (YKO) collection (including heterozygous and homozygous diploid knockouts for nearly every yeast ORF and haploid mutants for each mating type) have been used in pooled phenotypic profiling as well as on an individual basis to examine gene function in a wide variety of conditions. Fundamental discoveries on the nature of gene networks (21–24), genome-wide haploinsufficiency (25), drug target and mechanism of action (26, 27), and the essentiality of all genes in the genome (1) have been uncovered in *S. cerevisiae.*

Given the utility and success of the *S. cerevisiae* collection and its derivatives in related strains, targeted deletion efforts are being applied to other microorganisms, such as the fission yeast *Schizosaccharomyces pombe* (21, 28, 29) and the pathogenic yeast *Candida albicans* (30, 31). However, to successfully expand this methodology to all microorganisms of medical, industrial, or environmental significance would be a daunting task, as organism-specific deletion cassettes would have to be generated for each gene in each genome, and an efficient homologous recombination system (either endogenous or engineered) is required.

1.4. A Universal Approach to Generating Tagged Mutants

To bypass the need for homologous recombination and gene-by-gene deletions, we designed a generalizable method (Fig. 1) to create large numbers of individually archived, tagged mutants and applied it to the bacterium *Shewanella oneidensis* MR-1 and *C. albicans* (32, 33). Briefly, we chose transposon mutagenesis for its ability to rapidly generate large numbers of gene disruptions and combined it with *S. cerevisiae* DNA tag technology. The *S. cerevisiae* tags feature a pair of 20 base tags designed to have a similar melting temperatures, minimal cross-hybridization, and low levels of secondary structure that can occlude hybridization to an array (17).

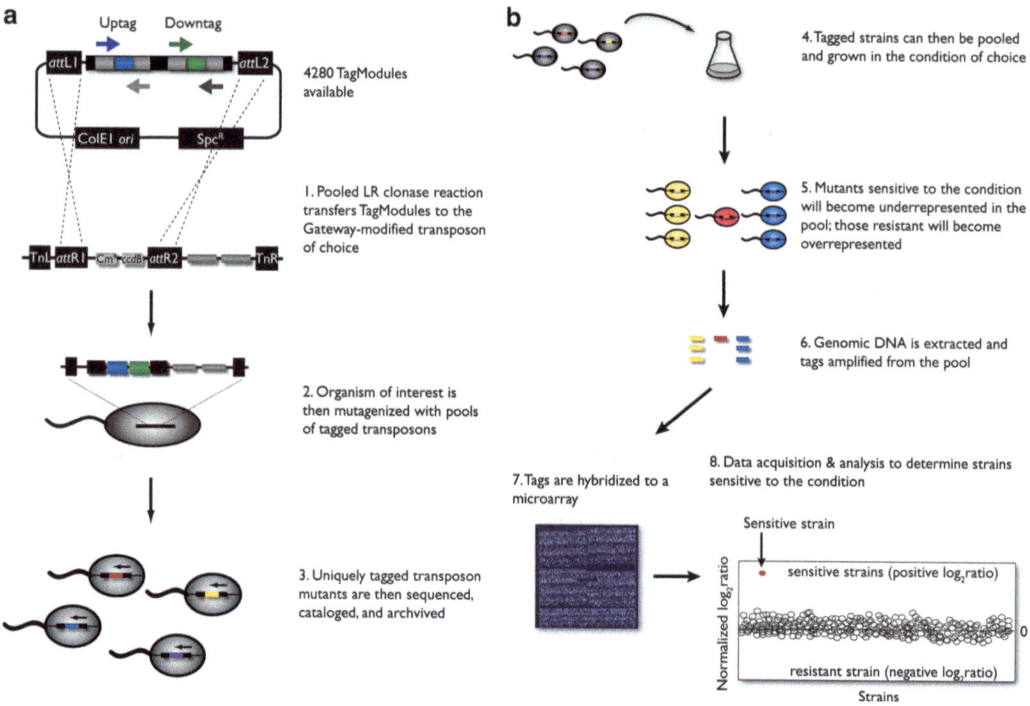

Fig. 1. Workflow for TagModule-based transposon mutagenesis of the organism of interest. (**a**). Each TagModule contains a unique uptag (*blue*) and a unique downtag (*green*) flanked by universal priming sites (*arrows*). The TagModule was cloned into the Gateway-compatible entry vector pCR8/GW/TOPO (Spc^R indicates spectinomycin resistance; ColE1 *ori* indicates the origin of replication) and is flanked by the *att*L1 and *att*L2 recombination sites (*black boxes*). To use the TagModules for tagged transposon mutagenesis, the transposon delivery plasmid is made Gateway-compatible by the addition of an *att*R1-*ccdB*-Cm^R-*att*R2 cassette (Step 1). Addition of LR clonase to a pool of TagModule entry clones and a transposon destination vector (Step 1) induces recombination at the *att* sites, replacing the Gateway conversion cassette with the TagModule. The resulting pool of tagged transposons can then be used to mutagenize the organism of interest (Step 2). TnL and TnR are the ends of the transposon, and each TagModule unique to a mutant is represented by a different color rectangle. To identify each mutant, individual mutants are sequenced (Step 3). (**b**). Uniquely tagged mutants can then be pooled and grown competitively in the experimental treatment of choice (Step 4). Strains lacking gene products necessary for growth in the treatment will grow more slowly and become underrepresented in the pool (e.g., *red* and *purple strains*); resistant strains will grow more quickly and be overrepresented (e.g., *blue* and *yellow strains*, Step 5). Following pooled growth, the genomic DNA is extracted and the uptags and downtags amplified with the common primers (Step 6). Hybridization of the tags to a microarray (Step 7) containing the tag complements (an alternative is tag counting via next-generation sequencing) yields information on the abundance of each tag, which is a proxy for the relative fitness of that mutant under that treatment condition (Step 8).

These tags are flanked by common priming sites that allow amplification of all tags simultaneously from a pool in a single PCR reaction. To leverage the supporting infrastructure and analysis tools generated for Affymetrix TAG4 arrays (e.g., optimization of screening and hybridization conditions and microarray processing), we used the original YKO bar codes. However, a portion of the *S. cerevisiae* tags contains sequence errors that degrade their hybridization performance (10, 32, 34). We, therefore, created a new, universal tag resource (termed TagModules) that is sequence-verified, quantitative, and reproducible as an in vitro tagging resource. Not only is the TagModule collection a source

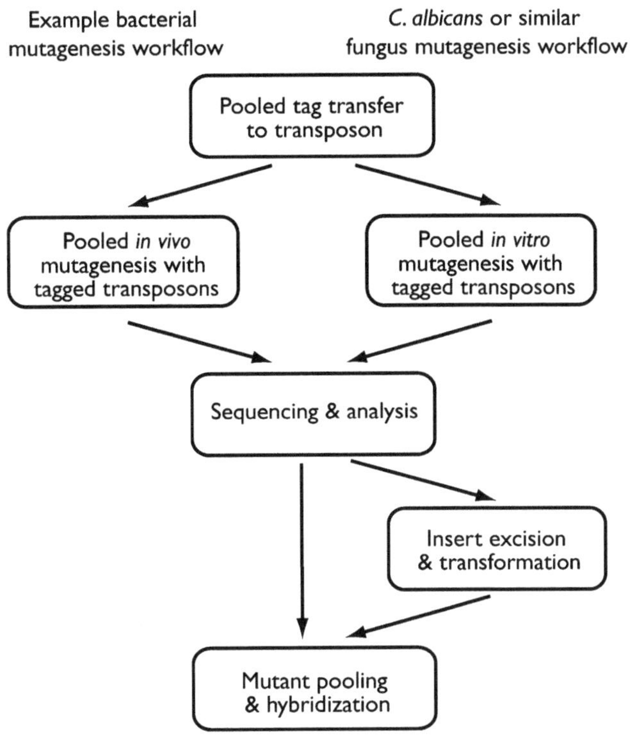

Fig. 2. Workflow for using the TagModules for genome mutagenesis is tailored to the organism of interest.

of tags that can be useful for any application requiring sample multiplexing – from the tagging of existing deletion/disruption collections to multiplexing high-throughput sequencing runs – when combined with transposon mutagenesis, these TagModules are adaptable to very different organisms to uncover genes necessary for growth in a variety of conditions (Fig. 2).

2. Materials

Catalog numbers, where available, are included in parentheses.

2.1. Commonly Used Reagents

1. Applicable antibiotics for selection (here, carbenicillin to a final concentration of 50 μg/mL (Sigma, C1613), kanamycin to 50 μg/mL (Sigma, K1876), spectinomycin to 10 μg/mL (Sigma, S0692), and chloramphenicol to 34 μg/mL (Sigma, C0378)).
2. LB liquid and solid agar medium in 100 × 15-mm Petri dishes and 245 × 245 × 18-mm Square BioDish XL Petri dishes (BD Falcon, 351040).
3. One Shot® MAX Efficiency® DH5α™-T1R Competent Cells (Invitrogen, 12297016).

4. TransforMax™ EC100™ Electrocompetent *E. coli* (Epicentre Biotechnologies, EC10010).
5. Antarctic Phosphatase (New England Biolabs, M0289L).
6. Quick Ligation Kit (New England Biolabs, M2200L).
7. QIAprep Spin Miniprep Kit (Qiagen, 27106).
8. HiSpeed Plasmid Maxi Kit (Qiagen, 12663).
9. QIAquick PCR Purification Kit (Qiagen, 28106).
10. Disposable pipetting reservoirs.
11. Multichannel pipettes (1000, 200, and 20 μL).
12. 5 and 50-mL centrifuge tubes.
13. QIAquick Gel Extraction Kit (Qiagen, 28704).
14. 10× TAE buffer (Sigma, T8280): bring to 1× by adding 100–900 mL ddH_2O.
15. Agarose, loading dye, and nucleic acid stain suitable for gel electrophoresis.
16. 1 Kb Plus DNA Ladder (Invitrogen, 10787026).
17. Taq DNA Polymerase with Standard Taq (Mg-free) Buffer (New England Biolabs, M0320L).
18. Deoxynucleotide Solution Mix (New England Biolabs, N0447L).
19. 25 mM $MgCl_2$ (Sigma, 63036).
20. YPD broth: Mix 10 g yeast extract (Sigma, Y1625), 20 g Bacto™ peptone (BD Biosciences, 211677), 20 g dextrose (Sigma, D9434), and 1 L ddH_2O to a 1-L bottle. Autoclave.
21. 50% glycerol. To make 1 L, add 500 mL ddH_2O to 500 mL 100% glycerol (Sigma, G5516) and autoclave.
22. 96- and 384-Well Deep Well Plates (Axygen Scientific, P-2ML-SQ-C-S & P-384240SQCS) and 1.1-mL Deep Well Plate, Half Height, 96-Well PP (Phenix Research Products, MAX-9610).
23. AeraSeal Sterile Microporous Sealing Film (Phenix Research Products, LMT-AERAS-EX).
24. 96-well and 384-well PCR plates and seal film.
25. Plate roller for sealing multiwell plates (Sigma, R1275).
26. Method for filter-sterilization.
27. 30 and 37°C shaking incubators for growing bacterial and yeast on plates and in tubes.

2.2. Creating a Gateway-Compatible Transposon

1. Transposon of choice containing a selectable marker for use in the organism of choice. Here, we used the EZ-Tn5™ pMOD™-6<KAN-2/MCS> Transposon Construction Vector (Epicentre Biotechnologies, MOD7906), which contains a kanamycin-resistance marker for selection of transposon

insertions in *E. coli*. This vector was modified to contain the *UAU1* cassette for selection of Arg+ mutants in *C. albicans* (32, 33). We refer to this here as Tn5-*UAU1*. See Note 1 for selection of a transposon.

2. Gateway® Vector Conversion System with One Shot® *ccdB* Survival Cells with Reading Frame A, B, or C.1 (Invitrogen, 11828029).
3. *SnaB*I restriction enzyme, 10× NEBuffer 4, and 100× BSA (New England Biolabs, R0130L).
4. *BsrG*I restriction enzyme, 10× NEBuffer 2, and 100× BSA (New England Biolabs, R0575L).
5. LB agar plates containing chloramphenicol, kanamycin, and carbenicillin.

2.3. TagModule Transfer to Transposon

1. TagModule collection, created in (32). See reference for how to obtain this collection.
2. Gateway® LR Clonase® II enzyme mix (Invitrogen, 11791020).
3. Cell Scrapers, Sterile (Greiner Bio-One, 541070).
4. *PshA*I restriction enzyme, 10× NEBuffer 4, and 100× BSA (New England Biolabs, R0593L).
5. LB agar 245 × 245 × 18-mm plates containing kanamycin and carbenicillin.

2.4. Constructing a Genomic Library

1. pUC19 plasmid (Invitrogen, 15364-011).
2. Oligonucleotides "polylinker" 5′ -CCTAGGTCCGGAACTA GTGATATCGGCCGGCCACGCGT-3′, "polylinker_EcoRI_L" 5′- ATCGATCGGAATTCATCCCTAGGTCC-3′, and "polylinker_EcoRI_R" 5′- ATCGATCGGAATTCA TCACGCGTGGC -3′ (IDT DNA, standard desalted) at 100 μM.
3. *EcoR*I restriction enzyme, 10× NEBuffer EcoRI (New England Biolabs, R0101).
4. *EcoR*V restriction enzyme, 10× NEBuffer 3, 100× BSA (New England Biolabs, R0195L).
5. *Spe*I restriction enzyme, 10× NEBuffer 2, 100× BSA (New England Biolabs, R0133).
6. QIAGEN Genomic-tip 100/G (Qiagen, 10243).
7. 50 mg/mL Zymolyase®-20 T (Amsbio, 120491-1): Dissolve 1 g into 20 mL 1 M sorbitol (add 91 g d-sorbitol (Sigma, W302902) to 500 mL ddH$_2$O and filter-sterilize; store aliquots at −20°C).
8. Proteinase K (Qiagen, 19131).
9. LB agar 245 × 245 × 18-mm plates containing carbenicillin.

2.5. In Vitro Transposon Mutagenesis

1. EZ-Tn5™ Transposase (Epicentre Biotechnologies, TNP92110).

2.6. Sequence Identification of Inserts

1. Appropriate sequencing primer at 100 μM in ddH_2O (here, U1: 5′ATGCGATGTCCACGAGGTCTCT-3′, IDT DNA, standard desalted).
2. BigDye® Terminator v3.1 (Applied Biosystems, 4336935).
3. 5× sequencing buffer: 400 mM Tris–HCl, pH 9.0 (Teknova, T1090) and 10 mM $MgCl_2$ (Sigma, 63036) sterilized through a 0.22-μm filter.
4. Seqprep™ 384 Plasmid Prep Kit or Seqprep™ 96 HT Plasmid Prep Kit (Edge Biosystems).
5. Performa® DTR 96- or 384-Well Plates (Edge Biosystems).

2.7. High-Throughput Transformation Via Homologous Recombination

1. Seqprep™ 96 HT Plasmid Prep Kit (Edge Biosystems).
2. 50% polyethylene glycol. For 500 mL, measure 250 g polyethylene glycol 3350 (Sigma, P4338) into a bottle and add ddH_2O to 500 mL. Stir overnight to dissolve and autoclave. Discard after 6 months.
3. 1 M lithium acetate. For 500 mL, add 33 g lithium acetate (Sigma, 517992) to 500 mL ddH_2O. Filter-sterilize.
4. 6-well plates, sterile (Corning, 3335).
5. 50 mg/mL uridine. Mix 1 g uridine (Sigma, U3750) with 100 mL water and filter-sterilize. This makes a 500× stock solution. Discard after 1 month.
6. 100× Tris–EDTA (TE) buffer solution (Sigma, T9285). To make a 10× TE stock, add 50–450 mL ddH_2O and filter-sterilize.
7. 1× TE/0.1 M LiOAc. To make 500 mL, add 5 mL 100× TE and 50 mL 1 M LiOAc to 445 mL ddH_2O.
8. SC–Arg + uridine selection medium: 1.7 g yeast nitrogen base without amino acids and without ammonium sulfate (Sunrise Science Products), 0.75 g dropout mix (SC –Arg –Ura) (Sunrise Science Products), 20 g dextrose (Sigma, D9434), and ddH_2O to 1 L. If making agar plates, add 20 g agar (Sigma, A1296). Autoclave, cool to ~50°C and then add 2 mL 50 mg/mL uridine (final concentration 100 μg/mL). Mix well. To make 6-well agar plates, add ~2 mL medium to each well.
9. Sonicated salmon sperm DNA kit (Agilent Technologies, 201190) or equivalent carrier DNA, 10 mg/mL. Boil 2 min just prior to use and use immediately.
10. Transforming mix for one 96-well plate: 32 mL 50% polyethylene glycol, 4 mL 10× TE, and 4 mL 1 M LiOAc. Prepare freshly for each transformation.

2.8. Construction of Strain Pool

1. Sterile 500-mL flask.
2. 200 μL 12-strip PCR tubes.

2.9. Experimental Pool Growth

1. 48-well plates (Greiner, M9437) if growing cultures in plates.
2. Adhesive plate seals (ABgene, AB-0580) if growing cultures in plates.
3. 200-mL culture flask (if growing cultures in flasks).
4. Spectrophotometer capable of OD_{600} absorbance measurement.
5. Temperature-controlled shaker for 250-mL flasks or shaking spectrophotometer such as Tecan Genios Spectrafluor Plus (Tecan).
6. Safe-Lock Microcentrifuge Tube, 2 mL (Eppendorf, 0030 120.094) for sample storage.

2.10. Preparation for Hybridization

1. Up primer mix: Dissolve Uptag (5′-GATGTCCAC GAGGTCTCT- 3′) and Buptagkanmx4 (5′ biotin-GTC-GACCTGCAGCGTACG- 3′) each in ddH_2O at 100 μM, then mix at a 1:1 ratio for a final concentration of 50 μM each. Store at −20°C. (IDT DNA, standard desalting).
2. Down primer mix: Dissolve Dntag (5′-CGGTGTCGGT CTCGTAG- 3′) and Bdntagkanmx4 (5′ biotin-GAAAAC-GAGCTCGAATTCATCG-3′) each in ddH_2O at 100 μM, then mix at a 1:1 ratio for a final concentration of 50 μM each. Store at −20°C. (IDT DNA, standard desalting).
3. Genflex Tag 16 K Array v2 (Affymetrix, 511331).
4. Hybridization Oven 640 (Affymetrix, 800138).
5. GeneChip Fluidic Station 450 (Affymetrix, 00-0079).
6. GeneArray Scanner 3000 (Affymetrix, 00-0212).
7. Safe-Lock Microcentrifuge Tube, 0.5 mL (Eppendorf, 0030 123.301).
8. Boiling water bath with floating rack for 0.5-mL tubes.
9. Teeny Tough-Spots (Diversified Biotech, LTTM1000).
10. Denhardt's Solution, 50× concentrate (Sigma, D2532).
11. Streptavidin, R-phycoerythrin conjugate (SAPE) (Invitrogen, S866). Store at 4°C; do not freeze. Discard after 6 months.
12. B213 oligonucleotide: (5′ biotin-CTGAACGGTAGCATCT-TGAC- 3′, IDT DNA, standard desalting).
13. Mixed oligonucleotides: Dissolve each of the following eight oligos (IDT DNA, standard desalted) in ddH_2O at 100 μM: Uptag (5′-GATGTCCACGAGGTCTCT- 3′), Dntag (5′-CGGTGTCGGTCTCGTAG-3′), Uptagkanmx (5′- GTC GACCTGCAGCGTACG-3′), Dntagkanmx (5′-GAAAAC

GAGCTCGAATTCATCG-3′), Uptagcomp (5′-AGAGACCTCGTGGACATC-3′), Dntagcomp (5′-CTACGAGACCGACACCG-3′), Upkancomp (5′-CGTACGCTGCAGGTCGAC-3′), Dnkancomp (5′-CGATGAATTCGAGCTCGTTTTC-3′). Mix an equal volume of each of the eight oligonucleotides for a final concentration of 12.5 μM each.

14. 0.5 M EDTA (BioRad, 161-0729).
15. 10% Tween: (Sigma, T2700).
16. 12× MES stock: For 10 mL, dissolve 0.7 g MES free acid monohydrate (Sigma, M5287) and 1.93 g MES sodium salt (Sigma, M5057) in 8 mL molecular-biology-grade water (e.g., Sigma). After mixing well, check pH; it should range from pH 6.5–6.7. Add water to a total volume of 10 mL. Filter-sterilize and store at 4°C protected from light (e.g., wrap the tube in foil). Replace if solution becomes visibly yellow or after 6 months, whichever comes first.
17. 2× Hybridization buffer: For 50 mL, mix 8.3 mL 12× MES stock, 17.7 mL 5 M NaCl (Sigma, 71386), 4.0 mL 0.5 M EDTA, 0.1 mL 10% Tween 20 (vol/vol), and 19.9 mL filtered ddH_2O. Filter-sterilize.
18. Wash A: Mix 300 mL 20× SSPE (Sigma, S2015), 1 mL 10% Tween (vol/vol), 699 mL ddH_2O. Filter-sterilize.
19. Wash B: Mix 150 mL 20× SSPE (Sigma, S2015), 1 mL 10% Tween (vol/vol), and 849 mL ddH_2O. Filter-sterilize.
20. YeaStar genomic DNA kit (Zymo Research, D2002).

3. Methods

Briefly, the protocol involves (1) the creation of a Gateway-compatible transposon into which the TagModules can be transferred, (2) transfer of the TagModules into the transposon, (3) tagged mutagenesis of the organism of interest, and (4) identification of tagged, gene disrupted mutants (outlined in Fig. 1). These sequence-identified mutants can then be arrayed as a tagged mutant collection or used in pooled growth using the tags to multiplex experiments. We provide an organism-agnostic overview of this process in Fig. 2, but for purposes of illustrating this protocol, we have detailed the protocol for the creation of a tagged transposon mutant collection in *C. albicans* (outlined in Fig. 3), as well as the methodology for performing phenotypic screens with the collection in a pooled format (Fig. 4).

3.1. Creating a Gateway-Compatible Transposon

1. To create a blunt end for ligation of the Gateway conversion cassette, in a 50 μL reaction volume, digest 2 μg of the Tn5-*UAU1* transposon with *SnaB*I in the following reaction

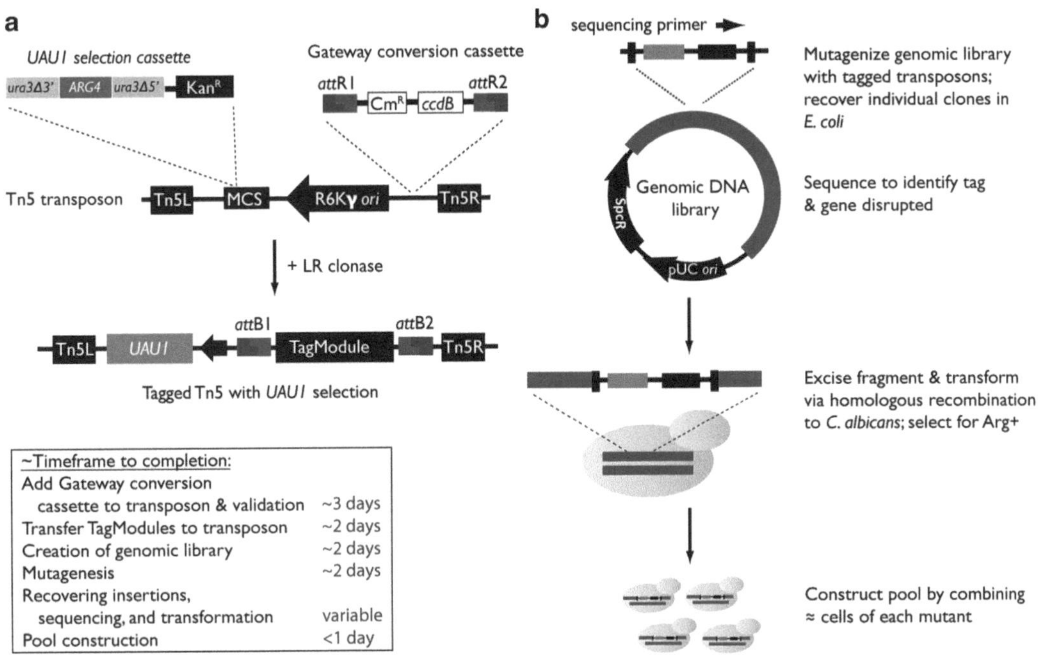

Fig. 3. Workflow for the construction of tagged transposon mutants in *C. albicans*. (**a**). Molecular engineering of the commercial Tn5 transposon for *C. albicans* mutagenesis: Gateway conversion cassette (containing the *ccdB* selection gene and chloramphenicol-resistance gene Cm^R), *UAU1* marker cassette for the transformation of transposon insertions in *C. albicans*, and kanamycin-resistance gene (Kan^R) for selection of in vitro transposon insertions in *E. coli*. Following the LR clonase reaction using a pool of TagModules, the TagModules are transferred into the Tn5 transposon, resulting in a pool of tagged transposons. (**b**). A *C. albicans* genomic library is mutagenized in vitro with pools of tagged transposons. Individual insertion events into the genomic library are then recovered in *E. coli* and sequenced (*arrow*) to identify the gene disrupted and the linked tag. Desired gene disruption events are then selected and excised from the genomic library and transformed via homologous recombination into the genome of *C. albicans* strain BWP17, selecting for Arg+ mutants. Homologous recombination is mediated by genomic DNA flanking the transposon insertion. Finally, pools are constructed by combining equivalent amounts of cells of each mutant.

conditions: 1× NEBuffer 4, 1× BSA, and 2 μL of *SnaB*I (10 U) for 1 h at 37°C. Following the reaction, clean up the reaction with the QIAquick PCR Purification Kit according to manufacturer's instructions to. See Note 2 regarding placement of the Gateway conversion cassette within the transposon.

2. Phosphatase-treat the blunt ends under the following conditions: in 30 μL volume, 1× Antarctic Phosphate buffer and 3 μL Antarctic Phosphatase (15 U) for 2 h at 37°C. Again, following the reaction, clean up the reaction with the QIAquick PCR Purification Kit.
3. Set up the ligation under the following conditions: in 20 μL volume, 50 ng *SnaB*I-cut, phosphatase-treated Tn5-*UAU1*, 2 μL (10 ng) Reading Frame A, B, or C.1, 1× Quick Ligation Reaction buffer, and 1 μL Quick T4 DNA Ligase. Incubate for 10 min at room temperature.

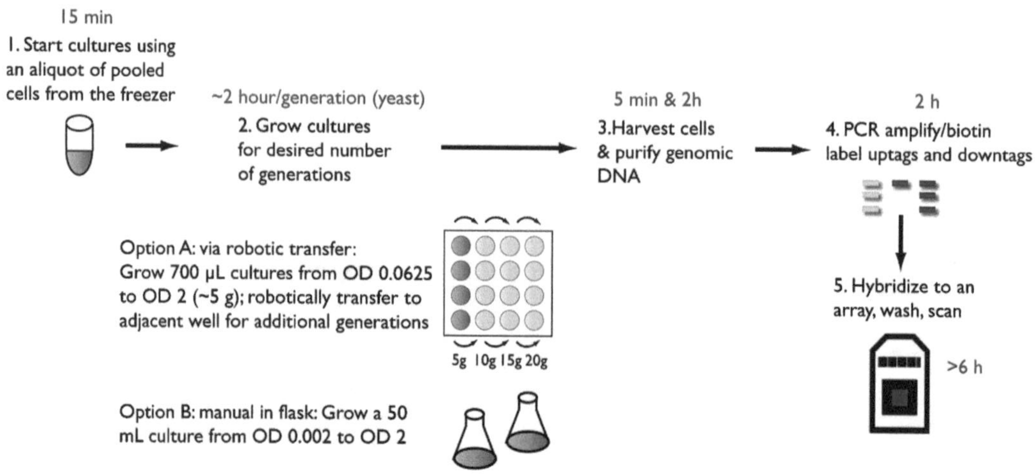

Fig. 4. Workflow for pooled growth assay and tag detection. Cultures are inoculated with thawed aliquots of pooled cells (Step 1), and then grown for the desired number of generations either robotically or manually (Step 2). Genomic DNA is then isolated from the harvested cells (Step 3), and uptags and downtags are independently amplified (Step 4) and hybridized to an array (Step 5).

4. Transform into One Shot® *ccdB* Survival Cells according to manufacturer's instructions and select on LB agar + chloramphenicol, kanamycin, and carbenicillin and grow at 37°C overnight.
5. Pick at least 12 individual colonies and inoculate to 5 mL liquid LB + chloramphenicol (additional antibiotics are optional) and grow at 37°C overnight. Following growth, extract the plasmids using the QIAprep Spin Miniprep Kit. Check for correct integration of the Gateway conversion cassette via digestion with *BsrGI* (reaction conditions: in 25 µL, 1× NEBuffer 4, 1× BSA, and 1 µL *BsrGI* (10 U) for 30 min at 37°C). *BsrGI* is present in both *att*R1 and *att*R2.
6. It is important to verify activity of the *ccdB* gene to reduce background transformants in the subsequent LR Clonase® transfer steps. Transform 1 µL of the correct plasmid each into One Shot® *ccdB* Survival Cells and One Shot® MAX Efficiency® DH5α™-T1R Competent Cells with the appropriate selection. An active *ccdB* gene would yield 10,000× more colonies when transformed into *ccdB* Survival Cells versus DH5α.

3.2. TagModule Transfer to Transposon

1. To make a single pool of the 4280 TagModules, use a 12-channel pipette to remove 5 µL from each well of the glycerol stock and pipette into a disposable reservoir. At this point, freeze a combined aliquot at −80°C for future growth. See Note 3 for recovering pools of TagModules or pools of tagged transposons.

2. Pour the combined liquid into a 50-mL centrifuge tube, centrifuge for 10 min at 1,200 × *g*, and pour off the supernatant. Using at least 4 spin columns, miniprep the pool of TagModules, combining after the final elution step.
3. To transfer the TagModules to the transposon, use 150 ng of the Gateway-converted Tn5-*UAU1* + Gateway and 300 ng of the TagModule pool in the Gateway® LR Clonase® II reaction according to manufacturer's guidelines. Electroporate 1 μL into TransforMax™ EC100™ Electrocompetent *E. coli* according to manufacturer's guidelines, and plate the entire reaction onto four 245 × 245 × 18-mm Square BioDish XL Petri dishes containing LB + agar with kanamycin and carbenicillin. We recommend performing two independent electroporations, each plated onto eight plates to obtain maximum coverage of the TagModules (target >20,000 colonies). Incubate at 37°C overnight.
4. Flood the plates with 25–50 mL (up to 1 mm in depth) LB medium + kanamycin + carbenicillin and, using a cell scraper, gently scrape all colonies off the agar plate. Collect with a pipette into two 50-mL centrifuge tubes. Vigorously shake the two centrifuge tubes on a rotary shaker for 1 h at 37°C to ensure that all the colonies are well suspended. Then, harvest the pools at 1,200 × *g* for 15 min and extract the plasmids from the cell pellets using the HiSpeed Plasmid Maxi Kit according to manufacturer's instructions. This final product contains a pool of TagModule-containing Tn5-*UAU1* transposons.
5. To prepare the pool of tagged transposons for mutagenesis, digest 10 μg of plasmid from Step 4 with *Psh*AI to excise the fragment containing only the mosaic ends, the selectable markers, and the TagModules from the pMOD™-6 backbone. Set up at least five reactions with 2 μg DNA and 20 U enzyme apiece with 1× NEBuffer 4 and 1× BSA in a total reaction volume of 50 μL. To ensure complete digestion, incubate for 3 h at 37°C. Add loading dye to 1× and load samples onto a 1% agarose gel with a 1 kb plus DNA ladder as a reference. Electrophorese at 150 V for ~1.5 h as appropriate. Excise the appropriate band (for our purposes, the upper ~5.8 kb band) with the QIAquick Gel Extraction Kit according to manufacturer's instructions.

3.3. Constructing a Genomic Library

If the mutagenesis is to be performed in vitro (see Note 4 for mutagenesis alternatives), a genomic library should be prepared from the organism of interest or purchased, if available. For the *C. albicans* mutagenesis, we used pUC19 as the library vector.

1. To add additional restriction sites to pUC19, clone in an additional polylinker by amplifying the synthesized oligonucleotide "polylinker" with two primers containing *EcoR*I restriction sites "polylinker_EcoRI_L" and "polylinker_EcoRI_R" in the following reaction conditions: 1 μL each oligonucleotide, 1× Standard Taq (Mg-free) Reaction Buffer, 200 μM dNTPs, 1.5 mM $MgCl_2$ in a 50 μL reaction volume. The PCR cycling conditions are as follows: 94°C 3′, 30 cycles of 94°C 30 s, 55°C 30 s, and 72°C 30 s, followed by 72°C for 3 . The resulting PCR product is digested with *EcoR*I in 1× EcoRI NEBuffer for 30 min at 37°C. The resulting *EcoR*I-cut PCR product is then gel-purified as described in Step 5 in Subheading 3.2.
2. pUC19 is then digested with *EcoR*I, phosphatase-treated, and purified as described in Steps 1 and 2 in Subheading 3.1.
3. "pUC19 + linker" is then generated by ligating the *EcoR*I-digested and phosphatase-treated pUC19 with the *EcoR*I-digested and gel-purified polylinker as described in Step 3 of Subheading 3.1. 2 μL of the ligation product is transformed into the One Shot® MAX Efficiency® DH5α™-T1R Competent Cells according to manufacturer's instructions and plated onto LB + agar + carbenicillin and grown at 37°C overnight.
4. To confirm the integration of the linker, individual colonies are picked and grown overnight in liquid LB + carbenicillin and miniprepped. Linearization of the vector by digestion by *EcoR*V or *Spe*I (to a 2.7 kb fragment) confirms integration.
5. To prepare genomic DNA, grow a culture of the strain of interest overnight. Here, we grew a 100 mL culture of *C. albicans* strain BWP17 (*ura3Δ::λimm434/ura3Δ::λimm434 his1::hisG/his1::hisG arg4::hisG/arg4::hisG* (35)) overnight in YPD, shaking at 30°C. Harvest cells by centrifugation at 1,200 ×*g* for 5 min. Resuspend the pellet in 4 mL of Zymolyase®-20 T and buffer Y1 (1 mL 50 mg/mL stock + 3 mL buffer Y1) in the QIAGEN Genomic-tip 100/G (Qiagen) kit, digest for 37°C for 30 min, and extract genomic DNA according to manufacturer's instructions.
6. Choose an appropriate restriction enzyme for digesting the genomic DNA; for this method, it should not cut the transposon, as the genomic fragment containing the transposon will be excised for transformation into the organism of interest. We use the example of *Spe*I, which is a 6-bp cutter. 4-bp cutters may require partial digestion to produce sufficiently large genomic fragments (see Note 5 for alternatives to cloning genomic libraries). Digest ~10 μg of genomic DNA with 10 U of *Spe*I in 1× NEBuffer 2 and 1× BSA for 1 h at 37°C and purify the digest with the QIAquick PCR Purification Kit.

We recommend running a 5 μL sample on a 1% agarose gel to confirm that average size ranges from 2–8 kb. If the samples appear overdigested, perform a partial digestion by shortening the digestion time and/or reducing the number of units of enzyme used. Concurrently, digest and phosphatase-treat pUC19+linker with *Spe*I as described in Steps 1 and 2 in Subheading 3.1.

7. Ligate the genomic DNA into the pUC19 + linker vector in the following reaction: 1× Quick Ligation buffer, 1 μL Quick T4 DNA Ligase, 100 ng *Spe*I digested and phosphatase-treated pUC19 + linker, and 1 μg of *Spe*I-cut genomic DNA. Ligate for 15 min at room temperature and perform two electroporations of 1 μL of the ligation mix into Transformax EC100 Electrocompetent *E. coli* according to manufacturer's instructions. Plate the two transformations to eight 245 × 245 × 18-mm Square BioDish XL Petri dishes containing LB + agar + carbenicillin. We strongly recommend also ligating a vector-only control to determine background. If the background is unduly high (e.g., greater than 1/100 of the colonies on the ligation plate), redigest and phosphatase-treat the pUC19+ linker vector.
8. As in Step 4 in Subheading 3.2, collect at least 20,000 colonies off the agar plate into liquid LB + carbenicillin medium and extract the plasmids using the HiSpeed Plasmid Maxi Kit (Qiagen) to make the final genomic library.

3.4. In vitro Transposon Mutagenesis

1. Add 200 ng of the pool of tagged transposons and 200 ng of the genomic library to EZ-Tn5™ 1× Reaction Buffer and 1 μL EZ-Tn5™ Transposase in a final volume of 10 μL. Incubate for 2 h at 37°C, then add 1 μL EZ-Tn5 10× Stop Solution and incubate for 10 min at 70°C to stop the reaction. Use 1 μL for electroporation into Transformax EC100 Electrocompetent *E. coli* according to manufacturer's instructions, and plate to LB + agar + kanamycin plates. Incubate overnight at 37°C. Extra reaction mix may be stored at −20°C and retransformed with small loss of efficiency.

3.5. Sequence Identification of Inserts

The subsequent steps are best performed robotically (e.g., using a Biomek FX or similar liquid handler workstation). See Note 6 on alternatives to sequencing individual insertions; this can also be performed in a pooled format.

1. Pick individual colonies from Subheading 3.3 either by hand or robotically and inoculate to 384- or 96-well half height plates containing liquid LB + kanamycin (500 μL for 96-well or 120 μL for 384-well). Seal the plates with aerated seal film and grow the plates shaking overnight at 37°C.
2. Add 40 μL of 50% glycerol to each well of a fresh 96- or 384-well deep well plate. Transfer 80 μL of the grown culture to

each well to make a frozen stock. At the same time, transfer 200 μL of the grown culture to a 96-well Seqprep plate (or equivalent high-throughput plasmid extraction system), if using 96-well plates. If using 384-well plates, inoculate the 384-well Seqprep plate and grow overnight, shaking at 37°C, according to manufacturer's instructions. Extract the plasmids according to manufacturer's guidelines, except incubate the plate at room temperature for at least 6 h or overnight following addition of the elution buffer; this step greatly increases the yield of sequenceable plasmid.

3. Prepare the sequencing reaction in a 96- or 384-well PCR plate as follows: 4 μL Seqprep plasmid, 0.032 μL 100 μM primer U1 (or applicable primer), 1 μL 5× reaction buffer, and 0.25–1 μL of 2. BigDye® Terminator v3.1, for a total volume of 5–6 μL. Run the reaction in a thermocycler with the following program: 50 cycles of 96°C 10 s, 50°C 5 s, 60°C 4 min. (see Note 7 for tips on optimizing the sequencing reaction).
4. Add 7.5 μL water to the sequencing reactions and clean up using 96- or 384-well PerformaDTR plates (or equivalent high-throughput sequence purification system) prior to capillary sequencing in an AB3730 DNA analyzer (Applied Biosystems) or method of choice.
5. Analyze each sequence to determine (1) the gene disrupted and (2) the TagModule associated with the disruption. For *C. albicans*, each sequence was parsed to blast the sequence up to the transposon junction against a list of the TagModule sequences to determine the TagModule identity. The region following the transposon junction was analyzed with BLAST-N against Assembly 21 of the *C. albicans* sequence. Percentage of gene disrupted was calculated as 1 − (# base pairs from transposon junction to gene start)/(total gene length). Sample-archived sequence data are shown in Fig. 5a.
6. Sort genes to maximize the number of unique TagModules and unique gene insertions. Return to the 96- or 384-well frozen stocks, pick the desired clones, and rearray them into a 96-well deep well plate containing 1 mL liquid LB + kanamycin. Cover with an aerated film seal and grow overnight shaking at 37°C.
7. Prep the plasmids containing transposon insertions with Seqprep 96 according to manufacturer's instructions and as in Step 2 of Subheading 3.5.
8. To digest the genomic DNA/transposon insertion plasmids, make a master mix of 10× NEBuffer 2, 100× BSA, and 0.5 μL/well (~5 U) of *Spe*I or appropriate enzyme as described and aliquot directly into the Seqprep plate from Step 7. Incubate overnight at 37°C to digest.

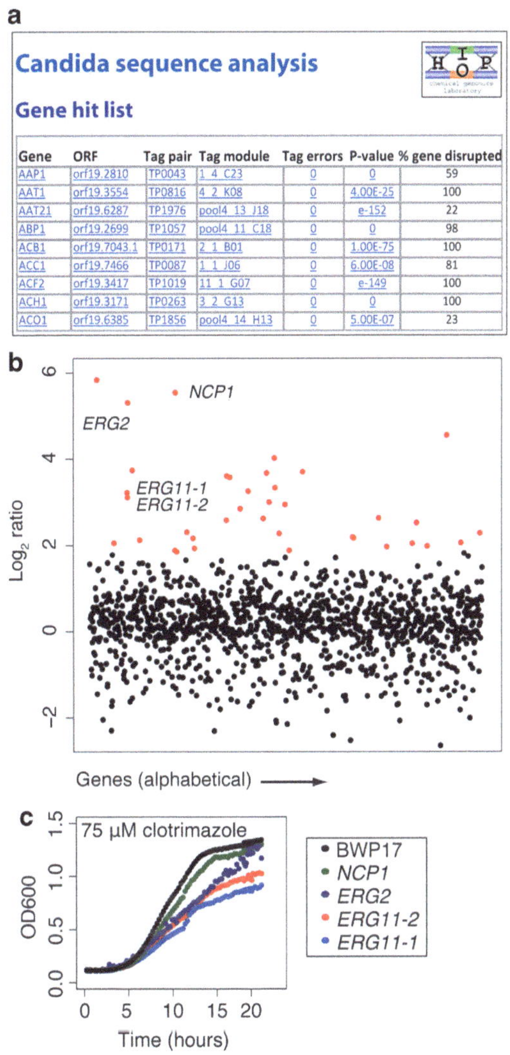

Gene	ORF	Tag pair	Tag module	Tag errors	P-value	% gene disrupted
AAP1	orf19.2810	TP0043	1 4 C23	0	0	59
AAT1	orf19.3554	TP0816	4 2 K08	0	4.00E-25	100
AAT21	orf19.6287	TP1976	pool4 13 J18	0	e-152	22
ABP1	orf19.2699	TP1057	pool4 11 C18	0	0	98
ACB1	orf19.7043.1	TP0171	2 1 B01	0	1.00E-75	100
ACC1	orf19.7466	TP0087	1 1 J06	0	6.00E-08	81
ACF2	orf19.3417	TP1019	11 1 G07	0	e-149	100
ACH1	orf19.3171	TP0263	3 2 G13	0	0	100
ACO1	orf19.6385	TP1856	pool4 14 H13	0	5.00E-07	23

Fig. 5. Sample data collected at certain points of the protocol. (**a**). Sample sequence data is collected and archived. Each sequence is analyzed to determine (1) the gene disrupted, and (2) the TagModule associated with the disruption. The percentage of each gene disrupted is calculated as 1 – (number of base pairs from transposon junction to gene start)/(total gene length). Genes can then be sorted to maximize the number of unique TagModules and unique gene insertions. (**b**). Sample data from screening results (adapted from (32)). The pool of tagged mutants was grown for 20 generations in the presence of clotrimazole and DMSO (control). Log_2 ratio (control intensity/treatment intensity) was calculated and plotted as a function of gene. Highly sensitive strains (*red*) included the known target of clotrimazole, ERG11p. Note that this assay frequently uncovers other sensitive mutants in addition to the compound's actual target. Generally, these mutants are those that act synthetically with the target, those that are part of a general stress/treatment response, or false positives that fail to confirm. (**c**). Example of confirmation data (adapted from (32) with permission from Oxford University Press). Results from the pooled growth assays can be validated by growing the strain in individual culture and compared against wild-type growth (*black*).

3.6. High-Throughput Transformation Via Homologous Recombination

1. Inoculate a 5-mL falcon tube with the transforming strain (here, BWP17) and grow overnight shaking at 30°C.
2. For one 96-well plate, inoculate 4 mL of the saturated overnight culture of BWP17 to 200 mL of YPD + 100 μg/mL uridine. Grow ~6 h, shaking at 30°C until exponential phase is reached (<1.5 OD_{600}).
3. Harvest cells by centrifugation in 50-mL Falcon tubes for 5 min at 1,200 × *g*. Wash pellet once with 25 mL 1× TE/0.1 M LiOAc. Resuspend the pellet in 10 mL 1× TE/0.1 M LiOAc.
4. To the wells of a 96-well half height plate using a multichannel pipette, add 15 μL freshly boiled 10 mg/mL sonicated salmon sperm DNA (ssDNA), 100 μL of the resuspended cells, and the entire digested plasmid prep from Step 8 in Subheading 3.5. Prepare the transforming mix and aliquot 700 μL to each well. Incubate overnight at room temperature.
5. Heat-shock the plate at 44°C for 1 h by placing in a water bath.
6. Pellet the cells by centrifuging the plate for 3 min at 300 × *g*.
7. Invert the plate and shake to remove the transforming mix supernatant. Gently blot dry on a stack of paper towels to remove excess liquid. Add 35 μL SC–Arg + uridine to the wells and resuspend by pipetting gently. For each well, plate the entire transformation reaction onto a well of a 6-well microtiter plate containing SC–Arg + uridine agar and glass beads to spread the mix. Incubate for 2–3 days at 30°C.
8. For each strain, pick two individual colonies and array them separately to 96-well deep well plates containing 1 mL SC–Arg + uridine liquid medium. Grow shaking at 30°C, overnight or until all wells are saturated. Freeze an aliquot (generally in duplicate) in 96- or 384-well plates as described in Step 2 of Subheading 3.5. The remainder can be used for pool construction (Subheading 3.7).
9. To validate integration, the strains can be checked individually via PCR confirmation with gene-specific primers (see Note 8 on integration). However, due to the large numbers of mutants that are typically generated with such an approach, an alternative is to check the TagModule activity when hybridized to a microarray.

(a) Construct two separate pools as described in the following Subheading, 3.7, using colony 1 and colony 2 from Step 8 in Subheading 3.6.

(b) Amplify the uptags and downtags separately and hybridize to a TAG4 array as described in Subheading 3.9.

(c) Calculate the median intensity of all unused tags on the array. Multiply by 3 to determine a background cutoff.

(d) For each pool, determine which strains are not represented in the pool.

(e) Individually pick colonies that have the highest signal from the collection of colony 1 or colony 2 transformants and rearray into a new set of 96- or 384-well plates. Combine these to make the final pool.

3.7. Construction of Strain Pool

1. Combine all the wells from the plates from Step 8 or 9 in Subheading 3.6 by transferring each well using a multichannel pipette to a reagent reservoir. Combine all reservoirs into a sterile flask and add 50% glycerol to a final concentration of 15%. Measure the OD_{600} of the pool and adjust (by dilution or centrifugation and resuspension) to 50 OD_{600}/mL with medium containing 15% glycerol. Mix well and aliquot ~50 μL into PCR strip tubes. Cover strips and store at −80°C indefinitely. Do not refreeze aliquots once thawed.

3.8. Experimental Pool Growth

This procedure (with timeline) is outlined in Fig. 4.

1. Break off individual tubes of pooled cells as needed and thaw on ice. If not using robotics, skip to Step 4 for manual cell growth.
2. Immediately dilute the pool into medium with drug or condition of choice, inoculating at an OD_{600} of 0.0625 in a total volume of 700 μL in a 48-well plate. Include at least one appropriate solvent control on the plate. For growth beyond five generations, fill adjacent wells with medium or condition of choice, but do not add cells. These wells will be used by the robotics for inoculation at five generation intervals (up to ~2 OD_{600} for yeast). Seal with a plastic plate seal; if the condition requires aerobic growth (e.g., nonfermentable carbon sources), pinprick the holes in the membrane seal toward the side of each well.
3. Grow the cells in a spectrophotometer, shaking at 30°C with an experimentally determined optimal shaking regimen. Part of the cell suspension can be harvested by the robot and saved on a cold plate on the robotic deck at user-defined generation time points (http://med.stanford.edu/sgtc/technology/access.html, for details contact C. Nislow or G. Giaever).
4. For manual cell growth, inoculate a 50 mL culture at a starting OD_{600} of 0.002 in a 250-mL culture flask. Grow in a shaking incubator at 30°C at 250 rpm. After cells reach a final OD_{600} of 2.0 (for yeast), they will have undergone ~10 generations of growth.
5. Following growth for the desired number of generations, harvest at least ~2 OD_{600} of cells into Safe-Lock Microcentrifuge Tubes.

3.9. Preparation for Hybridization and Hybridization

See Note 9. Alternatives to array hybridization.

1. Purify genomic DNA from ~2 OD_{600} of cells with the YeaStar kit according to manufacturer's instructions (Protocol I if using yeast DNA), or another suitable method specific to the organism of interest. If using the YeaStar kit, elute the DNA with 300 μL of 0.1× TE instead of the 60 μL of 1× TE specified in the protocol. The genomic DNA can then be stored indefinitely at −20°C.
2. Set up two PCR reactions for each sample, one for the uptags and the other for the downtags, with the following reaction conditions: 33 μL ddH_2O, 6 μL 10× PCR buffer without $MgCl_2$, 3 μL 50 mM $MgCl_2$, 1.2 μL 10 mM dNTPs, 1.2 μL 50 μM Up or Down primer mix, 0.6 μL 5 U/μL Taq polymerase, ~0.1 μg genomic DNA in 15 μL. Total volume is 60 μL. Thermocycle under the following conditions: 94°C 3 min, 30 cycles of 94°C 30 s, 55°C 30 s, 72°C 30 s and then 72°C 3 min and hold at 4°C. Check the resulting PCR products on a gel; a 60-bp product for both PCRs is expected. The PCR products can then be stored at −20°C indefinitely or used immediately.
3. Set the hybridization oven temperature to 42°C. Then, set up a boiling water bath and a slushy ice bucket.
4. To pre-wet the arrays, fill the arrays with 90 μL 1× hybridization buffer and incubate in the hybridization buffer at 42°C and 20 rpm for 10 min.
5. Prepare 90 μL of hybridization mix per sample, plus one extra, as described: 75 μL 2× hybridization buffer, 0.5 μL B213 control oligonucleotide (0.2 fm/μL), 12 μL mixed oligonucleotides (12.5 pm/μL), 3 μL 50× Denhardt's solution) in lock-top 0.5-mL tubes.
6. Add 30 μL uptag PCR and 30 μL downtag PCR to 90 μL hybridization mix for a total volume of 150 μL. Boil for 2 min and set on ice for at least 2 min. Briefly spin the tubes prior to use.
7. Remove the hybridization buffer from the arrays and add 90 μL hybridization/PCR mix. To prevent evaporation, cover the array gaskets with a Tough-Spot. Hybridize for 16 h at 42°C, 20 rpm. Ideally, keep this hybridization time constant for samples that are part of the same dataset, for consistency.
8. Freshly prepare 600 μL biotin labeling mix per sample plus an extra, as follows: 180 μL 20× SSPE, 12 μL 50× Denhardt's, 6 μL 1% Tween 20 (vol/vol), 1 μL 1 mg/mL streptavidin-phycoerythrin, 401 μL ddH_2O. Aliquot 600 μL into 2-mL tubes. Remove Tough-Spots from chips.
9. Remove hybridization mix from the arrays and fill chips with 90 μL Wash A. Prime the Affymetrix fluidics station.

10. Wash the arrays using an Affymetrix fluidics station according to the manufacturer's instructions, using "Gene-Flex_Sv3_450" protocol with the following modifications: 1 extra step with Wash A (1 cycle, 2 mixes) before staining, Wash B temperature 42°C instead of 40°C, stain at 42°C in hybridization oven instead of 25°C. It is also possible to perform the posthybridization wash, the biotin staining, and the post-staining wash manually (see p. 396 in C. Nislow and G. Giaever (36)). Following fluidics operations, run the fluidics station "SHUTDOWN_450" protocol.
11. Following the wash, check the arrays for air bubbles. If present, add 90 μL Wash A and remove until the bubbles disappear. If there are any marks or smudges on the array surface, clean the glass window with isopropanol and a lint-free tissue. Replace Tough-Spots on the gaskets to prevent evaporation and place the arrays in the scanner.
12. Scan in an Affymetrix GeneArray scanner at an emission wavelength of 560 nm.

3.10. Array Analysis

1. *Outlier masking*. Affymetrix TAG4 array contains five replicate features for each tag probe dispersed across the array so that outlier features can be identified and discarded prior to averaging intensity values for each tag.
   (a) For each array feature, examine the surrounding 5 × 5 features. If >13/25 probes in this region differ from their trimmed replicate mean (the mean of the middle three replicates, excluding the highest and lowest replicates) by more than 10%, this probe should be discarded.
   (b) Once these outlier-dense regions have been identified, pad them by including all probes within a 5-probe radius, as defined by $((x_1 - x_2)^2 + (y_1 - y_2)^2)^{1/2} < 6$ where x_1, x_2, y_1, and y_2 are the x and y coordinates for the two features.
   (c) Discard features for which standard deviation of feature pixels/mean feature pixels >0.3. The standard deviation is included in the .cel file for Affymetrix arrays.
   (d) After removal of outliers, average all unmasked replicates by calculating intensity values for each tag.
2. *Saturation correction*. Owing to feature saturation, the signal on the TAG4 array is not linearly related to tag concentration. Thus, this saturation can be corrected, or the degree of sensitivity or resistance will be underestimated for strains with bright tags. This is an optional correction; see ref. (37) if necessary.
3. *Array normalization*. For each array, the uptags and downtags should be normalized separately; the efficacy of the separate PCR reactions will affect their array intensities. *To quantile*

normalize a set of arrays, rank the values obtained from each array for uptags and downtags in the order of increasing intensity. For each rank, assign the tag at that rank to the median of all values for all arrays at that rank.

To mean normalize a set of arrays, for each set of uptags and downtags, divide by the mean. Multiply each tag set by the mean across all arrays to return the tag intensities to approximately their original value.

4. *Removing unusable tags*. Tags with low-intensity values in the control set will give poor-quality results when making comparisons with experimental values. To calculate an intensity value threshold for which to exclude tags:
   (a) Using any treatment–control pair, calculate $\log_2((i_c - \text{bg})/(i_t - \text{bg}))$ for each tag, where i_c is the control intensity, i_t is the treatment intensity, and bg is the mean intensity of the unassigned tag probes.
   (b) Pair the uptag and downtag ratios by strain, and for each tag pair, take the minimum intensity for the two tags in the two samples. Sort the ratio pairs by this minimum intensity.
   (c) Use a sliding window of size 50 on the ranked ratio pairs, starting with the lowest intensity pairs. Calculate the correlation of uptag and downtag ratio pairs within the window. Also calculate the average of the minimum intensities calculated in the previous step.
   (d) Slide the window by 25 pairs and repeat the previous step until all pairs have been crossed.
   (e) Plot the average minimum intensity versus the uptag–downtag correlation for all windows.
   (f) Choose an intensity threshold; generally we use the intensity value where the correlation first reaches 80% of its maximum level. Flag and remove from use any tags below this cutoff in all samples as unusable.
5. *Calculating sensitivity scores for control-treatment comparisons*. In methods in which a large set of common control arrays (>10) are used with one or two replicate arrays per experiment, z-scores are generally preferable, although $\log_2$ ratios are a more direct way of quantitating strain sensitivity.

   To use $\log_2$ ratios as a metric of sensitivity. For each strain, calculate $\log_2((\mu_c - \text{bg})/(\mu_t - \text{bg}))$, where μ_c is the mean intensity for the control samples, μ_t is the mean intensity for the treatment samples, and bg is the mean intensity of the unassigned probes. Strains with a positive $\log_2$ ratio are sensitive to the treatment, and those that are resistant have negative $\log_2$ ratios.

To use z-score as a metric of sensitivity. For each strain, average all usable tags and calculate the mean of the controls and the standard deviation of the controls. Calculate a z-score: $(\mu_c - t)/\sigma_c$, where t is the treatment intensity for that strain, and a p-value by fitting a t-distribution with $n_c - 1$ degrees of freedom to all scores for the experiment, where n_c is the number of control arrays.

3.11. Validation of Array Data

As with any screen, the methods above will generate multiple candidates that are sensitive to a particular treatment, and therefore, validation is a necessary step. The choice of which candidates to confirm is somewhat arbitrary, but in general, ranking the most sensitive strains by the $\log_2$ ratio or z-score and then testing the top candidates to confirm provides sufficient sampling. Most simply, this involves growing the strains individually with and without treatment, and comparing their growth rate with a control strain. Failure to confirm can be attributed to biological or technical reasons; for example, cross-contamination can occur between wells on a storage plate, or the strain may be incorrectly archived. In some cases, tags can cross-hybridize on a microarray, creating false negatives. We provide an example of data generated via the screen in Fig. 5b, c.

4. Notes

1. *Selecting a transposon*. Here, we used the Epicentre Biotechnologies pMOD-series transposons, which are amenable to an in vitro or an in vivo mutagenesis strategy. However, virtually any transposon that works in vivo or in vitro for either bacteria or fungi can be used so long as it is receptive to cloning modifications.
2. *Placement of the Gateway conversion cassette within the transposon*. Typically, the gene disrupted by the transposon is identified by sequencing outward into the genome with a primer at or adjacent to the transposon junction. Because at least one tag of a TagModule needs to be sequenced concurrently with at least ~50 bp of the flanking genomic sequence (this estimate depends on the amount of genomic DNA needed to BLAST-n a DNA sequence to a gene), ideally the TagModule should be placed as close to the transposon junction as possible. As even high-quality Sanger sequencing reads generally do not exceed 800 bp, placement of the TagModule should not exceed 100–200 bases away from the transposon junction, or the tag and the gene disrupted may have to be determined through two separate sequencing reactions.

3. *Recovering pools of TagModules or pools of tagged transposons.* If an aliquot of pooled TagModules or pooled tagged transposons are recovered from a frozen stock, inoculate at least 10^5 cells in liquid medium and then grow overnight at 37°C and plate onto selective medium. Alternatively, plate directly onto selective medium to recover a minimum of 50,000 colonies to obtain adequate coverage of TagModules. These guidelines apply to recovering pools of genomic libraries.
4. *Mutagenesis alternatives.* We used the EZ-Tn5 mutagenesis system from Epicentre Biotechnologies, using it for in vitro mutagenesis of a genomic library. However, this transposon can also be used in vivo via transposomics, in which the prepared, excised transposon is incubated with the Tn5 transposases in the absence of Mg^{2+}, rendering the transposases inactive. These transposomes can then be electroporated into the cells of interest, and once in the cellular environment, the transposons activate and integrate the transposon into the genomic DNA. Other transposons/transposase systems can be transferred via bacterial conjugation or expressed endogenously on a plasmid.
5. *Alternatives to cloning genomic libraries.* There are many alternatives to cloning genomic libraries, including using different restriction enzymes (make certain, however, that the enzyme used to excise the genomic fragment does not cut the transposon). Additional restriction enzymes can also be added to the "polylinker" or different vectors can be used; we have had good success with *Xba*I. Alternatively, one can shear the genomic DNA and blunt-ligate it into the library vector. Another possibility is to include an R6Kγ ori (provided on some pMOD transposons, Epicentre Biotechnologies) or similar origin on the transposon of choice. Then, restriction-digested and self-ligated genomic DNA fragments can be mutagenized without needing to be cloned into a backbone vector. This would, in principle, improve the mutagenesis efficiency as the transposon will not integrate into the vector backbone.
6. *Alternatives to sequencing individual insertions.* One can combine a smart-pooling approach with a next-generation sequencing platform as described in (38), which can greatly accelerate identification of mutants.
7. *Optimizing the sequencing reaction.* Cycling the sequencing reaction for up to 99 cycles greatly improves yield (although these reactions need to be run overnight). Additives such as DMSO and betaine also improve reaction efficiency, as does the addition of BigDye® Terminator v3.1 (up to 4 μL/reaction). If the mutagenesis is performed in vivo, two-step arbitrary PCR (39) can be used to isolate a PCR fragment

spanning the TagModule + flanking genomic DNA, which can then be sequenced directly.

8. *On integration*. In vivo mutagenesis systems can bypass the additional step in which mutagenized genomic DNA is excised and transformed via homologous recombination into the organism of interest. However, should in vitro mutagenesis + transformation be required, verification of integration is an important step. This can be accomplished by PCR verification using one of the TagModule up- or downtag primers (or their reverse complements, depending on the orientation and the location within the transposon into which the Gateway conversion cassette is cloned) plus a primer designed to fall into the flanking genomic region of the gene targeted. While this does not test for ectopic integration, transformation efficiencies via homologous recombination generally exceed 97+% with 60 bp of flanking homology in *S. cerevisiae* (33) and 100 bp in *C. albicans* (40). We estimate that the genomic DNA flanking the transposon insertion exceeds 1 kb given that genomic DNA was size-selected prior to cloning into the library vector. Alternatively, the entire pool can be sequenced en masse using next-generation sequencing so that the tagged transposons can be mapped to a genomic location as described (10).

9. *Alternatives to array hybridization*. With the costs of high-throughput sequencing decreasing, using high-throughput sequencing as a readout of tag abundance rather than array hybridization is becoming feasible (10). In this way, amplified PCR product can be measured directly as "counts" rather than as signal intensity as hybridized to an array. This would eliminate false negatives and positives stemming from tag cross-contamination, saturation or tag representation problems arising from very high or very low signal intensities. Furthermore, multiple experiments can be combined prior to sequencing by the addition of a 4–8 bp DNA index. As the tags themselves are 20 bp, a single, 2-step read of 24–28 bases is then sufficient, allowing a large number of experiments to be multiplexed.

Concluding comments. Here, we outline a protocol that, with modest modification, can easily be adapted to a wide range of microorganisms to create tagged mutant collections. These collections can be used as individuals or in a pooled format, although the true potential lies in the tags' ability to permit experimental multiplexing. The TagModules are adaptable to virtually any transposon or mutagenesis system, including start-to-stop deletions. While the gold standard for determining gene function is to phenotype a deletion allele of a gene, this method becomes cumbersome if applied to nonmodel organisms or collections of metagenomes. It is clearly cost- and resource-prohibitive to take the one-by-one deletion approach.

We emphasize that while we have reported a protocol on the tagged transposon mutagenesis for the pathogenic yeast *C. albicans*, a very similar protocol could be adapted to a wide variety of unicellular fungi. When modified, this protocol works well in bacteria (32), and currently collection for a number of additional genomes, fungal and bacterial, are under construction.

Acknowledgments

We thank A. Deutschbauer, G. Giaever, R. St.Onge, U. Schlecht, and R. Davis for discussions and advice. C.N. is supported by grants from the National Human Genome Research Institute (Grant Number HG000205), RO1 HG003317, CIHR MOP-84305, and Canadian Cancer Society (#020380). J.O. was supported by the Stanford Genome Training Program (Grant Number T32 HG00044 from the National Human Genome Research Institute) and the National Institutes of Health (Grant Number P01 GH000205).

References

1. Hillenmeyer M. E., Fung E., Wildenhain J., Pierce S. E., Hoon S., Lee W., Proctor M., St Onge R. P., Tyers M., Koller D., Altman R. B., Davis R. W., Nislow C., and Giaever G. (2008) The chemical genomic portrait of yeast: uncovering a phenotype for all genes. *Science* **320**, 362–365.
2. Giaever G., Chu A. M., Ni L., Connelly C., Riles L., Veronneau S., Dow S., Lucau-Danila A., Anderson K., Andre B., Arkin A. P., Astromoff A., El-Bakkoury M., Bangham R., Benito R., Brachat S., Campanaro S., Curtiss M., Davis K., Deutschbauer A., Entian K. D., Flaherty P., Foury F., Garfinkel D. J., Gerstein M., Gotte D., Guldener U., Hegemann J. H., Hempel S., Herman Z., Jaramillo D. F., Kelly D. E., Kelly S. L., Kotter P., LaBonte D., Lamb D. C., Lan N., Liang H., Liao H., Liu L., Luo C., Lussier M., Mao R., Menard P., Ooi S. L., Revuelta J. L., Roberts C. J., Rose M., Ross-Macdonald P., Scherens B., Schimmack G., Shafer B., Shoemaker D. D., Sookhai-Mahadeo S., Storms R. K., Strathern J. N., Valle G., Voet M., Volckaert G., Wang C. Y., Ward T. R., Wilhelmy J., Winzeler E. A., Yang Y., Yen G., Youngman E., Yu K., Bussey H., Boeke J. D., Snyder M., Philippsen P., Davis R. W., and Johnston M. (2002) Functional profiling of the *Saccharomyces cerevisiae* genome. *Nature* **418**, 387–391.
3. Hensel M., Shea J. E., Gleeson C., Jones M. D., Dalton E., and Holden D. W. (1995) Simultaneous identification of bacterial virulence genes by negative selection. *Science* **269**, 400–403.
4. Badarinarayana V., Estep P. W., 3 rd, Shendure J., Edwards J., Tavazoie S., Lam F., and Church G. M. (2001) Selection analyses of insertional mutants using subgenic-resolution arrays. *Nat Biotechnol* **19**, 1060–1065.
5. van Opijnen T., Bodi K. L., and Camilli A. (2009) Tn-seq: high-throughput parallel sequencing for fitness and genetic interaction studies in microorganisms. *Nat Methods* **6**, 767–772.
6. Hidalgo-Grass C., Ravins M., Dan-Goor M., Jaffe J., Moses A. E., and Hanski E. (2002) A locus of group A *Streptococcus* involved in invasive disease and DNA transfer. *Mol Microbiol* **46**, 87–99.
7. Polissi A., Pontiggia A., Feger G., Altieri M., Mottl H., Ferrari L., and Simon D. (1998) Large-scale identification of virulence genes from *Streptococcus pneumoniae*. *Infect Immun* **66**, 5620–5629.

8. Sassetti C. M., Boyd D. H., and Rubin E. J. (2001) Comprehensive identification of conditionally essential genes in mycobacteria. *Proc Natl Acad Sci USA* **98**, 12712–12717.
9. Groh J. L., Luo Q., Ballard J. D., and Krumholz L. R. (2005) A method adapting microarray technology for signature-tagged mutagenesis of Desulfovibrio desulfuricans G20 and *Shewanella oneidensis* MR-1 in anaerobic sediment survival experiments. *Appl Environ Microbiol* **71**, 7064–7074.
10. Smith A. M., Heisler L. E., Mellor J., Kaper F., Thompson M. J., Chee M., Roth F. P., Giaever G., and Nislow C. (2009) Quantitative phenotyping via deep barcode sequencing. *Genome Res.* **19**, 1836–1842.
11. Claus H., Frosch M., and Vogel U. (1998) Identification of a hotspot for transformation of *Neisseria meningitidis* by shuttle mutagenesis using signature-tagged transposons. *Mol Gen Genet* **259**, 363–371.
12. Hava D. L., and Camilli A. (2002) Large-scale identification of serotype 4 *Streptococcus pneumoniae* virulence factors. *Mol Microbiol* **45**, 1389–1406.
13. Ross-Macdonald P., Coelho P. S., Roemer T., Agarwal S., Kumar A., Jansen R., Cheung, K. H., Sheehan A., Symoniatis D., Umansky L., Heidtman M., Nelson F. K., Iwasaki H., Hager K., Gerstein M., Miller P., Roeder G. S., and Snyder M. (1999) Large-scale analysis of the yeast genome by transposon tagging and gene disruption. *Nature* **402**, 413–418.
14. Davis D. A., Bruno V. M., Loza L., Filler S. G., and Mitchell A. P. (2002) *Candida albicans* Mds3p, a conserved regulator of pH responses and virulence identified through insertional mutagenesis. *Genetics* **162**, 1573–1581.
15. Uhl M. A., Biery M., Craig N., and Johnson A. D. (2003) Haploinsufficiency-based large-scale forward genetic analysis of filamentous growth in the diploid human fungal pathogen *C.albicans. EMBO J* **22**, 2668–2678.
16. Castano I., Kaur R., Pan S., Cregg R., Penas Ade L., Guo N., Biery M. C., Craig N. L., and Cormack B. P. (2003) Tn7-based genome-wide random insertional mutagenesis of Candida glabrata. *Genome Res* **13**, 905–915.
17. Shoemaker D. D., Lashkari D. A., Morris D., Mittmann M., and Davis R. W. (1996) Quantitative phenotypic analysis of yeast deletion mutants using a highly parallel molecular bar-coding strategy. *Nat Genet* **14**, 450–456.
18. Hartwell L. H., Culotti J., Pringle J. R., and Reid B. J. (1974) Genetic control of the cell division cycle in yeast. *Science* **183**, 46–51.
19. Lashkari D. A., DeRisi J. L., McCusker J. H., Namath A. F., Gentile C., Hwang S. Y., Brown P. O., and Davis R. W. (1997) Yeast microarrays for genome wide parallel genetic and gene expression analysis. *Proc Natl Acad Sci USA* **94**, 13057–13062.
20. Goffeau A., Barrell B. G., Bussey H., Davis R. W., Dujon B., Feldmann H., Galibert F., Hoheisel J. D., Jacq C., Johnston M., Louis E. J., Mewes H. W., Murakami Y., Philippsen P., Tettelin H., and Oliver S. G. (1996) Life with 6000 genes. *Science* **274**, 546, 563–547.
21. Costanzo M., Baryshnikova A., Bellay J., Kim Y., Spear E. D., Sevier C. S., Ding H., Koh J. L. Y., Toufighi K., Mostafavi S., Prinz J., St. Onge R. P., VanderSluis B., Makhnevych T., Vizeacoumar F. J., Alizadeh S., Bahr S., Brost R. L., Chen Y., Cokol M., Deshpande R., Li Z., Lin Z.-Y., Liang W., Marback M., Paw J., San Luis B.-J., Shuteriqi E., Tong A. H. Y., van Dyk N., Wallace I. M., Whitney J. A., Weirauch M. T., Zhong G., Zhu H., Houry W. A., Brudno M., Ragibizadeh S., Papp B., Pal C., Roth F. P., Giaever G., Nislow C., Troyanskaya O. G., Bussey H., Bader G. D., Gingras A.-C., Morris Q. D., Kim P. M., Kaiser C. A., Myers C. L., Andrews B. J., and Boone C. (2010) The Genetic Landscape of a Cell. *Science* **327**, 425–431.
22. Tong A. H., Evangelista M., Parsons A. B., Xu H., Bader G. D., Page N., Robinson M., Raghibizadeh S., Hogue C. W., Bussey H., Andrews B., Tyers M., and Boone C. (2001) Systematic genetic analysis with ordered arrays of yeast deletion mutants. *Science* **294**, 2364–2368.
23. Pan X., Yuan D. S., Xiang D., Wang X., Sookhai-Mahadeo S., Bader J. S., Hieter P., Spencer F., and Boeke J. D. (2004) A robust toolkit for functional profiling of the yeast genome. *Mol Cell* **16**, 487–496.
24. Schuldiner M., Collins S. R., Thompson N. J., Denic V., Bhamidipati A., Punna T., Ihmels J., Andrews B., Boone C., Greenblatt J. F., Weissman J. S., and Krogan N. J. (2005) Exploration of the function and organization of the yeast early secretory pathway through an epistatic miniarray profile. *Cell* **123**, 507–519.
25. Deutschbauer A. M., Jaramillo D. F., Proctor M., Kumm J., Hillenmeyer M. E., Davis R. W., Nislow C., and Giaever G. (2005) Mechanisms of haploinsufficiency revealed by genome-wide profiling in yeast. *Genetics* **169**, 1915–1925.
26. Giaever G., Flaherty P., Kumm J., Proctor M., Nislow C., Jaramillo D. F., Chu A. M., Jordan M. I., Arkin A. P., and Davis R. W. (2004)

Chemogenomic profiling: identifying the functional interactions of small molecules in yeast. *Proc Natl Acad Sci USA* **101**, 793–798.

27. Lum P. Y., Armour C. D., Stepaniants S. B., Cavet G., Wolf M. K., Butler J. S., Hinshaw J. C., Garnier P., Prestwich G. D., Leonardson A., Garrett-Engele P., Rush C. M., Bard M., Schimmack G., Phillips J. W., Roberts C. J., and Shoemaker D. D. (2004) Discovering modes of action for therapeutic compounds using a genome-wide screen of yeast heterozygotes. *Cell* **116**, 121–137.
28. Kennedy P. J., Vashisht A. A., Hoe K. L., Kim D. U., Park H. O., Hayles J., and Russell P. (2008) A Genome-Wide Screen of Genes Involved in Cadmium Tolerance in *Schizosaccharomyces pombe. Toxicol. Sci.* **106**, 124–139.
29. Zuin A., Gabrielli N., Calvo I. A., Garcia-Santamarina S., Hoe K.-L., Kim D. U., Park H.-O., Hayles J., Ayte J., and Hidalgo E. (2008) Mitochondrial Dysfunction Increases Oxidative Stress and Decreases Chronological Life Span in Fission Yeast. *PLoS ONE* **3**, e2842.
30. Xu D., Jiang B., Ketela T., Lemieux S., Veillette K., Martel N., Davison J., Sillaots S., Trosok S., Bachewich C., Bussey H., Youngman P., and Roemer T. (2007) Genome-wide fitness test and mechanism-of-action studies of inhibitory compounds in *Candida albicans. PLoS Pathog* **3**, e92.
31. Noble S. M., French S., Kohn L. A., Chen V., and Johnson A. D. (2010) Systematic screens of a *Candida albicans* homozygous deletion library decouple morphogenetic switching and pathogenicity. *Nat Genet* **42**, 590–598.
32. Oh J., Fung E., Price M. N., Dehal P. S., Davis R. W., Giaever G., Nislow C., Arkin A. P., and Deutschbauer A. (2010) A universal TagModule collection for parallel genetic analysis of microorganisms. *Nucleic Acids Res* **38**, e146.
33. Oh J., Fung E., Schlecht U., Davis R. W., Giaever G., St.Onge R. P., Deutschbauer A., and Nislow C. (2010) Gene annotation and drug target discovery in *Candida albicans* with a tagged transposon mutant collection. *PLoS Pathog in press.*
34. Elson S. L., Noble S. M., Solis N. V., Filler S. G., and Johnson A. D. (2009) An RNA Transport System in *Candida albicans* Regulates Hyphal Morphology and Invasive Growth. *PLoS Genet* **5**, e1000664.
35. Wilson R. B., Davis D., and Mitchell A. P. (1999) Rapid Hypothesis Testing with *Candida albicans* through Gene Disruption with Short Homology Regions. *J. Bacteriol.* **181**, 1868–1874.
36. Nislow C., and Giaever G. (2007) Chemical Genomic Tools for Understanding Gene Function and Drug Action, In *Yeast Gene Analysis* (Stansfield, I., and Stark, J. R., Eds.) 2nd ed., pp 387–414, Elsevier Ltd.
37. Pierce S. E., Davis R. W., Nislow C., and Giaever G. (2007) Genome-wide analysis of barcoded *Saccharomyces cerevisiae* gene-deletion mutants in pooled cultures. *Nat Protoc* **2**, 2958–2974.
38. Han T. X., Xu X. Y., Zhang M. J., Peng X., and Du L. L. (2010) Global fitness profiling of fission yeast deletion strains by barcode sequencing. *Genome Biol* **11**, R60.
39. Jacobs M. A., Alwood A., Thaipisuttikul I., Spencer D., Haugen E., Ernst S., Will O., Kaul R., Raymond C., Levy, R., Chun-Rong L., Guenthner D., Bovee D., Olson M. V., and Manoil C. (2003) Comprehensive transposon mutant library of *Pseudomonas aeruginosa. Proc Natl Acad Sci USA* **100**, 14339–14344.
40. Gola S., Martin R., Walther A., Dunkler A., and Wendland J. (2003) New modules for PCR-based gene targeting in *Candida albicans*: rapid and efficient gene targeting using 100 bp of flanking homology region. *Yeast* **20**, 1339–1347.

Chapter 15

Global Strain Engineering by Mutant Transcription Factors

Amanda M. Lanza and Hal S. Alper

Abstract

Cellular hosts are widely used for the production of chemical compounds including pharmaceutics, fuels, and specialty chemicals. Strain engineering focuses on manipulating and improving these hosts for new and enhanced functionalities including increased titers and better bioreactor performance. These tasks have traditionally been accomplished using a combination of random mutation, screening and selection, and metabolic engineering. However, common metabolic engineering techniques are limited in their capacity to elicit multigenic, complex phenotypes. These phenotypes can also include nonpathway-based traits such as tolerance and productivity. Global transcription machinery engineering (gTME) is a generic methodology for engineering strains with these complex cellular phenotypes. In gTME, dominant mutant alleles of a transcription-related protein are screened for their ability to reprogram cellular metabolism and regulation, resulting in a unique and desired phenotype. gTME has been successfully applied to both prokaryotic and eukaryotic systems, resulting in improved environmental tolerances, metabolite production, and substrate utilization. The underlying principle involves creating mutant libraries of transcription factors, screening for a desired phenotype, and iterating the process in a directed evolution fashion. The successes of this approach and details for its implementation and application are described here.

Key words: Complex phenotype, Transcription machinery, gTME, Engineered phenotype, Sigma factor, TATA binding protein, Metabolic engineering

1. Introduction

Many prokaryotic and eukaryotic cellular systems are attractive hosts for the sustainable production of chemicals, fuels, and pharmaceuticals. Using classical metabolic engineering methodologies, these hosts are typically engineered for improved pathway fluxes and high product yields. These traditional approaches typically focus on one-gene-at-a-time perturbations. As a result, these tools are quite limited in their ability to elicit complex cellular phenotypes. Examples of complex phenotypes include chemical tolerances, faster growth rates, morphology, and higher bioconversion rates. In addition to

James A. Williams (ed.), *Strain Engineering: Methods and Protocols*, Methods in Molecular Biology, vol. 765, DOI 10.1007/978-1-61779-197-0_15, © Springer Science+Business Media, LLC 2011

these global phenotypes, rerouting metabolites through to products (especially toward secondary metabolites) can also constitute a complex phenotype requiring multiple gene modifications. Complex phenotypes differ from metabolic phenotypes in part due to the number of genes regulating these traits. Metabolic phenotypes are often controlled by a handful of genes, whereas no singular gene or pathway is responsible for complex, multigenic traits (1, 2). The improvement of complex cellular phenotypes has been a long standing goal of the food and biotechnology industry well before the advent of recombinant DNA technology (3) and continues to be actively researched (4).

Global transcription machinery engineering (gTME) is a generic methodology for engineering complex phenotypes. The major premise of this approach is that introducing dominant mutant alleles of generic transcription-related proteins can reprogram gene networks and cellular metabolism. By linking this mutant protein expression with phenotype selection and screening, it is possible to identify mutants eliciting novel, complex cellular phenotypes. However, since the linkage between mutant protein sequence and function is difficult to predict de novo, this method is typically reduced to practice by creating mutant libraries of transcription factors and selecting for the ones that improve a desired phenotype. Since its development in 2006, this method has been implemented in *Escherichia coli*, *Saccharomyces cerevisiae*, and *Lactobacillus plantarum* and used to isolate strains with improved complex and metabolic phenotypes including high ethanol tolerance (5, 6), overproduction of lycopene (6), and improved xylose fermentation (7). Thus, gTME is a useful tool for strain development that can be applied on its own or in conjunction with other cellular engineering techniques. The description of this method below begins by comparing gTME to alternative methods, and then provides the theoretical framework for the approach, followed by highlights of recent successes. This extended introduction is followed by a detailed description of the gTME method for both a generic phenotype and the specific example of improved ethanol tolerance in yeast.

1.1. Alternative Methods for Engineering Complex Cellular Phenotypes

Several rational and combinatorial methods exist for strain development. Most traditional, classical approaches utilize a mutation-selection strategy. The most common chemical mutagenic agent is *N*-methyl-*N*′-nitro-*N*-nitrosoguanidine (NTG), an alkylating agent that causes a GC to AT transition at the DNA level (3). Other treatments include radiation, UV rays, chemical treatment, and biological treatment such as phage or transposons (3). The extent and efficacy of mutation depends not only on the type of treatment but also on the dose, exposure time, type of damage, and conditions after treatment (3). Chemical mutagenesis is limited, however, in its ability to engineer strains with complex phenotypes. It is a slow and laborious process, making screening of large libraries hard to implement (4). Each round of mutation may result in incremental improvement, necessitating an iterative process. Additionally, it has been shown that the phenotypic diversity achieved by NTG mutagenesis is much lower than that of gTME (8). Despite limitations of chemical mutagenesis, it has been widely and successfully used. A major application of this method is in antibiotic production, where strains producing penicillin in excess of 50 g/L have been isolated (4). Nevertheless, one particular challenge of this approach for complex phenotypes is the extreme library size needed to screen the low-probability event of multiple genes being impacted.

As the understanding of cellular metabolism and pathways has increased, a more rational approach to strain engineering has evolved. In this approach, the "bottlenecks," or limiting steps in a pathway, are first identified. In some cases, this may be a single enzyme or network branch point, but often it is distributed among several enzymatic steps. In the case of distributed flux control, efforts are focused first on those enzymes with the most direct impact on desired product (9). The rate-limiting steps are then altered using metabolic engineering techniques along with an appropriate genetic modification. Limiting enzymes can be overexpressed to increase flux, whereas competing reactions can be minimized by knockout or knockdown of those genes. Additionally, heterologous genes can be cloned and expressed to augment an organism's natural chemistry, thereby creating new metabolic pathways and end products.

This rational strain engineering method can be iteratively applied to a particular host and used to engineer a strain with an improved phenotype. This method has been widely applied for the production of small molecules, food additives, metabolites, and polymer precursors (4). As a specific example, this approach has recently been used to develop a synthetic pathway for the biological production of glucaric acid in *E. coli*, resulting in titers in excess of 1 g/L (10). While metabolic engineering has significant utility, this method is best applied to a well-defined system where the genes contributing to a phenotype are both identified and well understood (4). For less characterized phenotypes, or those phenotypes involving more complex genetic interactions, metabolic engineering cannot be rationally applied. Additionally, experimental limitations in vector construction, transformation, and expression constrain the number of genes that can simultaneously be manipulated (6).

More recently, genomics-inspired approaches have been used to engineer strains with complex phenotypes. Approaches such as parallel gene trait mapping (11), multi-SCale Analysis of Library Enrichments (12, 13), shotgun genomics (14, 15), transposon mutagenesis (16–19), and genome shuffling (20, 21) have all been developed to map genotype to phenotype. These approaches are quite powerful, as they select for traceable changes to a cell that may be easily identified and transferred. As an example, multi-SCale Analysis of Library Enrichments, or SCALE, can be used to quickly and efficiently screen and analyze genomic libraries in surrogate hosts (22). These libraries contain different size DNA fragments, or scales, cloned into expression vectors and are grown competitively under selective conditions associated with the desired phenotypes. This method was initially validated in *E. coli* to identify genes associated with Pine-sol resistance, a complex tolerance phenotype (11) that would be very difficult or impossible to engineer using either a classical or rational approach to strain engineering.

1.2. Global Transcription Machinery Engineering Theoretical Framework

Global transcription machinery engineering is an alternative approach for imparting the multigenic changes necessary to elicit complex phenotypes in industrially relevant strains. Transcriptional profiling of cells expressing mutant transcription factors reveals that these factors can elicit the differential expression of hundreds of genes (5) compared to wild-type cells. Thus, the phenotypes enabled by gTME are not limited to a single gene or single pathway like the traditional approaches discussed above. Moreover, this broad action increases the likelihood of engineering complex phenotypes resulting from the concerted manipulation of several genes.

To accomplish this reprogramming, the method of gTME can be broken down into four basic steps with two optional steps depending on the end goal and desired level of phenotype improvement (Fig. 1):

- First, a target gene involved in key transcription activity is selected. Suitable targets include generic transcription factors known to interact directly or indirectly with hundreds of genes.
- Second, a mutant library (often comprising roughly 10^5–10^6 different mutant versions of this selected protein) is created.
- Third, the mutated gene library is expressed in the host cell. Coexpression of the mutant and endogenous, chromosomal version creates a screen for dominant mutant phenotypes. In this regard, the wild-type version allows for maintenance of crucial cellular functions whereas the mutant version can impart phenotype-specific cellular reprogramming.
- Fourth, mutant proteins imparting the desired reprogramming and thus phenotype of interest can be identified using trait-specific screening and selection. Owing to the use of large libraries, a high-throughput screen is almost always necessary.

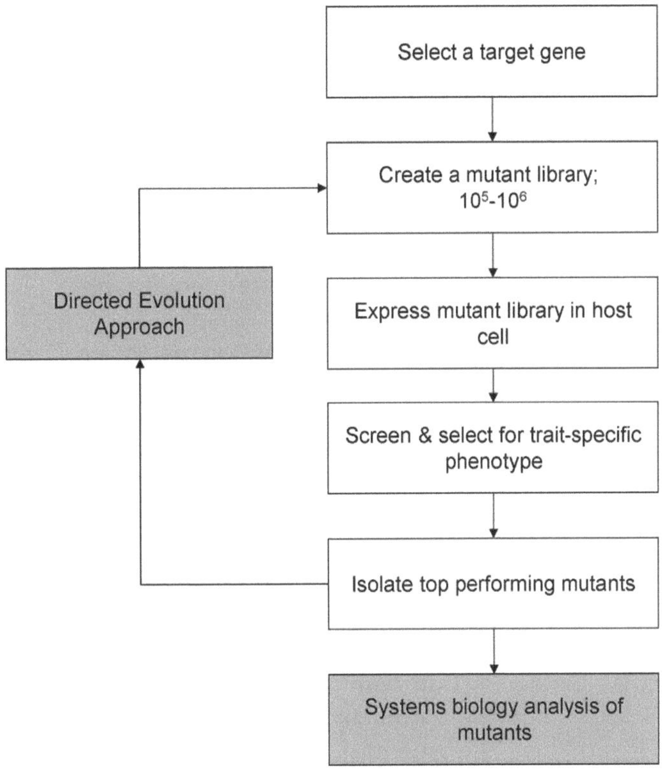

Fig. 1. General Methodology for gTME. The general framework for the gTME approach is depicted. Shaded boxes highlight optional steps depending on the goal and scope of the project.

- Fifth, the process (Steps 2–4) can be reiterated using a directed evolution approach since the phenotype of interest is linked to a single mutant protein of interest.
- Sixth, it is possible to use a systems biology approach for understanding the underlying reprogramming and mechanism. This step is not necessary if the end goal is simply to obtain an improved strain.

One limitation of gTME is that it can only access latent cellular potential and cannot introduce de novo function to a cell. This limitation arises from the mode of action, namely, altering gene expression, not protein function. Specifically, while mutant factors result in gene expression changes across the cell, the functionality of those genes is fixed. Thus, unlike chemical mutagenesis, there is no potential to evolve other enzymes by strictly using a gTME approach. This is a major difference between gTME and other methods for engineering phenotype, such as chemical mutagenesis, metabolic engineering, and metagenomics. However, these methods can be combined to effectively modify existing cellular mechanisms and introduce new functionality through heterologous DNA.

1.3. Examples of Applications and Future Directions of gTME

Transcription machinery engineering has been applied to several prokaryotes and to the eukaryote yeast. In *E. coli* and *L. plantarum*, sigma factors including σ^{70} (encoded by *rpoD*) and σ^{38} (encoded by *rpoS*), as well as the RNA polymerase itself, encoded by *rpoA*, (23), have been targets for gTME (6, 8, 24), whereas in yeast strains the focus has been on *taf25* and *spt15* (5, 7). Most of the mutant transcription factors identified from these varied screens either contained a small number of mutations (on the order of 1–5) or had large modifications (such as significantly truncated proteins).

In bacterial systems, the *rpoD* gene encoding the main sigma-70 factor has been a target of choice. In one study, a single mutant library of *rpoD* was developed and enabled the screening of three improved phenotypes: improved tolerance to ethanol, multiple, simultaneous phenotypes, and lycopene overproduction (6). In the first case, an *E. coli* mutant with a 6 h doubling time in 60 g/L of ethanol was identified. Tolerance to ethanol and sodium dodecyl sulfate (SDS) was selected to explore the ability of gTME to impart multiple, simultaneous phenotypes. Independent selection for an ethanol and an SDS mutant, followed by coexpression of both mutants, was found to be superior to either sequential or simultaneous selection strategies (6). In the third case, development of a metabolite overproducing mutant started with a pre-engineered parental strain that had previously been optimized for lycopene accumulation. This strain was then further engineered using both gTME and traditional gene knockouts. A single round of gTME performed better than a single gene knockout, and three distinct gene knockouts were necessary to achieve similar increases in lycopene production as a single round of gTME. In this case, gTME was a more efficient approach for isolating a metabolite overproducing strain, and by combining the two methods, a strain able to produce more than 7 mg/L of lycopene was isolated. Thus, these results illustrate the utility of gTME as a method for traditional strain engineering.

A separate study in *L. plantarum* isolated mutants with increased tolerance to low pH and high lactic acid concentration, both of which are industrially relevant conditions (8). In the case of lactic acid tolerance, a single amino-acid substitution was made to *rpoD*, whereas low pH tolerance was a result of four amino-acid substitutions and a truncation of the protein (8). A third study looked at improved hyaluronic acid (HA) production in recombinant *E. coli* by targeting both *rpoD* and *rpoS*. The top-performing mutant accumulated 560 mg/L of HA (24).

The majority of gTME work in yeast has focused on the TATA-binding protein encoded by the *SPT15* gene in *S. cerevisiae*. The first study looked at two mutant libraries for *spt15* and *taf25* for improved ethanol tolerance in high glucose fermentations (5). The *spt15* mutant outperformed the *taf25* mutant, giving a 12-fold improvement in optical density over a control strain when grown in 100 g/L of glucose and 6% ethanol by volume. Subsequent transcriptional profiling identified major gene targets of the mutant *spt15-300*, which when knocked out resulted in a loss of capacity for the ethanol-tolerant phenotype (5).

Spt15 mutant libraries for ethanol tolerance were also constructed and tested in a yeast diploid and Kyokai 7, an industrial strain for sake production (unpublished). This industrial strain was observed to have superior ethanol tolerance compared to both laboratory diploid and haploid strains. Using gTME, an s*pt15* mutant was isolated that conferred improved growth rates on Kyokai 7 with ethanol concentrations as high as 9% by volume. Previously identified s*pt15* mutants for laboratory strains were not effective in Kyokai 7, whereas the new mutant identified by screening in the industrial strain has negligible impact on *S. cerevisiae*. This indicates the strain specific behavior of *spt15* mutants and, therefore, the importance of screening mutants in an environment relevant to the desired phenotype.

A third study screened an *spt15* mutant library for xylose utilization in *S. cerevisiae* (7). The control strain was able to grow on glucose but not able to utilize xylose. A single strain was isolated that was able to grow modestly on 50–150 g/L of xylose as well as mixed sugar cultures. However, the mutant strain had decreased ethanol yield compared to the control strain, even when grown on the preferred glucose carbon source, indicating further strain improvements are necessary before this could be applied to an industrial biomass process (7).

Despite its infancy, the technique of global transcription machinery engineering has demonstrated an ability to engineer strains exhibiting complex cellular phenotypes in both prokaryotes and simple eukaryotes. These phenotypes often result from dramatic reprogramming of innate gene expression, resulting in significant shifts in metabolism and ultimately cellular performance. More importantly, these multiple, simultaneous changes are being conducted by a single, generic target. Thus, this approach transforms complex phenotype elicitation into the directed evolution of a single protein. This methodology could also be employed to identify genes involved in a particular pathway or phenotype that is not well understood. gTME represents an important alternative to more traditional methods such as chemical mutagenesis or single-gene based metabolic engineering. Using the method of gTME in conjunction with techniques such as metabolic engineering, directed evolution, and cell adaptation could be a powerful combination facilitating the isolation of complex, cellular phenotypes.

2. Materials

2.1. Selection of Gene Target

1. Genome sequence or (at minimum), sequence of target gene.
2. Genomic DNA or cDNA from the host organism. The Promega Wizard Genomic DNA Purification Kit works well for most cell types.
3. Standard PCR reagents and gene-specific primers.

2.2. Construction of Mutant Library

1. Broad-host expression vector compatible with both host organism and *E. coli*, containing a promoter of the appropriate strength.
2. Restriction enzymes and T4 DNA ligase for basic cloning.
3. Sequencing primers and mutagenesis primers.
4. Error-prone PCR kit or reagents.
5. Column cleanup kit for DNA fragments.
6. Competent *E. coli* cells.
7. Large petri dishes (150 × 10 mm) and LB-agar supplemented with an appropriate antibiotic.
8. Sterile plate scrapers.
9. 30% glycerol by volume, filter-sterilized.
10. LB media and culture tubes.
11. Plasmid extraction kit.
12. Device for measuring optical density.
13. Incubator for 37°C growth of *E. coli* cells.

2.3. Mutant Expression in Host Strain

1. Competent cells for the host organism and transformation reagents.
2. Large petri dishes (150 × 10 mm) and media supplemented with agar and the appropriate selection media for the host organism.
3. Sterile plate scrapers.
4. Incubator for growth of host cells.

2.4. Phenotype Selection Strategies

1. Selection conditions and proper flasks/tubes.
2. Incubator.
3. Plasmid recovery kit for the host organism.

2.5. A Directed Evolution Approach

1. Same reagents as Subheading 2.2.

2.6. Analysis of Selected Mutants

1. Sequencing primers.
2. Sequence alignment software, such as ClustalW.
3. Whole cell RNA extraction kit.
4. Global microarrays specific to host organism.

3. Methods

As described above, the basic gTME paradigm can be described by four main steps and two optional steps as follows: (1) selecting a target of interest, (2) creating a mutant library of the target, (3) expressing the mutant library in a host strain, (4) selecting for the phenotype of interest, (5) reiterating the process using directed evolution, and (6) evaluating the mutant using a systems biology approach. The detailed procedure for carrying out gTME depends significantly on the desired phenotype and cellular host. Thus, to provide adequate description of this method, each of the six steps is described in two ways in this section. First, this method is described generically. Second, to more clearly illustrate the methodology, the specific example of improved ethanol tolerance in yeast is selected as a case study (5).

3.1. Selection of Gene Target

3.1.1. Generic Methodology

1. Selection of a suitable gene target is the first step in gTME. In order to be most effective at generically impacting metabolism, global regulators of basic transcription have traditionally been selected. As a starting point, it is useful to consider a high-confidence target, or a gene in which dominant mutations have previously been shown to exist. To date, transcription factors associated with the basic RNA polymerase system such as *rpoD* and *rpoS,* as well as the polymerase itself, *rpoA,* have been targets in prokaryotic systems (6, 8, 23, 24). In yeast, *taf25* and *spt15* have been selected as targets (5, 7). Additional potential targets in yeast include the three RNA polymerases and approximately 75 general transcription factors (5). In order to obtain optimal results, it may be necessary to select and test multiple targets, as the ideal target may be phenotype specific (see Note 1).
2. The sequence of the target gene should be obtained from sequence databases or de novo sequencing. For many model organisms, transcription factors and associated proteins have been sequenced and characterized. This information greatly facilitates target selection. However, in other microbes of interest, these genes may not have been studied and genome sequence annotation may be limited. In these cases, it may be necessary to select a target gene by

comparing to better understood organisms using sequence homology.

3.1.2. Ethanol Tolerance Example

1. Two known transcription factors, *TAF25* and *SPT15*, were selected as targets.
2. The genome sequence of BY4741 is known and sequences for both genes were found using the Saccharomyces Genome Database.

3.2. Construction of a Mutant Library

3.2.1. Generic Methodology

1. First, a suitable expression vector must be found to clone the target gene selected in the steps above. The expression vector should contain an origin of replication, selectable marker, and promoter region for the host organism (described further in Step 2 below), as well as an antibiotic marker and origin of replication for *E. coli* (or another selected host for routine plasmid propagation).
2. A suitable promoter used to express the mutant transcription factor must be selected. The strength of the promoter can have an impact on gTME selection and screening, as well as the magnitude of the dominant phenotype, and should be considered as part of vector selection. It has previously been shown that a relatively strong constitutive promoter can be used with low copy number plasmids (see Note 2). Choice of promoter strength also depends on the ploidy of the strain.
3. Primers should be designed to amplify the gene of interest from the genome. Restriction enzyme sites should be appended to both primers to allow for cloning into the selected expression plasmid.
4. Genomic DNA is extracted for the desired host organism. The target gene is amplified from genomic DNA using a polymerase chain reaction (see Note 3) and the primers designed above. This product should be cloned directly after the vector's promoter using basic molecular biology techniques.
5. This completed plasmid should be sequence confirmed to ensure the gene was cloned in-frame and does not contain any sequence errors. This plasmid serves as the control vector for experiments as well as the starting point for library construction.
6. A set of mutagenesis primers should be designed to amplify and mutate the target gene. To ensure the maximum amount of mutation, these primers are designed to have nearly complete homology to the vector (see Note 4).
7. A mutant library using error-prone PCR (epPCR) should be constructed using the control vector as a template. This mutant library can be constructed using various epPCR techniques including: mutant polymerases, nucleotide analogues,

or increased Mg and Mn concentrations (25). Typically, nine 50 μL reactions are performed to obtain enough pooled DNA.

(a) Error-Prone PCR by mutant polymerases

- This is the most common method for producing mutated copies of the target gene. An epPCR reaction is assembled identically to standard PCR, except it uses a polymerase with an increased likelihood for incorporating mismatched base pairs. Mutation frequency can be controlled by changing the template concentration, where less initial template increases the mutation frequency (see Note 5).

(b) Error-Prone PCR by nucleotide analogues

- Another option is to use pyrimidine analogues that are recognized by polymerases and incorporated in the place of dTTP and dCTP (26). Individual base-pair substitution rates range from 1^{-2} to 4.4^{-2}, whereas overall mutation rates were found to be as high as 1.9^{-1}.

(c) Error-Prone PCR by Mg/Mn concentration

- Increased concentrations of $MgCl_2$ and the addition of $MnCl_2$ have both been shown to increase the mutation rate of a standard PCR reaction with the use of traditional enzymes such as *Taq* (27). The presence of either compound helps stabilize mismatched base pairs. $MgCl_2$ will typically be added at a concentration up to 7 mM (a 4.7-fold increase from a typical reaction). $MnCl_2$ is not typically present in PCR reactions and can be added at a concentration of 0.5 mM just prior to thermocycling, as it can precipitate.

8. Pooled DNA from the error-prone PCR step should be purified using standard molecular biology techniques. This mixture should be digested with the appropriate restriction enzymes (see Note 6).
9. The fragments and the vector can be ligated (usually overnight), then transformed into a bacterial host such as DH10β or DH5α using standard techniques. Typically, around 10–15 standard transformation reactions are performed using 2.5 μL of the ligation mixture for each transformation.
10. Cells are then plated onto large LB-agar plates (see Note 7) supplemented with the appropriate antibiotic and grown overnight at 37°C.
11. Colonies are counted from the plates after around 20 h of incubation to determine the library size (see Note 8).
12. The colonies are then pooled using sterile media and a sterile cell scraper to create a liquid library of the cells.

13. Portions of this liquid library are stored as glycerol stocks (in 15% glycerol by volume).
14. The remainder of this library is harvested for plasmids using methods such as Qiagen Mini-prep spin kit. Often, around 20 samples from the library are harvested for plasmids.
15. Final plasmids are pooled and DNA concentration is measured using a spectrophotometer.

3.2.2. Ethanol Tolerance Example

1. The p416-TEF-mut2 plasmid was selected as an expression vector. This supports bacterial and yeast replication. The prokaryotic and eukaryotic selection markers are ampicillin and uracil, respectively. This is a lower-copy number CEN-based plasmid.
2. The TEF-mut2 promoter was selected to drive mutant gene expression. This is a constitutive promoter found to give 7% expression compared to the native TEF promoter, from which it was derived (28).
3. Primers were designed with *Nhe*I and *Sal*I restriction sites. For TAF25: TCGAGTGCTAGCAAAATGGATTTTGAG GAAGATTACGAT and CTAGCGGTCGACCTAACGATAA AAGTCTGGGCGACCT, for SPT15: TCGAGTGCTAGC AAAATGGCCGATGAGGAACGTTTAAAGG and CTAG CGGTCGACTCACATTTTTCTAAATTCACTTAGCACA.
4. Yeast genomic DNA was extracted using the Promega Wizard Kit and a standard haploid laboratory strain of yeast, BY4741. The TAF25 and SPT15 genes were amplified from gDNA using *Taq* polymerase and digested with *Nhe*I and *Sal*I.
5. The resulting plasmids were sequence verified to ensure the correct sequences of TAF25 and SPT15.
6. Generic Mutagenesis primers were designed based on the p416 vector and *Nhe*I and *Sal*I restriction sites. GCATA GCAATCTAATCCAAGTTTTCTAGAATG and ATAACTA ATTACATGACTCGAGGTCGACTTA were chosen as the two mutagenesis primers.
7. Mutagenesis was performed using the GeneMorph II Random Mutagenesis Kit from Stratagene. Varied template concentrations (using the p416-TEF-mut2-TAF25 or SPT15 based plasmids) were selected to enable low, medium, and high mutation rates.
8. The fragments were purified using Qiagen PCR cleanup kit and digested overnight at 37°C using *Nhe*I and *Sal*I. The vector was digested with *Xba*I and *Sal*I overnight.
9. The fragments were ligated with the vector overnight at 16°C using T4 DNA ligase. The mixture was transformed into competent DH5α.

10. The transformation mixture was plated on LB-agar plates supplemented with 100 μg/mL of ampicillin and grown overnight at 37°C.
11. Colonies were counted and library size was determined to be 10^5.
12. The colonies were scraped off of plates to create a liquid library.
13. A portion of the liquid library was stored in a 15% glycerol mixture at −80°C.
14. Plasmids were extracted from the library using a Qiagen Mini-prep spin kit.
15. The plasmids were pooled and the concentrations were around 300 ng/μL.

3.3. Mutant Expression in Host Strain

3.3.1. Generic Methodology

1. Extracted plasmids are library-transformed into the host strain using a method suitable for the organism.
2. The transformation mixture is plated on a solid media containing a selectable marker specific to the host strain.
3. Colonies are then scraped off the plates using sterile media to create a liquid library of the cells; glycerol stocks can be made at this point.

3.3.2. Ethanol Tolerance Example

1. Plasmids containing mutant libraries for TAF25 and SPT15 were transformed into competent BY4741 using the standard Geitz's lithium acetate protocol. A total of 1 μg of plasmid was used for each transformation mixture.
2. The transformation mixture was plated onto dropout media (a total of 48 150×10 mm plates were used) lacking uracil. The plates were incubated for 2–3 days at 30°C.
3. Colonies were scraped from the plates into a liquid media to proceed directly with phenotypic selection.

3.4. Phenotype Selection Strategies

3.4.1. Generic Methodology

gTME requires a large library size to effectively isolate mutants conferring a desired phenotype. Because of this library size, a high-throughput screen for the isolation of desired mutants is essential. The screening method and the specific conditions will vary depending on the phenotype that is being isolated (see Note 9). Common phenotypes studied to date include tolerance to environmental conditions, metabolite production, and multigenic phenotypes. Screening for each of these traits should be uniquely approached and may require multiple rounds of library construction and refinement before identifying mutations conferring the desired phenotype.

1. Screening for improved tolerance (see Note 10)
   (a) Identify an environmental condition for which tolerance in the host organism is desired.
   (b) Expose the mutant library to gradually harsher conditions, selecting against those mutants unable to grow in the environment. Typically, starting values at or above the minimum inhibitory concentration of wild-type cells are used for rapid selection.
   (c) As cells from the library grow, this initial selection phase can be repeated by diluting and growing the surviving culture under the same environmental condition.
   (d) The environmental condition can be made increasingly harsher, thereby reducing the surviving population.
   (e) After several rounds (typically 5–10), this liquid culture can then be plated on solid media for the selection of individual clones.
   (f) Individually selected clones can then be tested again for growth under the environmental conditions (typically, 20–50 clones are isolated from plates for initial testing).
   (g) Finally, the plasmid DNA should be extracted and retransformed, and the resulting clones are again tested for growth under the environmental condition (see Note 11).
2. Screening for metabolite production (see Note 12)
   (a) Develop a high-throughput screen to detect the compound of interest. The most common screens are colorimetric, enzymatic, or a direct measurement of the compound.
      - Colorimetric screening, when possible, is ideal because it is fast and requires minimal labor. Candidates can be screened by eye or using a spectrophotometer (see Note 13).
      - Enzymatic screens are another option if the metabolite of interest interacts with enzymes whose products or cofactors result in a detectable photometric shift (see Note 14).
      - For those metabolites that cannot be screened by either a color change or enzymatic method, it may be necessary to directly measure metabolite concentration. If the metabolite is secreted, stationary phase culture can be pelleted and tested using an instrument such as a YSI biochemical analyzer, HPLC, or GC-MS. Additionally, the product of interest could be stained for or indirectly linked to a fluorescent protein (such as GFP) and detected using FACS.
   (b) Select individual colonies from the library and grow in 96-well plates (or on solid media, depending on the screen).

(c) Screen each clone for metabolite production and compare performance to the control strain.

(d) Isolate top-performing clones and repeat the screen using 5-mL culture tubes to confirm earlier results.

(e) For those clones exhibiting sufficient metabolite production, extract the plasmid DNA, retransform it, and once more test the clone for metabolite production (see Note 11).

3. Selection of multigenic phenotypes

(a) When dealing with more than one phenotype, the order of selection is important (see Note 15).

(b) Starting from the full mutant library, develop and carry out a selection strategy for the first phenotype.

- Identify and isolate top-performing clones.
- Repeat the screening to confirm earlier results.
- Extract the plasmid DNA, retransform it, and conduct a final test of the phenotype.

(c) Starting from the full mutant library, develop and carry out a selection strategy for the second, and any subsequent, phenotype.

(d) Coexpress combinations of the top-performing mutants for each phenotype in the host organism.

(e) Evaluate these combinations for performance under both conditions of interest.

3.4.2. Ethanol Tolerance Example

1. Yeast strains with an increased tolerance to ethanol were selected using a tolerance screen.

(a) Increased ethanol tolerance was selected as the desired phenotype.

(b) Selection was initially performed in YSC-URA media supplemented with 100 g/L of glucose and 5% ethanol by volume. 30 mL of culture in 30 × 115 mm closed top tubes were grown vertically at 30°C.

(c) The *taf25* library was subcultured four times under these conditions.

(d) The *spt15* library was subcultured twice under the initial condition and then twice at 120 g/L glucose and 6% ethanol.

(e) The mixture was plated on YSC-URA and 20 surviving colonies selected for both *taf25* and *spt15* mutations.

(f) These selected clones were grown in overnight cultures and assayed for growth rate in 60 g/L glucose and 5% ethanol. Improvement in growth performance was determined by OD, as compared to the control.

(g) Plasmids were extracted from the clones showing improved growth. These plasmids were retransformed and the ethanol-tolerant phenotype once more validated. The *spt15-300* mutant (containing the F177S, Y195H, and K218R mutations) was found to impart the most significantly improved phenotype.

3.5. A Directed Evolution Approach to gTME

3.5.1. Generic Methodology

Directed evolution is a protein engineering algorithm by which the fitness of a protein can be enhanced through iterative mutagenesis and selection. Such an approach can be applied to gTME for increased library diversity and improved phenotypes. Unlike traditional approaches, the directed evolution algorithm is applicable to gTME, since the phenotype improvement is linked to a mutant protein. Directed evolution can be performed in multiple ways. First, the top performing mutants isolated in Subheading 3.4 Phenotype selection strategies can be remutagenized (following the methods of Subheading 3.2 Construction of a mutant library), and then selection can be repeated. A previous study merging directed evolution and gTME showed that after two rounds of mutagenesis, the fold improvement in phenotype was incremental (6). Alternatively, it is possible to isolate a subset of the top-performing mutants and perform gene shuffling to create a new, diverse library (see Note 16). For the case of iterative mutagenesis, the directed evolution method would entail the following:

1. Isolate a top-performing mutant following Subheading 3.4. This mutant serves as the starting point for the construction of a new library. Proceed as described in Subheading 3.2, starting with Step 7 and using the mutant target gene as the template for mutagenesis.
2. Express the new mutant library in a host strain (see Subheading 3.3).
3. Select for an improved phenotype (see Subheading 3.4). During the selection phase, the new mutant library should be compared to both the wild-type, endogenous target gene and the mutant target gene isolated after initial library construction.
4. This process can be iterated more than once to obtain further improved phenotypes.

3.5.2. Ethanol Tolerance Example

For the case of improved ethanol tolerance in yeast by mutant *spt15*, a directed evolution approach was not used. However, this approach could have been performed by using *spt15-300* as a template for creating a mutagenesis library.

3.6. Analysis of Selected Mutants

3.6.1. Generic Methodology

1. Sequence analysis of the target gene

   After isolating a mutant cell line conferring a desired phenotype, it may be of interest to determine the mutation(s) associated with those changes. Identifying the locations and types of mutations that contribute to a specific phenotype can provide useful information about the functionality of transcription-related proteins.

(a) The plasmid carrying the mutant target gene is recovered from the cell line.

(b) The target gene is sequenced using standard technology and primers both upstream and downstream of the gene.

(c) The sequence is compared with the endogenous target gene to identify mutations.

- Previous studies have found mutations ranging from single amino-acid substitutions to domain truncations (see Note 17).

2. Transcription profiling

Mutant transcription factors are selected for their ability to indirectly regulate transcription of hundreds of genes; therefore, a desired phenotype is likely the result of unique interactions between that target gene and other genomic loci. An analysis of the cell's transcriptome, compared to the control cell line, can identify both upregulated and downregulated genes.

(a) The control and mutant cell lines are grown under desired media conditions until mid-log phase.

(b) Whole cell RNA is extracted. The extraction method varies depending on the cell type.

(c) The whole cell RNA is subjected to a global microarray analysis (see Note 18).

(d) The genes most impacted by the mutant transcription factor can be further probed using traditional knockouts and overexpression.

3.6.2. Ethanol Tolerance Example

1. Sequence analysis of *spt15-300* ethanol-tolerant mutant

(a) The *spt15-300* plasmid was extracted using the Zymoprep yeast plasmid miniprep kit and transformed into *E. coli* DH5α. This plasmid was isolated and sequenced using forward primer TCACTCAGTAGAACGGGAGC and reverse primer AATAGGGACCTAGACTTCAG.

(b) The sequence of *spt15-300* and the wild-type gene were compared using ClustalW. *Spt15-300* was found to contain three point mutations, each resulting in an amino-acid residue change.

2. Transcription profiling of *spt15-300* ethanol-tolerant mutant

(a) The control strain and the *spt15-300* mutant were grown in YSC-URA medium supplemented with 100 g/L of glucose and 5% ethanol to an OD of 0.4–0.5.

(b) Whole cell RNA was extracted using the Ambion RiboPure Yeast RNA extraction kit.

(c) Microarray services were contracted through Ambion using Affymetrix yeast 2.0 arrays. Arrays were run in triplicate with biological replicates.

(d) Over one hundred genes were found to be diversely expressed in the presence of the *spt15* mutant and increased ethanol concentration. One hundred eleven genes were found to be upregulated, with only 21 downregulated. Twelve of the most highly expressed genes in mutant were individually deleted. These deletions were then found to result in a loss of mutant capacity (5), indicating their individual importance in ethanol tolerance as well as confirming that such a phenotype relies on the concerted expression of multiple genes.

4. Notes

1. Previous studies have shown that different gene targets elicit different phenotypic effects. For example, both *taf25* and *spt15* mutant libraries were screened for improved ethanol tolerance, but the *spt15* mutants outperformed the *taf25* mutants, demonstrating that different members of the transcription machinery have differential influence over phenotypic responses (5). A similar effect was seen with *rpoD* and *rpoS* libraries (24). Thus, for a given phenotype, it may be necessary to select and test multiple target genes before finding one that has the desired impact on cellular metabolism.
2. Many gTME studies have successfully used low-copy plasmids coupled together with relatively high constitutive promoters (5, 6, 24). An alternative option is to use the native promoter for the target gene; however, expression may not be constitutive and may be subject to regulation. In a higher ploidy strain, promoter strength must be increased to provide a more profound dominant phenotype (unpublished).
3. While *Taq* polymerase is the most common enzyme choice, many other polymerases are available and optimized for different conditions including long amplicons, high-fidelity, and high GC-content templates.
4. The last nucleotides of the 5′ primer should be ATG and the last nucleotides of the 3′ primer should be TAA (or another stop codon). This will allow all possible residues within the target gene to be subjected to mutation in the next step.

5. Stratagene's GeneMorph II Random Mutagenesis kit recommends 560 ng, 280 ng, and 28 ng for low, medium, and high mutation rates, respectively (8). A typical error rate for low mutation is 0–4.5 mutations per kilobase, 4.5–9 for medium mutation, and 9–16 for high mutation.
6. The mutant library can be digested with *Dpn*I to remove copies of the circular template DNA, reducing the likelihood that the control plasmid is present in the transformation mixture and, thus, increasing the diversity of the library.
7. Because of the large library size, it is ideal to use 150 × 10 mm plates for library transformation mixtures. This will ensure a better distribution of transformants across the plate, allow for antibiotic selection, and facilitate selection of individual colonies. Typically, around 300–500 μL of transformation mixture is plated onto each plate.
8. A typical desirable library size is 10^5–10^6 (5, 6, 8).
9. In general, high-throughput selections are sensitive to environmental conditions including media composition, temperature, and selection stringencies, and there exists a dependency between isolated mutants and the conditions under which they are identified (3, *unpublished*). This phenomenon is especially important when applying gTME to industrial strains. Decisions such as minimal or complex media and shaking rate, for example, should be made in advance and maintained throughout selection to ensure that isolated mutants perform consistently.
10. Screening for improved tolerance to environmental conditions is perhaps the most straightforward of selection strategies because undesired mutants are unable to compete and, therefore, die off. However, the specifics of the screen depend on the phenotype. For example, when identifying ethanol-tolerant mutants, an initial concentration of 50 g/L of ethanol was used (6). In another study attempting to identify mutants tolerant to low pH, the library was initially exposed to either pH of 4.60 or 3.85 (8).
11. This final step is important for confirmation that the improved phenotypic condition is a result of the mutated transcription-related gene and not other genomic mutations that could have occurred to a particular cell.
12. Metabolite production can be a more difficult phenotype to screen for because the concentration of the desired compound must be directly or indirectly measured, and it is not possible to select against colonies lacking the desired phenotype.

13. A colorimetric screen was used to detect improved lycopene production in yeast. The colonies were initially screened for a reddish hue (6).
14. An enzymatic assay was successfully used to screen for glucuronic acid production. The enzyme uronate dehydrogenase converts glucuronate to glucarate, resulting in an accumulation of NADH, detectable at 340 nm (29), whereas NAD+ is not.
15. In a study looking for mutants tolerant to both ethanol and SDS, four distinct search strategies were examined: isolation of ethanol-tolerant mutants followed by SDS tolerance, isolation of SDS-tolerant mutants followed by ethanol tolerance, simultaneous selection of ethanol and SDS, or independent selection of ethanol and SDS mutants followed by coexpression of these mutants. This final search strategy was found to impart the most significant phenotype overall (6). A similar effect was found when isolating mutants adapted to high lactic acid and low pH, although the two phenotypes were not additive (8).
16. Gene shuffling is an alternative to directed evolution in which portions of various genes are recombined to increase genetic diversity (20, 21).
17. A variety of different mutations have been shown to confer unique phenotypes. For example, a single, nonsynonymous mutation to *rpoD* was found to significantly increase growth of bacteria in high lactic acid concentrations (8). In other cases, three or more amino-acid substitutions were found to confer improved phenotypes (5, 8). In several cases, top-performing mutant sigma factors were found as a result of truncation (6, 24). These truncations resulted in a loss of conserved regions and because sigma factors are involved in many cellular processes, it is hypothesized that loss of these regions can generate significant metabolic improvements. Alternatively, the majority of the point mutations in an *rpoD* mutant conferring high HA yield occurred in the nonconserved region of the gene (24).
18. Microarray studies can be contracted through a service such as Asuragen or Cogenics and should be run in triplicate and include biological references for statistical confidence. Microarray analysis not only will identify large sets of genes associated with a particular phenotype but also can be used to identify a small subset of genes most closely linked to the phenotype (i.e., largest change in expression compared to the control).

References

1. Stephanopoulos G., Alper H., and Moxley J. (2004) Exploiting biological complexity for strain improvement through systems biology. *Nat Biotechnol* **22**, 1261–1267.
2. Tyo K. E., Alper H. S., and Stephanopoulos G. N. (2007) Expanding the metabolic engineering toolbox: more options to engineer cells. *Trends Biotechnol* **25**, 132–137.
3. Parekh S., Vinci V. A., and Strobel R. J. (2000) Improvement of microbial strains and fermentation processes. *Appl Microbiol Biotechnol* **54**, 287–301.
4. Patnaik R. (2008) Engineering complex phenotypes in industrial strains. *Biotechnol Prog* **24**, 38–47.
5. Alper H., Moxley J., Nevoigt E., Fink G. R., and Stephanopoulos G. (2006) Engineering yeast transcription machinery for improved ethanol tolerance and production. *Science* **314**, 1565–1568.
6. Alper H., and Stephanopoulos G. (2007) Global transcription machinery engineering: A new approach for improving cellular phenotype. *Metabolic Engineering* **9**, 258–267.
7. Liu H., Yan M., Lai C., Xu L., and Ouyang P. (2010) gTME for improved xylose fermentation of *Saccharomyces cerevisiae*. *Appl Biochem Biotechnol* **160**, 574–582.
8. Klein-Marcuschamer D., and Stephanopoulos G. (2008) Assessing the potential of mutational strategies to elicit new phenotypes in industrial strains. *Proc Natl Acad Sci U S A* **105**, 2319–2324.
9. Stephanopoulos G., and Sinskey A. J. (1993) Metabolic engineering--methodologies and future prospects. *Trends Biotechnol* **11**, 392–396.
10. Moon T. S., Yoon S. H., Lanza A. M., Roy-Mayhew J. D., and Prather K. L. (2009) Production of glucaric acid from a synthetic pathway in recombinant *Escherichia coli*. *Appl Environ Microbiol* **75**, 589–595.
11. Gill R. T., Wildt S., Yang Y. T., Ziesman S., and Stephanopoulos G. (2002) Genome-wide screening for trait conferring genes using DNA microarrays. *Proc Natl Acad Sci U S A* **99**, 7033–7038.
12. Gall S., Lynch M. D., Sandoval N. R., and Gill R. T. (2008) Parallel mapping of genotypes to phenotypes contributing to overall biological fitness. *Metab Eng* **10**, 382–393.
13. Warnecke T. E., Lynch M. D., Karimpour-Fard A., Sandoval N., and Gill R. T. (2008) A genomics approach to improve the analysis and design of strain selections. *Metab Eng* **10**, 154–165.
14. Jin Y. S., and Stephanopoulos G. (2007) Multi-dimensional gene target search for improving lycopene biosynthesis in *Escherichia coli*. *Metab Eng* **9**, 337–347.
15. Kang M. J., Lee Y. M., Yoon S. H., Kim J. H., Ock S. W., Jung K. H., Shin Y. C., Keasling J. D., and Kim S. W. (2005) Identification of genes affecting lycopene accumulation in *Escherichia coli* using a shot-gun method. *Biotechnol Bioeng* **91**, 636–642.
16. Hemmi H., Ohnuma S., Nagaoka K., and Nishino T. (1998) Identification of genes affecting lycopene formation in *Escherichia coli* transformed with carotenoid biosynthetic genes: candidates for early genes in isoprenoid biosynthesis. *J Biochem* **123**, 1088–1096.
17. Alper H., Miyaoku K., and Stephanopoulos G. (2005) Construction of lycopene-overproducing *E. coli* strains by combining systematic and combinatorial gene knockout targets. *Nat Biotechnol* **23**, 612–616.
18. Winterberg K. M., Luecke J., Bruegl A. S., and Reznikoff W. S. (2005) Phenotypic screening of Escherichia coli K-12 Tn5 insertion libraries, using whole-genome oligonucleotide microarrays. *Appl Environ Microbiol* **71**, 451–459.
19. Badarinarayana V., Estep P. W., 3rd, Shendure J., Edwards J., Tavazoie S., Lam F., and Church G. M. (2001) Selection analyses of insertional mutants using subgenic-resolution arrays. *Nat Biotechnol* **19**, 1060–1065.
20. Patnaik R., Louie S., Gavrilovic V., Perry K., Stemmer W. P., Ryan C. M., and del Cardayre S. (2002) Genome shuffling of Lactobacillus for improved acid tolerance. *Nat Biotechnol* **20**, 707–712.
21. Wang Y., Li Y., Pei X., Yu L., and Feng Y. (2007) Genome-shuffling improved acid tolerance and L-lactic acid volumetric productivity in *Lactobacillus rhamnosus*. *J Biotechnol* **129**, 510–515.
22. Lynch M. D., Warnecke T., and Gill R. T. (2007) SCALEs: multiscale analysis of library enrichment. *Nat Methods* **4**, 87–93.
23. Klein-Marcuschamer D., Santos C. N., Yu H., and Stephanopoulos G. (2009) Mutagenesis of the bacterial RNA polymerase alpha subunit for improvement of complex phenotypes. *Appl Environ Microbiol* **75**, 2705–2711.
24. Yu H., Tyo K., Alper H., Klein-Marcuschamer D., and Stephanopoulos G. (2008) A high-throughput screen for

hyaluronic acid accumulation in recombinant *Escherichia coli* transformed by libraries of engineered sigma factors. *Biotechnol Bioeng* **101**, 788–796.

25. Arnold F. H., and Georgiou G. (2003) *Directed evolution library creation : methods and protocols*, Humana Press, Totowa, N.J.
26. Zaccolo M., Williams D. M., Brown D. M., and Gherardi E. (1996) An approach to random mutagenesis of DNA using mixtures of triphosphate derivatives of nucleoside analogues. *J Mol Biol* **255**, 589–603.
27. Cadwell R. C., and Joyce G. F. (1992) Randomization of genes by PCR mutagenesis. *PCR Methods Appl* **2**, 28–33.
28. Alper H., Fischer C., Nevoigt E., and Stephanopoulos G. (2005) Tuning genetic control through promoter engineering. *Proc Natl Acad Sci U S A* **102**, 12678–12683.
29. Yoon S. H., Moon T. S., Iranpour P., Lanza A. M., and Prather K. J. (2009) Cloning and characterization of uronate dehydrogenases from two pseudomonads and *Agrobacterium tumefaciens* strain C58. *J Bacteriol* **191**, 1565–1573.

Chapter 16

Genomic Promoter Replacement Cassettes to Alter Gene Expression in the Yeast *Saccharomyces cerevisiae*

Andreas Kaufmann and Michael Knop

Abstract

Promoter substitutions are frequently used to regulate the expression of genes in a specific manner such as for their conditional expression or for their overexpression. Chromosomal integration of a regulatable promoter upstream of an open reading frame (ORF) by homologous recombination using PCR-based gene targeting is straightforward and enables stable alterations of the genome. Furthermore, together with the promoter exchange, the target proteins can be tagged N-terminally with an epitope or a fluorescent protein.

Expression levels can be constitutively lowered or increased by using promoters of different strengths. Reversible regulation of gene expression at the level of transcription can be achieved by using either regulatable yeast-endogenous promoters (e.g., *GAL1-10*) or heterogeneous promoters with synthetic transcription factors (e.g., *TetO*). To regulate gene expression at the translational level, insertion of tetracycline-binding aptamers into the 5′ untranslated region (5′ UTR) of target genes can be used.

Key words: *Saccharomyces cerevisiae*, Yeast, Promoter replacement, Gene expression, N-terminal tagging, *GAL1-10*, Tetracycline, Aptamer, 5′ untranslated region, PCR-based gene targeting

1. Introduction

Controlled gene expression is a powerful method to analyze gene and protein function. Classically, endogenous promoters of *Saccharomyces cerevisiae* have been used to study the consequences of altered gene expression or protein depletion on particular aspects of cell physiology. Additionally, applications in protein production and synthetic biology benefit from the ability to control gene expression. Promoters with different strengths, such as the *ADH1*, *TEF1*, and *CYC1* promoters, are commonly used to constitutively lower or increase expression levels (1). By contrast, the *GAL1-10* promoter (2) allows regulation of gene expression

James A. Williams (ed.), *Strain Engineering: Methods and Protocols*, Methods in Molecular Biology, vol. 765, DOI 10.1007/978-1-61779-197-0_16, © Springer Science+Business Media, LLC 2011

as a function of the available carbon source: glucose for repression and galactose for induction of gene expression. Truncated variants of the *GAL1-10* promoter, *GALL* and *GALS*, enable controlling expression levels (3). Other regulatable promoters include the *MET3* and *MET25*, and *CUP1-1* promoters (3, 4), which are induced in the absence of methionine and in the presence of copper ions, respectively.

However, regulation of these promoters always interferes with cellular metabolism due to changes in growth media, and in many cases regulation is not tight enough to completely shut off transcription of essential genes. An improvement was the introduction of heterologous tetracycline-regulated promoters, which are either inducible or repressible (Tet-On/Off) (5–7). A clear advantage of this system is that promoter regulation with doxycycline, a derivative of tetracycline, does not interfere with the yeast cellular metabolism (8). On the contrary, regulation requires the presence of the tetracycline-regulated activators and repressors (6, 9), which require specific strain backgrounds or additional manipulations of the strains in use.

A novel concept for conditional yeast gene expression is the insertion of a genomically encoded aptamer (a short stretch of RNA that adopts a three-dimensional confirmation and binds a specific target molecule) into the 5′ UTRs of target mRNAs (10). This aptamer has a strong binding affinity for tetracycline (11) and thus inhibits translation when bound to tetracycline. In contrast to transcriptional regulation, such aptamer-based synthetic riboswitches rely on a direct RNA-ligand interaction, and they are, therefore, strain and growth medium independent.

In many yeast expression systems, the ORF of the target gene has to be cloned into vectors (1, 5, 7, 12, 13) that provide additional control over expression levels via their copy number. Plasmids containing the 2-μm inverted repeats are maintained in high copy number up to 200 plasmids per cell (14). By contrast, centromere-containing vectors are maintained at a low copy number of one to a few plasmids per cell (14). Both types of plasmids require continuous selection; otherwise, they will be lost from growing populations. Chromosomal integration of the expression construct using integrative plasmids enables stable, selection-free culturing of the modified yeast strains (15, 16). Additionally, chromosomal integration often yields tandem arrays of integrated plasmids and thus clones with specific expression levels can be selected for. For the purpose of heterologous protein production in yeast, plasmid-based strategies are favored. By contrast, to alter the expression of endogenous genes, e.g., in the course of functional studies, PCR-based gene targeting can be used to replace endogenous promoters of genes with promoters possessing the desired properties. This method requires so-called cassettes or modules that combine such promoters with a selection marker. Over the time, a broad range of cassettes have been generated that enable the genomic substitution of an endogenous promoter of a gene with a promoter of choice (Figs. 1 and 2, Table 1) (9, 10, 17–19). Additionally, several of these cassettes are also available in combination with different tags (e.g., HA or GFP), which enable the expression of N-terminally tagged proteins under the control of a specific promoter. Alternatively, new cassettes can be prepared easily and enable tailor-made manipulations of the gene of interest. The construction of new cassettes following a generic cloning strategy provides the advantage that amplification primers for existing cassettes will be compatible with the new ones (Fig. 3).

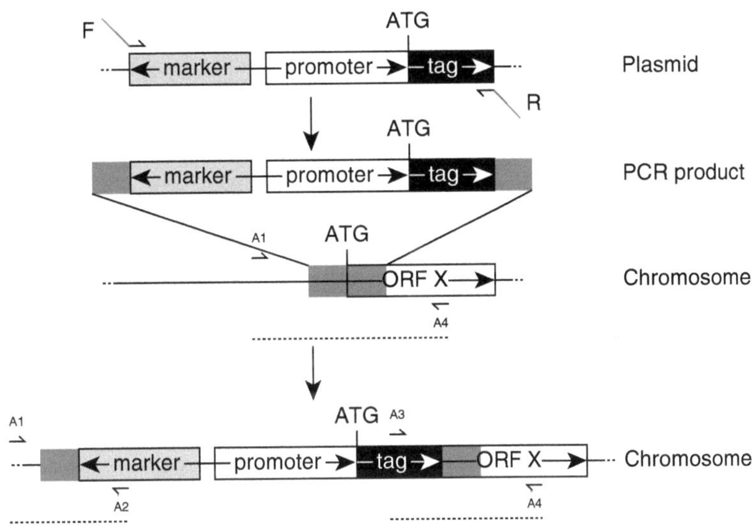

Fig. 1. Principle of promoter replacement cassettes. Genomic promoter replacement cassettes consist of a promoter of choice (regulatable or constitutive), a selectable heterologous marker (auxotrophic or dominant), and, optionally, an epitope tag (e.g., 3HA or GFP). The plasmid-encoded cassette is PCR-amplified using long oligonucleotides (F and R) consisting of a 20 bases plasmid-annealing site (*black arrows*) and a 45–55 bases homology region to the target locus (*gray tails*). The PCR product is transformed into yeast cells where it integrates by homologous recombination into the chromosome immediately upstream and downstream of the start codon (ATG) of the target open reading frame (ORF X) guided by its flanking homology regions (*gray boxes*). The ATG is usually deleted and, if a tag is used, an in-frame fusion of the tag and the ORF X is created. Cells that have been transformed with a cassette are selected for the presence of the marker gene. Analytical colony PCR with four short oligonucleotides (A1–A4) is used to verify the correct integration of the promoter replacement cassette into the target locus. The PCR products (*dashed lines*) of the primer pairs A1/A2 and A3/A4 are only generated in correctly transformed cells, but not in wild-type cells, and verify the presence of the cassette at the target locus (Fig. 4). In haploid wild-type cells, the A1/A4 PCR product is much shorter than in correctly transformed cells (e.g., 0.4 and 2.4 kb, respectively).

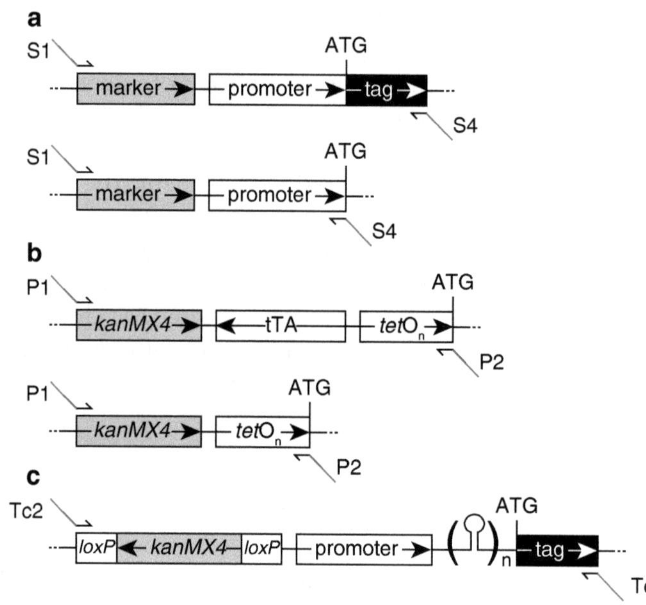

Fig. 2. Different types of promoter replacement cassettes. (**a**) The pYM plasmid collection (18) combines four different constitutive promoters (*ADH, CYC1, TEF*, and *GPD*) and five regulatable promoters (*GAL1, GALL, GALS, CUP1-1*, and *MET25*) with two dominant markers (*kanMX4* and *natNT2*) and three epitope tags (3HA, yeGFP, and ProA). All cassettes can be amplified using the S1 and S4 primers (Table 2). Similar cassettes exist (17, 19) offering other markers (e.g., *HIS3MX6*) and tags (e.g., GST), but different amplification primers have to be used. (**b**) The doxycycline-regulatable promoter replacement cassettes (6, 9) are only available with the *kanMX4* marker and without tags. They should be used in a strain background with a genomically encoded Ssn6-based repressor (6) to lower basal expression levels under repressing conditions and, for the second cassette shown, with a genomically integrated tTA activator (9). ***n***: 2 or 7 *tetO* repeats. (**c**) The cassettes for tetracycline aptamer-mediated regulation (10) combine constitutive promoters (*ADH1* and *TDH3*) with the recyclable *loxP-kanMX4-loxP* marker (36) and two epitope tags (3HA and 6HA). Inhibition of translation of target mRNAs occurs upon tetracycline binding to 1–3 (***n***) aptamers introduced into the 5′UTRs.

2. Materials

2.1. Primer Design for Genomic Integration

1. Computer platform-independent free plasmid editor software: ApE (http://www.biology.utah.edu/jorgensen/wayned/ape/).
2. Yeast genomic DNA sequence: SGD (http://www.yeastgenome.org/).
3. Plasmid DNA sequences: EUROSCARF (http://web.uni-frankfurt.de/fb15/mikro/euroscarf/).

Table 1
Selected cassettes for promoter substitutions

Promoter	on/off	Tag[a]	Marker[a]	Primers[b]	Size[c]	Plasmid[d]	Reference
CUP1-1	$CuSO_4$/–	–	*kanMX4*	S1/S4	1,990	pYM-N1	(18)
CUP1-1	$CuSO_4$/–	*3HA*	*natNT2*	S1/S4	1,980	pYM-N3	(18)
ADH	constitutive	–	*kanMX4*	S1/S4	2,987	pYM-N6	(18)
ADH	constitutive	*3HA*	*natNT2*	S1/S4	2,977	pYM-N8	(18)
CYC1	constitutive	–	*kanMX4*	S1/S4	1,816	pYM-N10	(18)
CYC1	constitutive	*3HA*	*natNT2*	S1/S4	1,806	pYM-N12	(18)
GPD	constitutive	–	*kanMX4*	S1/S4	2,143	pYM-N14	(18)
GPD	constitutive	*3HA*	*natNT2*	S1/S4	2,133	pYM-N16	(18)
TEF	constitutive	–	*kanMX4*	S1/S4	1,932	pYM-N18	(18)
TEF	constitutive	*3HA*	*natNT2*	S1/S4	1,922	pYM-N20	(18)
GAL1	Gal/Glc	–	*kanMX4*	S1/S4	1,978	pYM-N22	(18)
GAL1	Gal/Glc	*3HA*	*natNT2*	S1/S4	1,968	pYM-N24	(18)
GAL1	Gal/Glc	*GST*	*natMX6*	F4/R4	2,500	pFA6a-natMX6-PGAL1-GST	(17)
GAL1	Gal/Glc	*GST*	*HIS3MX6*	F4/R4	2,612	pFA6a-HIS3MX6-PGAL1-GST	(19)
GALL	Gal/Glc	–	*kanMX4*	S1/S4	1,951	pYM-N26	(18)
GALL	Gal/Glc	*3HA*	*natNT2*	S1/S4	1,941	pYM-N28	(18)
GALS	Gal/Glc	–	*kanMX4*	S1/S4	1,935	pYM-N30	(18)
GALS	Gal/Glc	*3HA*	*natNT2*	S1/S4	1,925	pYM-N32	(18)

(continued)

Table 1 (continued)

Promoter	on/off	Tag[a]	Marker[a]	Primers[b]	Size[c]	Plasmid[d]	Reference
MET25	–/Met	*3HA*	*natNT2*	S1/S4	1,892	pYM-N36	(18)
$tetO_7$[e]	–/doxycycline	–	*kanMX4*	1/2	4,000	pCM225	(6)
$tetO_7$[f]	–/doxycycline	–	*kanMX4*	P1/P2	2,200	pCM325	(9)
ADH1[g]	–/tetracycline	tc3-3HA	*loxP-kanMX4*	Tc1/Tc2	2,347	pADH1-tc3-3× HA	(10)
TDH3[g]	–/tetracycline	tc3-6HA	*loxP-kanMX4*	Tc1/Tc2	2,540	pTDH3-tc3-6× HA	(10)

[a]For more combinations of promoters, tags, and markers refer to the original publications
[b]F/R primer names for the PCR amplification of the cassette. For primer sequences see Table 2
[c]Size of the PCR product in base pairs (bp)
[d]All plasmids are available for noncommercial use through EUROSCARF (http://web.uni-frankfurt.de/fb15/mikro/euroscarf/) except the pFA6a-HIS3MX6-PGAL1-GST, which has to be requested directly from the authors (19)
[e]Host strain CML276 (*MATa ura3-52 leu2Δ1 his3Δ200 GAL2* CMVp(*tetR′-SSN6*)::*LEU2*), which contains the tetracycline-regulated Ssn6-based repressor, should be used to achieve lower basal levels in the presence of tetracycline (6)
[f]Host strain CML476 (CML276 *trp1::tTA*), which contains the tetracycline-regulated tTA activator, must be used (9)
[g]Constitutive promoter. Regulation occurs via RNA aptamer–tetracycline interaction (10)

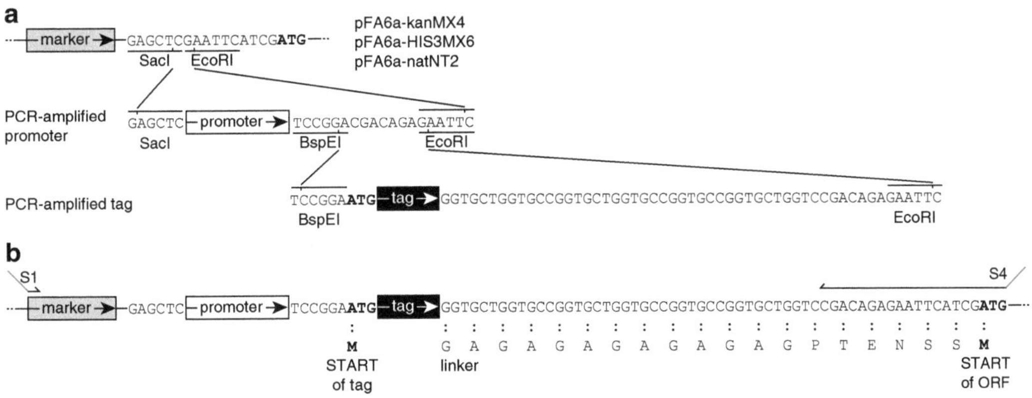

Fig. 3. Cloning strategy to construct novel cassettes for promoter substitutions and N-terminal tagging. (**a**) The promoter of choice is PCR-amplified with oligonucleotides, which add a SacI site at the 5′-end and BspEI-*CGACAGA*-EcoRI at the 3′-end of the promoter, and cloned into the vector carrying the marker. The tag is then PCR-amplified with oligonucleotides, which add BspEI-*ATG* at the 5′-end and a Gly-Ala-linker sequence followed by *CCGACAGA*-EcoRI at the 3′-end, and cloned into the vector carrying the marker and the promoter. Depending on the PCR template of the tag the linker sequence can be omitted from the oligonucleotide. Both cassettes, i.e., with and without tag, can be amplified using the S1/S4 primer annealing sites and can be used for gene targeting. (**b**) Final cassette with marker, promoter and tag. The Gly-Ala-linker sequence and the S4 annealing site are shown. Start codons (ATG) used for cassettes with or without promoters are indicated in bold letters.

2.2. PCR Amplification of Cassettes

1. Plasmid template: 10–50 ng/μl plasmid DNA (Table 1).
2. High-fidelity, high-processivity polymerase and 5× PCR buffer: VELOCITY DNA Polymerase (Bioline, Luckenwalde, Germany) or Herculase II Fusion DNA Polymerase (Stratagene, Santa Clara, CA, USA).
3. F and R primers: HPLC-purified oligonucleotides. 100 μM stock and 10 μM working solutions stored stock at −20°C.
4. dNTPs: 10 mM each of dATP, dTTP, dGTP, and dCTP stored at −20°C.
5. $MgCl_2$: 50 mM $MgCl_2$ stock solution.
6. DMSO: Dimethyl sulfoxide (spectroscopy grade).
7. Betaine: 5 M stock solution stored at 4°C.

2.3. Competent Frozen Yeast Cells

1. Background yeast strain, e.g., wild type S288C (see Note 1).
2. YPD: 10 g/l yeast extract, 20 g/l peptone. Sterilized by autoclaving. After autoclaving, sterile glucose is added to a final concentration of 2%. Stored at room temperature.
3. Glucose: 200 g/l D-(+)-Glucose. Sterilized by autoclaving and stored at room temperature.
4. SORB: 100 mM Lithium acetate, 10 mM Tris (from 1 M stock, pH 8; adjust pH with HCl), 1 mM EDTA (from 0.5 M stock, pH 8; adjust pH with NaOH), 1 M D-(−)-Sorbitol

(extra pure for microbiology). Sterilized by filtering through a sterile 0.2-µm membrane and stored at room temperature.

5. Carrier DNA: 10 mg/ml Salmon Sperm DNA. DNA is boiled for 5 min, chilled in an ice–water slurry and stored at –20°C. If thawed on ice, carrier DNA can be used 3–4 times before it becomes necessary to boil it again.
6. Sterile water.
7. DMSO: see Subheading 2.2, item 6.

2.4. Transformation of Yeast Cells and Selection for Transformants

1. PEG: 100 mM Lithium acetate, 10 mM Tris (see Subheading 2.3, item 4), 1 mM EDTA (see Subheading 2.3, item 4), 40% w/v PEG3350. Sterilized by filtering through a sterile 0.2-µm membrane and stored at room temperature for several months.
2. SC dropout medium: 6.7 g/l yeast nitrogen base without amino acids, ~2 g/l amino acid dropout mix that lacks the relevant amino acid (QBioGene/MP Biomedicals, Solon, OH, USA or see (20)), 2% w/v glucose (see Subheading 2.3, item 3). Sterilized by filtering through a sterile 0.2-µm membrane and stored at room temperature. For agar plates, 2× SC dropout medium is mixed with an equal volume of freshly autoclaved 40 g/l Bacto Agar solution before the agar has solidified.
3. YPD: see Subheading 2.3, item 2. For agar plates, add 20 g/l Bacto Agar before autoclaving.
4. Geneticin/G-418: 200 mg/ml (100% potency) stock solution in water, sterilized by filtering through a sterile 0.2-µm membrane, and stored in aliquots at –20°C. After autoclaving, the medium is allowed to cool to approximately 55°C prior to the addition of 200 mg/l Geneticin to the medium.
5. Hygromycin B: 100 mg/ml stock solution in water, sterilized by filtering through a sterile 0.2-µm membrane, and stored in aliquots at –20°C. After autoclaving, the medium is allowed to cool to approximately 55°C prior to the addition of 300 mg/l Hygromycin B to the medium.
6. Nourseothricin/ClonNAT: 100 mg/ml stock solution in water. Sterilized by filtering through a sterile 0.2-µm membrane and stored in aliquots at –20°C. After autoclaving, the medium is allowed to cool to approximately 55°C prior to the addition of 100 mg/l ClonNAT to the medium.
7. DMSO: see Subheading 2.2, item 6.

2.5. Validation of the Genomic Integration by Analytical Colony PCR

1. A1–A4 primers: desalted oligonucleotides. 100 µM stock and 10 µM working solution stored stock at –20°C.
2. Polymerase and PCR buffer: Taq DNA Polymerase and 10× PCR buffer.

3. dNTPs: 2 mM each of dATP, dTTP, dGTP, and dCTP stored at −20°C.
4. Betaine: see Subheading 2.2, item 7.

2.6. Altering Gene Expression Levels in Yeast Cell Cultures

1. YPD: see Subheading 2.3, item 2.
2. SC raffinose: as SC dropout medium (see Subheading 2.4, item 2), but with all amino acids and 2% w/v alpha-D-Raffinose (research grade) (see Note 2) instead of glucose.
3. SC glucose: as SC dropout medium (see Subheading 2.4, item 2), but with all amino acids.
4. Glucose: see Subheading 2.3, item 3.
5. SC-Met dropout medium: as SC dropout medium (see Subheading 2.4, item 2), but with amino acid dropout mix lacking methionine.
6. Galactose: 200 g/l D-(+)-Galactose (research grade) (see Note 2). Sterilized by autoclaving and stored at room temperature.
7. $CuSO_4$: 100 mM $CuSO_4$ stock solution. Sterilized by filtering through a sterile 0.2-μm membrane and stored at room temperature.
8. YPF: as YPD, but with 2% w/v fructose instead of glucose.
9. Doxycycline: 5 mg/ml doxycycline hydrochloride in 50% ethanol stock solution. Stored at −20°C. After autoclaving, the medium is allowed to cool to approximately 55°C prior to the addition of 1–50 μg/ml doxycycline to the medium (see Note 3).
10. Tetracycline: 10 mg/ml tetracycline hydrochloride in 70% ethanol stock solution. Stored at −20°C. After autoclaving, the medium is allowed to cool to approximately 55°C prior to the addition of 25–250 μg/ml tetracycline to the medium (see Note 4).

2.7. Yeast Cell Protein Extracts

1. NaOH/βMe: 1.85 M NaOH, 7.5% β-mercaptoethanol. Store NaOH at room temperature and add β-mercaptoethanol freshly before use.
2. TCA: 55% w/v trichloroacetic acid. Stored in the dark at room temperature.
3. HU-Buffer: 8 M urea, 5% w/v SDS, 200 mM sodium phosphate buffer pH 6.8, 1 mM EDTA, 100 mM DTT, with 0.1% w/v bromophenol blue as coloring and pH indicator. Store without DTT at −20°C. Store 1 M DTT stock solution in 1-ml aliquots at −20°C.
4. Sodium phosphate buffer: For 100 ml of 1 M sodium phosphate buffer pH 6.8, mix 46.3 ml and 53.7 ml of 1 M stocks of Na_2HPO_4 and NaH_2PO_4.

5. Liquid nitrogen (optional).
6. Ice-cold water.

3. Methods

3.1. Primer Design for Genomic Integration

1. Import the genomic DNA sequence of the target open reading frame (ORF) plus 1 kb upstream of the start codon and the plasmid DNA sequence of the cassette template into the plasmid editor software.
2. Design the forward primer (F) for the PCR amplification of the cassette: 45–55 bases upstream of the start codon of the gene including the ATG, followed by the forward primer annealing sequence of the plasmid used as template (Table 2, Fig. 1).
3. Design the reverse primer (R) for the PCR amplification of the cassette: the reverse complement of 45–55 bases downstream of the start codon of the gene (excluding the ATG), followed by the reverse primer annealing sequence of the plasmid used as template (Table 2, Fig. 1).
4. Besides the two long primers (F and R) bearing the target homology regions, a set of four short, 18–22 nucleotides-long primers (A1–A4, Fig. 1) are used to analyze the correct integration of the cassette into the genome. Design the locus-specific primers A1 and A4 such that they anneal 200–300 bp upstream and downstream of the start codon of the ORF, respectively, and design the marker-specific primers A2 and

Table 2
Primer sequences for cassette amplification

Forward primer (F)		Reverse primer (R)		
Name	**Annealing sequence 5′-3′**	**Name**	**Annealing sequence 5′-3′**	**Reference**
S1[a]	CGTACGCTGCAGGTCGAC	S4	CATCGATGAATTCTCTGTCG	(18)
F4	GAATTCGAGCTCGTTTAAAC	R4	ACGCGGAACCAGATCCGATT	(17, 19)
P1[b]	CGTACGCTGCAGGTCGACGG	P2	ATAGGCCACTAGTGGATCTG	(6, 9)
Tc1	AAGCTTCGTACGAGCGTAATC	Tc2	CATAGGCCACTAGTGGATCTG	(10)

[a]The annealing site for the S1 primer was chosen initially for the *kanMX4* and *HIS3MX6* deletion cassettes (29, 33) and was later used in the EUROFAN project (37)

[b]Same annealing site as the S1 primer but extended by GG at the 3′-end

A3 such that they anneal 200–300 bp downstream of the 5′-end and upstream of the 3′-end of the cassette, respectively (see Note 5).

3.2. PCR Amplification of Cassettes

The PCR amplification of the gene targeting cassettes with long oligonucleotides can cause problems, since the primer annealing sites can lead to self-annealing, the high GC content of the *natNT2* marker, and poor primer quality (18) (see Note 6). To circumvent these problems, different PCR conditions have been used (17, 18, 22). We present here one particular condition, using a DNA polymerase with a 3′-5′ proofreading exonuclease activity and enhanced processivity, which works reliably to amplify most cassettes, even templates longer than 4 kb (Fig. 4).

1. To a thin-walled PCR tube on ice, add in the following order 26.75 μl water, 5 μl dNTPs, 2.5 μl of primer F, 2.5 μl of primer R, 2 μl $MgCl_2$, 1 μl template DNA, and 10 μl 5× PCR Buffer. Briefly vortex and spin down the PCR mix.
2. Use the following program on a thermocycler: (1) 95–97°C, 2 min, (2) 95–97°C, 30 s, (3) 64°/68°C, 30 s, (4) 72°C, 30 s/kb, (5) repeat steps 2–4. 30 times, (6) 72°C, 5 min, (7) 4°C, indefinitely (see Note 7).
3. Put the PCR tube into the thermocycler, start the program, add 0.25 μl polymerase and mix by pipetting up and down several times.
4. After the PCR has finished, analyze 5 μl of the reaction by standard agarose gel electrophoresis (20).

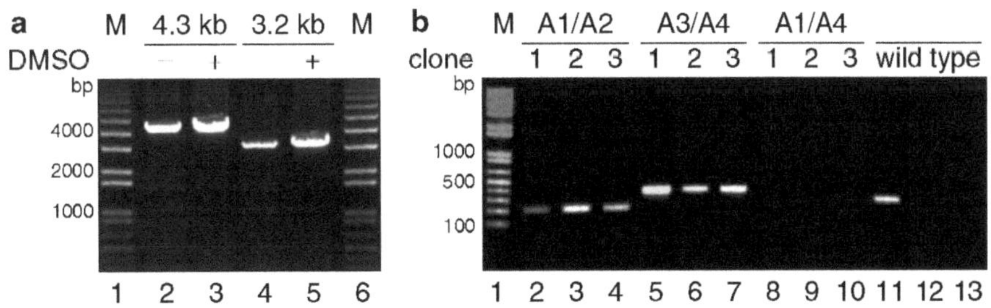

Fig. 4. Cassette amplification and validation of genomic integration by PCR. (**a**) PCR amplification of the 4.3 kb natNT2-ADH-mCherry-sfGFP and 3.2 kb natNT2-TEF-mCherry-sfGFP cassettes from pMaM96 and pMaM97, respectively (Matthias Meurer, personal communication). High-fidelity, high-processivity DNA polymerases reliably amplify gene targeting cassettes, even templates longer than 4 kb. Addition of DMSO or betaine can eliminate secondary structure formation and base-pair composition dependence of DNA melting and, thus, can increase product yield (32), especially when amplifying GC-rich templates such as the natNT2 maker. (**b**) The formation of both new DNA junctions at the 5′ and 3′-end of the integrated cassette (lane 2–4: primers A1/A2 and lane 5–7: primers A3/A4, respectively) is validated by analytical colony PCR in three independent clones. The primer pair A1/A4 only yields a product in a control PCR with DNA from a wild-type strain without the modification (lane 11), but not with DNA from transformed clones (lane 8–10), whereas the primer pairs A1/A2 and A3/A4 yield no product for a wild-type strain (lane 12 and 13, respectively).

3.3. Competent Frozen Yeast Cells

Choose a background yeast strain (see Note 1) that is suitable for the selection marker and expression system that you plan to use. Yeast transformation using frozen competent cells is based on the lithium acetate/polyethylene glycol method (23) with some modifications (see Note 8). For basics in manipulation and growth of yeast cells, please refer to Amberg et al. (21).

1. Inoculate 5 ml YPD with yeast cells and incubate the culture overnight at 30°C in a rotary shaker at 200 rpm.
2. Determine the cell titer by measuring the optical density of the culture at 600 nm (OD_{600}) in a spectrophotometer or by using a hemocytometer (see Note 9).
3. Inoculate 50 ml prewarmed 2× YPD with 2.5×10^8 cells and grow to approximately 2×10^7 cells/ml at 30°C in a rotary shaker at 200 rpm (see Note 10).
4. Transfer the yeast culture to a 50-ml centrifuge tube, harvest the cells by centrifugation (500 × *g*, 5 min), and wash the cell pellet once with 25 ml sterile water.
5. Resuspend the cell pellet in 1 ml SORB, transfer the suspension to a 1.5-ml centrifuge tube, and pellet the cells again by centrifugation (1,500 × *g*, 2 min).
6. Remove SORB completely by aspiration, resuspend the cells in a total volume of 360 µl of SORB, and add 40 µl of carrier DNA (0°C).
7. Prorate the cell suspension into individual tubes (e.g., as 50 µl aliquots, at room temperature) and store the tubes at −80°C (see Note 11).

3.4. Transformation of Yeast Cells and Selection for Transformants

When analyzing unknown or essential genes in haploid yeast strains, be careful not to repress its gene expression already during transformation and selection. In such cases, it is better to target the gene in a diploid cell, which can be sporulated (21) to obtain a haploid strain.

1. Add 5 µl of the unpurified PCR product into a 1.5-ml tube (see Note 12), add 50 µl of competent yeast cells and mix the suspension well.
2. Add 330 µl of PEG and 42 µl DMSO (final concentration ~10%), mix thoroughly, and place the tube in a 42°C water bath for 20 min (see Note 8).
3. Pellet the cells by centrifugation (1,500 × *g*, 2 min), remove the supernatant, and resuspend the cells in 100 µl liquid medium (see next step).
4. If auxotrophic markers are used for selection of transformants (e.g., *HIS3MX6*), resuspend the cells in liquid SC dropout medium (synthetic complete medium lacking the relevant

amino acid) and directly spread the cells on SC dropout agar plates. If dominant antibiotic resistance markers are used (e.g., *kanMX4*, *hphNT1*, or *natNT2* for the selection on Geneticin/G418, Hygromycin B, or Nourseothricin/ClonNAT, respectively), resuspend the cells in 3 ml YPD and allow them to recover at 30°C for 4 h to overnight while shaking. Cells are then harvested by centrifugation and spread on a YPD agar plate containing the selective antibiotic.

5. Incubate the plates at 30°C for approximately 2 days until transformed colonies become visible (see Note 13).
6. Pick at least three transformed colonies with a sterile toothpick and streak out for single cells on a fresh selective plate and incubate again until colonies become visible.

3.5. Validation of the Genomic Integration by Analytical Colony PCR

Analytical colony PCR (Figs. 1 and 4) on whole yeast cells, which are directly added to the PCR, allows quick validation of chromosomal alterations that occurred after transforming cells with linear DNA fragments (see Note 14).

1. For each transformed cassette and strain test at least three independent clones, i.e., three colonies originating from individual transformed colonies after streaking out for single cells.
2. Master mix 1–3: for each of the three primer combinations to be tested, i.e., A1/A2, A3/A4, A1/A4, prepare one master mix on ice consisting of 3 μl dNTPs, 2 μl 10× PCR buffer, 1 μl forward primer, 1 μl reverse primer, 3 μl betaine, and 10 μl water per reaction.
3. Master mix 4: prepare one master mix on ice consisting of 0.1 μl polymerase, 1 μl 10× PCR buffer, and 8.9 μl water per reaction.
4. Distribute 20 μl of the master mix 1–3 into individual PCR tubes.
5. Slightly touch a yeast colony with a sterile 10-μl pipette tip and transfer the cells to a PCR tube containing master mix 1. Repeat this for master mix 2 and 3 for the same colony and then for the other colonies to be tested.
6. Briefly vortex and centrifuge each PCR tube.
7. Use the following program on a thermocycler: (1) 96°C, 10 min, (2) 50°C, indefinitely, (3) 94°C, 30 s, (4) 50°C, 30 s, (5) 72°C, 30 s, (6) repeat steps 3–5, 35–40 times, (7) 4°C, indefinitely.
8. Start the program, at step 2 add 10 μl of master mix 4 to each PCR tube and mix by pipetting up and down several times and continue the program.

9. After the PCR has been finished, analyze 5 µl of each reaction by standard agarose gel electrophoresis (Fig. 4) (20).

3.6. Altering Gene Expression Levels in Yeast Cell Cultures

3.6.1. Constitutive Expression

1. Dilute a stationary yeast culture with YPD and grow to early log phase (OD_{600} of 0.4–0.6) (see Note 9).
2. Prepare yeast cell extracts as described in Subheading 3.7 to check the protein expression level (Fig. 5).

3.6.2. Induced Expression from GAL Promoters

1. Grow yeast cells overnight to stationary phase in SC raffinose medium. The carbon source must be one that does not repress expression of the *GAL* promoters, which is strongly repressed by glucose.
2. Dilute the cultures to an OD_{600} of 0.05–0.1 with SC raffinose medium and grow to early log phase (OD_{600} of 0.4–0.6).
3. Add sterile galactose to the cell culture at a final concentration of 2% to induce target gene expression for 90 min. To the negative control, add glucose at a final concentration of 2% to repress gene expression.
4. Prepare yeast cell extracts as described in Subheading 3.7 to check the protein expression level (Fig. 5).

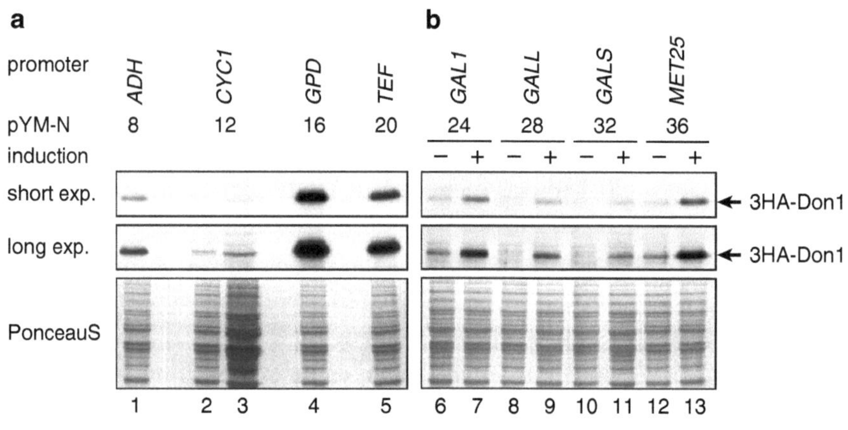

Fig. 5. Altered expression of *DON1* using different promoter replacement cassettes. The promoter of the gene *DON1* was exchanged for eight different promoters in combination with an N-terminal 3HA tag. The names of the pYM plasmids (18) used to amplify the promoter replacement cassettes are indicated. Cultures were grown into exponential growth phase. Immunoblot detection of the 3HA tag was done with the mouse monoclonal antibody 16B12 (Covance, Emeryville, CA, USA) and horseradish peroxidase-coupled goat anti-mouse IgG (H+L) (Jackson ImmunoResearch Laboratories, West Grove, PA, USA). Equal protein load was verified by staining the blots with Ponceau S. Two different exposures are shown to highlight the differences in promoter strength. (**a**) Constitutive promoters: *GPD* (lane 4) and *TEF* (lane 5) induce very strong protein expression; the *ADH* promoter (lane 1) is weaker; whereas the *CYC1* promoter (lane 2) is very weak. In the latter case, 3HA-Don1 was only detected with a fivefold protein load (lane 3). (**b**) Inducible promoters: induction was performed by adding 1% glucose (–) or 1% galactose (+) to YP Raffinose medium (all *GAL* promoters) or by washing and transferring the culture to SC-Met medium (*MET25* promoter). Cells were induced for 90 min. The inducible promoters differed in strength and the very strong *MET25* and the strong *GAL1* were slightly leaky when uninduced (lanes 6 and 12). Figure adapted from (18) with permission of John Wiley & Sons, Ltd.

3.6.3. Induced Expression from the MET25 Promoter

1. Dilute a stationary yeast culture with SC glucose medium containing methionine and grow to early log phase (OD_{600} of 0.4–0.6).
2. Harvest the cells by centrifugation (500 × *g*, 5 min), wash the cell pellet with SC-Met dropout medium, centrifuge again, and discard the supernatant.
3. Resuspend cells in SC-Met dropout medium to induce target gene expression for 90 min.
4. Prepare yeast cell extracts as described in Subheading 3.7 to check the protein expression level (Fig. 5).

3.6.4. Induced Expression from the CUP1-1 Promoter

1. Dilute a stationary yeast culture with SC glucose medium and grow to early log phase (OD_{600} of 0.4–0.6).
2. Add $CuSO_4$ to a final concentration of 100–300 μM to the culture (see Note 15) to induce target gene expression for 2–3 h.
3. Prepare yeast cell extracts as described in Subheading 3.7 to check the protein expression level (Fig. 5).

3.6.5. Promoter Shut Off Using Doxycycline-Regulated Promoters

1. Dilute a stationary yeast culture with YPD (or SC medium) and grow to an OD_{600} of 1–2.
2. Dilute the culture to an OD_{600} of 1 and make 4–5 tenfold serial dilutions in growth medium.
3. From each serial dilution, spot 5 μl onto not too wet plates without (control) and with 1–50 μg/ml doxycycline (see Note 3).
4. Allow 15–20 min to absorb the drops and observe growth differences after incubating the plates for 2–3 days at the desired temperature (see Note 4).

3.6.6. Inhibition of Translation Using Tetracycline Aptamer-Mediated Regulation

1. Dilute a stationary yeast culture with YPD or YPF, depending whether the *ADH1* or the *TDH3* promoter is used, respectively, and grow to an OD_{600} of 1–2.
2. Dilute the culture to an OD_{600} of 1 and make 4–5 tenfold serial dilutions in growth medium.
3. From each diluted culture, spot 5 μl onto not too wet plates without (control) and with 25–250 μg/ml tetracycline.
4. Allow 15–20 min for absorption of the drops and observe growth differences after incubating the plates for 2–3 days at the desired temperature.

3.7. Yeast Cell Protein Extracts

To check for expression of the targeted gene whole cell protein extracts are analyzed by SDS-PAGE and immunoblotting (Fig. 5). If no specific antibody is available, a promoter replacement cassette that introduces a tag at the N-terminus of the protein could be used. The NaOH/TCA method for yeast cell protein extracts (24, 25) is simple, fast, and reproducible, and small culture volumes are usually sufficient to check for protein expression by Western blot. For purification of proteins, other methods should be used, e.g., cell lysis by glass beads (26).

1. Pipette a sample corresponding to 0.5–3 OD_{600} of cells of an appropriate yeast cell culture (see Subheading 3.6) into a centrifuge tube on ice.
2. Harvest the cells by centrifugation (4°C); freeze the cell pellet in liquid nitrogen and store the sample at −80°C (optional).
3. Resuspend the cell pellet in 1 ml of ice-cold water, add 150 μl ice-cold NaOH/βMe, mix quickly, and incubate on ice for 15 min.
4. Add 150 μl ice-cold TCA, mix quickly, and incubate on ice for 10 min.
5. Centrifuge (16,100 × *g*, 10 min, 4°C), remove the supernatant, centrifuge again, and remove all traces of the supernatant.
6. Add 30–100 μl HU-Buffer per OD_{600} and denature the proteins for 10 min at 65°C on a thermomixer (see Note 16).
7. Centrifuge the samples to pellet cell debris (16,100 × *g*, 5 min, room temperature) and analyze aliquots corresponding to 0.1–0.5 OD_{600} of cells by SDS-PAGE (27), followed by immunoblotting (see Note 17) using standard procedures (28).

3.8. Cloning Strategy to Construct New Cassettes

Many cassettes for gene targeting in yeast were constructed in a modular way, i.e., markers, tags, and promoters were cloned using the same restriction sites of the pFA6a plasmid backbone. Primer annealing sequences were also kept constant, at least to some degree (6, 9, 17–19, 29). In practice, this means that the same set of four gene-specific primers can be used to delete a gene, exchange its promoter and tag it N- and C-terminally. The same applies when new modules become available, e.g., improved fluorescent proteins or novel methods to regulate protein expression (10). Figure 3 displays a cloning strategy to construct novel cassettes for N-terminal tagging and promoter substitutions based on the commonly used pFA6a marker plasmids and S1/S4 primer annealing sites.

4. Notes

1. In the SGD (http://wiki.yeastgenome.org/index.php/Commonly_used_strains), you will find further information on commonly used yeast strains and sources.
2. It is important to use special research-grade raffinose and galactose, since glucose contamination can lead to repression of *GAL* promoters.

3. Doxycycline concentrations ranging from 1 to 5 μg/ml are usually sufficient for optimal inhibition of expression, although concentrations up to 50 μg/ml can be used without significant alteration of growth (30).
4. To efficiently regulate the target gene, the tetracycline concentration has to be adjusted to the expression strength of the strong *TDH3* and the weaker *ADH1* promoter and the abundance of the target protein (10).
5. Whereas the sequence of the cassette-amplification oligonucleotides (F and R) is determined by the sequence upstream and downstream of the start codon, the analytical oligonucleotides (A1–A4) should be designed according to general primer design rules (for an overview of rules and available bioinformatics tools, see (31)).
6. Test the individual primers in combination with established primers to identify a faulty primer.
7. Use 64 and 68°C as annealing temperature for the Herculase II Fusion and Velocity DNA polymerase, respectively. If there is little or no PCR product, lower the annealing temperature stepwise to 60°C and increase the elongation time to 45 s. For GC-rich templates increase the denaturation temperature to 97°C and add 5% DMSO or 0.5 M betaine to the reaction (32).
8. See also the Gietz Lab Yeast Transformation Home Page (http://home.cc.umanitoba.ca/~gietz/) for additional yeast transformation protocols, e.g., addition of DMSO is not necessary if the cells are incubated for 40 min (instead of 20 min) at 42°C.
9. The cell titer is determined by measuring the absorption of the cell suspension at a wavelength of 600 nm, which probes for light scattering by the cells in the suspension. Dense cultures should be diluted because OD_{600} measurements are only linear in a small range between approximately 0.1 and 0.5. For many yeast strains, 1 ml of a cell suspension with 1 OD_{600} corresponds to 10^7 cells, but the relation between cell titer and OD_{600} varies greatly between different yeast strains, growth phases (i.e., stationary versus log phase), and spectrophotometers. To standardize OD_{600} measurements, they should be calibrated using a hemocytometer.
10. It is important to allow the cells to complete at least two divisions (this will take 3–5 h); however, transformation efficiency remains constant for 3–4 cell divisions (23). Adjust the volumes according to the cell number.
11. Simply place the tubes in a storage box into the −80°C freezer. Do not snap-freeze the cells in liquid nitrogen, since this will decrease viability.

12. 5 μl of one PCR reaction are usually sufficient for transformation of S288c- or W303-derived strains. For some other strain backgrounds (such as SK-1), a tenfold higher amount of DNA is required. For this purpose, ethanol-precipitate the PCR product and dissolve it in 1/10 of the original volume in water (18).
13. Selection for positive transformants on plates containing antibiotics often requires replica plating of the transformants after 2 days, presumably because of the high background of transiently transformed cells, which makes it difficult to recognize the correct transformants (25, 33).
14. In special cases, where a PCR product above 800 bp is expected, alternative methods involving extraction of genomic DNA prior to the PCR should be considered (34).
15. Some strains may exhibit a different sensitivity to $CuSO_4$. Therefore, a preliminary experiment may be needed to determine the tolerable $CuSO_4$ concentrations for a specific strain.
16. If the buffer capacity of the HU-buffer is not high enough to neutralize the remaining traces of the trichloroacetic acid (yellow color), add ~1 μl of 2 M Tris-base.
17. For unknown reasons, TCA-treated proteins require a longer blotting transfer time than nontreated proteins (about 1.5–2-fold the normal time) (35).

Acknowledgments

The authors would like to thank Constanze Kaiser and Matthias Meurer for critical reading of this manuscript and valuable improvements of the described methods. This work was supported by the Novartis Foundation.

References

1. Mumberg, D., Mailer, R., and Funk, M. (1995) Yeast vectors for the controlled expression of heterologous proteins in different genetic backgrounds. *Gene* **156**, 119–122.
2. Johnston, M., and Davis, R. W. (1984) Sequences that regulate the divergent GAL1-GAL10 promoter in *Saccharomyces cerevisiae*. *Mol. Cell. Biol.* **4**, 1440–1448.
3. Mumberg, D., Muller, R., and Funk, M. (1994) Regulatable promoters of *Saccharomyces cerevisiae*: comparison of transcriptional activity and their use for heterologous expression. *Nucleic Acids Res.* **22**, 5767–5768.
4. Etcheverry, T. (1990) Induced expression using yeast copper metallothionein promoter, in *Gene Expression Technology* (Goeddel, D. V.), pp. 319–29. Elsevier.
5. Belli, G., Gari, E., Piedrafita, L., Aldea, M., and Herrero, E. (1998) An activator/repressor dual system allows tight tetracycline-regulated gene expression in budding yeast. *Nucleic Acids Res.* **26**, 942–947.
6. Belli, G., Gari, E., Aldea, M., and Herrero, E. (1998) Functional analysis of yeast essential genes using a promoter-substitution cassette and the tetracycline-regulatable dual expression system. *Yeast* **14**, 1127–1138.

7. Gari, E., Piedrafita, L., Aldea, M., and Herrero, E. (1997) A set of vectors with a tetracycline-regulatable promoter system for modulated gene expression in *Saccharomyces cerevisiae*. *Yeast* **13**, 837–848.
8. Wishart, J. A., Hayes, A., Wardleworth, L., Zhang, N., and Oliver, S. G. (2005) Doxycycline, the drug used to control the tet-regulatable promoter system, has no effect on global gene expression in *Saccharomyces cerevisiae*. *Yeast* **22**, 565–9.
9. Yen, K., Gitsham, P., Wishart, J., Oliver, S. G., and Zhang, N. (2003) An improved tetO promoter replacement system for regulating the expression of yeast genes. *Yeast* **20**, 1255–1262.
10. Kötter, P., Weigand, J. E., Meyer, B., Entian, K., and Suess, B. (2009) A fast and efficient translational control system for conditional expression of yeast genes. *Nucleic Acids Res.* **37**, e120.
11. Müller, M., Weigand, J. E., Weichenrieder, O., and Suess, B. (2006) Thermodynamic characterization of an engineered tetracycline-binding riboswitch. *Nucleic Acids Res.* **34**, 2607–17.
12. Gao, C. Y., and Pinkham, J. L. (2000) Tightly regulated, beta-estradiol dose-dependent expression system for yeast. *BioTechniques* **29**, 1226–31.
13. Quintero, M. J., Maya, D., Arévalo-Rodríguez, M., Cebolla, A., and Chávez, S. (2007) An improved system for estradiol-dependent regulation of gene expression in yeast. *Microb. Cell Fact.* **6**, 10.
14. Schneider, J. C., and Guarente, L. (1991) Vectors for Expression of Cloned Genes in Yeast: Regulation, Overproduction, and Underproduction, in *Guide to Yeast Genetics and Molecular Biology* (Guthrie, C., and Fink, G. R.), pp. 373 388. Academic Press.
15. Sikorski, R. S., and Hieter, P. (1989) A system of shuttle vectors and yeast host strains designed for efficient manipulation of DNA in *Saccharomyces cerevisiae*. *Genetics* **122**, 19–27.
16. Taxis, C., and Knop, M. (2006) System of centromeric, episomal, and integrative vectors based on drug resistance markers for *Saccharomyces cerevisiae*. *Biotechniques* **40**, 73–78.
17. Van Driessche, B., Tafforeau, L., Hentges, P., Carr, A. M., and Vandenhaute, J. (2005) Additional vectors for PCR-based gene tagging in *Saccharomyces cerevisiae* and *Schizosaccharomyces pombe* using nourseothricin resistance. *Yeast* **22**, 1061–8.
18. Janke, C., Magiera, M. M., Rathfelder, N., Taxis, C., Reber, S., Maekawa, H., Moreno-Borchart, A., Doenges, G., Schwob, E., Schiebel, E., and Knop, M. (2004) A versatile toolbox for PCR-based tagging of yeast genes: new fluorescent proteins, more markers and promoter substitution cassettes. *Yeast* **21**, 947–962.
19. Longtine, M. S., McKenzie 3rd, A., Demarini, D. J., Shah, N. G., Wach, A., Brachat, A., Philippsen, P., and Pringle, J. R. (1998) Additional modules for versatile and economical PCR-based gene deletion and modification in *Saccharomyces cerevisiae*. *Yeast* **14**, 953–961.
20. Sambrook, J., and Russell, D. W. (2001) Molecular Cloning - A Laboratory Manual, 3 ed., p. 2344. Cold Spring Harbor Laboratory Press.
21. Amberg, D. C., Burke, D., and Strathern, J. N. (2005) Methods in yeast genetics: a Cold Spring Harbor Laboratory course manual, p. 230. Cold Spring Harbor Laboratory Press.
22. Goldstein, A. L., Pan, X., and McCusker, J. H. (1999) Heterologous URA3MX Cassettes for Gene Replacement in *Saccharomyces cerevisiae*. *Yeast* **15**, 507–511.
23. Daniel Gietz, R., and Woods, R. A. (2002) Transformation of yeast by lithium acetate/single-stranded carrier DNA/polyethylene glycol method, in *Guide to Yeast Genetics and Molecular and Cell Biology - Part B* (Guthrie, C., and Fink, G. R.), pp. 87–96. Academic Press.
24. Riezman, H., Hase, T., van Loon, A. P., Grivell, L. A., Suda, K., and Schatz, G. (1983) Import of proteins into mitochondria: a 70 kilodalton outer membrane protein with a large carboxy-terminal deletion is still transported to the outer membrane. *EMBO J* **2**, 2161–2168.
25. Knop, M., Siegers, K., Pereira, G., Zachariae, W., Winsor, B., Nasmyth, K., and Schiebel, E. (1999) Epitope tagging of yeast genes using a PCR-based strategy: more tags and improved practical routines. *Yeast* **15**, 963–972.
26. Conzelmann, A., Riezman, H., Desponds, C., and Bron, C. (1988) A major 125-kd membrane glycoprotein of *Saccharomyces cerevisiae* is attached to the lipid bilayer through an inositol-containing phospholipid. *EMBO J* **7**, 2233–40.
27. Laemmli, U. K. (1970) Cleavage of structural proteins during the assembly of the head of bacteriophage T4. *Nature* **227**, 680–5.
28. Harlow, E., and Lane, D. (1999) Using antibodies: a laboratory manual, p. 495. Cold Spring Harbor Laboratory Press.
29. Wach, A., Brachat, A., Alberti-Segui, C., Rebischung, C., and Philippsen, P. (1997) Heterologous HIS3 marker and GFP reporter modules for PCR-targeting in *Saccharomyces cerevisiae*. *Yeast* **13**, 1065–1075.

30. Ariño, J., and Herrero, E. (2003) Use of Tetracycline-Regulatable Promoters for Functional Analysis of Protein Phosphatases in Yeast, in *Protein Phosphatases* (S. Klumpp, a. J.), pp. 347–358. Academic Press.
31. Abd-Elsalam, K. A. (2003) Minireview Bioinformatic tools and guideline for PCR primer design. *African Journal of Biotechnology* **2**, 91–95.
32. Frackman, B. S., Kobs, G., Simpson, D., and Storts, D. (1998) Betaine and DMSO: Enhancing Agents for PCR. *Promega Notes* **65**, 27.
33. Wach, A., Brachat, A., Pohlmann, R., and Philippsen, P. (1994) New heterologous modules for classical or PCR-based gene disruptions in *Saccharomyces cerevisiae. Yeast* **10**, 1793–1808.
34. Maeder, C. I., Maier, P., and Knop, M. (2007) A Guided Tour to PCR-based Genomic Manipulations of *Saccharomyces cerevisiae* (PCR-targeting), in *Yeast Gene Analysis* (Stansfield, I., and Stark, M. J.) 2nd., pp. 55–78. Elsevier.
35. Wright, A. P., and Hartley, B. S. (1989) Extraction and rapid inactivation of proteins from *Saccharomyces cerevisiae* by trichloroacetic acid precipitation. *Yeast* **5**, 51–53.
36. Güldener, U., Heck, S., Fiedler, T., Beinhauer, J., and Hegemann, J. H. (1996) A new efficient gene disruption cassette for repeated use in budding yeast. *Nucleic Acids Res.* **24**, 2519–2524.
37. Dujon, B. (1998) European Functional Analysis Network (EUROFAN) and the functional analysis of the *Saccharomyces cerevisiae* genome. *Electrophoresis* **19**, 617–24.

Part III

Strain Engineering Other Industrially Important Microbes

Chapter 17

Microbial Genome Analysis and Comparisons: Web-Based Protocols and Resources

Medha Bhagwat and Arvind A. Bhagwat

Abstract

Fully annotated genome sequences of many microorganisms are publicly available as a resource. However, in-depth analysis of these genomes using specialized tools is required to derive meaningful information. We describe here the utility of three powerful publicly available genome databases and analysis tools. Protocols outlined here are particularly useful for performing pairwise genome comparisons between closely related microorganisms to identify similarities and unique features, for example to identify genes specific to a pathogenic strain of *Escherichia coli* compared to a nonpathogenic strain.

Key words: Pairwise genome comparisons, Bioinformatics tools, Microbial genome resources

1. Introduction

The recent outburst of whole-genome sequencing has marked the beginning of a new age in strain engineering. Powerful rationale based alterations of important microbial strains can be undertaken with the help of annotated genomes and genomic comparisons (1, 2). *Escherichia coli* based bacterial strain engineering strategies have come a long way from repeated cycles of random mutation selection. Recent innovations such as scar-free targeted gene deletions or point mutations of single amino-acid residues, and now synthetic genomes, have much more power and precision. In spite of these advances, in-depth genome sequence analysis is needed to derive meaningful information for genetic engineering of a model organism. Comparison of whole-genome sequences of multiple strains is not a trivial task, and to this end several sophisticated tools are available (3). This chapter

James A. Williams (ed.), *Strain Engineering: Methods and Protocols*, Methods in Molecular Biology, vol. 765,
DOI 10.1007/978-1-61779-197-0_17, © Springer Science+Business Media, LLC 2011

describes three main microbial genome resources: the National Center for Biotechnology Information (NCBI) Genome Database, Integrated Microbial Genomes (IMG), and BioCyc/MetaCyc. There are many additional resources that are not covered in this chapter, and a detailed list can be found at http://microbialgenomics.energy.gov/databases.shtml, a site maintained by the US Department of Energy.

2. Materials

Availability of the fully annotated genome sequence of the organism of interest or phylogenetically closest organism, a broadband Internet connection, a Web browser such as Internet Explorer, Firefox, Chrome, or similar ones.

3. Methods

3.1. NCBI's Genome Database (http://www.ncbi.nlm.nih.gov/sites/genome)

This database provides access to the sequences and annotations for archaea, bacteria, eukaryotes, plasmids, viruses, and viriods (4). It provides a link to the Microbial Genomes Resources page, http://www.ncbi.nlm.nih.gov/genomes/MICROBES/microbial_taxtree.html, a central page for the prokaryotic (bacterial and archaeal) genomes. This page lists several resources and has a link to the Prokaryotic Projects page http://www.ncbi.nlm.nih.gov/genomes/lproks.cgi. The default page lists alphabetically the organisms with complete genome sequences. The page also has an ability to filter genomes by kingdom or group.

As of August 2010, there are 1,211 completely sequenced microbial genomes, out of which 88 are archaeal and 1,123 are bacterial. This database provides information about each genome such as its size in megabases, percent GC content, and links to NCBI's relevant reference sequences (RefSeq). It also offers calculated analyses of the genomes using tools such as TaxMap (taxonomic distribution of protein homologs), ProtTable (tabular information about all encoded proteins and their sequences), COG Table (distribution of protein functions by Clusters of Orthologus Groups functional categories), BLAST (search against the genome nucleotide or protein sequences using Basic Local Alignment Search Tool), CDD search (conserved domains in the proteins identified by searching against the Conserved Domain Database), GenePlot (pairwise genome comparison of protein homologs), TaxPlot (comparison of proteins from a genome to proteins from two different genomes), gMap (genome nucleotide sequence comparison), FTP (access to genome nucleotide, protein, and RNA sequence files), and Publications (publications in PubMed) (see Note 1). The "Genomes in progress" tab lists 3,410 microbial genome sequencing projects in progress (52 archaeal and 3,358 bacterial). The "Organism Info" tab lists all microbial organisms for which complete genome sequence with analysis, assembled sequence, or unfinished sequence is available (see Note 2). This page also lists information about the organisms such as habitat, gram strain, shape, motility, salinity, and pathogenicity.

From the Prokaryotic Projects page, one could find information about *E. coli* complete genome sequences in the Complete Genomes tab. Scroll down the page to *E. coli*. Links to complete genome sequences for a number of strains are available. A table shows that genome size ranges from 4.6 Mb to 5.86 Mb. Let us explore one of the strains, *O157:H7 str. TW14359*. Click on the name of the strain in the organism column. The resulting page (http://www.ncbi.nlm.nih.gov/genomeprj/30045) lists detailed information about *E. coli* in general and information specific for this strain, and provides links to sequencing projects of all *E. coli* strains and *E. coli*-related resources outside NCBI.

The strain was isolated from spinach during the *E. coli* outbreak in 2006 and is pathogenic in humans causing hemorrhagic colitis. This page enumerates differences between pathogenic and nonpathogenic strains of *E. coli* and the particular features causing pathogenicity. Many strains of *E. coli* can cause disease by attaching to the host cell and introducing toxins that disrupt normal cellular processes. The genomes of pathogenic strains compared to nonpathogenic strains contain regions called pathogenicity islands (PAIs), which include genes for virulence proteins such as a type III secretion system, the locus of enterocyte effacement, numerous toxins and adhesins, fimbrial gene clusters and iron uptake systems. Such extra regions have usually been acquired by integration or transposition bacteriophage or plasmid DNA. The protocol below demonstrates how to identify such regions that may have been horizontally transferred.

A table is provided on the page with links to genome and plasmid overview, protein and RNA annotations, and some analysis tools. Click on the link NC_013008 under the "RefSeq" column. The next page provides an overview of the genome sequence, detailed annotations, access to multiple analysis tools, and the SequenceViewer to visualize the genome assembly (Fig. 1).

We use TaxMap tool to identify regions of the O157:H7 str. TW14359 genome that may have been horizontally transferred from bacteriophages and, thus, may be part of a pathogenicity island.

1. Click on the TaxMap link for O157:H7 str. TW14359. The resulting page (http://www.ncbi.nlm.nih.gov/sutils/taxik.cgi?gi=24828) displays a taxonomic distribution of protein

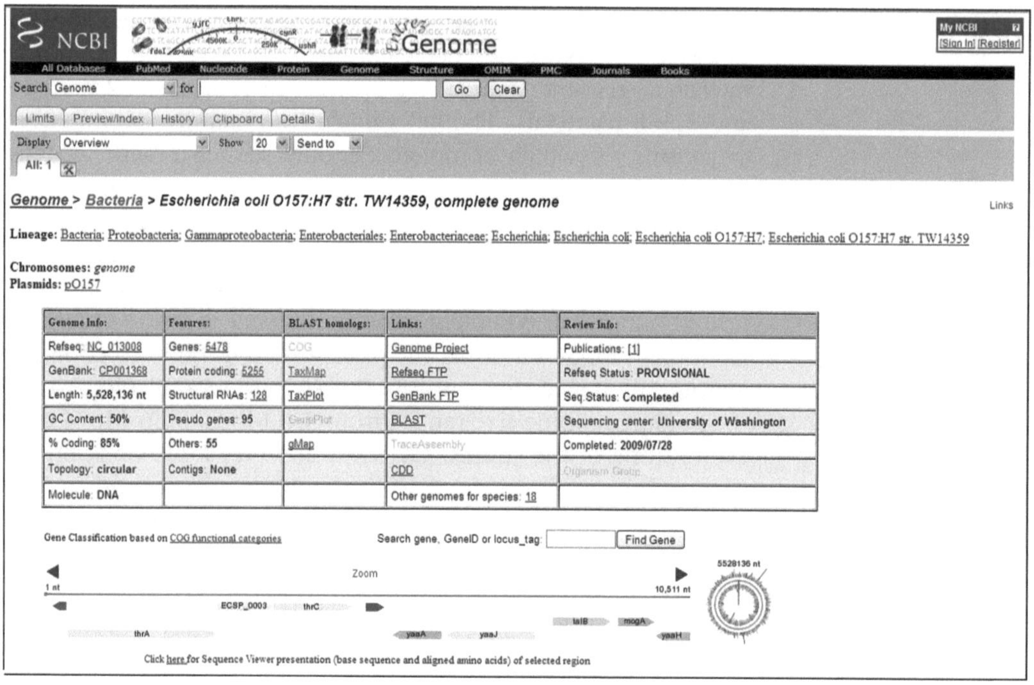

Fig. 1. Overview of *E. coli* 0157:H7 str. TW14359 genome sequence and annotation with access to analysis tools and Sequence Viewer to visualize the genome assembly and annotation.

homologs, excluding proteins from *E. coli*, after comparing each protein from the strain against all proteins from eukaryotes, eubacteria, viruses, and archaea. The top chart shows circles representing genes in their order on the genome. They are color-coded based on the taxonomic distribution of their best homolog (except *E. coli*): gray for virus, yellow for archaea, and blue for eubacteria. There are stretches of genes most similar to viruses indicating that the region may have been acquired from viruses/phages.

2. A click on one such region displays a table listing similarity scores of proteins in that region to the best protein from each domain, eukaryotes, eubacteria, viruses, and archaea. The majority of the proteins encoded by the genes colored in gray are most similar to bacteriophage BP-933 W proteins as shown in the result table.

3.2. Integrated Microbial Genomes (http://img.jgi.doe.gov/w)

Integrated microbial genomes (IMG) has been developed by the Department of Energy Joint Genome Institute (5). It is updated quarterly and contains all publicly available genomes from three domains of life – archaea, bacteria, and eukarya – along with plasmids and viruses. The resource provides access to genome sequences ("Find Genomes" tab), annotated information such as genes ("Find Genes" tab) and functions ("Find Functions" tab), and tools for comparison with other genomes ("Compare Genomes" tab). There are 5,648 genomes from all domains in complete, finished or draft form. The "Find Genomes" tab lets the user to search the genome by organism name or browse the entire list.

We use the Phylogenetic Profilers tool to obtain a list of genes found specifically in pathogenic strains compared to a nonpathogenic strain of *E. coli*. This tool lets you find a list of genes in one organism with homologs in another organism but do not have homologs in a third organism. We use this tool to find genes that are common to two pathogenic strains of *E. coli* (O157:H7 str. TW14359 and O157:H7 Sakai) with no homologs in a nonpathogenic strain (K-12 substr. MG1655).

1. Select the Find Genes tab and then click on the Phylogenetic Profilers option.
2. Select the link for Single Genes. Another option is to identify gene cassettes common in two organisms.
3. Select the radio button in the first column "Find Genes In" for *E. coli* O157:H7 str. TW14359, second column "With Homologs In" for *E. coli* O157:H7 Sakai and the third column "Without Homologs In" for *E. coli* str. K-12 substr. MG1655 (see Note 3).
4. Click on the "Go" button at the bottom of the page.

   Results as of August, 2010 show that 1,022 genes of O157:H7 str. TW14359 have homologs in O157:H7 Sakai, but do not have homologs in K-12 substr. MG1655. The Summary Statistics table provides access to functional classifications of these proteins based on COG, Enzyme, Pfam etc. The larger table on the page provides additional information such as identifier, length, and each of the functional classifications (see Note 4).

5. From the Summary Statistics table, select the COG functional category to study proteins specific to this pathogenic strain (Fig. 2).
6. Let us further analyze some COG categories such as Intracelluar trafficking that may have proteins associated with pathogenicity. Clicking on the number to the right of the "Intracellular trafficking, secretion and vesicular transport category" label lists proteins such as fimbrial proteins, adhesins, type III secretory proteins, hemolysin activator protein, etc. (see Note 5).

   These proteins are essential for pathogenicity of *E. coli* (6). For example, adhesin molecules of uropathogenic *E. coli* recognize mannose groups on the bladder epithelium and these specialized adhesin-containing fimbriae are required for colonization of the urinary tract. Synthesis of fimbriae often involves

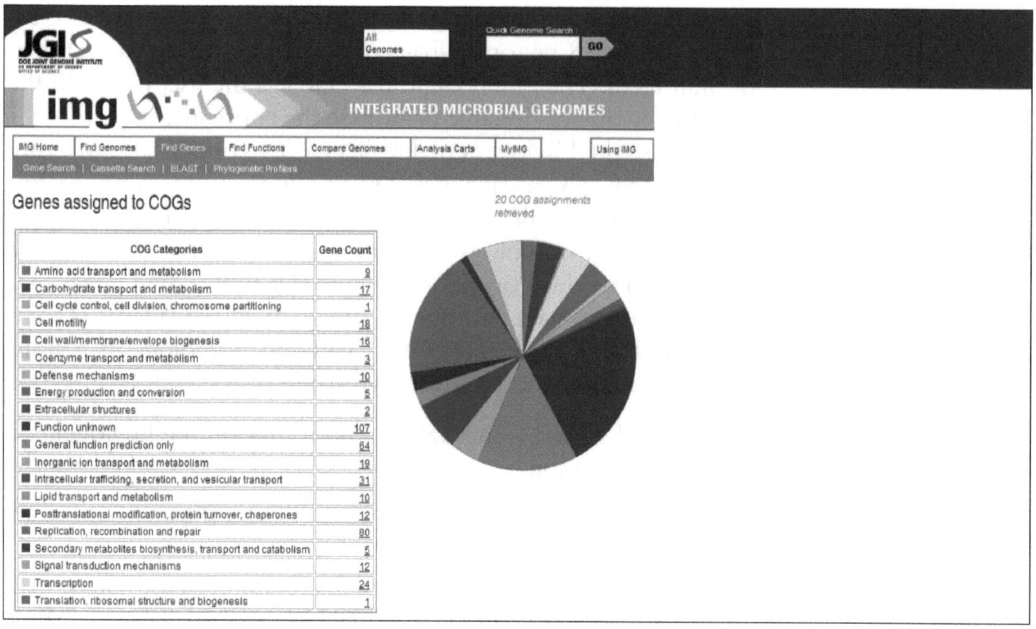

Fig. 2. COG functional categories of proteins unique in *E. coli* 0157:H7 str. TW14359 with homologs in *E. coli* 0157:H7 Sakai, but not in *E. coli* K-12 substr. MG1655.

complex secretion and assembly machinery (type III secretory proteins), which also excretes substances such as hemolysin that are toxic to host cells.

7. From the protein list obtained after selecting "Intracellular trafficking, secretion and vesicular transport" option, select the protein labeled "putative adhesin" with gene_id 644924025.

8. This leads you to a Gene Detail page with information about the gene such as links to DNA and protein sequence, function and domain, neighboring genes (neighborhood) and conserved neighborhood. Click on the "Show ortholog neighborhood regions" link to get the output depicted in Fig. 3. The result shows that this gene is only present in other pathogenic O157 strains. The result also displays some of the differences among various pathogenic strains. For example, rhsA protein in rhs element, shown by a long arrow, is present in other pathogenic O157:H7 strains, but absent in str. TW14359.

3.3. MetaCyc and Biocyc (http://metacyc.org and http://BioCyc.org)

MetaCyc is a database of nonredundant, curated, and experimentally elucidated metabolic pathways. It contains more than 1,500 pathways, from over 1,900 different organisms, involved in both primary and secondary metabolism, as well as associated compounds, enzymes, and genes. EcoCyc is a database dedicated to the bacterium

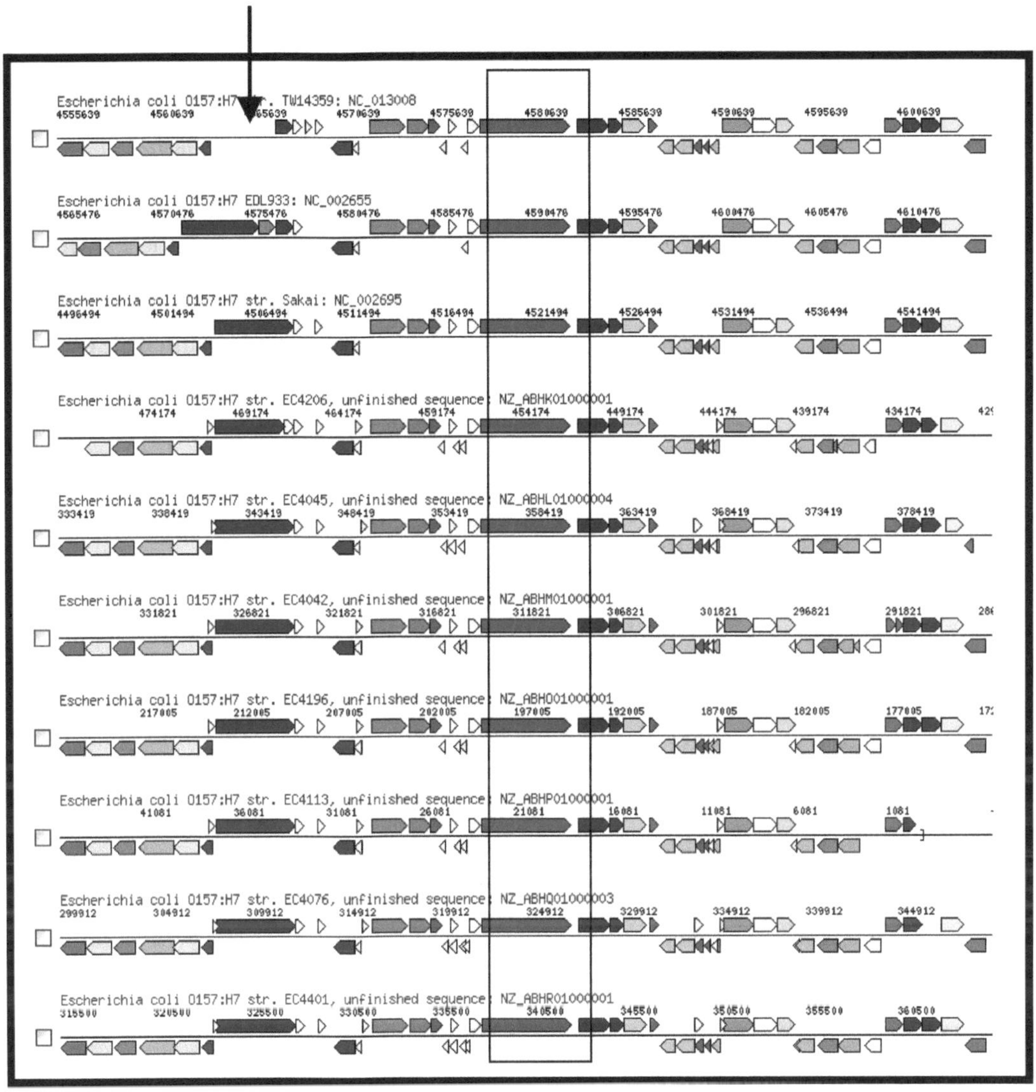

Fig. 3. Ortholog neighborhood region output for the *E. coli* O157:H7 str. TW14359 putative adhesin with gene_id 644924025. The protein is indicated by a rectangle in all genomes. Absence of rhsA protein in rhs element in the O157:H7 str. TW14359 strain is indicated by an *arrow*.

E. coli K-12 MG1655, providing literature-based curation of the entire genome, metabolic pathways, transporters, and transcription regulation (7). BioCyc contains several Pathway/Genome Databases (PGDBs), tools for navigating and analyzing these databases and is arranged in three tiers (see Notes 6 and 7) (8).

The following example demonstrates effective use of some of the tools at BioCyc. Let us begin by querying a specific topic important for microbial food safety. The acid tolerance of food-borne pathogens helps them survive human gastric challenge before they colonize the intestine (9, 10). *Escherichia coli* O157:H7 has three acid resistance systems. One of them uses an arginine decarboxylation system. Although *Shigella* and *E. coli* are closely related, this system was considered to be absent in *Shigella* (10–12). We examine if genome-based comparative tools could be used to identify genes related to arginine dependent acid resistance in *Shigella*.

1. Access the BioCyc Web page http://biocyc.org/.
2. Type "acid resistance" in the top right-hand corner window and click on the "Quick Search" button (see Note 8). The result lists two major acid resistance pathways that are dependent upon the availability of amino acids, namely, glutamate and arginine (10, 13).
3. Click on the "arginine dependent acid resistance link". The result page takes us to the page shown in Fig. 4. One can opt to seek "more details" (detailed structure and biochemistry of the enzymatic reactions) or "less details" (overview of the pathway).
4. Click on the "Species Comparison" button to study whether this pathway is present in other species.
5. Select three enteric pathogens associated with food-borne infections: *E. coli O157:H7 EDL933*, *Shigella dysenteriae Sd197*, and *Salmonella enterica serovar Typhimurium str LT2* (keeping the box for K-12 substr. MG1655 checked).
6. Click on the Submit button at the bottom of the page.
7. The results page includes a table listing color-coded information about the pathway and genes involved in this pathway (see Note 9).

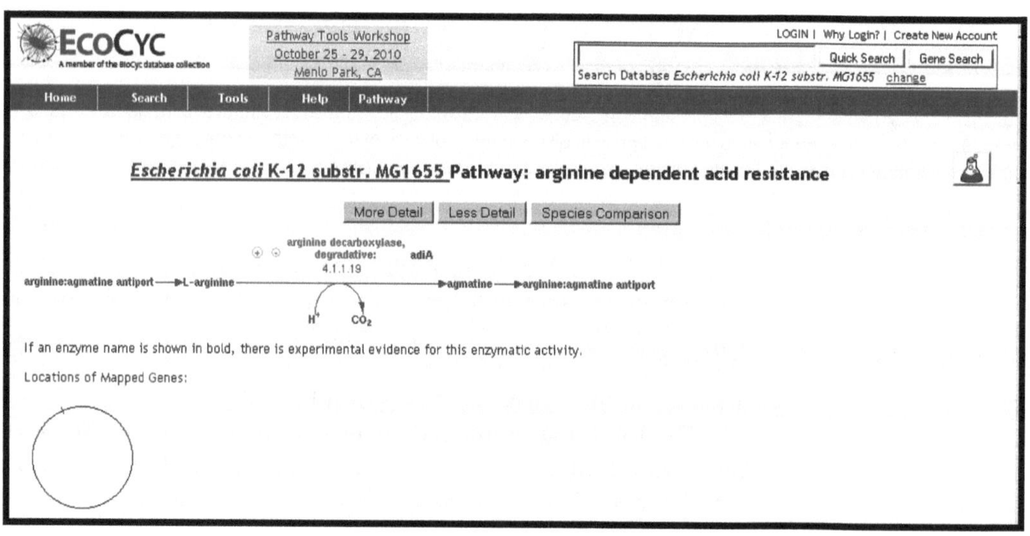

Fig. 4. EcoCyc pathway output for "arginine dependent acid resistance" query.

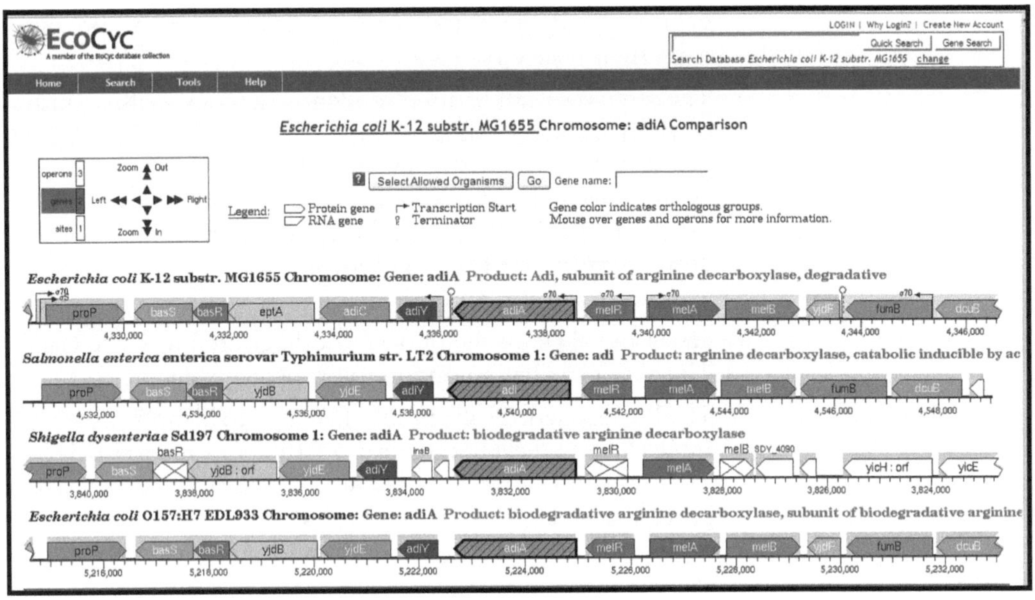

Fig. 5. Comparison of *adiA* locus in the multigenome browser. *Hash marks* show the gene of interest, adiA.

8. In the *E. coli* row and the Operons column, click on the *adiA* gene arrow to access the Web page to get further details about this gene.
9. Since our focus is on cross-species comparison, click on the "Align in Multi-Gene Browser" (Fig. 5).

All three components of arginine dependent acid resistance (*adiA*, arginine decarboxylase; *adiY*, AraC-type transcriptional regulator; and *adiC/yjdB*, arginine-agmatine transporter) are present in near-identical manner in all the four bacterial species (*see* **Note 10**). Although this pathway was originally considered to be absent in *Shigella* and *Salmonella*, it was later discovered that an arginine dependent acid resistance pathway is indeed operative in *Salmonella* in response to different physiological stimuli (14).

4. Notes

1. Note the tools legend given at the top of the page. T – TaxMap, P – ProtTable, C – COG Table, L – BLAST, S – CDD search, G – GenePlot, X – TaxPlot, M – gMap, F – FTP, R – Publications.
2. This page is a good starting point to access information about microbial genome sequencing projects listed at the NCBI site. Detailed information about the completed

genomes and genomes being sequenced can be obtained from their respective tabs.

3. The same tool can be used to identify genes specific to either of the pathogenic strains. For example, to identify unique genes in O157:H7 str. TW14359 compared to O157:H7 Sakai, select the radio button for "Find Genes In" for O157:H7 str. TW14359 and the radio button for "Without Homologs In" for O157:H7 Sakai.
4. To export the information to a file, select any or all of these genes and click on the button "Add Selected to Gene Cart". The next page has an option to export to Excel file.
5. The Phylogenetic Profiler tool when used to identify proteins in Intracelluar trafficking category in K12 compared to O157:H7 str. TW14359 shows the presence of periplasmic proteins and general secretary pathway proteins, which are different from the ones found in pathogenic strains needed for virulence.
6. Tier 1 databases have received manual literature-based curation. Tier 2 and Tier 3 databases contain computational predictions of metabolic pathways, genes coding for missing enzymes, and operons with moderate or no curation, respectively.
7. The downloadable version of BioCyc includes the Pathway Tools software, which provides more speed and power than the BioCyc used on the Web. Multiple database configurations are available for installation with the software including multiple *E. coli*, *Shigella*, *Bacillus*, *Mycobacterium*, and mammalian genomes.
8. The "Gene Search" option works for gene name, partial or full EC number of an enzyme, and UniProt identifier.
9. The color coding in the Evidence Glyph column indicates the evidence for enzymes and reactions in the pathways. See the Web page for a detailed key.
10. The default organisms are those which have been selected in BioCyc Web page; however, one can select different organisms by selecting "Select allowed organisms" and making appropriate choices.

References

1. Bhagwat M., and Aravind L. (2007) in *Methods in Molecular Biology: Comparative Genomics-I* (Bergman, N. H., Ed.) pp 177–186, Humana Press, Totowan, NJ.
2. Wheeler D., and Bhagwat M. (2007) in *Methods in Molecular Biology: Comparative genomics-I* (Bergman, N. H., Ed.) pp 149–176, Humana Press, Totowan, NJ.
3. Bhagwat A. A., and Bhagwat M. (2008) Methods and tools for comparative genomics of foodborne pathogens. *Foodborne Pathogens and Disease* **5**, 487–497.
4. Sayers E. W., Barrett T., Benson D. A., Bolton E., Bryant S. H., Canese K., Chetvernin V., Church D. M., DiCuccio M., Federhen S., Feolo M., Geer L. Y., Helmberg W., Kapustin

Y., Landsman D., Lipman D. J., Lu Z., Madden T. L., Madej T., Maglott D. R., Marchler-Bauer A., Miller V., Mizrachi I., Ostell J., Panchenko A., Pruitt K. D., Schuler G. D., Sequeira E., Sherry S. T., Shumway M., Sirotkin K., Slotta D., Souvorov A., Starchenko G., Tatusova T. A., Wagner L., Wang Y., Wilbur W. J., Yaschenko E., and Ye J. (2010) Database resources of the National Center for Biotechnology Information Nucl. Acids Res. **38**, D5–D16.

5. Markowitz V. M., Chen I. M. A., Palaniappan K., Chu K., Szeto E., Grechkin Y., Ratner A., Anderson I., Lykidis A., Mavromatis K., Ivanova N. N., and Kyrpides N. C. (2009) The integrated microbial genomes system: an expanding comparative analysis resource. *Nucl. Acids Res.* **38**, 1–9.
6. Manning S. D., Motiwala A. S., Springman A. C., Qi W., Lacher D. W., Ouellette L. M., Mladonicky J. M., Somsel P., Rudrik J. T., Dietrich S. E., Zhang W., Swaminathan B., Alland D., and Whittam T. S. (2008) Variation in virulence among clades of *Escherichia coli* O157:H7 associated with disease outbreaks. *Proc. Natl. Acad. Sci. USA.* **105**, 4868–4873.
7. Keseler I. M., Bonavides-Martinez C., Collado-Vides J., Gama-Castro S., Gunsalus R. P., Johnson D. A., Krummenacker M., Nolan L. M., Paley S., Paulsen I. T., Peralta-Gil M., Santos-Zavaleta A., Shearer A. G., and Karp P. D. (2009) EcoCyc: A comprehensive view of *Escherichia coli* biology. *Nucl. Acids Res.* **37**, D464-470.
8. Caspi R., Altman T., Dale J. M., Dreher K., Fulcher C. A., Gilham F., Kaipa P., Karthikeyan A. S., Kothari A., Krummenacker M., Latendresse M., Mueller L. A., Paley S., Popescu L., Pujar A., Shearer A. G., Zhang P., and Karp P. D. (2010) The MetaCyc database of metabolic pathways and enzymes and the BioCyc collection of pathway/genome databases. *Nucl. Acids Res.* **38**, D473-479.
9. Bhagwat A. A. (2006) in *Microbiology of Fresh Produce* (Matthews, K. R., Ed.) pp 121–165, American Society for Microbiology, Washington, D. C.
10. Foster J. W. (2004) *Escherichia coli* acid resistance: tales of an amateur acidophile. *Nat. Rev. Microbiol.* **2**, 898–907.
11. Lampel K. A., and Maurelli A. T. (2001) in *Food Microbiology* (Doyle, M. P., Beuchat, L. R., and Montville, T., Eds.) pp 247–261, ASM Press, Washington, D. C.
12. Foster J. W. (2000) in *Bacterial Stress Responses* (Storz, G., and Hengge-Aronis, R., Eds.) pp 99–115, ASM Press, Washington, D.C.
13. Bhagwat A. A., Chan L., Han R., Tan J., Kothary M., Jean-Gilles J., and Tall B. D. (2005) Characterization of enterohemorrhagic *Escherichia coli* strains based on acid resistance phenotypes. *Infect. Immun.* **73**, 4993–5003.
14. Kieboom J., and Abee T. (2006) Arginine-dependent acid resistance in *Salmonella enterica* serovar Typhimurium. *J. Bacteriol.* **188**, 5650–5653.

Chapter 18

Plasmid Artificial Modification: A Novel Method for Efficient DNA Transfer into Bacteria

Tohru Suzuki and Kazumasa Yasui

Abstract

Bacterial transformation is an essential component of many molecular biological techniques, but bacterial restriction-modification (R-M) systems can preclude the efficient introduction of shuttle vector plasmids into target bacterial cells. Whole-genome DNA sequences have recently been published for a variety of bacteria. Using homology and motif analyses, putative R-M genes can be identified from genome sequences. Introducing DNA methyltransferase genes into *Escherichia coli* cells causes subsequently transformed plasmids to be modified by these enzymes.

We propose a new method, designated Plasmid Artificial Modification (PAM). A PAM plasmid encoding the modification enzymes expressed by the target bacterial host is transformed into *E. coli* (PAM host). Propagation of a shuttle vector from the PAM host to the target bacterium ensures that the plasmid will be modified such that it is protected from restriction endonuclease digestion in the target bacterium. The result will be a higher transformation efficiency.

Here, we describe the use of PAM and electroporation to transform *Bifidobacterium adolescentis* ATCC15703. By introducing two genes encoding modification enzymes, we improved transformation efficiency 10^5-fold.

Key words: Transformation, Restriction-modification system, Plasmid Artificial Modification, Electroporation, Genome sequence, DNA methyltransferase

1. Introduction

Following the recent innovation of whole-genome sequencing technology, vast amounts of bacterial sequence information have become available (1). As of August 2010, 1,136 whole-genome bacterial sequences have been published, and over 4,800 sequencing projects are in progress. This enormous amount of data has been used inefficiently in molecular biological studies because reverse genetic tools such as convenient shuttle vectors, efficient

James A. Williams (ed.), *Strain Engineering: Methods and Protocols*, Methods in Molecular Biology, vol. 765, DOI 10.1007/978-1-61779-197-0_18,

transformation methods, gene knockouts and random mutagenesis techniques, and so on, have not been readily available (2). To address this issue, we have developed a simple method that can establish transformation techniques in bacteria for which the genome sequence is available.

To construct a transformation system for a new host, many researchers initially attempt to perform electroporation because of its convenience (3–5). In the case of *E. coli* K-12, the electroporation efficiency reaches 10^{10} colony forming units (cfu)/μg of pUC19 DNA when using suitable, competent cell selection. However, the transformation efficiency is typically lower in other bacterial species. In many cases, very few or no colonies are obtained by electroporation.

1.1. Limiting Factors of Transformation Efficiency

The potential limiting factors of transformation efficiency are (1) the physical barrier of cell surface structures such as the membrane, cell wall, or exopolysaccharides; (2) electric and/or oxidative stress during the electroporation procedure; (3a) stability of the plasmid replicon in the target host; and (3b) plasmid DNA digestion by host nucleases. We focused on the third point. Generally, bacterial cells express many endo- and exonucleases. Among these enzymes, the restriction-modification (R-M) enzymes are the most critical component protecting bacteria from invasion by foreign DNA (6, 7).

1.2. Principles of PAM

Restriction enzymes recognize and cleave within specific 4–8 bp DNA sequences; however, these enzymes do not recognize the same cleavage sites when the sites are modified by sequence-specific DNA methyltransferases (7). DNA methylation prevents restriction enzyme digestion of the host's own DNA. Most bacteria express specific R-M systems, which act as barriers against the invasion of foreign DNA by infecting phages, conjugative plasmids, or other mobile DNA elements. According to REBASE (1), 88% of bacterial genomes encode one or more R-M systems, and 43% encode four or more (Fig. 1). Multiple R-M systems, acting in concert to prevent the incorporation of foreign DNA, make it difficult to apply reverse genetics techniques. From a whole-genome sequence, it is not difficult to identify the gene encoding the modification enzyme because it usually flanks its cognate restriction enzyme. In addition, specific motifs indicative of DNA methylases have been well reported (8). It was speculated that if all, or at least some, of the modification enzymes expressed by a target bacterium were to be expressed in *E. coli*, then a plasmid transformed into *E. coli* would be modified as if it was replicated in the target bacterium. Such a plasmid would be protected from cleavage by restriction enzymes during transformation into the target bacterium. The result would be greatly improved transformation efficiency. We term this approach Plasmid Artificial Modification (PAM; Fig. 2) (9).

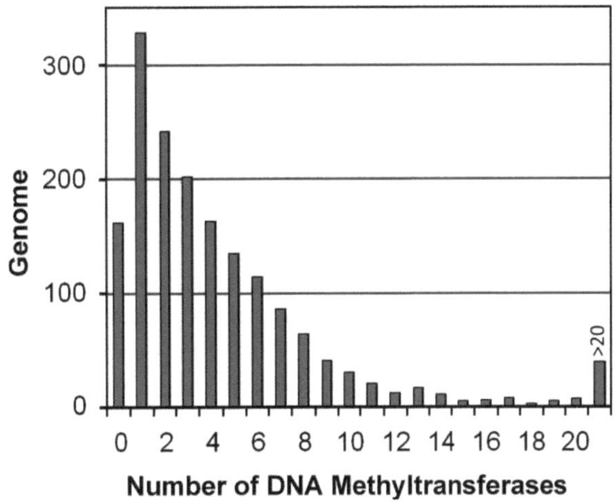

Fig. 1. Distribution of DNA methyltransferase genes in bacterial genomes. Data (1,200 Bacteria, 91 Archaea) were taken from REBASE Genomes (http://tools.neb.com/~vincze/genomes/) on August 1, 2010.

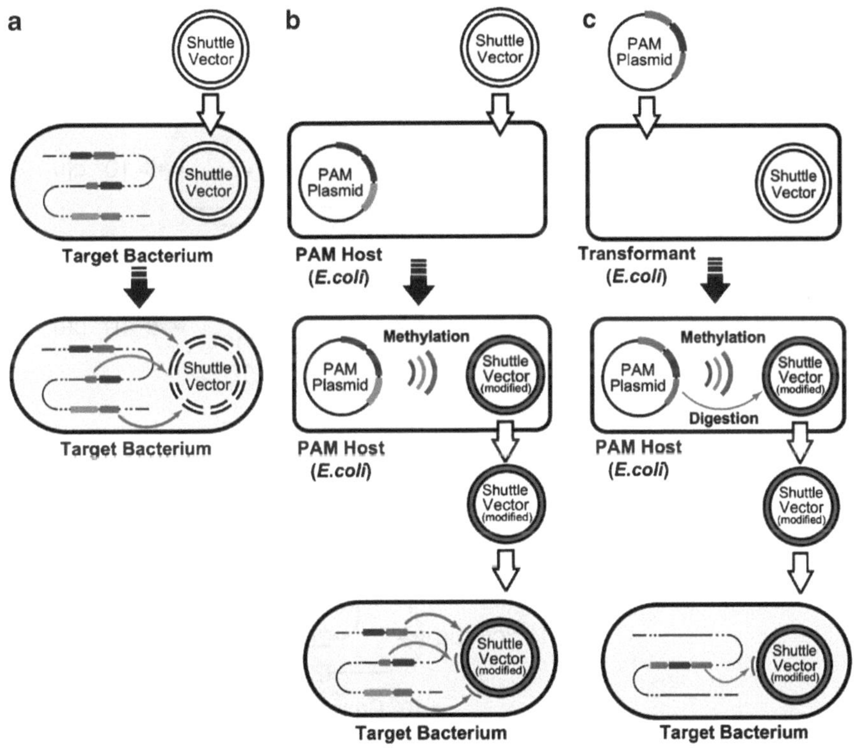

Fig. 2. The PAM concept. *Panel A*: The conventional method for the transformation of bacteria. The introduced shuttle vector is degraded by a restriction enzyme of the target bacterium. A small amount of vector survives and replicates in the target bacterium. *Panel B*: A PAM plasmid expressed by *E. coli* (the PAM host) carries all the modification methylase genes expressed by the target bacterium. A shuttle vector plasmid is introduced into the PAM host and is methylated by the appropriate modification enzymes. The shuttle vector is then isolated and introduced into the target host by electroporation. The vector plasmid is protected from host restriction enzymes and yields a higher transformation efficiency. *Panel C*: The R-M system is a complicated structure composed of a gene cluster that may include subunits or unknown accessory genes. The PAM plasmid, containing the known modification gene(s) and the uncharacterized components, is introduced into an *E. coli* transformant harboring a shuttle vector. Restriction enzyme digestion occurs, but some copies of the plasmid survive in the PAM host. The plasmid is then isolated and introduced into the target bacterium (reproduced from (9) with permission from Oxford Journals).

This is not a novel concept. It was first suggested in the 1950s by W. Arber in his Nobel-prized work (6). Researchers who investigate R-M systems know that it is essential to clone both the restriction enzyme and the methyltransferase genes into *E. coli*. Elhai et al. reported the synergistic effect of three methyltransferases on conjugation efficiency in *Anabaena* spp. (10). Our contribution is the systemization of PAM for genome-sequenced bacteria. Toward this end, we have applied PAM to types I and II R-M systems in three *Bifidobacterium* strains and one *Lactococcus* strain. We obtained efficiency increase of 7–10^5-fold in our electroporation experiments (9).

1.3. Goal of Improved Transformation Efficiency

How efficient a transformation is necessary for a molecular biology experiment? For a simple plasmid transformation, an efficiency of 10^3 cfu/μg plasmid DNA is minimal and translates to approximately 10^2 transformant colonies resulting from a typical electroporation experiment (Fig. 3a). After constructing

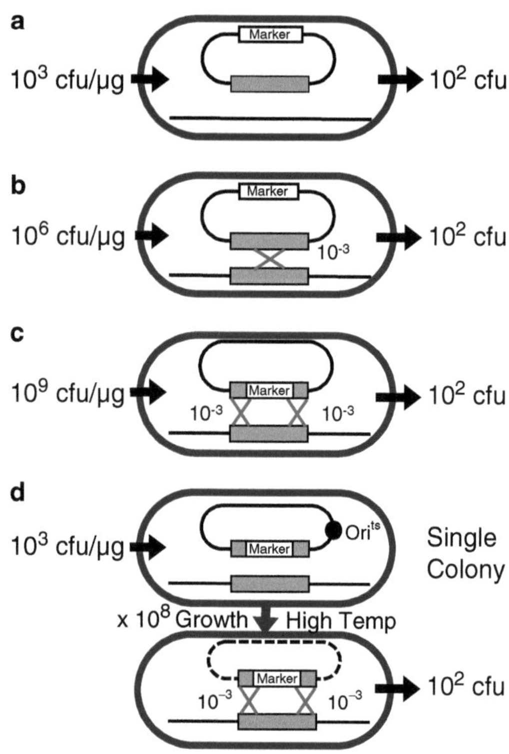

Fig. 3. Schematic representation of required transformation efficiencies. (**a**) Simple transformation, (**b**) single-crossover recombination, (**c**) double-crossover recombination, (**d**) double-crossover recombination using a plasmid with a temperature-sensitive origin of replication (ori^{ts}). The estimate cfu values (*left*: cfu/μg; *right*: cfu obtained) are calculated using typical condition of electroporation, 0.1 μg Plasmid DNA.

the plasmid in *E. coli*, a few colonies are sufficient for an electroporation experiment if the spontaneous mutation rate is relatively low. However, a 10–100-fold higher efficiency is desired if longer DNA fragments, such as >10 kbp, are to be introduced. Still higher efficiencies are required if the transformation process involves homologous recombination to achieve gene knockout or replacement. Biswas et al. examined the relationship between homologous fragment length and integration frequency using *L. lactis*. These authors suggested that a homologous fragment length of about 1 kb was associated with an integration frequency of 10^{-3} to 10^{-4} integrations per cell (11). This suggests that $>10^3$ cfu are expected in a single-crossover homologous recombination experiment. As a result, 10^6 cfu/μg or higher efficiencies are needed for single-crossover homologous recombination (Fig. 3b), and 10^9 cfu/μg or higher efficiencies are needed for double-crossover recombination (Fig. 3c). These higher efficiencies are reached by only a few target hosts, such as *E. coli*. If a vector with a temperature-sensitive origin of replication is available (12), an efficiency of 10^3 cfu/μg is sufficient for all experiments (Fig. 3a, d; Table 1).

The first attempt in a transformation experiment should be conventional electroporation as described in Subheadings 3.6 and 3.7. If a sufficient transformation efficiency is not obtained, use PAM to improve the efficiency.

Table 1
Transformation efficiencies required in the molecular biological experiments

Purpose	Description	Minimal goal (CFU/μg DNA)	Sufficient goal
Transformation	Electroporation (Fig. 3a)	10^3	10^5
Shotgun cloning	Electroporation	10^7	10^9
Homologous recombination	Single crossover (Fig. 3b)	10^6	10^8
Homologous recombination	Double crossover (Fig. 3c)	10^9	10^{11}
Homologous recombination	Double crossover Ori^{ts} shuttle vector (Fig. 3d)	10^3	10^5

2. Materials

2.1. Selection of R-M Systems from Database

1. REBASE: Information regarding R-M systems is available at the New England Biolabs website (http://rebase.neb.com/rebase/rebase.html).
2. MiGAP: The automated annotation service, Microbial Genome Annotation Pipeline, is available at the MiGAP website (https://migap.lifesciencedb.jp/mgap/jsp/index.jsp) (13).

2.2. Escherichia coli Strains and Shuttle Vector

1. *Escherichia coli* HST08: F$^-$, *endA1*, *supE44*, *thi-1*, *recA1*, *relA1*, *gyrA96*, *phoA*, φ80*lacZ*Δ*M15*, Δ(*lacZYA-argF*)*U*169, Δ(*mrr-hsd*RMS-*mcr*BC), Δ*mcrA*, λ$^-$ (Takara Bio).
2. TOP10: F$^-$, *mcrA*Δ(*mrr-hsd*RMS-*mcr*BC), φ80*lacZ*Δ*M15*, Δ*lacX74*, *nupG*, *recA1*, *araD139*, Δ(*ara-leu*)*7697*, *galE15*, *galK16*, *rpsL* (Str^r), *endA1*, λ$^-$ (Invitrogen).
3. DM1: F$^-$, *dam*–13::Tn9(Cm^r), *dcm*, *mcr*B, *hsd*R$^-$M$^+$, *gal1*, *gal2*, *ara*, *lac*, *thr*, *leu*, *tonr*, *tsxr*, Su^0, λ$^-$ (Invitrogen).
4. pKKT427: A *Bifidobacterium–E. coli* shuttle vector (10). A modified pBRASTA101 replicon (pTB6). This Spectinomycin resistant (Sp^R) shuttle vector was constructed by the modification of a previously reported shuttle vector, pBRASTA101 (14, 15), a composite plasmid of pUC18, and a multiple-cloning site (MCS).
5. pBAD33: Plasmid pBAD33 (p15A ori, Cm^r, *araBAD* promoter–*rrnB* terminator, *araC*) (Fig. 6) reported by Guzman et al. (16).

2.3. Construction of PAM Plasmid with the In-Fusion In Vitro Cloning Technique

1. 1.0 U/μl *KOD*-Plus-DNA Polymerase (*KOD*-plus): DNA polymerase from the hyperthermophilic Archaeon *Thermococcus kodakaraensis* KOD1, which exhibits excellent PCR fidelity and efficiency (17). The enzyme solution contains two types of anti-KOD DNA polymerase antibodies that inhibit polymerase and 3′–5′ exonuclease activity, thus allowing for hot-start PCR (Toyobo Biologics).
2. NucleoSpin Extract II Kit (Clontech cat# 740609.50 and 740609.250).
3. 10× Buffer for *KOD*-Plus (10× reaction buffer): available with the polymerase.
4. 2 mM dNTPs: available with the polymerase.
5. 25 mM $MgSO_4$: available with the polymerase.
6. Primers: sequences described below are for amplification of BAD1233 and BAD1283 (see Fig. 7). For other genes, replace the coding region (Capitalized) with the corresponding regions from the target gene. Each primer is reconstituted at 10 μM in water.

PMT1-F: 5′-gggctagcgaattcg ATGAGCAAGGAAATCAA AGT-3′

PMT1-R: 5′-gatccccgggtaccgTTACCGTTTCGAATCGTTGT-3′

PMT2-F: 5′-gcaggcatgcaagctATGATAAATAACCGGGAGTA-3′

PMT2-R: 5′-caaaacagccaagctTCATTCCTTGCTAGCATCAA-3′

OMT-F′: 5′-tcgaaacg ATGATAAATAACCGGGAGT-3′

OMT-R′: 5′-gttattta TCAT CGTTTCGAATCGTTGT-3′

7. SYBR Gold: SYBR Gold nucleic acid gel stain (Invitrogen, cat# S-11494). Store at ≤−20°C, desiccate, protect from light. Stable for 1 year. Dilute 10,000× with TE buffer before use.
8. Blue Light Transilluminator: Safe Imager (Invitrogen, cat# G6600).
9. *Hinc*II: Restriction endonuclease *Hinc*II. 10× reaction buffer (NEBuffer 3) and BSA (100×) are available from New England Biolabs or other suppliers.
10. *Escherichia coli* HST08 chemically competent cells: prepare competent cells as described previously (Takara Bio, cat# 9128).
11. In-Fusion Dry-Down PCR cloning kit (Clontech).
12. SOC broth: 2.0 g tryptone, 0.5 g yeast extract, 0.5 g NaCl, 1.0 mL 250 mM KCl. Adjust pH to 7.0 with NaOH, bring to 100 mL, then autoclave (120°C, 15 min). Cool, then add autoclaved 0.5 mL 2.0 M $MgSO_4$ and 2.0 mL 1.0 M glucose. Dispense into sterilized 1.5 mL microtubes.
13. Chloramphenicol (1,000× Cm): Dissolve 20 mg/mL chloramphenicol (Cm) in ethanol. Store at −20°C.
14. LB broth: 10 g tryptone, 5.0 g yeast extract, 10 g NaCl in 900 mL deionized water. Adjust pH to 7.0 with 1 M NaOH. Bring to 1 L and autoclave 120°C for 20 min.
15. LB (Cm) broth: After autoclaving 100 mL LB, cool to <50°C, then add 100 μL of 1,000× Cm.
16. LB (Cm) agar: Add 1.5 g agar to 100 mL LB broth. Autoclave 120°C for 20 min. Cool to <50°C, then add 100 μL of 1,000× Cm. Pour 25 mL each into 10 cmø Petri dish.
17. QIAprep Spin Miniprep Kit (Qiagen, cat# 27104).
18. Gel electrophoresis reagents: 1% agarose gel (12 cm) and buffer, standard 5× sample buffer.
19. TE buffer: 10 mM Tris–HCl, 1 mM EDTA, pH 8.0.

2.4. Construction of PAM Host

1. LB agar: Add 1.5 g agar to 100 mL LB broth. Autoclave 120°C for 20 min. Pour 25 mL each into 10 cmø Petri dish.
2. LB (Cm) agar: Add 1.5 g agar to 100 mL LB broth. Autoclave 120°C for 20 min. Cool to <50°C, then add 100 μL of 1,000× Cm. Pour 25 mL each into 10 cmø Petri dish.

3. 0.1 M PIPES (pH 7.0): Add 3.024 g PIPES and 0.4 g NaOH into 40 mL water. Fill up to 100 mL. Add 40–60 mL 0.1 M NaOH to adjust pH to 7.0 at 20°C.
4. $CaCl_2$ solution: 60 mM $CaCl_2$, 10 mM PIPES (pH 7.0 at 20°C). Dissolve 0.882 g $CaCl_2 \cdot H_2O$ in 99 mL water. Add 1 mL 0.1 M PIPES (pH 7.0). Autoclave at 110°C for 10 min.
5. Glycerol (20%): Measure 20 g glycerol. Fill to 100 mL with water. Dispense as 1 mL aliquots into microtubes.

2.5. Preparation of Shuttle Vector from PAM Host

1. Spectinomycin (1,000× Sp): Dissolve 150 mg/mL Spectinomycin in ethanol. Store at −20°C.
2. LB (Cm,Sp) agar: Add 1.5 g agar to 100 mL LB broth. Autoclave 120°C for 15 min. Cool to <50°C, then add 100 μL of 1,000× Cm and 100 μL of 1,000× Sp. Pour 25 mL each into 10 cmø Petri dish.
3. LB (Cm,Sp) broth: Add 100 μL of 1,000× Cm and 100 μL of 1,000× Sp to 100 mL LB broth.
4. 10% Arabinose: 10 g L-arabinose (Sigma-Aldrich, cat# A3256) in 100 mL water. Sterilize at 120°C for 10 min.
5. QIAprep Spin Miniprep Kit (Qiagen, cat# 27104).

2.6. Preparation of Competent Cells for Electroporation

Here, the transformation method for *Bifidobacterium* is described. Culture conditions and media should be customized to the target bacterial strain.

1. 2% l-cysteine: Dissolve 3.51 g l-cysteine·HCl·H_2O (MW = 175.63) in 100 mL water. Sterilize by filtration (0.22 μm pore). Store at −20°C.
2. 34% sodium ascorbate: Dissolve 34 g sodium l-ascorbate in 100 mL water. Sterilize by filtration (0.22 μm pore). Store at −20°C.
3. MRS broth: prepare 10.0 g casein peptone, tryptic digest, 10.0 g meat extract, 5.0 g yeast extract, 20.0 g glucose, 1.0 g Tween 80, 2.0 g K_2HPO_4, 5.0 g sodium acetate, 2.0 g diammonium citrate, 0.2 g $MgSO_4 \cdot 7H_2O$, 50.0 mg $MnSO_4 \cdot H_2O$, 1.0 L distilled water. Heat with frequent agitation, and boil for 2–3 min to completely dissolve. Check the medium pH within 6.0–6.5. Autoclave at 120°C for 15 min. This medium is also commercially available (Lactobacilli MRS, Difco).
4. MRS + AC broth: Same as MRS broth above but supplemented with 0.34% ascorbic acid and 0.02% l-cysteine. Add 1 mL 34% sodium ascorbate and 1 mL 2% l-cysteine to 100 mL MRS broth.
5. Sucrose buffer: prepare 1.712% (w/v) sucrose, 1 mM ammonium citrate. Adjust pH to 6.0 with citric acid. Sterilized by autoclaving (110°C, 10 min), then store under anaerobic conditions using an Anaeropack at 4°C.

6. MRS agar: Add 1.5 g agar to 100 mL MRS broth. Autoclave at 120°C for 15 min. Dispense 25 mL per 100 mM diameter Petri dish.

2.7. Transformation of Cells by Electroporation

1. Electroporation equipment: MicroPulser electroporator (Bio-Rad, cat# 165–2100) or equivalent.
2. Electroporation cuvette: 0.2 cm gap (Bio-Rad, cat# 165–2086).
3. MRS + C broth: Same as MRS broth above but supplemented with 0.02% l-cysteine. Add 1 mL 2% l-cysteine to 100 mL MRS broth.
4. MRS + AC (Sp) agar: The same as MRS broth above but containing 1.5% agar, supplemented with 0.34% ascorbic acid, 0.02% l-cysteine, and 150 μL/mL Sp. Add 1.5 g agar to 100 mL MRS broth. Autoclave at 120°C for 15 min. Cool to <50°C, then add 1 mL 34% sodium ascorbate, 1 mL 2% l-cysteine, and 1 mL 1,000× Sp. Dispense 25 mL per 10 cmø Petri dish. Store under anaerobic conditions.
5. Anaeropack: acts as an oxygen absorber/CO_2 generator. Pouch-Anaero, including O_2 absorber bag and pouch, is available from Mitsubishi Gas Chemical.

3. Methods

3.1. Cloning Strategy of DNA Methyltransferases

Once the whole-genome sequence of a given target bacterium has been completed and submitted to a public DNA database (DDBJ/EMBL/NCBI), the data should become available in REBASE shortly thereafter. The REBASE database focuses only on bacterial R-M systems (http://tools.neb.com/~vincze/genomes/). Most methyltransferases can be easily identified by conserved sequence motifs. If the genome sequence of interest is unpublished, a pipeline genome annotation service such as MiGAP may provide information about the associated R-M system. The protein sequence datasets of R-M systems also are available on the FTP server of REBASE, which can be used for in-house BLAST searches.

R-M systems have been categorized into four types, Type I–Type IV (8). From a homology analysis, it is possible to group related R-M clusters. Each type requires a different cloning strategy. In the case of the simple Type II system (Fig. 4a) in which the R gene is homologous to Type II R genes, and the R-M genes are arranged side-by-side, only the M gene should be cloned into a PAM plasmid. If R and M genes are located farther apart in the genome, the genes may be pseudogenes and should be validated by RT-PCR or some other method. In the case of Type I R-M systems (Fig. 4c), it is sufficient to clone the *hsdM* and *hsdS* genes encoding the methyltransferase and specificity subunits, respectively (9). For a Type IV R-M system (Fig. 4e), it is not necessary to clone any methyltransferase, but it is important to use a host *E. coli* that lacks methyltransferase (e.g., *dam*⁻ and *dcm*⁻ strains), such as DM1.

For the expression of an exogenous gene in an *E. coli* cell, a promoter, ribosome binding site, and terminator must also be present. Usually, methyltransferases are expressed at a low level, and 100 bp of 5′- and 3′-flanking regions are cloned with the structural gene, including the original promoter and terminator (Fig. 5a). In the case of species that are evolutionarily distant from *E. coli*, it is

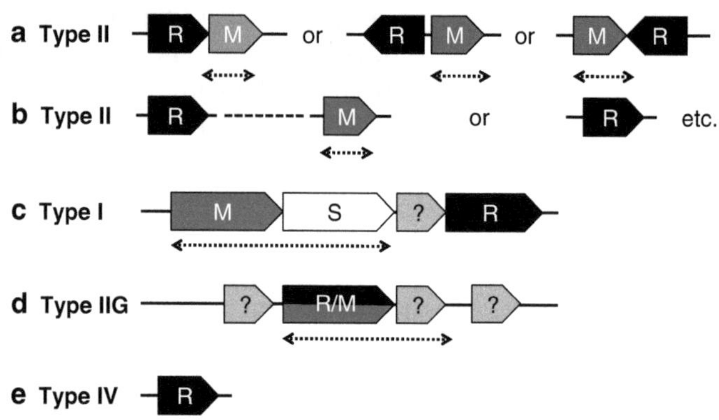

Fig. 4. Cloning targets of PAM. (**a**) For Type II R-M systems in which R and M genes exist in a cluster, clone the M gene. (**b**) If R and M genes are separated in the genome sequence, consider these to be pseudogenes until validated. (**c**) In the Type I R-M system, the M, S (specificity subunit), and R subunits are clustered. Select the M and S subunits for coexpression in the PAM host. (**d**) The R and M catalytic domains are fused in the Type IIG system. In this case, select the coding sequence and flanking genes which form an operon with R/M. (**e**) For Type IV systems, it is not necessary to clone M, but a *dam*⁻, *dcm*⁻ strain must be selected. *Dotted lines* indicate cloning targets.

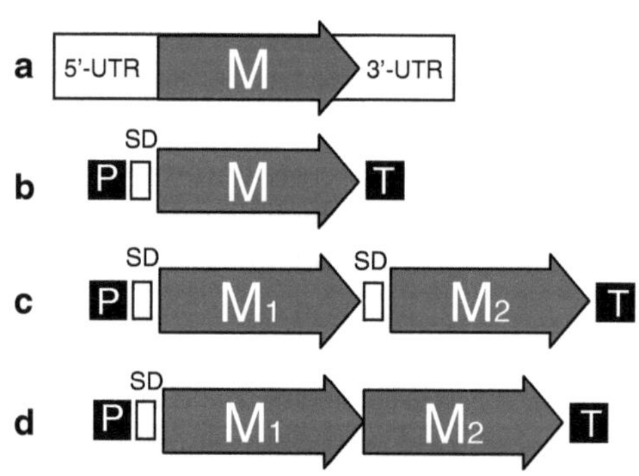

Fig. 5. Design of expression unit of DNA methyltransferase. (**a**) Clone the methylase (M) gene, including the 5 - and 3 -UTR (approximately 100 bp) regions that contain the original promoter and terminator. (**b**) Clone the coding region of the gene between the promoter and terminator. (**c**) Clone two (or more) genes as an operon. (**d**) Clone as in (**c**) but without the internal ribosome binding site (SD).

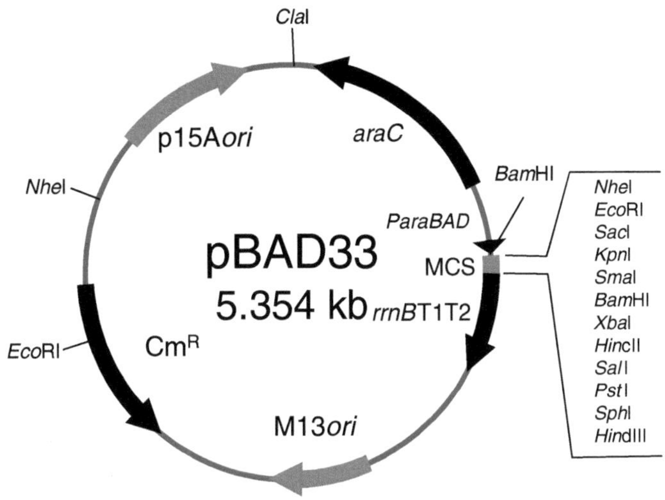

Fig. 6. Plasmid map of pBAD33 (16). The MCS exists between the P*araBAD* promoter and the *rrnB* transcription terminator. The chloramphenicol acetyl transferase gene (Cm) acts as a selection marker, and p15A ori, which does not show incompatibility with pMB1 (ColE1) ori, was derived from pACYC184.

necessary to clone an expression unit including a promoter, a ribosome-binding site (Shine-Delgarno [SD]), an MCS, and a transcription terminator (Fig. 5b). Plasmid pBAD33 is a suitable vector for PAM experiments (Fig. 6) (16). It consists of a controllable promoter from the *araBAD* operon, which is induced by the addition of L-arabinose into the culture medium, and an *rrnB* terminator. To construct the operon, insert SD sequences (AGGAGG) before each initiation codon (ATG) (Fig. 5c). Alternatively, an initiation codon for the second gene may be inserted next to the stop codon of the first gene (Fig. 5d). In the case of *B. adolescentis* ATCC15703, the latter method was utilized (9).

3.2. Selection of Plasmid Vectors and E. coli Strains

A plasmid vector capable of replicating in the target cell is needed. Ideally, a shuttle vector between *E. coli* and a closely related species should be used, but certain broad-range vectors also may be used. If no working plasmid is available, it is necessary to construct a novel vector that may result from a survey of cryptic plasmids in the host. If a shuttle vector between *E. coli* and the target strain is available and exhibits both a temperature-sensitive replication origin (ori^{ts}) and a suitable selection marker, only 10^3 cfu/μg efficiency is needed. Using error-prone PCR of the replication origin, it is not difficult to obtain an ori^{ts} (Rep^{ts}) mutant plasmid (14). After simple transformation, the cells are grown at the non-permissive temperature for plasmid replication. A few cells will grow, signaling that the selection marker was integrated into the chromosome by homologous recombination (Fig. 3c, d).

To construct a PAM plasmid, we employed plasmid vector, pBAD33 (18), which has the inducible promoter of the *araBAD* operon; the regulatory gene, *araC*, of *E. coli*; and the Cm-resistant gene. It also has a replication origin from plasmid p15A, which is compatible to the ColE1 origin. It allows for the coexistence of pKKT427 or other pUC18-19 derived shuttle vectors.

When cloning a foreign DNA methylase gene into *E. coli*, we typically use DH10B or TOP10 cells in which the Type I R-M system's *mcrA*, *mcrBC*, and *mrr* genes have been deleted. However, in some cases, DH5α cells can achieve higher transformation efficiencies. From the genotype record, the difference between the R-M systems is the expression of *hsdMS*. It is also possible that the target host recognizes the *E. coli* K-12 Type I system. However, an unknown limiting factor may exist in these *E. coli* strains. *E. coli* K-12 also expresses *dam* and *dcm* DNA methylase genes, and it is possible that the target host expresses unknown Type IV enzymes, which digest GmATC or CmCWGG, the recognition sequences of *dam* and *dcm*, respectively. Mutants for these genes, such as DM1, serve as alternative PAM hosts.

3.3. Construction of PAM Plasmid Using the In-Fusion In Vitro Cloning System

The PAM system requires cloning of multiple DNA methyltransferase genes. We employ two cloning strategies. The first is the In-Fusion Dry-Down PCR Cloning kit. This technique allows for the fusion of a PCR fragment with the homologous ends of a linearized vector (19). The Primer sequences are designed by using a computer software, Primer 3 (20) or equivalent. At the 5′-ends of both PCR primers, 15-bp extensions are added that precisely matched the ends of the linearized vector (see Note 1). When the vector is combined with the insert, the In-Fusion enzyme converts the double-stranded extensions into single-stranded DNA and fused these regions to the corresponding ends of the linearized vector (Fig. 7).

1. Amplify target DNA fragments (e.g., BAD_1233 and BAD_1283) by colony direct PCR using primers as described in Fig. 7.
   Mix 5 μL 10× reaction buffer, 5 μL dNTPs, 2 μL $MgSO_4$, 1.5 μL each primers, and 34 μL water in a 0.2-mL microtube. Add 1 μL *KOD*-plus, then mix gently.
2. Pick a fresh colony with a toothpick and dip it into each reaction mixture for a very short period. Fewer than 10^3 cells are needed to perform the amplification, and an excess amount of cells will inhibit the PCR reaction.
3. Put the reaction tubes into a thermal cycler at 95°C for 3 min for denaturation and DNA polymerase activation. Next, run 35 cycles each at 95°C for 30 s to denature, 60°C for 30 s to anneal, and 68°C for 60 s to elongate. Finally, complete the annealing at 68°C for 3 min.

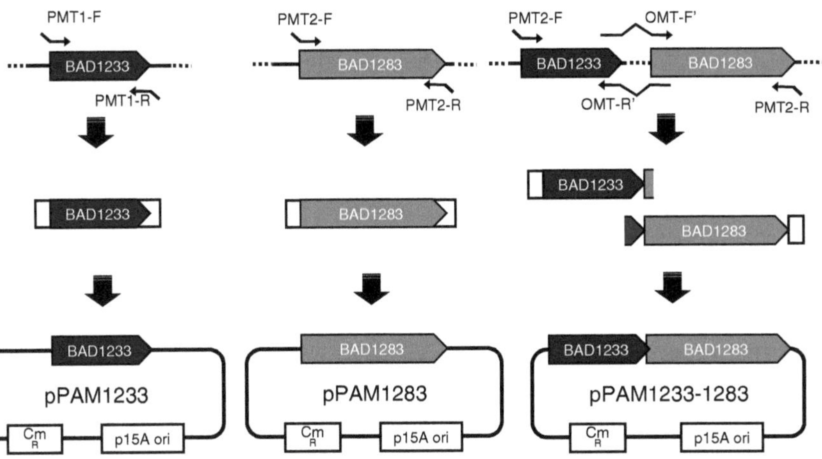

Fig. 7. Construction of pPAM plasmids. *Panel A*: Putative methyltransferase genes were amplified by PCR using the primers listed in Subheading 2.3. PCR products were joined by in vitro homologous recombination to plasmid vector pBAD33, which had been cleaved by *Hinc*II, using the In-Fusion Dry-Down PCR cloning kit (Clontech) to obtain pPAM plasmids.

4. Check the amplification reaction by applying an aliquot (5 μL) of the PCR products to an agarose gel electrophoresis. If the PCR products correspond to more than one band on the gel, consider redesigning the PCR primers.
5. Add 10 μL sample buffer to the remaining PCR products, then apply to a 1% agarose gel electrophoresis (12 cm) and electrophoreses for 2 h (~ 5 V/cm).
6. Stain the gel with SYBR Gold for 30 min.
7. Observe the DNA on a Blue Light Transilluminator. Excise the DNA band from the agarose gel using a disposable razor.
8. Purify the DNA using a NucleoSpin Extract II Kit.
9. To digest 1 μg plasmid, pBAD33, add 10 μL 10× reaction buffer and 1 μL BSA (100×). Fill to 100 μL with water. Add 5 μL (50 U) of *Hinc*II, mix gently, and incubate for 3–6 h.
10. Isolate digested plasmid as described in steps 5–8.
11. Mix 50–200 ng of each PCR fragment with 100 ng linearized vector (2:1 molar ratio) in 10 μL water.
12. Add 10 μL of vector + insert DNA + H_2O (from step 11) to each In-Fusion Dry-Down pellet. Mix well by pipetting up and down.
13. Incubate the reaction for 15 min at 37°C, followed by 15 min at 50°C, then place the tube on ice.
14. Dilute the In-Fusion reaction mixture with 40 μL TE buffer and mix well.
15. Thaw one vial of frozen HST08 chemically competent cells on ice. Tap tube gently to ensure that the cells are suspended.

16. Add 2.5 μL of the diluted reaction mixture to the cells. Mix gently to ensure even distribution of the DNA solution. Incubate the tube on ice for 30 min.
17. Heat-shock the cells in a water bath at 42°C for 45 s, then place them on ice for 1 min.
18. After a heat shock, add 450 μL of SOC medium to the cells, then incubate at 37°C for 60 min while shaking at 250 rpm.
19. Take 50 μL from each transformation, bring the volume to 100 μL with SOC medium, and spread on separate LB (Cm) plates.
20. Centrifuge the remaining mix at 6,000 rpm (3000 × *g*) in a microfuge for 5 min, resuspend the cells in 100 μl fresh SOC, and spread the remainder of the transformation mix on a separate LB (Cm) plate.
21. Incubate plates at 37°C overnight.
22. Pick individual, isolated colonies. Isolate plasmid DNA using a standard method (e.g., QIAprep spin miniprep kit). To determine the presence of the insert, analyze DNA by restriction digestion or PCR screening.
23. Select some clones and sequence the inserts to confirm the genes are correctly cloned into the vector.

3.4. Construction of PAM Host

1. Streak TOP10 cells (stored at −80°C) on an LB agar plate and grow for single colonies at 20°C or room temperature.
2. Pick up one colony (ca. 2 mmφ) with a plastic loop, then suspend in 50 μl of ice-cold $CaCl_2$ solution.
3. Add 10 μL PAM plasmid DNA (e.g., pPAM1233-1283; 100 ng/μL) to the cells, gently.
4. Place it in ice-water for 30 min.
5. Put the tube in a 42°C water-bath for 60 s, then chill in ice-water for 30 s.
6. Add 250 μL of SOC.
7. Incubate for 1 h at 37°C.
8. Plate 20 × 100 μl aliquots on LB (Cm)-agar plates.
9. Incubate overnight at 37°C.
10. Pick up some colonies with toothpicks, dip into 3 mL LB (Cm), and then culture overnight at 37°C (see Note 2).
11. Put 1 mL of culture into 1 mL glycerol (20%), then store at −80°C. Use the remaining culture for plasmid preparation.

3.5. Preparation of Shuttle Vector from PAM Host

When the PAM host is successfully obtained (Subheading 3.4), introduce a shuttle vector. Here, the case of a *Bifidobacterium*–*E. coli* shuttle vector (pKKT427) is described. In the pBAD33 plasmid, the insert genes are strongly expressed in the presence of L-arabinose, but occasionally, induction reduces the growth rate. Therefore, two culture conditions with arabinose (induced) or without arabinose (repressed) are tested.

1. Plate *E. coli* TOP10 cells harboring pBAD1233-1283 on an LB (Cm) plate and grow to single colonies at 20°C or room temperature.
2. Perform the same transformation procedure as in steps 2–10 in Subheading 3.4, but substitute the Sp^R shuttle vector (e.g., pKKT427) in step 3 and LB (Cm, Sp) plate and broth in steps 8 and 10.
3. A clone, which is resistant to both Sp and Cm, is inoculated into two tubes containing 5 mL LB (Cm, Sp).
4. After overnight cultivation, add 50 μL 10% arabinose to one tube, and incubate for 3 h.
5. Prepare plasmids using a QIAprep Spin Miniprep Kit (see Subheading 2.3). Elute plasmid DNA with 50 μL water to prevent arcing upon electroporation.

3.6. Preparation of Competent Cells for Electroporation

Here, a transformation method optimized for *B. adolescentis* ATCC15703 (9) and *B. longum* 105-A is described (14). However, almost the same procedure, except for culture media and conditions, is applicable for other bacteria such as lactic acid bacteria or other anaerobic bacteria.

1. Inoculate a microspatula (~10 μL) of −80°C freezer-stocked cells into 15 mL MRS + AC medium in a screw-capped glass tube. Cultivate for 2 days at 37°C (see Note 3).
2. Inoculate a single colony of the cells into 10 mL MRS + AC. Incubate anaerobically at 37°C to an OD_{660} of ~0.5–0.6.
3. Transfer 10 mL of culture to an ice-cold tube. Harvest the cells by centrifugation at 5,000 × *g* for 15 min at 4°C. Decant the supernatant and resuspend the cell pellet in 10 mL of ice-cold sucrose buffer.
4. Harvest the cells by centrifugation at 5,000 × *g* for 15 min at 4°C. Decant the supernatant and resuspend the cell pellet in 5 mL of ice-cold sucrose buffer.
5. Harvest the cells by centrifugation at 5,000 × *g* for 15 min at 4°C. Carefully decant approximately 4 mL of the supernatant. The remainder is about 1 mL.
6. Mix the cell suspension by pipetting.
7. Transfer 50 μL of the suspension to an Eppendorf tube on ice.
8. Use the electrocompetent cells immediately because the transformation efficiency will decrease quickly with time.

3.7. Transformation of the Bacteria by Electroporation

1. Add 10 pg to 0.5 μg shuttle vector plasmid DNA into the tube containing electrocompetent cells on ice.
2. Incubate the tube for 15 min on ice.
3. Set the electroporation apparatus to 2.5 kV and 25 μF. Set the pulse controller to 200 Ω (14) (see Note 4). Transfer the DNA and cells into an ice-cold cuvette. Place the cuvette in the electroporation device. Tap the solution to ensure that the cells and DNA sit at the bottom of the cuvette.
4. Apply the pulse by pushing the button or flipping the switch.
5. As quickly as possible after the pulse, add 1 mL MRS + C (not MRS + AC) into the electroporation cuvette, then transfer to a 1.5-mL Eppendorf tube. Incubate 3 h anaerobically at 37°C.
6. Transfer 10 μL of the culture into 990 μL of MRS + C medium, then mix vigorously. Repeat dilution to obtain the appropriate fold dilution (e.g., ×1, ×100, ×10,000).
7. Spread 100 μL onto an MRS + AC (Sp) plate, then pouch in an Anaero-Pack.
8. Incubate 2 days under anaerobic conditions using an Anaero-Pack.
9. Count colonies and calculate cfu by multiplication with the fold dilution. The expected transformation efficiency improvement in *Bifidobacteria* using PAM plasmids is summarized in Table 2.

Table 2
Comparison of electroporation efficiencies in *Bifidobacteria* using PAM

Donor host	Recipient	Efficiency[a] (CFU/μg DNA)
TOP10	*B. adolescentis* ATCC15703	$1–3 \times 10^0$
TOP10/pPAM1233	*B. adolescentis* ATCC15703	$4–6 \times 10^4$
TOP10/pPAM1283	*B. adolescentis* ATCC15703	$1–2 \times 10^4$
TOP10/pPAM1233-1283	*B. adolescentis* ATCC15703	$0.9–4 \times 10^5$
B. adolescentis ATCC15703	*B. adolescentis* ATCC15703	9×10^4
TOP10	*B. longum* 105-A	1.5×10^6 to 5×10^6
B. longum 105-A	*B. adolescentis* ATCC15703	6×10^3 to 8×10^3

[a]After electroporation, the cells were diluted ×1 or ×100 with MRS, plated on MRS-AC agar, supplemented with 150 μg/mL Sp, and incubated at 37°C under anaerobic conditions

4. Notes

1. In (9), we adopted overlap extension PCR to join two insert fragments, then connected these to the vector. Here, we described single-step cloning with two inserts and one vector, using the In-fusion cloning system, which has given the same result.
2. If transformants are not obtained by this convenient chemically competent cell method, try the high-efficiency method (21) or the electroporation method as described previously (3).
3. In some cases of anaerobic bacteria, the addition of ascorbic acid assists the bacterial growth rate and reduces lag-time under anaerobic conditions. However, reduce the ascorbic acid under aerobic conditions, because it reacts with oxygen to produce H_2O_2 (22). Thus, avoid adding Na-ascorbate or ascorbic acid to medium when it is to be used under aerobic conditions.
4. To optimize the electroporation conditions, consider optimizing any of the following (23): (1) cell growth medium, (2) growth phase at OD_{660}, (3) harvest wash solution, (4) prepulse incubation (min), (5) electroporation temperature (°C), (6) electroporation medium, (7) cell density, (8) volume of cells, (9) DNA concentration, (10) DNA resuspension buffer, (11) volume of DNA, (12) cuvette gap, (13) voltage field strength (kV/cm), (14) capacitor (μF), (15) resistor (Ω), (16) time constant (ms). After the pulse, try optimizing the (1) outgrowth medium, (2) outgrowth temperature, (3) length of incubation, (4) selection conditions. The conditions of previous studies have been summarized for over 60 strains, including, Gram-positive and -negative bacteria, and are available in the Bio-Rad Bulletin D035552 (http://www3.bio-rad.com/cmc_upload/Literature/210246/Bulletin_D035552.pdf).

Acknowledgments

The author would like to thank Professor M. Shimizu-Kadota for useful discussion. This work was partly supported by the Grant-in-Aid for Scientific Research on Priority Areas in Applied Genomics from the Ministry of Education, Culture, Sports, Science, and Technology of Japan.

References

1. Roberts R.J., Vincze T., Posfai J., Macelis D. (2010) REBASE – a database for DNA restriction and modification: enzymes, genes and genomes. *Nucl. Acids Res,* **38**, D234–D236.
2. Schweizer H.P. (2008) Bacterial genetics: past achievements, present state of the field, and future challenges. *BioTechniques,* **44**, 633–641.
3. William J., Dower W.J., Miller J.F., Ragsdale C.W. (1988) High efficiency transformation of *E. coli* by high voltage electroporation. *Nucl. Acids Res,* **16**, 6127–6145.
4. Calvin N.M., Hanawalt P.C. (1988) High-efficiency transformation of bacterial cells by electroporation. *J. Bacteriol,* **170**, 2796–280.
5. Miller J.F. (1994) Bacterial transformation by electroporation *In* Bacterial pathogenesis. Part A. Identification and regulation of virulence factors. *Methods in Enzymology,* **235**, 375–385.
6. Arber W., Linn S. (1969) DNA modification and restriction. *Annu. Rev. Biochem,* **38**, 467–500.
7. Tock M.R., Dryden D.T. (2005) The biology of restriction and anti-restriction. *Curr. Opin. Microbiol,* **8**, 466–472.
8. Roberts R.J. et al. (2003) A nomenclature for restriction enzymes, DNA methyltransferases, homing endonucleases and their genes. *Nucl. Acids Res.* **31**, 1805–1812.
9. Yasui K., Kano Y., Tanaka K., Watanabe K., Shimizu-Kadota M., Yoshikawa H., Suzuki T. (2009) Improvement of bacterial transformation efficiency using plasmid artificial modification. *Nucl. Acids Res.* **37**, e3. doi: 10.1093/nar/gkn884
10. Elhai J., Vepritskiy A., Muro-Pastor A.M., Flores E., Wolk C.P. (1997) Reduction of conjugal transfer efficiency by three restriction activities of *Anabaena* sp. strain PCC 7120. *J. Bacteriol,* **179**, 1998–2005.
11. Biswas I., Gruss A., Ehrlich S.D., Maguin E. (1993) High-efficiency gene inactivation and replacement system for gram-positive bacteria. *J. Bacteriol,* **175**, 3628–3635.
12. Hashimoto-Gotoh T., Sekiquchi M. (1977). Mutations of temperature sensitivity in R plasmid pSC101. *J. Bacteriol,* **131**, 405–412.
13. Sugawara H., Ohyama A., Mori H., Kurokawa K. (2009) Microbial Genome Annotation Pipeline (MiGAP) for diverse users. The 20th International Conference on Genome Informatics (GIW2009) Poster and Software Demonstrations (Yokohama), S001-1–2.
14. Matsumura H., Takeuchi A., Kano Y. (1997) Construction of *Escherichia coli-Bifidobacterium longum* shuttle vector transforming *B. longum* 105-A and 108-A. *Biosci. Biotechnol. Biochem,* **61**, 1211–1212.
15. Tanaka K., Samura K., Kano Y. (2005) Structural and functional analysis of pTB6 from *Bifidobacterium longum. Biosci. Biotechnol. Biochem,* **69**, 422–425.
16. Guzman L.M., Belin D., Carson M.J., Beckwith J. (1995) Tight regulation, modulation, and high-level expression by vectors containing the arabinose PBAD promoter. *J. Bacteriol,* **177**, 4121–4130.
17. Mizuguchi H., Nakatsuji M., Fujiwara S., Takagi M., Imanaka T. (1999) Characterization and application to hot start PCR of neutralizing monoclonal antibodies against *KOD* DNA polymerase. *J. Biochem,* **126**, 762–768.
18. Novick R.P. (1987) Plasmid incompatibility. *Microbiol. Rev,* **51**, 381–395.
19. Zhu B., Cai G., Hall E.O., Freeman G.J. (2007) In-fusion assembly: seamless engineering of multidomain fusion proteins, modular vectors, and mutations. *Biotechniques,* **43**, 354–359.
20. Rozen S., Skaletsky H.J. (2000) *Primer3 on the WWW for general users and for biologist programmers.* Humana Press, Totowa, NJ.
21. Inoue H., Nojima H., Okayama H. (1990) High efficiency transformation of *Escherichia coli* with plasmids. Gene, **96**, 23–28.
22. Richter H.E., Loewen P.C. (1981) Induction of catalase in *Escherichia coli* by ascorbic acid involves hydrogen peroxide. *Biochem. Biophys. Res. Commun,* **100**, 1039–1046.
23. Mercenier A., Chassy B.M. (1988) Strategies for the development of bacterial transformation systems. *Biochimie,* **70**, 503–517.

Chapter 19

Broad-Host-Range Plasmid Vectors for Gene Expression in Bacteria

Rahmi Lale, Trygve Brautaset, and Svein Valla

Abstract

This chapter provides methods and insights into the use of broad-host-range plasmid vectors useful for expression of genes in a variety of bacteria. The main focus is on IncQ, IncW, IncP, and pBBR1-based plasmids which have all been used for such applications. The specific design of each vector is adapted to its use, and here we describe, as an example, a protocol for construction (in *Escherichia coli*) of large insert DNA libraries in an IncP type vector and transfer of the library to the desired host. The genes of interest will in this case have to be expressed from their own promoters and the libraries will be screened by a method that best fits the functions of the gene or gene clusters searched for.

Key words: Broad-host-range, RK2, Plasmid, XylS/*Pm*, Conjugation

1. Introduction

Plasmids represent one of the most important biological tools used in recombinant DNA technology by serving as vectors for DNA cloning, gene modification, and delivery and expression of genes in bacteria. They can be easily isolated, genetically modified, and transferred into bacterial cells in modified versions to construct recombinant cells with desired new functions. Most plasmids used in biotechnology are double-stranded, circular, extra-chromosomal DNA elements that can replicate in a cell independent from the host chromosome. Naturally occurring plasmids vary in size from less than 1 to more than a 1,000-kb, and are generally dispensable. Such plasmids are virtually never used directly in biotechnology, but are modified to facilitate all the varying and specialised needs associated with this technology.

James A. Williams (ed.), *Strain Engineering: Methods and Protocols*, Methods in Molecular Biology, vol. 765,
DOI 10.1007/978-1-61779-197-0_19, © Springer Science+Business Media, LLC 2011

Plasmids typically encode only a few of the proteins needed for their presence in a host, therefore their maintenance heavily depended on the availability of other proteins, such as DNA polymerase, DNA ligase, DNA helicase, and others, expressed from chromosomal genes. Each plasmid minimally possesses an origin of vegetative replication (*ori*) and they often encode essential replication initiation protein(s). Note that the widely used ColE1-type plasmids do not encode such a protein. Naturally occurring plasmids often encode additional genes that are beneficial for the host under certain conditions, such as genes conferring antibiotic resistance and genes encoding enzymes for degradation of carbon sources that are not commonly distributed in nature. On the other hand, many naturally occurring plasmids have not yet been ascribed biological functions and are denoted as cryptic (1).

Plasmid copy number varies widely and is determined by the specific copy number control system associated with each plasmid type. Each such system determines a particular number of copies of the corresponding plasmid per cell or per host chromosome, varying from, for example one to two (low copy number plasmids) or up to several hundred (high copy number plasmids). The copy number is a parameter of profound importance relative to their use as vectors for gene expression, and it is common to introduce mutations in the copy number control system to modify the vectors for specific needs in biotechnology (2).

Plasmids can be categorised based on several features they possess, such as the mode of replication mechanism they use (theta or rolling circle) or whether they are transferable (mobile) or not. Some naturally occurring plasmids are self-transmissible (conjugative), while others are mobilizable only. Of particular relevance for this chapter is that some plasmids can replicate in more than one host (broad-host-range plasmids), while others display a narrow-host-range. Plasmids may also be categorised with respect to the incompatibility group they belong to (IncQ, IncP, IncW, IncN, IncC, IncU, etc.), and they are considered incompatible (belong to the same Inc group) if they cannot coexist stably in a host cell, reflecting that their replication systems are the same or similar (3, 4).

Some plasmids can carry large heterologous DNA fragments ranging up to 100 kb and above, and are useful for creating metagenomics libraries and for cloning of large gene clusters encoding complex metabolic pathways. If plasmids are to be used for high level protein expression a suitable promoter can be inserted to achieve efficient transcription of the gene of interest. A variety of promoters with different expression characteristics are available for such purposes, and the selection of promoter systems heavily depends on the host as well. For many purposes, the possibility of regulating the recombinant expression level is favourable, and accordingly inducible promoter systems are popular. A selection of

Table 1
Examples of bacteria in which broad-host-range plasmids with regulatable promoter systems have been used

Promoter system	Inducer	Replicon	Species	References
P_{tac}, lacI[Q]	IPTG	IncQ	*Escherichia coli*, *Pseudomonas putida*	(26)
P_{trc} (with the lac repressor)	IPTG	IncQ	*E. coli*, *Synechocystis*, *Nostoc* 7120, *N. punctiforme*	(27)
P_{l} (with the temperature-sensitive Cl repressor)	Heat	pBBR1 based	*E. coli*, *P. aeruginosa*	(28)
P_{alkB}	*n*-Alkanes	pKKPalk[a]	*E. coli*, *P. putida GPo12* and *P. fluorescens KOB2D1*	(29)
P_{BAD}	Arabinose	pBBR1	*Escherichia coli*, *P. aeruginosa*, *P. putida*, *Burkholderia cepacia*, *Xanthomonas campestris pv. phaseoli*	(11, 30–32)
P_{tac}	IPTG	IncQ	*E. coli*, *P. aeruginosa*, *Edwardsiella tarda*, *Vibrio* (*Listonella*) *anguillarum*	(33)
*Plac*UV5	IPTG	pBBR1 based	*E. coli*, *Rhodobacter capsulatus*	(34)
rhaP$_{BAD}$	Rhamnose	pBBR1 based	*E. coli*, *P. putida* KT2440	(35)
Pm	Benzoic acid derivatives	IncP	*Agrobacterium tumefaciens*, *Azotobacter vinelandii*, *E. coli*, *P. aeruginosa*, *P. denitrificans*, *P. fluorescens*, *P. putida*, *X. campestris*	(36–38)

[a]This group of plasmids have not been discussed in this chapter

regulatable promoter systems used in broad-host-range replicons are listed in Table 1. Typically, a DNA region consisting of multiple unique cloning sites is positioned close to the promoter enabling insertion of foreign gene or gene cluster downstream and under transcriptional control of the promoter in the vector. An alternative method used for recombinant gene expression is to apply plasmids that integrate themselves into the host chromosome. For such purposes, the so-called "suicide plasmids" are typically used as they cannot replicate autonomously in the relevant host. Chromosomal integration can be favourable when a low recombinant gene dosage is desirable, to reduce background expression (in the absence of inducer) or when problems with plasmid segregative instability are experienced (5). For high level protein production, the disadvantage of using chromosomal insertions is very often that sufficiently high expression levels may depend on a high gene dosage.

Different methods are available for transfer of plasmids to bacterial cells, including transformation of naked DNA to naturally or chemically induced competent cells, transduction, electroporation, conjugation, sonoporation, etc. (6). Plasmid-mediated antibiotic resistance is the dominating marker used to select bacterial cells that have received and can stably maintain the vector. Among the different transfer methods chemical transformation may be technically simplest, but often lead to poor transformation frequencies in many species, while electroporation methods can frequently overcome such problems. However, bacterial conjugation is often the most efficient process (leading to highest frequencies of transfer), and the DNA is then transferred from donor to recipient bacteria by a specialised multi-protein complex, termed the "conjugation apparatus." The corresponding genes have usually been integrated into the chromosome of the donor strain as the many genes required for conjugation would make the vector excessively large. Alternatively, the conjugation apparatus is located on a separate "helper plasmid" which can either be in a third strain or in the donor strain used in a conjugation, method referred as "tri-parental mating" (7). In both cases, the conjugation process requires a specific origin of transfer (often denoted as *ori*T) in the vector to be transferred.

Broad-host-range or promiscuous plasmids are a subgroup of all plasmids, and many different types have been reported. However, only a fraction of these have actually been heavily used in biotechnology, and in the following we, therefore, focus on four groups that are frequently cited in the scientific literature, IncQ, IncW, IncP, and pBBR1-based plasmids. All of them replicate in *E. coli*, in which most DNA manipulations are usually carried out. The biological reasons for their unique promiscuous properties are poorly understood.

IncQ plasmids are characterised by their small size, medium range copy number, and a very wide host-range, including many Gram-positive bacteria. The best known (natural isolates) are RSF1010, R300B, and R1162, which are considered to be similar if not identical (8). Numerous vectors have been developed from these backbones and they are heavily used, for example in *Pseudomonas* species. This group of plasmids are not naturally self-transmissible but they can be transferred to different bacterial species if a conjugative helper-plasmid is provided in *trans*. The plasmids require three plasmid-encoded proteins for replication (RepA-C) in addition to the vegetative replication origin.

The IncW promiscuous group have members with self- and non-transmissible features. The group is known to include the smallest known conjugative plasmids (9), and the most thoroughly studied members are pSa and pR388 (10). Their replicons consist of a gene encoding replication protein RepA, in addition to the vegetative origin of replication.

pBBR1-based plasmids are relatively small, and mobilizable (when the transfer functions are provided in *trans*). They encode a replication protein (Rep) that is essential for stable plasmid maintenance, even in the absence of antibiotic selection. Based on compatibility studies pBBR1 is found not to be a member of the broad-host-range IncC, IncP, IncQ, or IncW plasmids, and it may therefore represent a novel incompatibility group (11).

The naturally occurring IncP group plasmids are self-transmissible and relatively large. They are divided into four distinct subgroups: α, β, γ, and δ (12). Plasmids belonging to IncP have been shown to replicate in many Gram-negative bacterial species (Table 2) and are considered as the most promiscuous (for Gram-negative bacteria) of all plasmids known to-date (13).

Table 2
List of species in which extensively used broad-host-range plasmids are known to replicate

Plasmid group	Species	References
IncP	*Achromobacter parvulus*, *Acinetobacter* spp., *Aeromonas* spp., *Agrobacterium* spp., *Alcaligenes* spp., *Aliivibrio salmonicida*, *Anabaena* spp., *Azospirillum braziliense*, *Azotobacter* spp., *Bordetella* spp., *Caulobacter* spp., Enterobacteriaceae, *Gluconacetobacter xylinus*, *Haemophilus influenzae*, *Hypomycrobium* X, *Legionella pneumophila*, *Methylophilus methyltrophus*, *Methylococcus methanolicus*, *Methylosinus trichosporium*, *Myxococcus xanthus*, *Neisseria* spp., *Paracoccous denitrificans*, *Pseudomonas* spp., *Rhizobium* spp., *Rhodopseudomonas* spp., *Rhodospirillum* spp., *Shewanella* spp., *Thiobacillus* spp., *Xanthomonas campestris*	((14), unpublished data)
IncQ	*Acinetobacter calcoaceticus*, *Actinobacillus pleuropneumoniae*, *Actinomyces naeslundii*, *A. viscosus*, *Aerobacter aerogenes*, *Aeromonas hydrophila*, *Agrobacterium tumefaciens*, *Alcaligenes eutrophus*, *Azotobacter vinelandii*, *Brevibacterium methylicum*, *Caulobacter crescentus*, *Desulfovibrio vulgaris*, *Erwinia carotovora*, *E. chrysanthemi*, *Escherichia coli*, *Gluconacetobacter xylinus*, *Gluconobacter* spp., *Hyphomicrobium* spp., *Klebsiella aerogenes*, *K. pneumoniae*, *Methylophilus methylotrophus*, *Moraxella* spp., *Mycobacterium aurum*, *M. smegmatis*, *Paracoccus denitrificans*, *Pasteurella multocida*, *Porphyromonas gingivalis*, *Proteus mirabilis*, *Providencia* spp., *Pseudomonas* spp., *Rhizobium leguminosarum*, *R. meliloti*, *Rhodobacter sphaeroides*, *R. capsulatus*, *Rhodopseudomonas spheroides*, *Salmonella* spp., *Serratia marcescens*, *Streptomyces lividans*, *Synechococcus* spp., *Thiobacillus ferrooxidans*, *Vibrio salmonicida*, *Yersinia enterocolitica*, *Xanthomonas campestris*, *X. maltophilia*	(39–42)

(continued)

Table 2 (continued)

Plasmid group	Species	References
IncW	*Acinetobacter calcoaceticus*, *Aeromonas liquefaciens*, *A. salmonicida*, *Agrobacterium tumefaciens*, *A. rhizogenes*, *Alcaligenes eutrophus*, *Enterobacter* sp., *Erwinia amylovora*, *E. carotovora* subsp. *Carotovora*, *E. herbicola*, *E. rubrifaciens*, *E. stewartii*, *Escherichia coli*, *Klebsiella* spp., *Legionella pneumophila*, *Methylophilus methylotrophus*, *Myxococcus virescens*, *M. xanthus*, *Proteus rettgeri*, *P. mirabilis*, *Providencia stuartii*, *Pseudomonas* spp., *Rhizobium leguminosarum*, *R. trifolii*, *Salmonella enteritidis*, *S. typhimurium*, *S. ordonez*, *Serratia marcescens*, *Shigella* spp., *Vibrio cholerae*, *Xanthomonas campestris* pv. *campestris*, *X. campestris* pv. *malvacearum*, *Zymomonas mobilis*	(10, 43)
pBBR1 based	*Alcaligenes eutrophus*, *Bartonella bacilliformis*, *Bordetella* spp., *Brucella* spp., *Caulobacter crescentus*, *Escherichia coli*, *Gluconacetobacter xylinus*, *Paracoccous denitrificans*, *Pseudomonas fluorescens*, *P. putida*, *Rhizobium meliloti*, *R. leguminosarum* by. *viciae*, *Rhodobacter sphaeroides*, *Salmonella typhimurium*, *Vibrio cholerae*, *Xanthomonas campestris*	(44, 45)

The list is not exhaustive

These replicons require only *ori*V and the replication initiation protein TrfA in order to replicate in a bacterial cell. Systems helpful for increasing stability are also known for this plasmid group. Plasmid RK2, the best characterised member of the IncP group, harbours a well-defined origin of conjugative transfer (*ori*T), and small size replicons containing this origin have been demonstrated to be conjugatively transferable to and replicate in numerous Gram-negative bacterial species (14). They have also been reported to be transferable to Gram-positive bacteria, yeast, and even mammalian cells (15–17).

Five plasmids, R18, R68, RK2, RP1, and RP4 are commonly presented as separate members of the IncP group but could not be differentiated by restriction profile analysis nor by heteroduplex experiments (18, 19), and these are now referred to as the core "Birmingham" IncPα plasmids (13). Many IncP-derived vectors have been reported to replicate in a wide range of bacterial hosts, and numerous specialised vectors of this type have been reported in the scientific literature.

2. Materials

1. *Escherichia coli* strains: EPI300 cells harbouring the plasmid pRS44, S17-1 (ATCC 47055) cells harbouring the plasmid pRS48, EPI300-T1^{R} cells (Epicentre), chemically competent S17-1 (cell lines are available from the authors upon request).

2. Electrocompetent *Pseudomonas fluorescens.*
3. Luria-Bertani (LB) broth and agar: 10 g/L tryptone, 5 g/L yeast extract, 5 g/L NaCl. For agar plates, add 20 g/L agar. Autoclave to sterilise.
4. LB agar + antibiotics: add antibiotics kanamycin or tetracycline as indicated in each case to make LB + kanamycin and LB + tetracycline plates.
5. SOC broth: 20 g/L tryptone, 5 g/L yeast extract, 2 mL of 5 M NaCl, 2.5 mL of 1 M KCl, 10 mL of 1 M $MgCl_2$, 10 mL of 1 M $MgSO_4$, 20 mL of 1 M glucose (autoclave).
6. *Pseudomonas* isolation agar + antibiotics (PIA): prepared according to the manufacturer's (Difco) instructions. Add antibiotics kanamycin or tetracycline as indicated in each case to make PIA + kanamycin and PIA + tetracycline plates.
7. Antibiotics: Kanamycin (stock: 50 mg/mL in distilled water, sterile filter), tetracycline (stock: 10 mg/mL in 70% ethanol) used at the final concentrations 50, and 10 μg/mL, respectively.
8. Enzymes: *Bam*HI, *Sau*3AI, Calf Intestinal Alkaline Phosphatase (CIP), T4 DNA Ligase (400,000 U/μL) (New England Biolabs [NEB]), Fast-Link DNA Ligase (Epicentre).
9. Enzyme buffers: 10× NEB 1 and 3, 10× T4 DNA Ligase buffer (NEB), 10× Fast-Link Ligation Buffer (Epicentre).
10. 10× BSA (NEB).
11. 10 mM ATP (Roche).
12. 6× DNA loading buffer (3 mL glycerol, 25 mg bromophenol blue, distilled water to 10 mL).
13. Lambda Mix DNA Marker, 19 (Fermentas).
14. 1 mM L-arabinose.
15. 0.5 mM EDTA.
16. Electroporation cuvettes (0.2 cm).
17. 60% Glycerol solution (autoclave).
18. 0.6% Agarose gel (both standard and low melting point agarose) in 1× TBE buffer (2 mM EDTA [pH 8], 89 mM Tris, 89 mM boric acid).
19. Kits: QIAquick Gel Extraction kit (Qiagen), Wizard Plus Miniprep kit (Promega), GELase Agarose Gel-Digesting Preparation kit (Epicentre), Copy Control Fosmid Library Production kit (Epicentre).
20. Lab equipments: spectrophotometer, centrifuge, incubator, thermal block, gel electrophoresis unit and gel imaging system, Gene Pulser (Bio-Rad), shaking incubator.

3. Methods

This section describes one specific and recently developed protocol for the expression of genes in Gram-negative bacteria by use of a particular type of RK2-based (IncP) vector (pRS44, (20)) recently constructed in our group. This vector does not contain an inserted promoter to express the target genes, as the goal is to use the vector for functional screening of metagenomics libraries in which the nature of the inserts is unknown. However, we want to emphasise that we have also constructed numerous vectors based on the same replicon that can be used for high-level expression of specific target genes, using the *xylS/Pm* expression cassette. For this particular purpose, the vectors have so far been applied mainly in *E. coli*, but efforts to expand their applications to other species are ongoing.

In the protocol described below, we use pRS44 (Fig. 1) for insertion of foreign DNA which in principle may originate from any source, although we have so far mainly used it to construct libraries of DNA isolated directly from the environment. pRS44 is based on the commercially available vector pCC1FOS (Epicentre), and in addition it carries the *parDE* element from RK2, which is known to enhance segregative stability of RK2 replicons across species barriers (21, 22). It also contains the origin of conjugative transfer, *ori*T; enabling transfer of libraries to numerous hosts. The kanamycin resistance gene is used for selection in this particular vector. pRS44 contains two origins of replication, and one of them (the origin from plasmid F) is active in *E. coli* in which the library is constructed. It also contains the origin of vegetative replication (*ori*V) from RK2, but this origin is not active unless TrfA is provided in *trans*. Thus, during library construction and maintenance the library is kept under control of the low copy number F origin, a feature that may be important to ensure the stability of very large insert libraries. The F origin is narrow in its host-range, but after conjugative transfer to a non-*E. coli* host control of replication is switched to the broad-host-range RK2 origin. This is achieved by inserting the *trfA* gene into the chromosome of the recipient host prior to conjugative transfer of the library. For this purpose, a suicide vector pRS48 (Fig. 2) with *trfA* in a transoposon is used (20).

In addition to its function as a replication initiation protein, TrfA also controls RK2 replicon copy numbers and various *trfA* mutations and specify different copy numbers across species barriers. The copy number of the library plasmids in the foreign host can therefore be varied, depending on what particular *trfA* variant is inserted into the chromosome. This may be of significant importance in relation to expression and detection of new gene

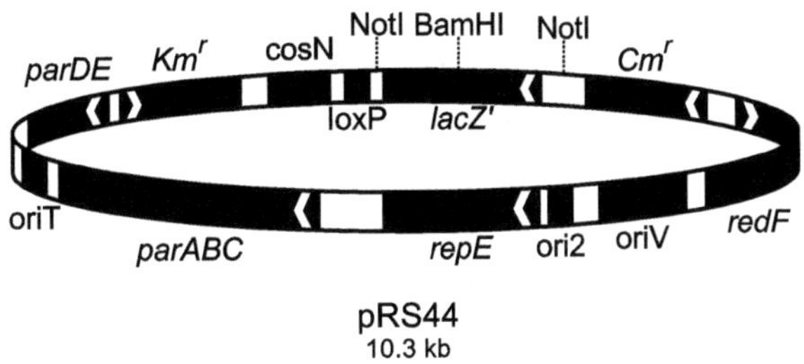

Fig. 1. Map of plasmid pRS44. pRS44 can replicate as a single copy replicon via *ori2* and *repE*, while *oriV* contributes to a medium copy number if its replication initiation protein TrfA is expressed in the same cell. pRS44 DNA can easily be prepared in large quantities in the *E. coli* strain EPI300 by expressing a mutant *trfA*-gene from an arabinose-induced promoter, as described by (46). *Bam*HI site can be used for introducing *Sau*3AI digested DNA. NotI sites can be used for insert size determination. *cosN* the site used in packaging of the environmental DNA library in bacteriophage lambda, *Cm* chloramphenicol resistance gene, *Km* kanamycin resistance gene, *oriT* origin of conjugative transfer, *lacZ* gene encoding beta-galactosidase, *loxP* bacteriophage P1 site for Cre-recombinase cleavage, *parDE* and *parABC* stabilisation elements from the RK2 plasmid.

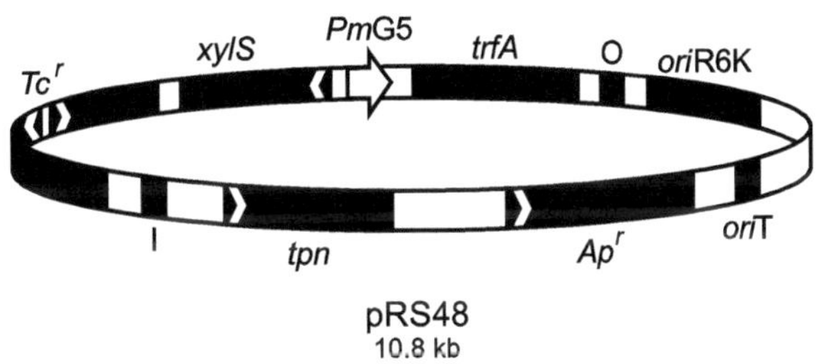

Fig. 2. Map of plasmid pRS48. The *trfA*-gene can be inserted into the chromosome of hosts of interest by the transposon present in this narrow-host-range plasmid. The inside and the outside ends of the transposon are marked I and O, respectively. *tpn* gene encoding the transposase, which is not a part of the transposon, *xylS* gene encoding activator of *PmG5* transcription in the presence of benzoic acid-type inducers, like *m*-toluate, *PmG5 Pm* promoter variant, *oriT* origin of conjugative transfer, *oriR6K* origin of replication, *oriT* origin of transfer, *Ap* ampicillin resistance gene, *Tc* tetracycline resistance gene.

functions in metagenomics and other applications, as has already been observed in our laboratory in experiments with *E. coli* (Aakvik, unpublished).

3.1. Construction of a DNA Library in Plasmid pRS44

The DNA to be cloned may originate from, in principal, any source and we have chosen not to describe how to isolate it as the protocol will be heavily dependent on the source. For environmental DNA there are typically problems with contaminating compounds that must be resolved (see, e.g.(23)).

3.1.1. Preparation of Vector DNA (pRS44) for Cloning

1. Streak *E. coli* strain EPI300 harbouring pRS44 from a frozen stock on a LB-agar plate containing kanamycin and incubate overnight at 37°C.

2. Pick a single colony with a sterile loop and inoculate 10 mL LB-broth with kanamycin in a 125-mL flask.
3. Add 1 mM L-arabinose to switch from single to high copy number. The basis for this is that the *E. coli* EPI300 strain carries a chromosomally integrated high copy variant of the *trfA* gene under the control of the AraC/P_{BAD} promoter, which can be induced with L-arabinose. The induction will result in the production of variant TrfA replication protein which subsequently activates replication from the *ori*V from single copy to high copy number, facilitating preparation of large quantities of vector DNA.
4. Grow cells overnight (16–20 h) with shaking (225–250 rpm) at 37°C.
5. Purify plasmid DNA with a Wizard Plus Miniprep kit by following the manufacturer's protocol.
6. Linearise 1 μg of pRS44 by digesting with BamHI. 1 μg of DNA is digested with 1 units of BamHI in 30 μL of 1× NEB 3 buffer containing 1× BSA and sterile distilled water, and incubated for 1 h at 37°C (see Note 1).
7. Dephosphorylate the linearised DNA by treatment with calf intestine alkaline phosphatase (CIP) to prevent self-circularisation. Add 0.5 u/μg CIP and incubate 1 h at 37°C.
8. Purify DNA by gel purification by QIAquick Gel Extraction kit by following the manufacturer's protocol.

3.1.2. Isolation and Purification of DNA to Be Cloned

9. Purify DNA from the desired source (see Note 2).
10. Carry out small-scale Sau3AI partial digestions with the purified DNA to determine the optimum conditions for degradation to the size 35–40 kb (see Note 3).
11. Prepare a 0.6% agarose gel in 1× TBE buffer.
12. Add 2 μL DNA loading buffer to each tube and load 20 μL of each sample to the wells. Load DNA marker (e.g., Lambda Mix Marker, 19) to the right- or left-most well (see the details below) to estimate the size of the digested DNA.
13. Run electrophoresis of the digested DNA samples at 5 V/cm until the bromophenol blue dye reaches the bottom of the gel.
14. Stain the gel and inspect under UV-light.
15. Define the optimum conditions to obtain the desired DNA size and carry out large-scale digestion of DNA in the amounts required. It is important to scale up volumes only, so that no concentrations are changed. A rule of thumb is that in the optimal sample the highest band intensity will be at twice the desired size (70–80 kb).

16. Run electrophoresis of the large-scale digested DNA with 0.6% LMP agarose. Ensure that the gel does not warm up during the run as this may melt the agarose.
17. Cut out the lane where the DNA marker is and stain it (e.g., ethidium bromide, SYBR safe or gelred, etc.).
18. Visualise the cut out lane with a ruler next to it. Note the run length of the area of interest and cut out the corresponding area from the rest of the unstained gel (see Note 4). Stain the remaining of the unstained gel to ensure that the DNA has been excised properly. It may also be wise to allow a small part of the sample to be stained to make sure digestion has worked as expected.
19. Purify the DNA from the excised agarose gel using the GELase kit by following manufacturer's instructions.
20. Determine the concentration of the purified DNA spectrophotometrically.

3.1.3. Construction of a DNA Library in E. coli

In this specific protocol described here the DNA is cloned by *in vitro* packaging in bacteriophage lambda. This is because transformation efficiencies are low for plasmids with big inserts and it is therefore easier to obtain many clones by lambda infection which is virtually 100% efficient.

21. Ligate the size-selected DNA to linearised pRS44 vector at a 10:1 molar ratio, respectively. Combine the following reagents in the order listed and mix thoroughly after each addition. Sterile water, 1 μL 10× Fast-Link Ligation Buffer, 1 μL 10 mM ATP, linearised pRS44 (0.5 μg/μL), insert DNA (0.25 μg of ≈40 kb DNA), 1 μL Fast-Link DNA Ligase, total reaction volume should be 10 μL (see Note 5). Incubate at 16°C for 4 h. Transfer the reaction to 65°C for 10 min to heat-inactivate the DNA ligase.
22. Package the ligation mixture (10 μL) and plate on EPI300-$T1^R$ plating cells by following the protocol of the Copy Control Fosmid Library Production kit.
23. Add ~2 mL of LB-broth to the plate with overnight grown infected EPI300-$T1^R$ cells, scoop, and resuspend them (see Note 6).
24. Transfer the cell suspension to the next plate (if more than one overnight plate was used) and repeat the resuspension process. Do this for as many plates as desired.
25. Transfer the final resuspension to a sterile tube and add sterile glycerol to a final concentration of 20%. Mix the solution and store aliquots of 100 μL (which would each constitute a library of the desired coverage) at −80°C.

3.2. Transfer of the DNA Library to Non-E. coli Hosts

Before the library can be transferred to a non-*E. coli* host the *trfA* gene has to be introduced to the chromosome of this host by transposon insertion. pRS48 carries the *trfA* gene and the plasmid is transferred by electroporation. Upon successful introduction and expression of the *trfA* gene the library can then be conjugally transferred. pRS48 is based on the plasmid mini-Tn5 Km (24). It harbours the replication origin *ori*R6K, whose function depends on the Pir protein for replication (24). pRS48 replicates in *E. coli* strain S17-1 λ *pir* which expresses the Pir protein chromosomally. The strain S17-1 λ *pir* also carries the RK2 *tra* genes in its chromosome, allowing this strain to be used as donor in conjugation. The expression of the *trfA* gene is under the control of the benzoic acid derivatives inducible positive regulator/promoter XylS/*Pm* system which has been shown to be active in several Gram-negative bacterial species (Table 1). Therefore, it is important to ensure that in the selected host for DNA library expression, the XylS/*Pm* system is functional.

3.2.1. Isolation of Plasmid pRS48

1. Isolation of pRS48 follows the same steps as the protocol described above for pRS44 with the exception that pRS48 encodes for tetracycline resistance thus cells should be inoculated to medium with tetracycline.

3.2.2. Transposon Insertion of the trfA Gene to Pseudomonas fluorescens by Electroporation

2. Place electroporation cuvettes, cells, and pRS48 on ice for at least 30 min prior to electroporation. Mix 1 μL (≈200 ng) pRS48 plasmid with 40 μL electrocompetent *P. fluorescens* cells. Mix gently. Set the electroporation device based on the manufacturer's guidelines for *Pseudomonas* (see Note 7).
3. Transfer the mixture to a 0.2-cm cuvette and tap gently to ensure the even distribution of the sample. Place the cuvette into the electroporation chamber. Avoid touching to the metal sides of the cuvette while handling.
4. Pulse once.
5. Add 900 μL of pre-warmth SOC medium immediately after electroporation and transfer the mixture to a test tube and incubate for 1 h at 30°C with shaking.
6. Transfer the cells on to PIA with tetracycline, and incubate at 30°C overnight.
7. Pick several transformants and inoculate for LB-broth with tetracycline for overnight growth and save as stock in 20% glycerol at –80°C.

3.2.3. Introduction of the DNA Library to E. coli S17-1 by Heat-Shock Transformation

The *E. coli* strain S17-1 harbouring the pRS44 DNA insert library will serve as a donor in conjugation. Therefore, the library has to be introduced into this host. The library construction could also be performed in S17-1 directly. However, the construction procedure is more efficient in the strain used for this purpose (see above), and since library construction is the critical step to obtain a library with the highest possible number of clones, the two-step procedure tends to give better results. Transformation to S17-1 is not limiting because excessive amounts of library DNA is available after library construction has been completed.

8. Inoculate 1% DNA library (100 μL aliquot from step 25 in Subheading 3.1) to 10 mL of kanamycin-supplemented LB-broth in a 125-mL flask.
9. Grow cells for 2 h with shaking (225–250 rpm) at 37°C.
10. Purify plasmid DNA with the Wizard Plus Miniprep kit by following the manufacturer's protocol.
11. Thaw an aliquot of frozen competent *E. coli* S17-1 cells on ice.
12. Add 1 μL of plasmid DNA to thawed cells and mix gently by tapping the tube with your finger. Return the tubes to ice (see Note 8).
13. Incubate the DNA/cell mixture on ice for 10 min.
14. Heat shock the cells in a 42°C water bath for 40 s, and return to ice.
15. Add 900 μL of SOC broth and incubate cells for 1 h at 37°C with shaking (225–250 rpm).
16. Plate the cells on LB-agar with kanamycin and incubate at 37°C overnight.
17. Scoop the transformants (see Note 6) and save 100 μL (for subsequent use, see below). Store the rest as stock in 20% glycerol at −80°C.

3.2.4. Conjugation

For conjugation, the donor and recipient strains have to be grown to early exponential phase without selection as transfer efficiencies have been found to be highest with such cells.

18. Inoculate 10 mL LB-broth without antibiotic with *P. fluorescens* in a 125-mL flask and incubate overnight at 30°C with shaking (225–250 rpm).
19. Inoculate 1% from the overnight grown *P. fluorescens* in a 10-mL LB-broth without antibiotic in a 125-mL flask and grow the cells to early exponential phase (OD_{540} 0.3–0.5) for about 4 h.
20. Inoculate *E. coli* S17-1 library (100 μL from step 17) to 10 mL LB-broth without antibiotic in a 125-mL flask and grow for 4 h.
21. Mix 2 mL culture from donor and recipient in a tube and centrifuge, 4,000 × *g* for 5 min.

22. Discard the supernatant and resuspend the cells in 100 μL of LB-broth and spot on an LB-agar plate without selection (see Note 9).
23. Incubate at 30°C overnight.
24. The next day scoop most of the bacteria from the mating spot with a sterile inoculation loop and resuspend in 1 mL LB-broth. This is the undiluted bacterial conjugation mixture. Make serial dilutions and plate out 100 μL [10^{-2}, 10^{-4}, and 10^{-6}] on PIA agar with kanamycin for selection. Incubate at 30°C for up to 48 h.
25. Transconjugants may be saved as stock in 20% glycerol at -80°C.

4. Notes

1. DNA fragments obtained after partial Sau3AI digestion can be ligated into BamHI restriction sites of the vector, as these two restriction enzymes generates compatible ends after digestion.
2. As indicated above DNA isolation procedures will not be described here as they are heavily dependent on the source. However, there are comprehensive method papers that can be followed if necessary (e.g. (23)).
3. Sau3AI recognises the four base sequence GATC in DNA, meaning that it typically generates very short DNA fragments (e.g. average 250 bp) if digestion is allowed to proceed to completion. Partial digestion can, however, be used to generate fragments of the desired size for this protocol, provided the original DNA molecules are sufficiently large (e.g. exceeding 100 kb). As the frequency of Sau3AI restriction sites (GC-content may vary) and concentration and size of the source DNA will vary there cannot be a fixed recipe for a partial digestion. Therefore, a small-scale test digestion has to be performed initially in order to determine the amount of enzyme needed to obtain the desired size of the partially digested DNA. The following protocol should be considered as a general guideline for partial digestion.

   Begin with 200 μL of DNA preparation (>0.1 μg/μL). Label ten Eppendorf tube vials 1–10. Add the following ingredients to the DNA solution: 20 μL 10× concentrated NEB buffer 1, 2 μL BSA solution (10 mg/mL), and mix. From this mixture dispense 40 μL to tube 1, and 20 μL to tubes 2–10. Ensure that from this point until incubation all tubes are kept on ice. Add 1 μL of Sau3AI at 10 μ/μL to tube 1 and mix. Transfer 20 μL from tube 1 and transfer to tube 2 and mix.

Repeat this process up to tube 9. Do not transfer to tube 10 as this will serve as a non-digested control. Incubate all ten tubes for 1 h at 37°C. Stop the digestion by adding 1 μL of 0.5 mM EDTA (modified protocol described in (25)). If small amounts of DNA is available the protocol may be scaled further down, but note that the digested DNA needs to be visualised after electrophoresis (see Subheading 3.1.2).

4. To avoid DNA damage the sample to be used is not subjected to ethidium bromide staining or UV-light exposure.
5. The number of clones obtained will vary depending on the quality of the insert DNA. Therefore, the number of ligation reactions can be increased and the ligation reaction can be scaled if necessary.
6. In order to collect all the cells on the plate add sterile liquid broth to the plate and tilt it towards yourself so that the liquid can be pooled at the bottom. By using a sterile glass rod scoop all the colonies and bring it to the pooled liquid. If the colonies stick to the glass rod they can be removed by a sterile pipette tip.
7. Depending on the electroporation device used the set-ups may vary, please consult the instrument manuals for the settings. The following conditions can be used for the Bio-Rad Gene Pulser (0.2 cm cuvette, 12.5 kV/cm, 100 Ω, 25 μF) (25).
8. Excess mixing and pipetting of the cells decrease transformation efficiency. The volume of cells used should be at least 20 times that of the DNA.
9. It is important to note the spread of the cells on the agar plate and just let the drop soak in before incubation.

References

1. Primrose S. B. and Twyman R. M. (2006) 'Basic biology of plasmid and phage vectors', *Principles of gene manipulation* (7th edn: Blackwell), 55–74.
2. del Solar G. and Espinosa M. (2000) Plasmid copy number control: an ever-growing story. *Mol. Microbiol.* **37**, 492–500.
3. Datta N. (1979) 'Plasmid classification: incompatibility grouping', in K. Timmis and A. Puhler (eds.), *Plasmids of Medical, Environmental and Commercial Importance* (Amsterdam: Elsevier/North Holland), 3–12.
4. Novick R. P. (1987) Plasmid incompatibility. *Microbiol. Rev.* **51**, 381–95.
5. Imanaka T. and Aiba S. (1981) A perspective on the application of genetic engineering: stability of recombinant plasmid. *Ann. N.Y. Acad. Sci.* **369**, 1–14.
6. Aune T. E. V. and Aachmann F. L. (2010) Methodologies to increase the transformation efficiencies and the range of bacteria that can be transformed. *Appl. Microbiol. Biotechnol.* **85**, 1301–13.
7. Figurski D. H. and Helinski D. R. (1979) Replication of an origin-containing derivative of plasmid RK2 dependent on a plasmid function provided in trans. *Proc. Natl. Acad. Sci. U. S. A.* **76**, 1648–52.
8. Haring V. and Scherzinger E. (1989) 'Replication proteins of the IncQ plasmid RSF1010', in C. M. Thomas (ed.), *Promiscuous Plasmids of Gram-Negative Bacteria* (Academic Press, London, United Kingdom.), 95–124.
9. Fernández-López R., et al. (2006) Dynamics of the IncW genetic backbone imply general trends in conjugative plasmid evolution. *FEMS Microbiol. Rev.* **30**, 942–66.

10. Valentine C. R. I. and Kado C. I. (1989) 'Moelcular Genetics of IncW plasmids', in C. M. Thomas (ed.), *Promiscuous Plasmids of Gram-Negative Bacteria* (Academic Press, London, United Kingdom.), 125–63.
11. Lefebre M. D. and Valvano M. A. (2002) Construction and evaluation of plasmid vectors optimized for constitutive and regulated gene expression in *Burkholderia cepacia* complex isolates. *Appl. Environ. Microbiol.* **68**, 5956–64.
12. Vedler E., Vahter M., and Heinaru A. (2004) The completely sequenced plasmid pEST4011 contains a novel IncP1 backbone and a catabolic transposon harboring *tfd* genes for 2,4-dichlorophenoxyacetic acid degradation. *J. Bacteriol.* **186**, 7161–74.
13. Schluter A., Szczepanowski R., Puhler A., and Top E. M. (2007) Genomics of IncP-1 antibiotic resistance plasmids isolated from wastewater treatment plants provides evidence for a widely accessible drug resistance gene pool. *FEMS Microbiol. Rev.* **31**, 449–77.
14. Thomas C. M. and Helinski D. R. (1989) 'Vegetative replication and stable inheritance of IncP plasmids', in C. M. Thomas (ed.), *Promiscuous Plasmids of Gram-Negative Bacteria* (Academic Press, London, United Kingdom.), 1–25.
15. Bates S., Cashmore A. M., and Wilkins B. M. (1998) IncP plasmids are unusually effective in mediating conjugation of *Escherichia coli* and *Saccharomyces cerevisiae*: involvement of the *tra2* mating system. *J. Bacteriol.* **180**, 6538–43.
16. Poyart C. and Trieu-Cuot P. (1997) A broad-host-range mobilizable shuttle vector for the construction of transcriptional fusions to beta-galactosidase in gram-positive bacteria. *FEMS Microbiol. Lett.* **156**, 193–98.
17. Waters V. L. (2001) Conjugation between bacterial and mammalian cells. *Nat. Genet.* **29**, 375–76.
18. Burkardt H. J., Riess G., and Puhler A. (1979) Relationship of group P1 plasmids revealed by heteroduplex experiments: RP1, RP4, R68 and RK2 are identical. *J. Gen. Microbiol.* **114**, 341–8.
19. Currier T. C. and Morgan M. K. (1981), Restriction endonuclease analyses of the incompatibility group P-1 plasmids RK2, RP1, RP4, R68, and R68.45. *Curr. Microbiol.* **5**, 323–27.
20. Aakvik T., et al. (2009) A plasmid RK2-based broad-host-range cloning vector useful for transfer of metagenomic libraries to a variety of bacterial species. *FEMS Microbiol. Lett.* **296**, 149–58.
21. Blatny J. M., Brautaset T., Winther-Larsen H. C., Karunakaran P., and Valla S. (1997) Improved broad-host-range RK2 vectors useful for high and low regulated gene expression levels in gram-negative bacteria. *Plasmid* **38**, 35–51.
22. Sia E. A., Roberts R. C., Easter C., Helinski D. R., and Figurski D. H. (1995) Different relative importances of the *par* operons and the effect of conjugal transfer on the maintenance of intact promiscuous plasmid RK2. *J. Bacteriol.* **177**, 2789–97.
23. Liles M. R., et al. (2009) Isolation and cloning of high-molecular-weight metagenomic DNA from soil microorganisms, *Cold Spring Harb. Protoc.* **2009**, pdb.prot5271.
24. de Lorenzo V., Eltis L., Kessler B., and Timmis K. N. (1993) Analysis of *Pseudomonas* gene products using lacIq/Ptrp-lac plasmids and transposons that confer conditional phenotypes. *Gene* **123**, 17–24.
25. Sambrook J. and Russel D. (2000) *Molecular cloning: a laboratory manual* (Cold Spring Harbor Laboratory Press, New York, N.Y.).
26. Bagdasarian M. M., Amann E., Lurz R., Rückert B., and Bagdasarian M. (1983) Activity of the hybrid trp-lac (tac) promoter of *Escherichia coli* in *Pseudomonas putida*. Construction of broad-host-range, controlled-expression vectors. *Gene* **26**, 273–82.
27. Huang H. H., Camsund D., Lindblad P., and Heidorn T. (2010) Design and characterization of molecular tools for a Synthetic Biology approach towards developing cyanobacterial biotechnology. *Nucleic Acids Res.* **38**, 2577–93.
28. Schofield D. A., et al. (2003) Development of a thermally regulated broad-spectrum promoter system for use in pathogenic gram-positive species. *Appl. Environ. Microbiol.* **69**, 3385–92.
29. Smits T. H., Seeger M. A., Witholt B., and van Beilen J. B. (2001) New alkane-responsive expression vectors for *Escherichia coli* and *Pseudomonas*. *Plasmid* **46**, 16–24.
30. Newman J. R. and Fuqua C. (1999) Broad-host-range expression vectors that carry the L-arabinose-inducible *Escherichia coli* araBAD promoter and the *araC* regulator. *Gene* **227**, 197–203.
31. Prior J. E., Lynch M. D., and Gill R. T. (2010) Broad-host-range vectors for protein expression across gram negative hosts. *Biotechnol. Bioeng.* **106**, 326–32.
32. Sukchawalit R., Vattanaviboon P., Sallabhan R., and Mongkolsuk S. (1999). Construction and characterization of regulated L-arabinose-inducible broad host range expression vectors in *Xanthomonas*. *FEMS Microbiol. Lett.* **181**, 217–23.
33. Singer J. T., et al. (2010) Broad-host-range plasmids for red fluorescent protein labeling of

gram-negative bacteria for use in the zebrafish model system. *Appl. Environ. Microbiol.* **76**, 3467–74.

34. Katzke N., et al. (2010) A novel T7 RNA polymerase dependent expression system for high-level protein production in the phototrophic bacterium *Rhodobacter capsulatus*. *Protein Expression Purif.* **69**, 137–46.
35. Jeske M. and Altenbuchner J. (2010) The *Escherichia coli* rhamnose promoter rhaP(BAD) is in *Pseudomonas putida* KT2440 independent of Crp-cAMP activation. *Appl. Microbiol. Biotechnol.* **85**, 1923–33.
36. Keil S. and Keil H. (1992) Construction of a cassette enabling regulated gene expression in the presence of aromatic hydrocarbons. *Plasmid* **27**, 191–99.
37. Mermod N., Ramos J. L., Lehrbach P. R., and Timmis K. N. (1986) Vector for regulated expression of cloned genes in a wide range of gram-negative bacteria. *J. Bacteriol.* **167**, 447–54.
38. Ramos J. L., Gonzäles-Carrero M., and Timmis K. N. (1988) Broad-host range expression vectors containing manipulated meta-cleavage pathway regulatory elements of the TOL plasmid. *FEBS Lett.* **226**, 241.
39. Davison J., Heusterspreute M., Chevalier N., Ha-Thi V., and Brunel F. (1987) Vectors with restriction site banks. V. pJRD215, a wide-host-range cosmid vector with multiple cloning sites. *Gene* **51**, 275–80.
40. Davison J. (2002) Genetic tools for pseudomonads, rhizobia, and other gram-negative bacteria. *BioTechniques* **32**, 386–8, 90, 92–4, passim.
41. Frey J. and Bagdasarian M. (1989) 'The Moelcular Biology of IncQ plasmids', in C. M. Thomas (ed.), *Promiscuous Plasmids of Gram-Negative Bacteria* (Academic Press, London, United Kingdom.), 79–94.
42. Labes M., Pühler A., and Simon R. (1990) A new family of RSF1010-derived expression and lac-fusion broad-host-range vectors for gram-negative bacteria. *Gene* **89**, 37–46.
43. Leemans J., et al. (1982) Broad-host-range cloning vectors derived from the W-plasmid Sa. *Gene* **19**, 361–64.
44. Antoine R. and Locht C. (1992) Isolation and molecular characterization of a novel broad-host-range plasmid from *Bordetella bronchiseptica* with sequence similarities to plasmids from gram-positive organisms. *Mol. Microbiol.* **6**, 1785–99.
45. Kovach M. E., et al. (1995) Four new derivatives of the broad-host-range cloning vector pBBR1MCS, carrying different antibiotic-resistance cassettes. *Gene* **166**, 175–76.
46. Wild J., Hradecna Z., and Szybalski W. (2002) Conditionally amplifiable BACs: switching from single-copy to high-copy vectors and genomic clones. *Genome Res.* **12**, 1434–44.

Chapter 20

A Simple Method for Introducing Marker-Free Deletions in the *Bacillus subtilis* Genome

Takuya Morimoto, Katsutoshi Ara, Katsuya Ozaki, and Naotake Ogasawara

Abstract

A genetic tool for introducing marker-free deletions is essential for multiple manipulations of genomes. We have developed a simple and efficient method for creating marker-free deletion mutants of *Bacillus subtilis* through transformation with recombinant PCR products, using the *Escherichia coli mazF* gene encoding an endoribonuclease that cleaves free mRNAs as a counterselection tool.

Key words: *Bacillus subtilis*, Marker-free mutant, *MazF*

1. Introduction

Antibiotic-resistance and other selectable marker genes are routinely used to create new strains of bacteria in which a particular genomic region is replaced and inactivated by a DNA fragment containing the marker gene (termed a marker cassette). However, the insertion of a constitutively expressed marker gene often alters the expression of adjacent genes, modulating the phenotypic consequences of inactivating the functions encoded in the deleted sequence. Furthermore, a genetic tool that introduces marker-free deletions is essential for multiple manipulations of genomes (1–4). One method for eliminating the marker cassette in primary transformants that has been widely used in work with *Escherichia coli* is placement of FLP recognition target (FRT) sites, which promote intramolecular homologous recombination, at both ends of the cassette using a helper plasmid encoding FLP recombinase (5). In *Bacillus subtilis*, a number of counterselection systems

James A. Williams (ed.), *Strain Engineering: Methods and Protocols*, Methods in Molecular Biology, vol. 765,
DOI 10.1007/978-1-61779-197-0_20,

have been developed that allow for the selection of cells that have lost the marker cassette from primary transformants (6–8).

Zhang and colleagues described a procedure that can be used in any genetic background to generate marker-free deletions in *B. subtilis*, utilizing the *E. coli mazF* gene, which encodes an endoribonuclease that cleaves free mRNAs, as a counterselection tool (9). We have created an improved and simplified variation of this method for constructing clean deletion mutants without remaining additional sequences that avoids the need for cloning *B. subtilis* genomic fragments in *E. coli*. In this procedure, illustrated schematically in Fig. 1, the *mazF*-encoding cassette is fused with the flanking sequences of the target region using recombinant polymerase chain reaction (PCR) (10). Upstream and downstream sequences (fragments A and B) of the flanking region to be deleted are amplified from the genomic DNA of the *B. subtilis*

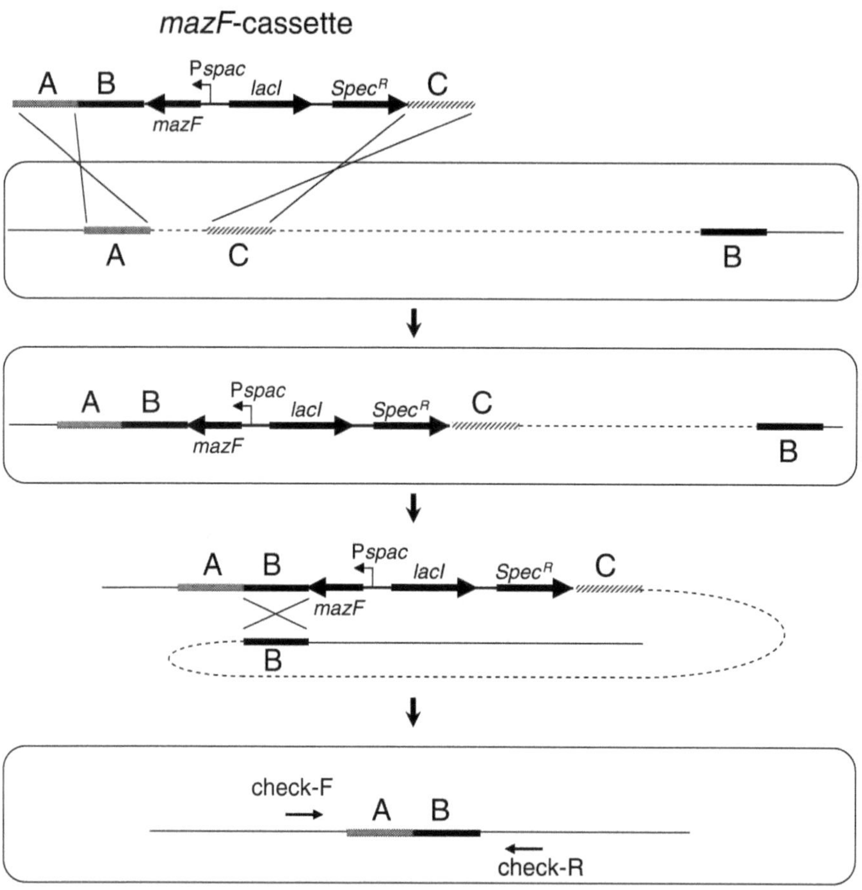

Fig. 1. Outline of the construction of marker-free deletion mutants. The outline of the procedure for introducing marker-free deletions is shown (*see* text for detail). *B. subtilis* cells are transformed with recombinant PCR product DNA in the order A – B – *mazF*-cassette – C. Cells in which the PCR product has been integrated into the target region to be deleted through homologous recombination (between fragment A and C loci) are selected by drug resistance in the absence of IPTG. Primary transformants are cultivated on IPTG-plates (i.e., *mazF* toxin-inducing conditions) to obtain cells in which the *mazF*-cassette has been excised by intramolecular homologous recombination at region B. Expected deletions in the resultant clones are confirmed by PCR using the primers check-F and check-R.

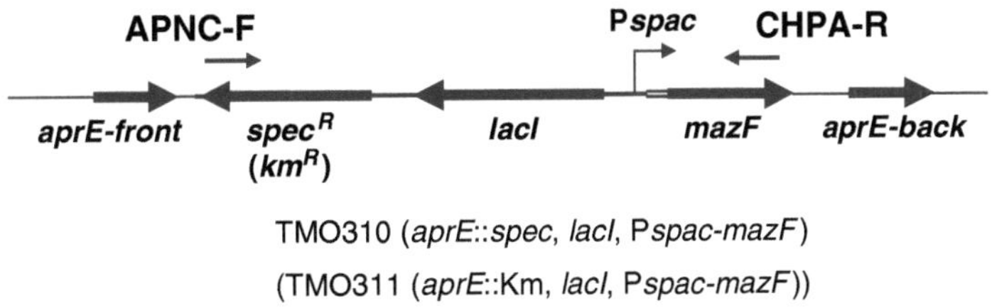

Fig. 2. Structure of *B. subtilis* strains harboring the *mazF*-cassette in the *aprE* locus. The structure of the *aprE* locus of TMO310 (168, *aprE*::*spec*^{R}, *lacI*, P*spac*-*mazF*) is shown. The spectinomycin-resistance gene (*spac*^{R}) is replaced with the kanamycin-resistance gene (*km*^{R}) in TMO311 (168, *aprE*::*km*^{R}, *lacI*, P*spac*-*mazF*). These strains do not grow on LB agar plates containing 100 μM IPTG. The *mazF*-cassette is amplified by PCR using the primers APNC-F and CHPA-R with TMO310 or TMO311 genomic DNA as a template.

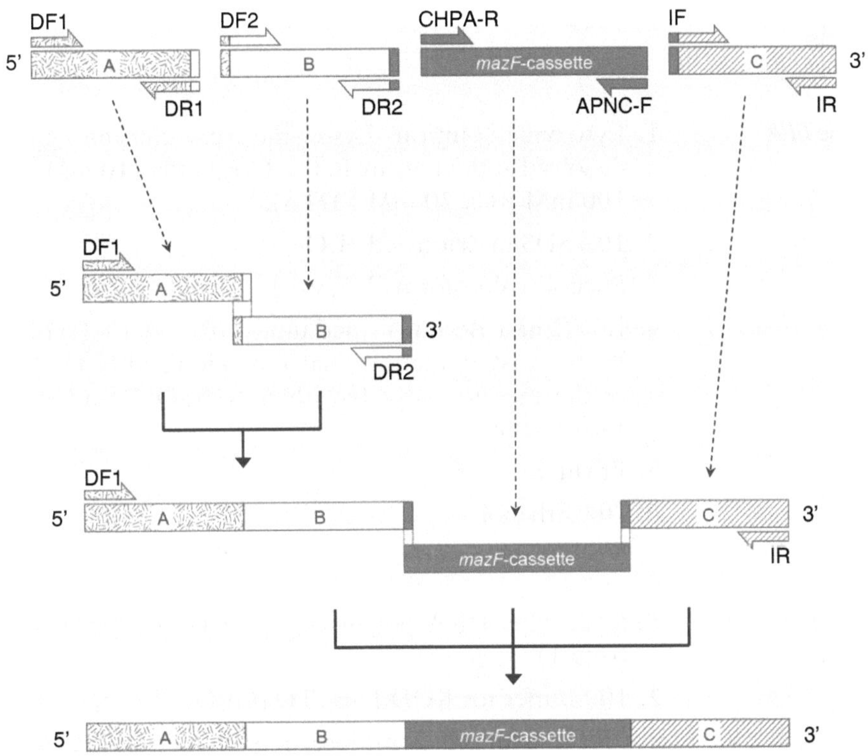

Fig. 3. Recombinant PCR to fuse fragments A, B, C and the *mazF*-cassette. The recombinant PCR scheme used to fuse fragments A, B, C and the *mazF*-cassette is shown. The 5′-ends of primers DR1 and DF2 contain overlapping sequences (10 bp of the 3′-end of fragment A and 10 bp of the 5′-end of fragment B). The 5′-ends of primers DR2 and IF contain 19-bp sequences of the 5′- and 3′-end, respectively, of the *mazF*-cassette.

strain to be manipulated. The *mazF*-cassette is amplified from the genomic DNA of *B. subtilis* strains that contain a drug-resistance gene and the *mazF* gene under the control of an IPTG-inducible *spac* promoter (Fig. 2). An internal sequence (fragment C) in the target region is also amplified. These PCR products are fused by recombinant PCR in the order A – B – *mazF*-cassette – C, as illustrated in Fig. 3, and integrated into the target region through homologous recombination between fragment A and C loci.

The resulting recombinants are selected for drug resistance in the absence of IPTG. Thereafter, the primary transformant is cultivated in the presence of IPTG (i.e., *mazF* toxin-inducing conditions), and clones in which the *mazF*-cassette has been excised by intramolecular homologous recombination at region B are selected (Fig. 1). We have routinely used this method to obtain marker-free deletions ranging from 8.5 to 128 kbp (10), but it is also likely to be applicable to smaller and larger deletions. Furthermore, because expression of the *mazF* gene is toxic to all bacteria (9), our method should be effective in any bacterium in which it is possible to introduce the *mazF*-cassette into the genome by double-crossover homologous recombination.

2. Materials

2.1. Genomic DNA Extraction

1. Lysozyme solution: Lysozyme from chicken egg white is dissolved at 0.5 mg/ml in the TKE buffer (10 mM Tris–HCl, 100 mM KCl, 20 mM EDTA).
2. 10% SDS in deionized H_2O.
3. Phenol/chloroform (1:1, v/v).
4. LB (Luria-Bertani) medium: 10% Bacto-Tryptone, 5% Bacto-yeast extract, 5% NaCl in deionized H_2O. Adjust pH of the medium to 7.0 with 5 N NaOH, and sterilize by autoclaving.
5. Ethanol.
6. 70% Ethanol.
7. Nuclease-free water.

2.2. Fragment PCR Reaction

1. KOD Plus DNA polymerase (1.0 U/μl) (TOYOBO) (see Note 1).
2. 10× Buffer for KOD Plus (TOYOBO).
3. dNTP solution: 2 mM each of dATP, dCTP, dGTP, and dTTP.
4. 25 mM $MgSO_4$.
5. Nuclease-free water.
6. Template genomic DNA:

   Genomic DNA (0.1 μg/ml) of the *B. subtilis* strain to be manipulated.

   Genomic DNA (0.1 μg/ml) of *B. subtilis* TMO310 or TMO311 strains (10) (see Note 2).
7. Primers for preparation of the *mazF*-cassette:

   10 μM APNC-F: 5′-CGACAGCGGAATTGACTCAAGC-3′.

10 μM CHPA-R: 5′-CGCGGATCCTACCCAATCAGTACGTTAATTTTG-3′.

8. Primers for amplification of *B. subtilis* genomic DNA Fragment A:

   10 μM DF1: 18–30-mer sequence of the 5′-end of region A.

   10 μM DR1: 20-mer complementary to the 5′-end of primer DF2 + 18–30-mer sequence of the 3′-end of region A.

9. Primers for amplification of *B. subtilis* genomic DNA Fragment B:

   10 μM DF2: 20-mer complementary to the 5′-end of primer DR1 + 18–30-mer sequence of the 5′-end of region B.

   10 μM DR2: 20-mer complementary to the 5′-end of primer CHPA-R (CTGATTGGGTAGGATCCGCG) + 18–30-mer sequence of the 3′-end of region B.

10. Primers for amplification of *B. subtilis* genomic DNA Fragment C:

    10 μM IF: 18-mer complementary to the 5′-end of primer APNC-F (GAGTCAATTCCGCTGTCG) + 18–30-mer sequence of the 5′-end of region C.

    10 μM IR: 18–30-mer sequence of the 3′-end of region C.

11. Primers for validation of the desired deletion (Fig. 1; see step 4 in Subheading 3.6):

    10 μM check-F: 18–30-mer sequence upstream of region A.

    10 μM check-R: 18–30-mer sequence downstream of region B.

2.3. Purification of PCR Products

1. 1% agarose gel containing 0.5 μg/ml ethidium bromide.
2. Wizard SV Gels and PCR cleanup kit (Promega) including the following solutions (see Note 3):

   Wizard SV Minicolumns.

   Membrane Binding Solution.

   Membrane Wash Solution.

   Nuclease-Free Water.

   Collection Tubes.

2.4. Recombinant PCR

1. Items 1–5 in Subheading 2.2.
2. Primer DF1 (see item 8 in Subheading 2.2).
3. Primer DR2 (see item 9 in Subheading 2.2).
4. Primer IR (see item 10 in Subheading 2.2).
5. Purified fragment A DNA (0.5 μg/ml).
6. Purified fragment B DNA (0.2–0.5 μg/ml).
7. Purified fragment C DNA (0.5 μg/ml).

8. Purified *mazF*-cassette DNA (1 μg/ml).
9. 20% PEG solution: 20% PEG8000 in 2.5 M NaCl.
10. 70% Ethanol.
11. Nuclease-free water.

2.5. Transformation of B. subtilis Cells

1. LB agar: LB medium (see item 4 in Subheading 2.1) containing 15 g/L agar. Sterilize by autoclaving and pour into 150 mm Petri plates.
2. 10× Salt: 2% $(NH_4)_2SO_4$, 14% K_2HPO_4, 6% KH_2PO_4, 1% trisodium citrate·2 H_2O, 0.2% $MgSO_4 \cdot 7\ H_2O$ in deionized H_2O. Sterilize the solution by autoclaving.
3. 100× Trace element (11): 0.073% $CaCl_2 \cdot 2H_2O$, 0.017% $ZnCl_2$, 0.0043% $CuCl_2 \cdot 2H_2O$, 0.0006% $CoCl_2 \cdot 6H_2O$, 0.0006% $Na_2MoO_4 \cdot 2H_2O$, 0.135% $FeCl_2 \cdot 4H_2O$, 0.001% $MnCl_2 \cdot 4H_2O$ in deionized H_2O. Sterilize the solution by filtration.
4. 50% glucose in deionized H_2O. Sterilize the solution by autoclaving.
5. 5% casein acid hydrolysate (Difco) in deionized H_2O. Sterilize the solution by autoclaving.
6. 5 mg/ml L-tryptophan in deionized H_2O. Sterilize the solution by filtration.
7. CI medium (10 ml): Mix 1 ml of 10× Salt, 0.5 ml of 50% glucose, 40 μl of 5% casein acid hydrolysate, 100 μl of 5 mg/ml L-tryptophan, 100 μl 100× trace element, and 8.26 ml sterile water.
8. CII medium (10 ml): Mix 1 ml of 10× Salt, 0.5 ml of 50% glucose, 20 μl of 5% casein acid hydrolysate, 10 μl of 5 mg/ml L-tryptophan, 100 μl 100× trace element, and 8.37 ml sterile water.
9. Recombinant PCR product (from step 4, Subheading 3.4.3).
10. Spec-plate: LB agar plate (see item 1 in Subheading 2.5) containing 100 μg/ml spectinomycin (*mazF* from TMO310).
11. Km-plate: LB agar plate containing 5 μg/ml kanamycin (*mazF* from TMO311).
12. Spec/IPTG-plate: LB agar plate containing 100 μg/ml spectinomycin and 1 mM IPTG (*mazF* from TMO310).
13. Km/IPTG-plate: LB agar plate containing 5 μg/ml kanamycin and 1 mM IPTG (*mazF* from TMO311).

2.6. Selection of Marker-Free Cells

1. Spec-plate: (see item 10 in Subheading 2.5) (*mazF* from TMO310).
2. Km-plate: (see item 11 in Subheading 2.5) (*mazF* from TMO311).

3. IPTG-plate: LB agar plate containing 1 mM IPTG.
4. Spec/IPTG-plate: (see item 12 in Subheading 2.5) (*mazF* from TMO310).
5. Km/IPTG-plate: (see item 13 in Subheading 2.5) (*mazF* from TMO311).
6. LB medium: (see item 4 in Subheading 2.1) containing 100 μg/ml spectinomycin or 5 μg/ml kanamycin as needed (see Fig. 2).
7. 50% Glycerol. Sterilize by autoclaving.

3. Methods

3.1. Genomic DNA Extraction from B. subtilis Cells

1. Inoculate 1 ml of LB medium containing the appropriate antibiotic (for selection of TMO310, TMO311, or a strain to be manipulated) with a single colony of *B. subtilis* and incubate with shaking overnight at 37°C.
2. Transfer 500 μl of the cell culture to 1.5-ml microcentrifuge tube and recover the cell by centrifugation at 16,000 × *g* for 2 min at 4°C.
3. Remove the supernatant by pipette.
4. Resuspend the cells in 100 μl of lysozyme solution by vortex and incubate for 10 min at 37°C.
5. Add 10 μl of 10% SDS to the suspension and mix by vortex.
6. Add 100 μl of phenol–chloroform solution. Mix by vortex and then centrifuge at 16,000 × *g* for 5 min at room temperature. Transfer 50 μl of the aqueous solution to a new 1.5-ml microcentrifuge tube.
7. Precipitate genomic DNA from supernatant by adding 125 μl of ethanol. Mix by vortex and then centrifuge at 16,000 × *g* for 5 min at 4°C.
8. Remove the supernatant by pipette and then add 400 μl of 70% ethanol to the pellet.
9. Recover the genomic DNA by centrifugation at 16,000 × *g* for 2 min at 4°C.
10. Carefully remove the supernatant by pipette and add 1 ml of 70% ethanol. After centrifugation for 1 min at 16,000 × *g*, remove the supernatant and then allow the remaining ethanol to evaporate.
11. Dissolve the genomic DNA pellet in 100 μl of H_2O.

3.2. Fragment PCR

3.2.1. Preparation of the mazF-Cassette

The *mazF*-cassette, containing *mazF* under the control of the *spac* promoter, a *lacI* gene controlling the *spac* promoter, and a drug-resistance gene (Fig. 2), is amplified by PCR using the primers APNC-F and CHPA-R with TMO310 or TMO311 genomic DNA as a template (see Note 2).

1. Set up the PCR reaction mix (100 μl) as follows:

10× Amplification buffer (TOYOBO)	10 μl
dNTPs solution (2 mM)	10 μl
$MgSO_4$ (25 mM)	6.4 μl
Primer APNC-F (10 μM)	1 μl
Primer CHPA-R (10 μM)	1 μl
B. subtilis TMO310 or TM311 genomic DNA (0.1 μg/ml)	1 μl
KOD Plus DNA polymerase (TOYOBO)	2 μl
H_2O	68.6 μl

2. Perform PCR reaction using the following cycling conditions:

   Denaturation at 94°C for 2 min

   30 cycles of incubation at 94°C for 15 s, 58°C for 30 s, and 68°C for 2 min

   Store at 4°C.

3. Fractionate 2 μl of the PCR reaction by agarose gel electrophoresis and verify if a PCR product of the predicted size is present.

3.2.2. Preparation of Genomic DNA Fragments A, B, and C (Fig. 1)

Usually, we amplify about 500 bp of DNA sequence flanking the region to be deleted (fragment A and B) and the internal sequence of the target region (fragment C) (see Note 4). The melting temperature (T_m) of primers is calculated using the equation $T_m = 2(A+T)+4(G+C)$. The G+C content of primers should be 40–60% ($T_m = 56–68$), and the T_m values for the primers in a pair should not differ by more than 5°C. An overlapping sequence is introduced into primer pairs, DR1 and DF2, DR2 and CHPA-R, and APNC-F and IF, to join PCR products by recombinant PCR, as illustrated in Fig. 3.

1. Set up the PCR reaction mix (100 μl) for each fragment (A, B and C) as follows:

10× amplification buffer (TOYOBO)	10 μl
dNTPs solution (2 mM)	10 μl
$MgSO_4$ (25 mM)	6.4 μl

Forward primer (10 μM)	1 μl
Reverse primer (10 μM)	1 μl
B. subtilis genome DNA (0.1 μg/ml)	1 μl
KOD Plus DNA polymerase (TOYOBO)	2 μl
H_2O	68.6 μl

2. Perform PCR reaction using the following cycling conditions:

   Denaturation at 94°C for 2 min

   30 cycles of incubation at 94°C for 15 s, 58°C for 30 s (see Note 5), and 68°C for 30 s

   Store at 4°C.

3. Fractionate 2 μl of the PCR reactions by agarose gel electrophoresis and verify if PCR products of the predicted sizes are present.

3.3. Purification of the mazF-Cassette and Fragments A, B, and C Using Wizard SV Gels and PCR Cleanup Kit

Removal of genomic DNA and primers used in PCR reactions is important for efficient joining of PCR products by recombinant PCR.

1. Fractionate PCR reaction mixtures (*mazF*, A, B, and C) by agarose gel electrophoresis.
2. Excise the gel slices containing PCR products using a clean razor blade and transfer to 1.5-ml microcentrifuge tube.
3. Add Membrane Binding Solution at a ratio of 10 μl of the solution per 10 mg of gel slice.
4. Incubate at 65°C until the gel slice is completely dissolved.
5. Place an SV Minicolumn in the Collection Tube, apply the dissolved gel mix to the column, and incubate for 1 min.
6. Centrifuge the SV Minicolumn assembly in a microcentrifuge at 10,000 × *g* for 30 s. Remove the SV Minicolumn from the Spin Column assembly and discard the liquid in the Collection Tube. Return the SV Minicolumn to the Collection Tube.
7. Wash the column by adding 600 μl of Membrane Wash Solution. Centrifuge the SV Minicolumn assembly for 30 s at 10,000 × *g*. Repeat the wash with 600 μl of Membrane Wash Solution.
8. Transfer the SV Minicolumn to a 1.5-ml microcentrifuge tube and centrifuge for 2 min at 16,000 × *g* to remove residual Membrane Wash Solution.
9. Transfer the SV Minicolumn to a 1.5-ml microcentrifuge tube. Apply 50 μl of Nuclease-Free Water to the center of

the column and incubate at room temperature for 1 min. Centrifuge for 1 min at 16,000 × *g* to collect purified DNA.

10. Measure the OD_{260} of the DNA solution using a UV spectrometer to calculate the concentration of DNA.

3.4. Recombinant PCR (Fig. 3)

3.4.1. Ligation of Fragments A and B

1. Set up the PCR reaction mix (100 μl) as follows:

10× Amplification buffer (TOYOBO)	10 μl
dNTPs solution (2 mM)	10 μl
$MgSO_4$ (25 mM)	6.4 μl
Primer DF1 (10 μM)	1 μl
Primer DR2 (10 μM)	1 μl
Purified fragment A DNA (0.5 μg/μl)	1 μl
Purified fragment B DNA (0.5 μg/μl)	1 μl
KOD Plus DNA polymerase (TOYOBO)	2 μl
H_2O	67.6 μl

2. Perform PCR reaction using the following cycling conditions:

   Denaturation at 94°C for 2 min

   30 cycles of incubation at 94°C for 15 s, 58°C for 30 s (see Note 5), and 68°C for 1 min

   Store at 4°C.

3. Purify the PCR product using Wizard SV Gels and PCR cleanup kit (Promega) as described in Subheading 3.3 above.

3.4.2. Ligation of Fragment A–B, *mazF*-Cassette and Fragment C

1. Set up the PCR reaction mix (100 μl) as follows:

10× Amplification buffer (TOYOBO)	10 μl
dNTPs solution (2 mM)	10 μl
$MgSO_4$ (25 mM)	6.4 μl
Primer DF1 (10 μM)	1 μl
Primer IR (10 μM)	1 μl
Purified fragment A–B DNA (0.4–1.0 μg/μl)	1 μl
Purified *mazF*-cassette (1 μg/μl)	0.5 μl
Purified fragment C DNA (0.2–0.5 μg/μl)	1 μl
KOD Plus DNA polymerase (TOYOBO)	2 μl
H_2O	67.1 μl

2. Perform PCR reaction using the following cycling conditions:

   Denaturation at 94°C for 2 min

   30–35 cycles of incubation at 94°C for 15 s, 58°C for 30 s, and 68°C for 4 min

   Store at 4°C.

3. Fractionate 2 μl of the PCR reaction by agarose gel electrophoresis and verify if a PCR product of the predicted size is present (see Note 6).

3.4.3. Purification of Recombinant PCR Product by PEG Precipitation

Because residual buffer salts in purified PCR products impair the efficiency of transformation of *B. subtilis* cells, we use PEG precipitation to purify the final recombinant PCR product.

1. Add an equal volume of 20% PEG solution to the PCR reaction mix in a 1.5-ml microcentrifuge tube. Vortex the mixture and store on ice for 15 min.
2. Pellet the PCR products by centrifugation at 16,000 × *g* for 15 min at 4°C.
3. Carefully remove the supernatant by pipette and add 1 ml of 70% ethanol. After centrifugation for 1 min at 16,000 × *g*, remove the supernatant and then allow the remaining ethanol to evaporate.
4. Dissolve the pellet in 50 μl of H_2O.

3.5. Transformation of B. subtilis Cells with Recombinant PCR Product Containing the mazF-Cassette

Bacillus subtilis cells are transformed with recombinant PCR product using competent cells, as described by Anagnostopoulos and Spizizen (12), with selection for drug resistance in the absence of IPTG.

1. Inoculate *B. subtilis* cells on an LB agar plate and incubate for ~10 h at 30°C.
2. Transfer cells to 2 ml of CI medium to yield a cell density of $OD_{600} = 0.1$.
3. Cultivate cells with shaking for 4 h at 37°C.
4. Centrifuge cell cultures at 6,000 × *g* for 5 min at room temperature and resuspend cells in 2 ml of CII medium.
5. Cultivate cells with shaking for 30 min at 37°C to induce competency.
6. Add 50 μl recombinant PCR product from step 4 in Subheading 3.4.3 to 500 μl competent cells and incubate with shaking for 90 min at 37°C.
7. Spread an appropriate volume of transformed competent cells on Spec- or Km-plate (as appropriate; see Fig. 2), and incubate at 37°C. Transformant colonies should appear within ~12 h. We usually use 100 μl cells to obtain ~10–30 colonies per plate.

8. To examine IPTG-sensitive growth attributable to the induction of *mazF*, transfer several colonies onto Spec- or Km-plates with and without IPTG and incubate at 37°C for 4–7 h. Select transformants that do not grow on Spec- or Km-IPTG plates (see Note 7).

3.6. Selection of Marker-Free Cells that Have Lost the mazF-Cassette

1. Inoculate the IPTG-sensitive transformants into 2 ml of LB liquid medium containing 100 μg/ml spectinomycin or 5 μg/ml kanamycin and incubate with shaking at 37°C overnight.
2. Spot and spread 5–10 μl of overnight culture on an IPTG-plate, followed by incubation at 37°C for 10–12 h to obtain IPTG-resistant cells produced by excision of the *mazF*-cassette by intramolecular homologous recombination at region B (Fig. 1) (see Note 8).
3. Inoculate cells from the IPTG-resistant colony into 2 ml of LB liquid medium containing 1 mM IPTG and incubate with shaking at 37°C overnight.
4. Mix 300 μl of overnight culture with 200 μl of 50% glycerol and store at -80°C. Extract genomic DNA from the remaining overnight culture (see Subheading 3.1) to verify deletion of the target sequence by PCR using the primers check-F and check-R (Fig. 1) (see Note 9).

4. Notes

1. Many types of thermostable DNA polymerases are provided by various suppliers. High-fidelity DNA polymerases are recommended in mutant construction to avoid introduction of undesired mutations. We routinely use KOD Plus polymerase (TOYOBO), which has strong proofreading activity and high processivity (13). In addition, the KOD Plus enzyme solution contains two antibodies that inhibit polymerase activity and prevent 3′–5′ exonuclease activity during the setup of PCR mixes. Thus, the PCR error rate of KOD Plus is approximately 80 times less than that of Taq DNA polymerase (13).
2. The nucleotide sequences of the *mazF*-cassette have been deposited in the NCBI/EMBL/DDBJ database under accession numbers AB526354 and AB526355; the TMO310 and TMO311 strains are available from us upon request.
3. Several DNA purification kits based on the ability of DNA to bind to silica membranes in the presence of chaotropic salts are provided by various manufacturers. We routinely use Wizard SV Gels and PCR cleanup kit (Promega).

4. A length of 500 bp is sufficient for efficient transformation of *B. subtilis* cells by homologous recombination. However, a shorter sequence (300 bp) could be used as an inner fragment (Fragment C) to obtain small deletions.
5. We usually set the annealing temperature at 58°C. If the PCR reaction fails, try changing the annealing temperature over a range of 50–64°C.
6. Usually, 50 ng/μl DNA is obtained by recombinant PCR. If the PCR ligation fails, yielding low or no product, try lowering the annealing temperature and/or increasing the amount of *mazF*-cassette in the PCR reaction.
7. Sometimes, IPTG-resistant clones appear during cultivation, possibly as a result of suppressor mutations that inactivate *mazF* activity. Take care not to select clones with subtle growth on Spec- or Km-IPTG plate for the following step.
8. We extract and store genomic DNA from remaining overnight cultures of primary transformants to use for introducing the deletion in *B. subtilis* strains with other genetic backgrounds.
9. Usually, cultivation of two candidate clones is enough to obtain marker-free mutants.

Acknowledgments

We are grateful to Shu Ishikawa for helpful discussions. This work is part of the subproject "Development of a Technology for the Creation of a Host Cell" included within the industrial technology project "Development of Generic Technology for Production Process Starting Productive Function" of the Ministry of Economy, Trade, and Industry, funded by the New Energy and Industrial Technology Development Organization (NEDO), Japan.

References

1. Forster A. C., and Church G. M. (2007) Synthetic biology projects *in vitro*. *Genome Res*, **17**, 1–6.
2. Kolisnychenko V., Plunkett G., 3 rd, Herring C. D., Feher T., Posfai J., Blattner F. R., and Posfai, G. (2002) Engineering a reduced *Escherichia coli* genome. *Genome Res*, **12**, 640–647.
3. Posfai G., Plunkett G., 3 rd, Feher T., Frisch D., Keil G. M., Umenhoffer K., Kolisnychenko V., Stahl B., Sharma S. S., de Arruda M., Burland V., Harcum S. W., and Blattner F. R. (2006) Emergent properties of reduced-genome *Escherichia coli*. *Science*, **312**, 1044–1046.
4. Hashimoto M., Ichimura T., Mizoguchi H., Tanaka K., Fujimitsu K., Keyamura K., Ote T., Yamakawa T., Yamazaki Y., Mori H., Katayama T., and Kato J. (2005) Cell size and nucleoid organization of engineered *Escherichia coli* cells with a reduced genome. *Mol Microbiol*, **55**, 137–149.
5. Datsenko K. A., and Wanner B. L. (2000) One-step inactivation of chromosomal genes

in *Escherichia coli* K-12 using PCR products. *Proc Natl Acad Sci USA*, **97**, 6640–6645.

6. Fabret C., Ehrlich S. D., and Noirot P. (2002) A new mutation delivery system for genome-scale approaches in *Bacillus subtilis*. *Mol Microbiol*, **46**, 25–36.
7. Morimoto T., Kadoya R., Endo K., Tohata M., Sawada K., Liu S., Ozawa T., Kodama T., Kakeshita H., Kageyama Y., Manabe K., Kanaya S., Ara K., Ozaki K., and Ogasawara N. (2008) Enhanced recombinant protein productivity by genome reduction in *Bacillus subtilis*. *DNA Res*, **15**, 73–81.
8. Liu S., Endo K., Ara K., Ozaki K., and Ogasawara N. (2008) Introduction of marker-free deletions in *Bacillus subtilis* using the AraR repressor and the *ara* promoter. *Microbiology*, **154**, 2562–2570.
9. Zhang X. Z., Yan X., Cui Z. L., Hong Q., and Li S. P. (2006) *mazF*, a novel counter-selectable marker for unmarked chromosomal manipulation in *Bacillus subtilis*. *Nucleic Acids Res*, **34**, e71.
10. Morimoto T., Ara K., Ozaki K., and Ogasawara N. (2009) A new simple method to introduce marker-free deletions in the *Bacillus subtilis* genome. *Genes Genet Syst*, **84**, 315–318.
11. Harwood C. R. and Archibald A. R. (1990) Growth, maintenance and general techniques, in *Molecular Biological Methods for Bacillus* (John Wiley & Sons, Chichester, New York, Brisbane, Toronto, Singapore), pp. 549.
12. Anagnostopoulos C., and Spizizen J. (1961) Requirements for Transformation in *Bacillus Subtilis*. *J Bacteriol*, **81**, 741–746.
13. Takagi M., Nishioka M., Kakihara H., Kitabayashi M., Inoue H., Kawakami B., Oka M., and Imanaka T. (1997) Characterization of DNA polymerase from *Pyrococcus* sp. strain KOD1 and its application to PCR. *Appl Environ Microbiol*, **63**, 4504–4510.

Chapter 21

Transposon-Mediated Random Mutagenesis of *Bacillus subtilis*

Adam C. Wilson and Hendrik Szurmant

Abstract

The depth of knowledge concerning its physiology and genetics make *Bacillus subtilis* an attractive system for strain engineering and analysis. Transposon-based mutagenesis strategies generate large libraries of mutant strains that can be used to investigate molecular mechanisms relevant in fundamental research or to generate desirable phenotypes in applied research. This section presents a mini-Tn10-based transposon mutagenesis system that is capable of genome-wide insertional mutagenesis in *B. subtilis* and related organisms. Using appropriately designed selections or screens, the desired strain phenotypes can be isolated from transposon mutant libraries. This transposon system then allows rapid identification of the genetic locus responsible for the desired phenotype, and, due to the natural competence of *B. subtilis*, the identified genotypic change can easily be confirmed as responsible for the phenotypic change.

Key words: Transposon, Mutagenesis, Mini-Tn10, *Bacillus subtilis*, Backcross, Plasmid rescue

1. Introduction

Bacillus subtilis is the leading model organism for understanding the genetics and physiology of Gram-positive bacteria. Aside from ubiquitous and essential bacterial processes, *B. subtilis* has the capability of developing competence for DNA-uptake, to perform chemotaxis and, perhaps most importantly, to divide asymmetrically to develop dormant spores. The substantial base of knowledge derived from decades of study on these and other phenomena has established *B. subtilis* as a popular organism for use in industrial and genetic engineering applications.

Random insertional disruption by transposition is an important tool for the investigation of biochemical pathways and the development of novel strains containing desirable phenotypes (1–3).

James A. Williams (ed.), *Strain Engineering: Methods and Protocols*, Methods in Molecular Biology, vol. 765,
DOI 10.1007/978-1-61779-197-0_21, © Springer Science+Business Media, LLC 2011

The section will address the use of the mini-Tn10 transposon for insertional disruption in *B. subtilis*, an organism in which mini-Tn10 transposition has been show to occur at high frequency with very little host sequence specificity (4).

Plasmid pAW016 carries all of the components necessary for random insertional disruption in *B. subtilis* (Fig. 1). pAW016 is a derivative of the pIC333 transposon delivery system (5), in which the Erythromycin-resistance cassette for plasmid selection has been replaced with a more effective chloramphenicol-resistance cassette (6). The transposable Tn10 element carried by pAW016 contains a selectable Spectinomycin-resistance marker as well as the Gram-negative pUC origin of replication, facilitating identification of insertion sites by plasmid rescue (see Subheading 3.5). The transposase enzyme is located in the plasmid backbone, outside the transposed sequence, resulting in stable insertion of the transposon. The transposase enzyme can be removed following transposition by exploiting the temperature-sensitive pWVO1 origin of replication, which is the only replication origin on the plasmid that is functional in *B. subtilis*. Following a shift to a non-permissive temperature, the plasmid carrying the transposase can no longer replicate, and removal of the transposase can be verified

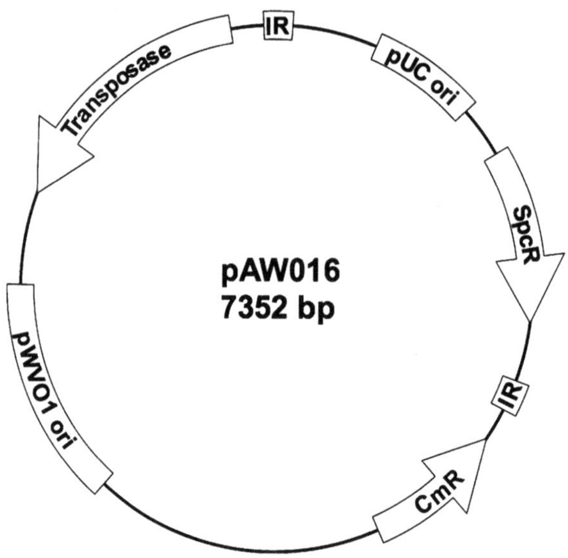

Fig. 1. Plasmid map of transposon deliver vector pAW016. The plasmid is a derivative of the pIC333 vector. The transposable Tn10 element is flanked by two inverted repeat sequences (IR). Between these sequences is the gene for spectinomycin resistance (Spc^R). Also within the transposable Tn10 element is the Gram-negative pUC origin of replication, which can be used for insertion sequence identification by plasmid rescue (Subheading 3.5). The heat-sensitive Gram-positive pWV01 origin of replication facilitates isolation of transposon mutants in the presence of spectinomycin selection and at an elevated temperature of 45°C. Since the transposase gene is located outside of the transposable element, transposon mutants are stable once a transposed strain is cured for the plasmid. (**a**) Chloramphenicol resistance marker (Cm^R) can be used as an initial selection when transforming the plasmid and can also serve as a counter selection tool to assure loss of the plasmid following transposition.

by loss of the chloramphenicol-resistance cassette encoded in the plasmid backbone. Figure 2 displays a schematic of the entire transposition protocol starting with plasmid transformation and ending with identification of the insertion sequence.

Plasmid pAW016 was developed for use in the related species *Bacillus anthracis*, but can function with high efficiency in *B. subtilis* and likely also in other Gram-positives. Additionally, the Mariner-based transposon delivery system developed in parallel with the

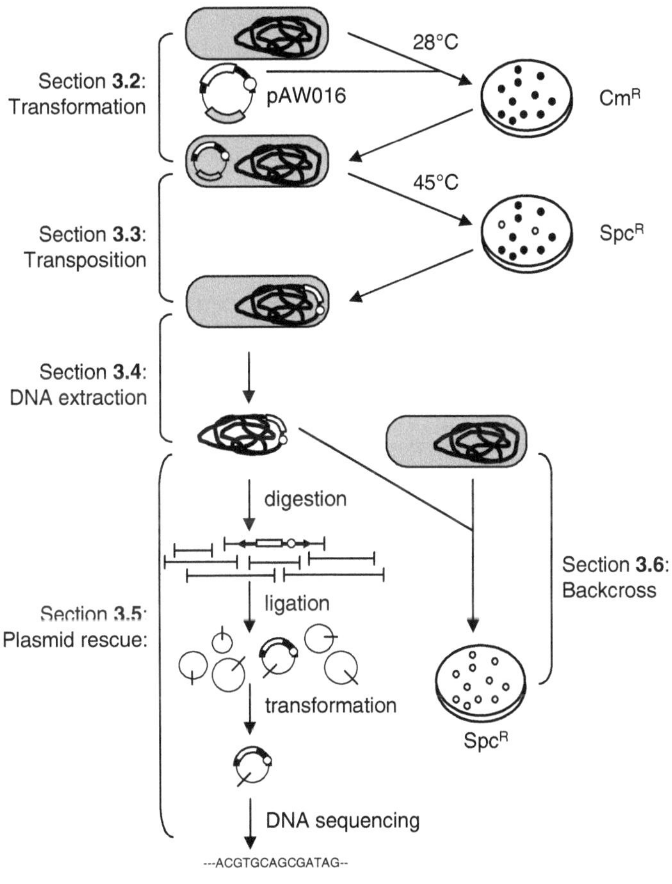

Fig. 2. Flowchart of the transposon mutagenesis approach in *Bacillus subtilis*. The transposition protocol includes five steps with detailed protocols available in the main text. As an initial step, a *B. subtilis* wild-type strain (*gray ovals*) is transformed with transposon delivery vector pAW016 (Subheading 3.1) at a growth temperature of 28°C and selection for chloramphenicol-resistant (Cm^R) colonies (Subheading 3.2). Next, a single colony is grown and transposon insertion mutants are selected by shifting the growth temperature to 45°C (Subheading 3.3). Spectinomycin-resistant (Spc^R) colonies with the desired phenotype (*white colonies for illustration*) are grown and genomic DNA is isolated (Subheading 3.4). The insertion sequence is identified by plasmid rescue (Subheading 3.5) involving a restriction digest of the genomic DNA followed by ligation and transformation of *E. coli* cells selecting for Spc^R. The pUC origin within the transposable element (*white circle*) facilitates *E. coli* propagation. Rescued plasmids can be isolated and sequenced for identification of the insertion sequence. Additionally, a backcross of genomic DNA into the wild-type *B. subtilis* strain is advisable to assure that the transposon insert is responsible for the observed phenotype (Subheading 3.6).

mini-Tn10 system uses many of the same components, and the protocols described in this section can be seamlessly applied to the Mariner-based system described in (6).

2. Materials

2.1. Preparation of Plasmid DNA for Transposon Delivery

1. pAW016, a plasmid carrying the mini-Tn10 transposon and transposase (6). The plasmid can be obtained by contacting the authors.
2. LB and LB-agar containing 10 μg/ml chloramphenicol: 10 g Bacto-tryptone, 5 g yeast extract, 10 g NaCl, distilled water to 1 l. Autoclave and store at room temperature. For LB-agar, add 15 g Bacto-agar per liter before sterilization. Allow to cool, add antibiotic to indicated concentration, and pour into sterile petri dishes.
3. *E. coli* strain SCS110 chemically competent cells or other *lacIq* carrying strain.
4. SOB Medium: 20 g Bacto-tryptone, 5 g yeast extract, 0.5 g NaCl, distilled water to 1 l. Autoclave and store at room temperature.
5. SOC Medium: 100 μl sterilized 1 M $MgCl_2$, 100 μl sterilized 1 M $MgSO_4$, 100 μl sterilized 2 M glucose, sterilized SOB Medium to 10 ml. Prepare immediately before use in a sterile 10–15 ml tube.
6. Chloramphenicol stock (10 mg/ml in ethanol) for *E. coli* selection at final concentration of 10 μg/ml.
7. QIAprep Spin Miniprep Kit (Qiagen, Valencia, CA).

2.2. Preparation of Competent B. subtilis Cells and Transformation with Transposon Delivery Vector

1. pAW016, prepared freshly as in Subheading 3.1.
2. *B. subtilis* strain JH642 (*trpC2 phe-1*) or any other competent *B. subtilis* strain available from the Bacillus Genetic Stock Center (http://www.bgsc.org).
3. LB-agar + 5 μg/ml chloramphenicol: see item 2 in Subheading 2.1 for preparation of LB-agar containing antibiotic.
4. 10× Spizizen salts: 20 g $(NH_4)_2SO_4$, 140 g K_2HPO_4, 60 g KH_2PO_4, 10 g sodium citrate ($Na_3C_6H_5O_7{\cdot}2H_2O$), 2 g $MgSO_4{\cdot}7H_2O$, distilled water to 1 l. Autoclave and store at room temperature.
5. GM1: 2 ml of medium: 200 μl 10× Spizizen salts, 10 μl 1 M $MgSO_4$, 20 μl 50% (w/v) glucose, 10 μl 5% (w/v) casein hydrolysate, 20 μl 5 mg/ml phenylalanine, 50 μl 2 mg/ml tryptophan, and 1,680 μl sterile distilled water.
6. GM2: 2 ml of medium: 200 μl 10× Spizizen salts, 10 μl 1 M $MgSO_4$, 20 μl 50% (w/v) glucose, 5 μl 5% (w/v) casein

hydrolysate, 2.5 μl 5 mg/ml phenylalanine, 6.25 μl 2 mg/ml tryptophan, and 1,700 μl sterile distilled water.

7. Chloramphenicol stock (see item 6 in Subheading 2.1) for *B. subtilis* selection at final concentration of 5 μg/ml.

2.3. Transposon Mutagenesis

1. LB and LB-agar + 50 μg/ml spectinomycin: see item 2 in Subheading 2.1 for preparation of LB-agar containing antibiotic.
2. Spectinomycin stock (100 mg/ml in H_2O, filter sterilized) for *B. subtilis* selection at final concentration of 50 μg/ml.
3. Glycerol freezing solution: 50% (w/v) glycerol in distilled water. Autoclave and store at room temperature.

2.4. Extraction of B. subtilis Genomic DNA

1. Lysis solution: 5 mg/ml lysozyme, 2 μg/ml RNase A, 100 mM EDTA (pH 8.0), 10 mM Tris–HCl (pH 8.0) in distilled water. Prepare fresh.
2. Tris-saturated phenol, pH 8.0. Phenol is equilibrated with 10 mM Tris–HCl, pH 8.0, 1 mM EDTA.
3. Chloroform.
4. Phenol/chloroform mixture at a volume to volume ratio of 1:1.
5. 100% reagent-grade ethanol.
6. 70% ethanol.
7. 10% SDS stock in distilled water.
8. TE Buffer: 10 mM Tris–HCl (pH 8.0), 1 mM EDTA in distilled water. Autoclave and store at room temperature.

2.5. Identification of Transposon Insertion Site by Plasmid Rescue

1. Restriction enzymes *Eco*RI, *Hin*dIII, and *Nsi*I and T4 DNA ligase.
2. Tris-saturated phenol, pH 8 (see item 2 in Subheading 2.4).
3. Chloroform (see item 3 in Subheading 2.4).
4. Phenol/chloroform mixture at a volume to volume ratio of 1:1 (see item 4 in Subheading 2.4).
5. 3 M Sodium acetate, adjusted with glacial acetic acid to pH 5.2.
6. 100% reagent-grade ethanol (see item 5 in Subheading 2.4).
7. 70% reagent-grade ethanol (see item 6 in Subheading 2.4).
8. Chemically competent *E. coli* stains DH5α or TG1. Electrically competent *E. coli* cells can also be used depending upon the investigator's preferences.
9. Spectinomycin stock for *E. coli* selection at final concentration of 100 μg/ml.
10. LB and LB-agar +100 μg/ml spectinomycin: see item 2 in Subheading 2.1 for preparation of LB-agar containing antibiotic.

11. QIAprep Spin Miniprep Kit (Qiagen, Valencia, CA).
12. Oligonucleotide primers mini-Tn1 (5′-GAGTCAGTGAGCGAGGAAGC-3′) and mini-Tn2 (5′-AGCCTGTCGGAATTGGTTTT-3′). Primers are outward-facing at either end of the transposon, allowing sequencing of insertion-flanking genome sequences from rescued plasmid.

2.6. Backcross to Confirm Linkage Between Transposon Insertion and Desired Phenotype

1. *B. subtilis* genomic DNA isolated as described in Subheading 3.4.
2. *B. subtilis* strain JH642.
3. All solutions are identical to those used for a *B. subtilis* transformation with plasmid DNA (see Subheading 2.3).
4. LB-agar + 50 μg/ml spectinomycin: see item 1 in Subheading 2.3.

3. Methods

3.1. Preparation of Plasmid DNA for Transposon Delivery

1. Transform chemically compsetent *E. coli* SCS110 with 10–20 ng of pAW016 plasmid DNA following manufacturer's instructions (see Note 1).
2. Allow cells to recover in SOC medium at 37°C for 1 h. Replication of the plasmid in *E. coli* at 37°C utilizes the pUC origin of replication located within the transposon sequence.
3. Plate 150 μl of transformation reaction on LB-agar containing 10 μg/ml chloramphenicol. Incubate overnight at 37°C (see Note 2).
4. The following day, pick up 25–50 colonies with an inoculating loop and use to inoculate 6 ml of LB containing 10 μg/ml chloramphenicol. Incubate liquid culture overnight at 37°C with aeration.
5. Extract plasmid from overnight culture with QIAprep Spin Miniprep Kit according to manufacturer's instructions.
6. Store plasmid at −20°C until further use.

3.2. Preparation of Competent B. subtilis Cells and Transformation with Transposon Delivery Vector

1. Heavily streak *B. subtilis* JH642 on LB-agar so as to produce a lawn of cells and incubate plate overnight at 37°C.
2. The following day, inoculate 2 ml of GM1 in a 18 × 150 mm test tube with a heavy loopful of bacteria (approximate starting $OD_{525nm} = 0.1$) from the overnight plate.
3. Incubate 4–6 h at 37°C with aeration until culture appears very dense with an OD_{600} just over 1.0.

4. Add 100 μl of dense culture (from step 3) to two prewarmed 18 × 150 mm test tubes containing 900 μl of GM2 per reaction.
5. Incubate both cultures for 1.5 h at 37°C with aeration.
6. Add 0.5–2.0 μg pAW016 from Subheading 3.1 (total volume should not exceed 10 μl of DNA solution) to the first tube. Do not add DNA to second tube, which will serve as negative control.
7. Incubate both tubes for one additional hour at 37°C with aeration.
8. Transfer cultures to 1.5 ml microcentrifuge tube and centrifuge at 6,000 × *g* for 2 min.
9. Decant most of the supernatant, leaving behind about 100 μl of liquid, and resuspend cells in remaining supernatant.
10. Plate cells on LB-agar containing 5 μg/ml chloramphenicol.
11. Incubate plates at 28°C for 18–24 h until chloramphenicol-resistant colonies are visible (see Note 3).

3.3. Transposon Mutagenesis

1. Prepare five to ten 18 × 150 mm test tubes containing 2 ml LB broth with 50 μg/ml spectinomycin and inoculate with chloramphenicol-resistant colonies (Subheading 3.2, step 11)
2. Incubate cultures overnight at 28°C with aeration. Cultures should appear turbid by 14 h (see Note 4).
3. Dilute cultures 1:50 in 18 × 150 mm test tubes containing LB broth with 50 μg/ml Spectinomycin and incubate at 28°C with aeration for 3 h, allowing overnight cultures to re-enter exponential growth phase.
4. Shift cultures to 45°C and incubate with aeration for an additional 5 h. Growth at nonpermissive temperatures inhibits replication of the transposon delivery vector and selects for growth of transposon mutants. These nonpermissive incubations can also be done at 37°C at a slightly reduced efficiency of transposon mutant selection.
5. Serially dilute cultures in sterile microcentrifuge tubes and plate dilutions on LB-agar containing 50 μg/ml spectinomycin.
6. Incubate plates overnight at nonpermissive temperature of 37°C or above (see Note 5).
7. The next day, streak candidate spectinomycin-resistant mutant colonies on LB-agar plates with 50 μg/ml spectinomycin to confirm mutant clones. Incubate overnight at 37°C or above, depending upon screening conditions. This step may be dispensable when generating a random mutant library (see Note 6).
8. Inoculate 8 ml LB broth containing 50 μg/ml spectinomycin with spectinomycin-resistant mutant colonies and incubate 8–12 h at 37°C with aeration.

9. Freeze two aliquots of culture as glycerol stocks at −80°C and use remainder of culture for genomic DNA extraction (see below). Final glycerol stock should be 15% (w/v) glycerol; typically 700 μl of bacterial culture to 300 μl of glycerol freezing solution.

3.4. Extraction of *B. subtilis* Genomic DNA

1. Transfer remaining culture (from Subheading 3.3, step 8) to a 15-ml conical centrifuge tube at centrifuge at 5,000 × *g* for 5 min at room temperature. Aspirate supernatant and retain pellet.
2. Resuspend the pellet in 600 μl lysis solution, transfer to a microcentrifuge tube, and incubate at 37°C for 30 min.
3. Add 20 μl of 10% SDS solution, incubate at 60°C, and gently mix until lysate clears.
4. Add 600 μl of 1:1 phenol/chloroform and incubate at 60°C for 30 min, mixing every 5 min.
5. Centrifuge in microcentrifuge tube at maximum speed for 5 min at room temperature. Transfer upper aqueous phase to a new microcentrifuge tube.
6. Add 500 μl of chloroform and mix vigorously (no vortexing) for 1 min.
7. Centrifuge in microcentrifuge tube at maximum speed for 5 min at room temperature. Transfer upper aqueous phase to a new microcentrifuge tube.
8. Add 800 μl of 100% ethanol and mix. The DNA appears as a white texture.
9. Remove supernatant with a pipetman, avoiding the DNA. If DNA is not easily visible, centrifuge at maximum speed for 1 min before aspirating the supernatant.
10. Add 500 μl 70% ethanol, mix, and centrifuge at maximum speed for 1 min at room temperature.
11. Aspirate the supernatant, dry the DNA under vacuum for 5 min, and dissolve the dried DNA in 200 μl TE buffer. The DNA should be at a concentration of 0.4–0.5 μg/μl.
12. Alternatively, genomic DNA can be extracted using a commercially available kit, such as the UltraClean Microbial DNA Isolation Kit (MoBio Laboratories, Carlsbad, CA).

3.5. Identification of Transposon Insertion Site by Plasmid Rescue

1. Assemble an *Eco*RI restriction digest containing 2 μg of genomic DNA in a 40-μl final reaction volume according to the manufacturer's instructions. For the occasional mutant that cannot be rescued following an *Eco*RI digest, digestions with the restriction enzymes *Hin*dIII and *Nsi*I often will result in plasmid formation.
2. Digest genomic DNA for 2–4 h at 37°C.

3. Heat inactivate restriction enzyme according to the manufacturer's instructions. If heat inactivation is not indicated, extract reaction with phenol/chloroform followed by chloroform extraction (see below, steps 5–9).
4. Add digested genomic DNA to a ligation reaction using T4 DNA ligase in a total reaction volume of 200 μl according to manufacturer's instructions. The larger reaction volume encourages self-ligation of restriction fragments. Incubate reaction overnight at 16°C.
5. Add 200 μl 1:1 phenol/chloroform to ligation reaction and vigorously mix for 1 min. Centrifuge tube at maximum speed for 5 min at room temperature. Transfer upper aqueous phase to a new microcentrifuge tube.
6. Add 200 μl chloroform and vigorously mix for 1 min. Centrifuge tube at maximum speed for 5 min at room temperature. Transfer upper aqueous phase to a new microcentrifuge tube.
7. Add 20 μl 3 M sodium acetate pH 5.2 and 500 μl 100% ethanol and vigorously mix for 1 min. Centrifuge tube at maximum speed for 15 min at 4°C. Carefully aspirate the supernatant with pipetman.
8. Add 500 μl 70% ethanol and vigorously mix for 1 min. Centrifuge tube at maximum speed for 5 min at 4°C.
9. Aspirate the supernatant, dry the DNA under vacuum for 5 min, and resuspend the precipitated DNA in 5–10 μl of distilled water.
10. Use resuspended DNA to transform chemically competent *E. coli* following manufacturer's instructions. *E. coli* strains DH5α and TG1 have both been transformed successfully by this method.
11. Plate transformation on LB-agar containing 100 μg/ml spectinomycin and incubate plates overnight at 37°C.
12. Streak candidate *E. coli* colonies on LB-agar plates with 100 μg/ml spectinomycin to confirm clones. Incubate overnight at 37°C (see Note 7).
13. Use confirmed spectinomycin-resistant *E. coli* colonies to inoculate 5 ml of LB broth with 100 μg/ml spectinomycin. Incubate overnight at 37°C.
14. Extract plasmid DNA from overnight *E. coli* cultures using the Qiagen QIAprep Spin Miniprep Kit according to manufacturer's instructions.
15. Sequence regions flanking the transposon insertion using sequencing oligonucleotide primers mini-Tn1 and/or mini-Tn2.

16. Identify insertion site by submitting sequence results in FASTA format to nucleotide BLAST (http://blast.ncbi.nlm.nih.gov/). Terminal portions of the transposon sequence are typically contained in the initial sequencing data, facilitating the precise determination of the transposon insertion site.

3.6. Backcross to Confirm Linkage Between Transposon Insertion and Desired Phenotype

1. Prepare competent *B. subtilis* strain JH642 as in Subheading 3.2, steps 1–5.
2. Add approximately 5.0 μg (usually 10 μl) of transposon-mutagenized *B. subtilis* genomic DNA from Subheading 3.4 (total volume should not exceed 10 μl of DNA solution) to the first tube. Do not add DNA to second tube, which will serve as negative control.
3. Incubate both tubes for one additional hour at 37°C with aeration.
4. Transfer cultures to 1.5 ml microcentrifuge tube and centrifuge at 6,000 × *g* for 2 min.
5. Decant most of the supernatant, leaving behind about 100 μl of liquid, and resuspend cells in remaining supernatant.
6. Plate cells on LB-agar containing 50 μg/ml spectinomycin.
7. Incubate plates at 37°C for 14 h or until spectinomycin-resistant colonies with the desired phenotype are visible (see Note 8).

4. Notes

1. Plasmid pAW016 is unstable due to the presence of the transposase enzyme. Transcription of the transposase is controlled by the synthetic tac promoter, a hybrid of the trp and lac promoters (7), which is repressible due to the presence of the lac operator. Propagation in the *E. coli* SCS110 strain, which carries the hyperactive lacIq, decreases transposase expression and increases plasmid stability. Despite this advantage, pAW016 remains somewhat unstable. When performing transposition, the use of freshly transformed plasmid, as outlined in this protocol, is strongly recommended. While glycerol stocks of *E. coli* harboring pAW016 are a useful backup, use of plasmid derived from the glycerol stocks should be a last resort.
2. pAW016 may be propagated in *E. coli* at 37°C. This ensures use of the pUC origin of replication because the pWVO1 origin of replication will not function at this nonpermissive temperature. As the pUC origin is contained in the transposon sequence and the chloramphenicol-resistance cassette is

located on the nontransposed plasmid backbone, incubation at 37°C in the presence of chloramphenicol promotes physical linkage of the transposon to the delivery plasmid and isolation of intact plasmid from *E. coli.*

3. Maintaining pAW016-transformed *B. subtilis* at temperatures permissive for replication through the pWVO1 origin is critical in these pretransposition steps. Incubation at temperatures above the nonpermissive temperature of approximately 35°C with selection can result in lack of growth or, in rare occasions, selection of unwanted mutants or irregular integration events. Because the tac promoter is not repressed in *B. subtilis*, low level transposition will occur immediately following transformation with pAW016. Though this has not caused significant problems, if premature transposition is a concern, verify the loss of most spectinomycin-resistance when freshly transformed cells are grown at 45°C. pAW016 plasmid DNA can also be extracted from transformed *B. subtilis* and checked by restriction digest to confirm retention and stability of the plasmid, though this additional check is rarely needed.
4. The length of culture at 28°C can be varied depending upon the structure of the experiment. Longer, multiday incubations and dilutions at 28°C will result in higher numbers of transposon insertion mutants per culture, but may also result in multiple transposon insertions per cell. Also consider that extended culture may result in clonal expansion of early mutants and the disproportionate representation of specific mutants in a library.
5. A small number (typically less than 1%) of spectinomycin-resistant clones are isolated at the nonpermissive temperature that have retained pAW016 plasmid sequences (and hence chloramphenicol-resistance) as a result of plasmid integration or secondary mutation that reduces temperature sensitivity of the plasmid. Retention or integration of nontransposon pAW016 sequences, which include the transposase-encoding gene, can result in multiple transposon insertions and/or changes in insertion sites. An unusually large number of spectinomycin-resistant colonies in Subheading 3.3, steps 6 and 7 is often indicative of an early occurring plasmid integration or retention event that may have been subsequently clonally amplified. Screening individual mutants of interest for retention of chloramphenicol-resistance can be valuable in eliminating plasmid-retention mutants before further analysis.
6. The nature of the screen or selection will influence the handling of the mutants. If a random transposon library is being generated, a number of transposon mutants corresponding to the investigator's target sample size can be immediately

isolated and stored from the initial mutant selection in Subheading 3.3, step 6 without the subsequent culture and extraction steps. It may be worthwhile to check the insertion site of a number of mutants as a quality control check before continuing with the library manipulation. Alternatively, Southern blotting of a subset of library mutants using a transposon-specific sequence as a probe will test the quality of the transposon library, including the presence of multiple insertions in single clones. If the transposon mutants are being immediately screened or selected for the desired phenotype, it becomes more important to confirm each candidate transposon mutant as outlined in Subheading 3.3, step 7 and Note 8 or by Southern blotting as mentioned above.

7. Factors such as inadequately spaced restriction sites or incompatibility of *B. subtilis* genetic sequences in *E. coli* can occasionally result in transposon mutants that cannot be analyzed by plasmid rescue. In these situations, an inverse PCR approach (8) using primers mini-Tn1 and mini-Tn2 (see item 12 in Subheading 2.5) has proven successful in identifying otherwise refractory insertion sites.
8. The natural ability of *B. subtilis* to uptake genomic DNA (natural competence) and integrate DNA via a double crossover homologous recombination event, allows for an easy yet powerful approach of assuring that a single transposition is indeed responsible for an observed phenotype. When transforming genomic DNA into the wild-type strain, an overwhelming majority of spectinomycin-resistant colonies should have the desired phenotypes. In those instances where secondary mutations have led to the observed phenotype, a large majority of spectinomycin-resistant colonies should have an otherwise wild-type appearance. Depending on the frequency with which secondary mutations are observed to be responsible for desired phenotypes, it might be more cost-effective to complete the backcross under Subheading 3.6 before sequencing DNA as per Subheading 3.5.

Acknowledgments

This work was supported partly by Grant GM19416 from the Institute of General Medical Sciences, and Grant AI055860 from the National Institute of Allergy and Infectious Diseases, National Institutes of Health, awarded to James A. Hoch.

References

1. Dartois V., Djavakhishvili T., and Hoch J.A. (1996) Identification of a membrane protein involved in activation of the KinB pathway to sporulation in *Bacillus subtilis*. *J Bacteriol.* **178**, 1178–1186.
2. Inaoka T. and Ochi K. (2007) Glucose uptake pathway-specific regulation of synthesis of neotrehalosadiamine, a novel autoinducer produced in *Bacillus subtilis*. *J Bacteriol.* **189**, 65–75.
3. Szurmant H., Nelson K., Kim E.J., Perego M., and Hoch J.A. (2005) YycH regulates the activity of the essential YycFG two-component system in *Bacillus subtilis*. *J Bacteriol.* **187**, 5419–5426.
4. Petit M.A., Bruand C., Janniere L., and Ehrlich S.D. (1990) Tn10-derived transposons active in *Bacillus subtilis*. *J Bacteriol.* **172**, 6736–6740.
5. Steinmetz M. and Richter R. (1994) Easy cloning of mini-Tn10 insertions from the *Bacillus subtilis* chromosome. *J Bacteriol.* **176**, 1761–1763.
6. Wilson A.C., Perego M., and Hoch J.A. (2007) New transposon delivery plasmids for insertional mutagenesis in *Bacillus anthracis*. *J Microbiol Methods.* **71**, 332–335.
7. de Boer H.A., Comstock L.J., and Vasser M. (1983) The tac promoter: a functional hybrid derived from the trp and lac promoters. *Proc Natl Acad Sci U S A.* **80**, 21–25.
8. Ochman H., Gerber A.S., and Hartl D.L. (1988) Genetic applications of an inverse polymerase chain reaction. *Genetics.* **120**, 621–623.

Chapter 22

Integrative Food Grade Expression System for Lactic Acid Bacteria

Grace L. Douglas, Yong Jun Goh, and Todd R. Klaenhammer

Abstract

Lactobacillus acidophilus NCFM is a probiotic microbe with the ability to survive passage to the gastrointestinal tract, interact intimately with the host and induce immune responses. The genome of NCFM has been determined and the bacterium is genetically accessible. Therefore, *L. acidophilus* has excellent potential for use as a vaccine delivery vehicle to express antigens at mucosal surfaces. Plasmids, commonly used to carry antigen encoding genes, are inherently unstable and require constant selection by antibiotics, which can be problematic for in vivo studies and clinical trials. Chromosomal expression of gene cassettes encoding antigens offers enhanced genetic stability by eliminating requirements for marker selection. This work illustrates the integration and inducible expression of the reporter gene *gusA3*, encoding a β-glucuronidase (GusA3), in the *L. acidophilus* chromosome. A previously described *upp*-counterselectable gene replacement system was used to direct insertion of the *gusA3* gene into an intergenic chromosomal location downstream of *lacZ* (LBA1462), encoding a β-galactosidase. The transcriptional activity of integrated *gusA3* was evaluated by GUS activity assays using 4-methyl-umbelliferyl-β-D-glucuronide (MUG) and was determined to be one to two orders of magnitude higher than the GusA3-negative parent, NCK1909. The successful chromosomal integration and expression of GusA3 demonstrate the potential of this method for higher levels of inducible gene expression in *L. acidophilus.*

Key words: Chromosomal gene insertion, Probiotic, β-Glucuronidase

1. Introduction

Lactobacillus acidophilus NCFM is a widely used probiotic microbe that has demonstrated the ability to adhere to the intestinal epithelium, induce immunomodulation, and inhibit pathogens (1, 2). The potential exists to expand the benefits of *L. acidophilus* by introducing and expressing beneficial genes, specifically bioactive proteins, enzymes, and vaccines. The ability of probiotic lactobacilli to survive passage through the stomach

James A. Williams (ed.), *Strain Engineering: Methods and Protocols*, Methods in Molecular Biology, vol. 765,
DOI 10.1007/978-1-61779-197-0_22, © Springer Science+Business Media, LLC 2011

and interact with dendritic cells at the intestinal mucosa to initiate immune responses has generated significant interest for their use as orally administered vaccine delivery vehicles. Expression of cloned antigen genes from probiotic bacteria to elicit an immune response in the gastrointestinal tract has shown great potential in recent years (3). However, due to genetic instability, genes carried on plasmids often require maintenance with antibiotic pressure that will be problematic during in vivo clinical trials. Expression of genes directly from the chromosome would eliminate plasmid stability issues and provide a food grade delivery system suitable for expression of antigens.

Meanwhile, the need to functionally analyze the molecular mechanisms enabling *L. acidophilus* NCFM to survive intestinal transit and interact with the intestinal mucosa has necessitated development of efficient gene inactivation and replacement systems (4, 5). These systems first select for a single-crossover homologous recombination using a pORI plasmid construct to generate isogenic mutants for *L. acidophilus* (4–6). Subsequently, selection for a second homologous recombination event enables replacement of a target gene with a deletion mutant allele, eliminating the plasmid backbone and associated issues with (1) unknown effects of antibiotic selection required to maintain inserts and (2) the ability to inactivate multiple genes in a single strain using the same targeting and selectable markers. The extensive screening required to identify double-crossover events limited the efficiency of traditional approaches. This was overcome by the development of a uracil phosphoribosyltransferase gene (*upp*)-based counterselectable marker in the pORI-based system to provide positive selection for excision of the integrated plasmid and selection of double-crossover recombinants. This approach has been used to obtain deletion mutants in *L. acidophilus* NCFM with minimal screening (5).

The *upp* counterselectable gene replacement system was used here for the first time to integrate and express a reporter gene in the chromosome of *L. acidophilus* NCFM. The reporter gene, *gusA3*, encoding a β-glucuronidase (GusA3), was successfully inserted into the chromosomal intergenic location downstream of *lacZ* (LBA1462), encoding a β-galactosidase, with no loss of DNA at the insertion site. The *lacZ* gene is induced and highly transcribed in the presence of lactose (7). This transcriptional activity was exploited for the expression of GusA3, measured by the hydrolysis of 4-methyl-umbelliferyl-β-D-glucuronide (MUG). The successful *gusA3* integration into a targeted location, for inducible expression by lactose, demonstrates the potential of this method for stable and efficient chromosomal integration and expression of genes that can potentially expand the beneficial applications of *L. acidophilus* and related commensal bacteria.

2. Materials

2.1. Bacterial Cultures and DNA Isolation

1. Strains (Table 1): Cloning host, NCK1831 (8); integration cloning vector, pTRK935 (5); *Lactobacillus acidophilus* NCFM NCK56 (1); background strains for the *upp*-based counterselection system, NCK1909 and NCK1910 (5); *gusA3* source, pTRK892 (7).
2. Chromosomal DNA isolation: ZR Bacterial Fungal DNA Miniprep Kit (Zymo Research Corporation, Orange, CA).
3. Plasmid isolation: Qiagen QIAprep Spin Miniprep Kit (Qiagen Inc., Valencia, CA).

2.2. Cloning and Gene Integration

1. Primers are designed manually or with Clone Manager Professional 9 (Sci-Ed Software, Cary, NC) and synthesized by Integrated DNA Technologies (IDT, Coralville, IA) or an alternative vendor. See Table 2 for primers for *gusA3* integration downstream of chromosomal *lacZ*.
2. Restriction enzyme digests: Restriction enzymes *Sal*I, *Not*I, and *Pvu*I (Roche Molecular Biochemical, Indianapolis, IN) with appropriate buffer as per manufacturer's instructions.

Table 1
Bacterial strains and plasmids used in this study

Strains and plasmids	Genotype/characteristics	References
E. coli strains		
NCK1831	EC101 RepA$^+$ host strain for pORI-based plasmids, Kn^r	(8)
NCK1911	EC101 harboring pTRK935	(5)
NCK1978	MC1061 harboring pTRK892	(7)
L. acidophilus strains		
NCK56	*L. acidophilus* NCFM	(1)
NCK1909	*L. acidophilus* NCFM Δ*upp* strain	(5)
NCK1910	NCK1909 harboring pTRK669, host for gene targeting insertion or replacement	(5)
Plasmids		
pTRK669	Temperature-sensitive helper plasmid, *repA*, Cm^r	(4)
pTRK892	Source of *gusA3*, Em^r	(7)
pTRK935	Counterselective integration vector with *upp* expression cassette, *lacZ*, Em^r	(5)

Table 2
Cloning and sequencing primers for *L. acidophilus* NCFM *gusA3* chromosomal integration

Primer name	Primer sequence[a]	5′ Restriction site
Cloning primers		
lacZF1	CAAG GTCGAC-TCTCGTCTGTATATTCTAAC	with *Sal*I
lacZR1	CAAG GCGGCCGC-CGAACGAAAATGTCCGGCCT	with *Not*I
lacZF2	CAAG GCGGCCGC-ACCTTATTTATTTGATCTACGG	with *Not*I
lacZR2	CAAG CGATCG-TCTATGAACGCAATATTCC	with *Pvu*I
lacZ_p892_GusF2	CAAG GCGGCCGC-TAAGAAGGCTGAATTCTAC	with *Not*I
lacZ_p892_GusR2	CAAG GCGGCCGC-TGAGCACGATTATTTG	with *Not*I
PCR screening primers		
pTRK935up	TGAAATACCGCACAGATG	
pTRK935down	ACACAGGAAACAGCTATG	
lacZscF1	CGCACAGATGCGTAAGGAG	
lacZscR1	CGCGATGAGTAACCGAACC	
lacZscF2	AATAAAGAGTTGCCTGATCCTGAAG	
lacZscR2	TAACGGTTTAAGCAGACAAAGTCAC	
lacZscF3	CGTCCTTATACTGGAACTTTAG	
lacZscR3	CCCAGGCTTTACACTTTATG	
lacZup	GAATCCATGAGTCGAAATATC	
lacZdown	TTTAGGGTCAAAGACTAAGG	

[a]Enzyme restriction sites are *underlined*

3. Polymerase Chain Reaction (PCR) for cloning: Primers, nucleotides, and High Fidelity DNA Polymerase (Roche) with appropriate buffer, according to manufacturer.
4. DNA purification after agarose gel electrophoresis: Zymo Gel DNA Recovery Kit followed by Zymo Clean and Concentrator-5 (Zymo Research Corporation).
5. Dephosphorylation of DNA: SuperSAP phosphatase (USB Corp., Cleveland, OH).
6. Ligation: T4 DNA ligase (New England Biolabs, Ipswich, MA) or Fast-Link DNA ligation kit (Epicentre Biotechnologies, Madison, WI).
7. Antibiotic stocks: Erythromycin (Em) 50 mg/ml in 70% ethanol; kanamycin (Kn) 40 mg/ml in deionized water, filter-sterilized through a Nalgene 0.45 μm filter; chloramphenicol (Cm) 10 mg/ml in 70% ethanol; penicillin G 10 mg/ml in deionized water, filter-sterilized through a Nalgene 0.45 μm filter. Store Em, Kn, and Cm stocks at −20°C. Store penicillin G at 4°C for 1–2 months.
8. *E. coli* EC101 growth medium: Prepare Brain Heart Infusion (BHI) (Becton, Dickinson and Company (BD), Sparks, MD)

broth in distilled water, autoclave at 121°C for 20 min, cool to ~55°C, and add Kn to a final concentration of 40 μg/ml.

9. *E. coli* EC101 chemical competent cell modified RF1 solution (9): Dissolve 10 mM potassium acetate, 50 mM magnesium chloride, 100 mM rubidium chloride, 10 mM calcium chloride, and 15% (wt/vol) glycerol in distilled water, adjust to pH 8.0, and filter-sterilize through a 0.45-μm filter. Store at 4°C for up to 3 years.
10. *E. coli* EC101 chemical competent cell modified RF2 solution (9): Dissolve 10 mM MOPS, 75 mM calcium chloride, 10 mM rubidium chloride, and 15% glycerol in distilled water, adjust to pH 6.5, and filter-sterilize through a 0.45-μm filter. Store at 4°C for up to 3 years.
11. *L. acidophilus* (NCK1910) growth medium: Prepare de Man Rogosa and Sharpe (MRS) (BD) broth in distilled water, autoclave, cool, and add Cm to a final concentration of 5 μg/ml.
12. NCK1910 competent cell buffer, 3.5× SMEB (10–12): Dissolve 1 M sucrose and 3.5 mM magnesium chloride in distilled water, adjust to pH 7, and filter-sterilize through a 0.45-μm filter. Store at 4°C for up to 3 years.
13. 5-Bromo-4-chloro-3-indolyl-β-D-galactopyranoside (X-gal): Dissolve 100 mg X-gal in 5 ml dimethyl formamide (20 mg/ml). Store in the dark at −20°C.
14. Isopropyl-β-D-thiogalactopyranoside (IPTG): Dissolve 1 g IPTG in 5 ml deionized water (200 mg/ml). Filter-sterilize through a 0.45-μm filter. Aliquot in 1-ml portions and store at −20°C.
15. Medium for selection of *E. coli* EC101 transformants: Prepare BHI medium both as broth and with 1.5% (wt/vol) agar in distilled water, autoclave, and add Em and Kn to a final concentration of 150 and 40 μg/ml, respectively.
16. Reagents for blue-white screening of transformants: Spread a mixture of 40 μl BHI broth, 40 μl X-gal, and 20 μl IPTG onto each plate and incubate at 37°C for 30 min prior to plating the transformants.
17. Medium for selection of NCK1910 transformants: Prepare MRS medium both as broth and with 1.5% agar in distilled water, autoclave, cool, and add both Em and Cm to a final concentration of 2 μg/ml.
18. NCK1910 plasmid integration selection medium for replica plating: Prepare MRS, both as a broth and with 1.5% agar, in distilled water, autoclave, cool, and add Em to a final concentration of 2 μg/ml. Prepare MRS agar (1.5%) in distilled water, autoclave, cool, and add Cm to a final concentration of 5 μg/ml.

19. Counterselection reagent for selection of double recombinants.

    5-fluorouracil (5-FU) stock solution: Prepare between 60 and 70 mg/ml 5-FU in dimethyl sulfoxide (DMSO). Store for up to 4 months at 4°C in the dark.

20. Counterselection medium for selection of double recombinants: *Lactobacillus* counterselection agar is a semi-defined medium with glucose (GSDM) (13) and 5-FU (5): Prepare 2% (wt/vol) glucose (dextrose), 0.1% (vol/vol) Tween 80, 0.2% (wt/vol) ammonium citrate, 0.5% (wt/vol) sodium acetate, 0.01 % (wt/vol) magnesium sulfate heptahydrate, 0.005% (wt/vol) manganese sulfate, 0.2% (wt/vol) dipotassium phosphate, 0.5% (wt/vol) yeast nitrogen base (BD), 1% (wt/vol) casitone (BD), and 1.5% agar in distilled water, autoclave, allow to cool, and add 5-FU stock solution to a final concentration of 100 μg/ml. Store plates for up to 2 months at 4°C in the dark.

21. Screening of transformants/recombinants (colony PCR): Choice-*Taq* Blue DNA polymerase (Denville Scientific Inc., Metuchen, NJ), standard PCR reagents.

22. 20–30% glycerol solution. Sterilize by autoclaving.

2.3. DNA Analysis

1. Agarose gels (0.8% wt/vol): Prepare fresh in 1× Tris–acetate-EDTA (0.04 M Tris–acetate, 0.001 M EDTA) (TAE) buffer (14).
2. Running buffer (1× TAE).
3. 1-kb Plus DNA Ladder (Invitrogen, Carlsbad, CA).
4. Sequencing primers are designed as described above (see item 1 in Subheading 2.2).

2.4. β-Glucuronidase Activity Assay (8, 15)

1. GUS Buffer: Prepare 100 mM sodium phosphate buffer (6 ml of 1 M disodium hydrogen phosphate and 44 ml of 1 M sodium dihydrogen phosphate in 450 ml distilled water, autoclaved) with 2.5 mM EDTA (pH 8.0) in distilled water, adjust to pH 6.0, and filter-sterilize through a 0.45-μm filter. Store at 4°C.
2. Stop Buffer: Prepare 0.2 M sodium carbonate and filter-sterilize through a 0.45-μm filter.
3. GUS/Stop Buffer: Prepare one part GUS buffer to four parts stop buffer. Store at 4°C.
4. MUG stock solution: Prepare 10 mM MUG in GUS buffer. Store at −20°C.
5. MUG working solution: Prepare 2 mM MUG in GUS buffer. Store at −20°C.
6. 4-Methylumbelliferone (4-MU) stock solutions: Prepare a 10 mM 4-MU stock in methanol. Dilute to 1,000 nM in GUS/Stop buffer. Store at −20°C.

7. Bovine serum albumin (BSA) stock solution: Prepare 12.5 mg/ml BSA in GUS buffer. Store at −20°C.
8. Semisynthetic medium (SSM) (16): Prepare 1% (wt/vol) bactopeptone (BD), 0.5% (wt/vol) yeast extract (BD), 0.2% (wt/vol) dipotassium phosphate, 0.5% (wt/vol) sodium acetate, 0.2% (wt/vol) ammonium citrate, 0.02% (wt/vol) magnesium sulfate heptahydrate, 0.005% (wt/vol) manganese sulfate, 0.1% (vol/vol) Tween 80, and 1% (wt/vol) of either glucose (dextrose) or lactose in distilled water and autoclave.
9. Bradford protein assay solution (Sigma) to measure protein concentration.
10. 0.1 mm glass beads (Biospec Products, Inc., Bartlesville, OK).
11. Costar flat-bottom 96-well plate (Corning Inc., Corning, NY).

3. Methods

3.1. Chromosomal Gene Integration and Expression

The following is an example of how a gene is inserted stably into the *L. acidophilus* NCFM chromosome using the *upp*-counterselective integration system, and then tested for gene expression levels. The chromosomal locus targeted to demonstrate integration and gene expression is located downstream of the *lacZ* gene encoding a β-galactosidase, which is inducible by lactose. The gene integrated into this loci is *gusA3*. To target an alternative transgene to a different chromosomal locus, new primers are designed using the design criteria described in Subheadings 3.2, 3.6, and 3.7 (see Table 2). *L. acidophilus* NCFM chromosomal DNA is isolated and used as a template for PCR amplification of target insertion regions.

3.2. Construction of a Plasmid for Insertion of gusA3 Downstream of lacZ (Fig. 1)

1. The *lacZ* gene is located on the complement strand of the *L. acidophilus* chromosome. Design two sets of primers, the first with 5′ restriction enzyme sites *Sal*I/*Not*I (*Sal*I in forward primer, *Not*I in reverse primer; Table 2) and the second with 5′ restriction enzyme sites *Not*I/*Pvu*I for directional triple ligation into the pTRK935 cloning vector (see Note 1). Primers of this design for targeting to *lacZ* are described in Table 2. Amplify the region immediately downstream of the *lacZ* gene (fragment 1) from *L. acidophilus* NCFM with High Fidelity DNA polymerase and the first set of primers, lacZF1/lacZR1 (623 bp) (Table 2). Amplify the 3′ end of the *lacZ* gene (fragment 2) with the second set of primers, lacZF2/lacZR2 (752 bp) to flank the target insertion location (see Note 2).
2. Co-electrophorese the PCR products alongside 2 μg of 1-kb Plus DNA ladder on a 0.8% agarose gel run at 110 V for 50 min, stain with ethidium bromide, and visualize under UV light. Gel-purify, and double digest the PCR products using

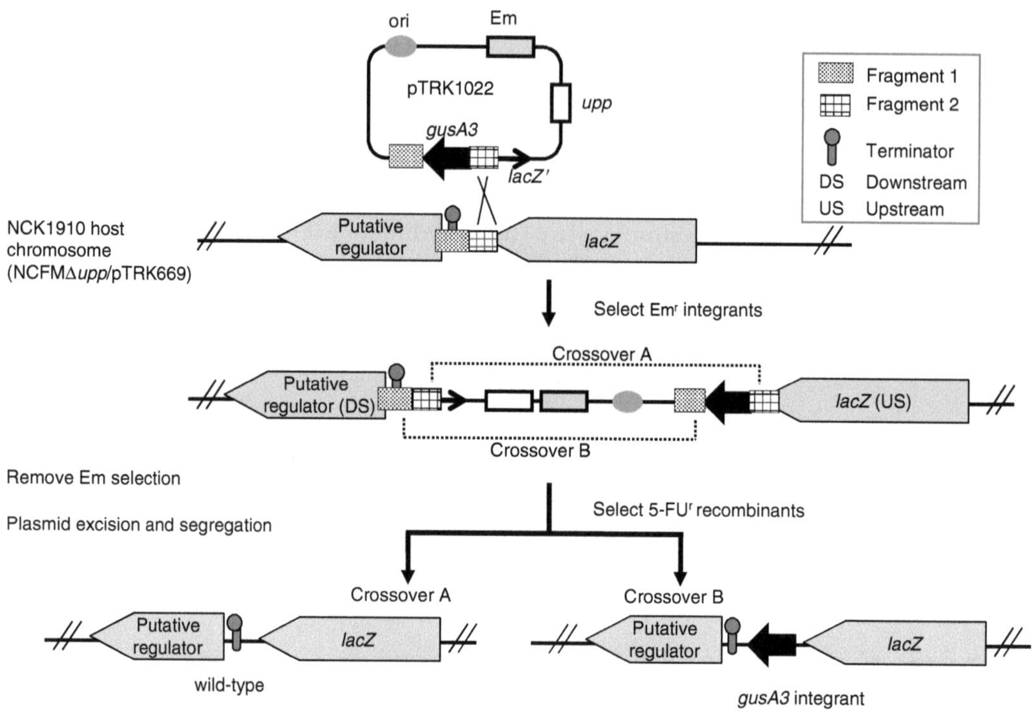

Fig. 1. Food grade chromosomal gene integration strategy. Single-crossover recombination of pTRK1022 occurs in the chromosomal *lacZ* region homologous to either fragment 1 or 2. Recombination via fragment 2 is shown. Removal of Em selection for the integrated plasmid facilitates a second recombination event and plasmid excision. The resultant double-recombinants can be either wild-type colonies or *gusA3* integrants. This figure is not drawn to scale.

*Sal*I/*Not*I (fragment 1) and *Not*I/*Pvu*I (fragment 2), respectively, for each primer set for 3 h at 37°C, according to manufacturer's instructions. Digest the pTRK935 cloning vector with *Pvu*I and *Sal*I and gel-purify the product.

3. Perform a triple ligation reaction (10 μl total) with 50 ng of vector at both 1:2:2 and 1:3:3 molar ratios of vector: insert1:insert2, per manufacturer's instructions. A control ligation containing only the digested cloning vector should also be included. Incubate the ligation mixture at 65–70°C for 10–15 min to inactivate the DNA ligase and then hold the samples on ice prior to transformation.

3.3. Preparation of E. coli EC101 Chemical Competent Cells, Modified from Hanahan (9)

1. Grow EC101 overnight in BHI broth with 40 μg/ml Kn at 37°C with aeration. Transfer the culture at 1% (vol/vol) inoculum into 100 ml of fresh BHI broth with 40 μg/ml Kn and grow as described above until the OD_{540nm} reaches about 0.4 (around 1–2 h).
2. Split the culture into 2–50 ml conical tubes, ice 20 min, and then centrifuge at 1,717×*g* for 10 min at 4°C. Discard the supernatant, resuspend each cell pellet in 20 ml of modified RF1 and hold for 20 min on ice (see Note 3). Repeat centrifugation

as described above, resuspend the cells all together in 3 ml of modified RF2 and hold for 20 min on ice.

3. Aliquot the cell suspension in 100 μl volumes into microcentrifuge tubes on ice, and store at −80°C for up to a year.

3.4. Plasmid Transformation into EC101, Modified from Hanahan (9)

1. Thaw EC101 competent cells on ice.
2. Transfer 5 μl of each ligation mixture from step 3 in Subheading 3.2 into separate microcentrifuge tubes containing 100 μl of competent EC101 cells and maintain on ice for 15–30 min. Additionally, add 5 μl of a mixture containing 50 ng of undigested plasmid to a tube of EC101 cells as a control for transformation efficiency.
3. Heat-shock the cells by placing each tube in a 42°C water bath for 2 min and return to ice for 2 min.
4. Recover each tube of transformants in 1 ml of prewarmed BHI medium at 37°C with aeration for 1–2 h and plate on BHI agar containing 150 μg/ml Em and 40 μg/ml Kn overlaid with X-Gal and IPTG. Colonies will be visible after 24–48 h.

3.5. Colony PCR Screening of EC101 Transformants (5)

1. The multiple cloning sites in pTRK935 are within the *lacZ* (*lacZ*-alpha) gene used for blue/white screening through alpha complementation in the Δ *lacZ E. coli* cloning hosts. Successful ligation products should produce white colonies whereas plasmids without an insert produce blue colonies. Choose white colonies for colony PCR screening for the *lacZ* recombinant plasmids and simultaneously grow each colony in BHI broth with 150 μg/ml Em and 40 μg/ml Kn (see Note 4).
2. Use pTRK935up and pTRK935down primers for colony PCR screening (Table 2). Use Choice-*Taq* Blue DNA Polymerase as per the manufacturer's instructions.
3. Co-electrophorese PCR products alongside 2 μg of 1-kb Plus DNA ladder on a 0.8% agarose gel run at 110 V for 30 min, stain with ethidium bromide, and visualize under UV light.
4. Select three positive clones containing the correct insert size, stock culture in 20–30% glycerol solution (final glycerol concentration 10–15%), and store at −80°C (see Note 5).
5. Isolate plasmids and sequence the inserts using pTRK935up and pTRK935down primers. Select a recombinant plasmid with the correct insert sequence (pTRK1021) for the subsequent *gusA3* insertions (Table 3).

3.6. Insertion of gusA3 into the NotI Site of the Cloned lacZ Insert of pTRK1021

1. Obtain primers to amplify the *gusA3* gene (lacZ_p892_GusF2/ lacZ_p892_GusR2) with 5′ *Not*I restriction sites for insertion into pTRK1021 (Table 2). PCR amplify the *gusA3* gene from pTRK892, using High Fidelity DNA polymerase.

Table 3
***E. coli* and *L. acidophilus* NCFM strains constructed for *gusA3* chromosomal integration and expression**

Host background	Description
E. coli EC101 strains	
NCK2137	Host of pTRK1021, pTRK935 recombinant plasmid containing *lacZ* fragments
NCK2138	Host of pTRK1022, pTRK1021 recombinant plasmid containing *gusA3*
L. acidophilus NCFM strains	
NCK2139	*gusA3* chromosomal insertion downstream of *lacZ*

2. Digest the amplified *gusA3* fragments with *Not*I and ligate into a similarly digested and SuperSAP phosphatase-treated pTRK1021. Perform the ligation at both 1:2 and 1:3 molar ratios, vector:insert as per manufacturer's instructions. Include a control ligation containing only digested pTRK1021 in the ligation. Inactivate the DNA ligase as described previously (step 3 in Subheading 3.2), and hold the samples on ice prior to transformation.
3. Repeat Subheading 3.4 and steps 1–4 in Subheading 3.5 following *gusA3* insertion into pTRK1021. All colonies will be white and can be randomly selected for screening (see Note 6).

3.7. Screening for Directional Orientation of gusA3 Insertion by PCR and DNA Sequencing

1. Design primers for sequencing the entire *lacZ::gusA3* (lacZ-scF1/R1, F2/R2, F3/R3) region in plasmids from selected transformants (Table 2). The *gusA3* gene is required to be in the same transcriptional direction as *lacZ* to enable *gusA3* expression. Before sequencing, confirm correct directional orientation with PCR using the lacZscF1/R2 primers. Transform a recombinant plasmid with the correct sequence, pTRK1022 (Table 3) into *Lactobacillus acidophilus* strain NCK1910 as described in Subheadings 3.8–3.10.

3.8. Preparation of NCK1910 Competent Cells (10–12)

1. Grow NCK1910 overnight in MRS broth with 5 μg/ml Cm at 37°C under static conditions in ambient atmosphere. Transfer the culture at a 2% (vol/vol) inoculum into 100 ml of fresh MRS broth with 5 μg/ml Cm and grow as described above until the OD_{600nm} reaches about 0.1–0.2 (around 3 h).
2. Add penicillin G to a final concentration of 10 μg/ml and incubate the culture for 1.5–2 h.

3. Split the culture into 2–50 ml conical tubes and centrifuge at 1,717 × *g* for 10 min at 4°C. Discard the supernatant.
4. Resuspend each pellet in 20 ml of cold 3.5× SMEB buffer by pipetting up and down gently (see Note 3). Centrifuge the culture again at 1,717 × *g* for 10 min at 4°C. Discard the supernatant.
5. Repeat step 4 two more times.
6. Resuspend the cells in a total of 1 ml of cold 3.5× SMEB for a 100-fold concentration. Competent NCK1910 cells must be made fresh each day prior to transformation and kept on ice.

3.9. Plasmid Transformation of NCK1910

1. Add around 500 ng of plasmid DNA to 200 μl of competent NCK1910 cells in a cold microcentrifuge tube and mix gently.
2. Transfer the cell/DNA mixture into a cold 0.2 cm electroporation cuvette and electroporate at 2.5 kV, 25 μFD, and 400 Ω. Return cells to ice for 2 min.
3. Recover cells in 1 ml of MRS broth at 37°C overnight and then plate on MRS agar with 2 μg/ml of both Em and Cm. Incubate plates under anaerobic conditions for 24–72 h.

3.10. Selection and Screening for Homologous Recombination Events in the NCK1910 Chromosome (Fig. 1)

1. Choose colonies and screen similarly to that described in steps 2 and 3 in Subheading 3.5. Propagate colonies chosen for PCR in MRS broth with 2 μg/ml of both Em and Cm overnight at 37°C under static conditions.
2. Stock three positive transformants in glycerol as described previously (step 4 in Subheading 3.5). Transfer two of these cultures three times at 1% inoculum (~30 generations) in MRS broth with 2 μg/ml Em in a 42°C water bath, which selects for the loss of the temperature-sensitive replication helper plasmid, pTRK669.
3. Dilute the cells and plate at 10^{-6} on MRS agar with 2 μg/ml Em and grow for 48 h at 37°C under anaerobic conditions.
4. Select around 50–100 colonies and replica plate on MRS agar with 2 μg/ml Em and MRS agar with 5 μg/ml Cm and grow for 48 h at 37°C under anaerobic conditions. The Em-resistant, Cm-sensitive colonies indicate integration of the plasmids into the targeted chromosomal loci via a single-crossover homologous recombination event. The *upp*-based counterselectable marker present on the integrated plasmid also renders the colonies sensitive to 5-FU.
5. Choose three Em-resistant, Cm-sensitive colonies, grow overnight in MRS broth with 2 μg/ml Em at 37°C, and stock in glycerol as described previously (step 4 in Subheading 3.5) (see Note 7). Transfer two of these cultures three times at 1%

inoculum in MRS broth without Em and incubate at 37°C. Dilute the cultures to 10^{-3} and plate on 5-FU GSDM plates for 24–72 h at 37°C under anaerobic conditions in the dark. Colonies that grow should have lost the plasmid backbone and its associated *upp* expression cassette through a double-crossover recombination event, leaving a mix of colonies that are either wild type or that contain a *gusA3* integrant allele (Fig. 1).

6. Screen 5-FU resistant colonies with PCR for the *gusA3* integration event using primers specific for upstream and downstream of the chromosomal integration region (lacZup/lacZdown) and Choice *Taq* Blue DNA Polymerase (Table 2) (see Note 8). Propagate colonies chosen for PCR in MRS broth overnight at 37°C under static conditions.
7. Stock three positive integrants in glycerol as previously described (step 4 in Subheading 3.5). Isolate genomic DNA and PCR amplify the region containing the integration with High Fidelity DNA polymerase and with the primers lacZup/lacZdown. Gel purify the products and sequence with the primers lacZup/lacZdown and lacZscR1/F2/R2/F3 (Table 2). Select an integrant with the correct sequence (NCK2139) for testing of GusA3 activity.

3.11. Analysis of β-Glucuronidase Activity, Modified from Duong (8) and Russell (15)

1. Grow NCK2139 to log phase (OD_{600nm} = 0.5–0.7) in MRS broth and collect 10 ml aliquots by centrifugation at 1,717 × *g* for 10 min at room temperature. Resuspend cultures in 10 ml SSM with 1% lactose for 1 h at 37°C to induce *lacZ* transcription, and subsequently *gusA3* transcription. For comparison, resuspend cultures in 10 ml SSM with 1% glucose. Use NCK1909, the background Δ*upp* strain, as a GusA3 negative control.
2. Prepare cell-free extracts (CFEs) by centrifugation of the cultures at 1,717 × *g* for 10 min at 4°C. Wash the pellets two times in 10 ml of cold GUS buffer. Resuspend the pellets in 0.5 ml of GUS buffer, transfer to prechilled 2 ml screw cap tubes with 0.5 g of 0.1 mm glass beads, and disrupt by bead beating for three 1-min cycles, with 1 min on ice between each cycle. After centrifugation at 8,600 × *g* for 2 min at room temperature, transfer CFEs to chilled microcentrifuge tubes and hold on ice.
3. Preparation of standards for Bradford assay (prepared fresh): Dilute the 12.5 mg/ml BSA stock to 0.25, 0.5, 0.75, and 1.0 mg/ml protein in GUS buffer to obtain a standard curve (Abs_{595nm} for 0–1.0 mg/ml protein).
4. Bradford assay: Add 10 μl of each CFE (diluted if necessary) to a Costar flat-bottom 96-well plate. Add 10 μl of each BSA standard to the 96-well plate. Then add 190 μl of Bradford

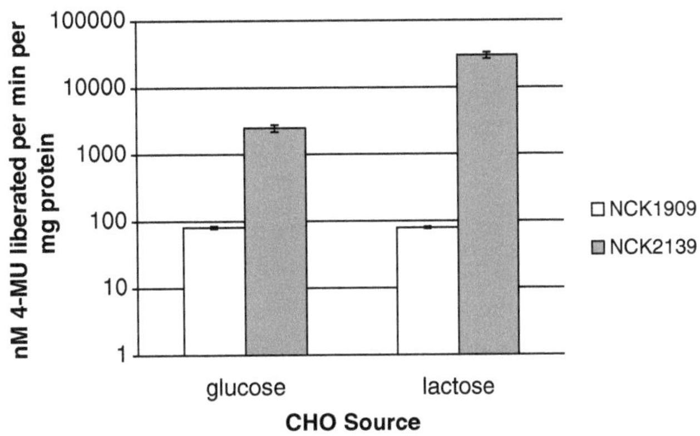

Fig. 2. GusA3 activity of *L. acidophilus gusA3* integrant (NCK2139, chromosomal insertion of *gusA3* downstream of the *lacZ* gene) in the presence of lactose (inducer) or glucose (control). NCK1909, reference strain. The data are the means +/– one standard deviation of two independent replicates.

assay solution to each well and incubate the plates for 5 min at room temperature. Measure Abs_{595nm} using a 96-well microtiter plate reader.

5. Preparation of standards for GUS assay (prepared fresh daily): Dilute the 1,000 nM 4-MU stock to 10, 50, 100, 150, 200, and 250 nM in GUS/Stop buffer to obtain a standard curve (0–250 nM 4-MU).
6. GUS assay: Serially dilute each CFE from 0.1 to 0.0001 mg/ml protein in GUS buffer (see Note 9). Transfer the diluted CFEs (100 μl) into clean microcentrifuge tubes, incubate for 1 min at 37°C, and vortex with 100 μl MUG working solution. After 5 min of incubation at 37°C, add 800 μl of stop buffer and vortex the samples. Then transfer 200 μl of each sample and 200 μl of each MU standard to a 96-well plate. Measure fluorescence intensity using a 96-well plate reader at 355 nm excitation and 460 nm emission.
7. Define GusA3 activity as nanomoles of 4-methylumbelliferone released per min per mg protein (Fig. 2).

4. Notes

1. Restriction enzyme sequences chosen for the 5′ end of each primer should not be present elsewhere within PCR products, and should only be present once in the cloning vector at the chosen insertion location. For convenience, select two enzymes with a compatible reaction buffer.
2. Both PCR fragments representing the upstream or downstream regions flanking the target insertion site should be

around 600–800 bp each for homologous recombination to efficiently take place.

3. To create a uniform cell suspension, it is easier to resuspend the cell pellet first in 10 ml of buffer, then add the remaining 10 ml of buffer.
4. Generally a successful clone is obtained after screening around 30 white colonies. However, more extensive screening may be required. Additionally, sometimes positive clones are blue, as a result of an insert that is cloned in-frame with the *lacZ* gene, so blue colonies may be screened if white colonies do not contain the recombinant plasmid.
5. Multiple cultures are stocked at each step to ensure a reserve culture is available if the final DNA sequence contains errors.
6. Blue-white screening with IPTG and X-gal will no longer be applicable for screening of positive clones at this step because the *lacZ* gene in pTRK935 has already been interrupted with the previous ligation.
7. Replica plating of 200 colonies without obtaining a Cm sensitive colony indicates that the selected chromosomal location may be essential and disruption by gene integration affects cell viability. In this case, an alternate gene integration site should be selected.
8. Usually a successful integrant is found after screening only 30 colonies, but additional colonies may be screened if necessary.
9. Different protein dilutions may be required to obtain a valid fluorescence measurement based on the transcriptional activity of the chromosomal integration location and the resulting expression of *gusA3*.

Acknowledgments

This work was supported in part by the North Carolina Dairy Foundation and Danisco USA, Inc. (Madison, WI). GD was supported by an NIH-Molecular Biotechnology Training Fellowship, and an IFT Graduate Scholarship. We are grateful to S. O'Flaherty, R. Sanozky-Dawes, E. Pfeiler, E. Durmaz, and J. Schroeter for comments and insightful discussions.

References

1. Sanders M. E., and Klaenhammer T. R. (2001) Invited review: the scientific basis of *Lactobacillus acidophilus* NCFM functionality as a probiotic. *J. Dairy Sci.* 84, 319–331.
2. Altermann E., Russell W. M., Azcarate-Peril M. A., Barrangou R., Buck B. L., McAuliffe O., Souther N., Dobson A., Duong T., Callanan M., Lick S., Hamrick A., Cano R., and Klaenhammer T. R. (2005) Complete genome sequence of the probiotic lactic acid bacterium *Lactobacillus acidophilus* NCFM. *Proc. Natl. Acad. Sci. U.S.A. 102*, 3906–3912.
3. Wells J. M., and Mercenier A. (2008) Mucosal delivery of therapeutic and prophylactic molecules using lactic acid bacteria. *Nat. Rev. Microbiol.* 6, 349–362.
4. Russell W. M., and Klaenhammer T. R. (2001) Efficient system for directed integration into the *Lactobacillus acidophilus* and *Lactobacillus gasseri* chromosomes via homologous recombination. *Appl. Environ. Microbiol.* 67, 4361–4364.
5. Goh Y. J., Azcarate-Peril M. A., O'Flaherty S., Durmaz E., Valence F., Jardin J., Lortal S., and Klaenhammer T. R. (2009) Development and application of a *upp*-based counterselective gene replacement system for the study of the S-layer protein SlpX of *Lactobacillus acidophilus* NCFM. *Appl. Environ. Microbiol.* 75, 3093–3105.
6. Maguin E., Duwat P., Hege T., Ehrlich D., and Gruss A. (1992) New thermosensitive plasmid for gram-positive bacteria. *J. Bacteriol.* 174, 5633–5638.
7. Duong T., Miller M. J., Barrangou R., Azcarate-Peril M. A., and Klaenhammer T. R. (2011) Construction of vectors for inducible and constitutive gene expression in *Lactobacillus*, *Microbial Biotechnology.* 4, 357–367.
8. Law J., Buist G., Haandrikman A., Kok J., Venema G., and Leenhouts K. (1995) A system to generate chromosomal mutations in *Lactococcus lactis* which allows fast analysis of targeted genes. *J. Bacteriol.* 177, 7011–7018.
9. Hanahan D. (1985) Techniques for transformation of *E. coli*, in *DNA cloning: a practical approach* (Glover, D. M., Ed.), pp 109–135, IRL Press Ltd., Oxford, England.
10. Luchansky J. B., Kleeman E. G., Raya R. R., and Klaenhammer T. R. (1989) Genetic transfer systems for delivery of plasmid deoxyribonucleic acid to *Lactobacillus acidophilus* ADH: conjugation, electroporation, and transduction. *J. Dairy Sci.* 72, 1408–1417.
11. Wei M.Q., Rush C. M., Norman J. M., Hafner L. M., Epping R. J., and Timms P. (1995) An improved method for the transformation of *Lactobacillus* strains using electroporation. *J. Microbiol. Methods* 21, 97–109.
12. Walker D. C., Aoyama K., and Klaenhammer T. R. (1996) Electrotransformation of *Lactobacillus acidophilus* group A1. *FEMS Microbiol. Lett.* 138, 233–237.
13. Kimmel S. A., and Roberts R. F. (1998) Development of a growth medium suitable for exopolysaccharide production by *Lactobacillus delbrueckii* ssp. *bulgaricus* RR. *Int. J. Food Microbiol.* 40, 87–92.
14. Sambrook J., Fritsch E. F., and Maniatis T. (1989) *Molecular Cloning, A Laboratory Manual*, Vol. 1, 2 ed., Cold Spring Harbor Laboratory Press, New York.
15. Russell W. M., and Klaenhammer T. R. (2001) Identification and cloning of *gusA*, encoding a new beta-glucuronidase from *Lactobacillus gasseri* ADH. *Appl. Environ. Microbiol.* 67, 1253–1261.
16. Barrangou R., Altermann E., Hutkins R., Cano R., and Klaenhammer T. R. (2003) Functional and comparative genomic analyses of an operon involved in fructooligosaccharide utilization by *Lactobacillus acidophilus. Proc. Natl. Acad. Sci. U.S.A.* 100, 8957–8962.

Chapter 23

ClosTron-Mediated Engineering of *Clostridium*

Sarah A. Kuehne, John T. Heap, Clare M. Cooksley, Stephen T. Cartman, and Nigel P. Minton

Abstract

The genus *Clostridium* is a diverse assemblage of Gram positive, anaerobic, endospore-forming bacteria. Whilst certain species have achieved notoriety as important animal and human pathogens (e.g. *Clostridium difficile*, *Clostridium botulinum*, *Clostridium tetani*, and *Clostridium perfringens*), the vast majority of the genus are entirely benign, and are able to undertake all manner of useful biotransformations. Prominent amongst them are those species able to produce the biofuels, butanol and ethanol from biomass-derived residues, such as *Clostridium acetobutylicum*, *Clostridium beijerinkii*, *Clostridium thermocellum*, and *Clostridium phytofermentans*. The prominence of the genus in disease and biotechnology has led to the need for more effective means of genetic modification. The historical absence of methods based on conventional strategies for "knock-in" and "knock-out" in *Clostridium* has led to the adoption of recombination-independent procedures, typified by ClosTron technology. The ClosTron uses a retargeted group II intron and a retro-transposition-activated marker to selectively insert DNA into defined sites within the genome, to bring about gene inactivation and/or cargo DNA delivery. The procedure is extremely efficient, rapid, and requires minimal effort by the operator.

Key words: Clostridia, ClosTron, Gene knock-out, Group II intron, Modular shuttle vectors, Retro-transposition-activated marker, FLP recombinase

1. Introduction

The genus *Clostridium* is composed of a diverse collection of Gram positive, anaerobic bacteria, unified by their ability to form endospores (1). Certain species have achieved notoriety as important animal and human pathogens. Notable examples include *Clostridium difficile*, *Clostridium botulinum*, *Clostridium tetani*, and *Clostridium perfringens* (2). The vast majority of the genus are, however, entirely benign, and are able to undertake all

James A. Williams (ed.), *Strain Engineering: Methods and Protocols*, Methods in Molecular Biology, vol. 765,
DOI 10.1007/978-1-61779-197-0_23, © Springer Science+Business Media, LLC 2011

manner of useful biotransformations. Prominent amongst them are saccharolytic species able to produce organic acids and solvents, together with those organisms capable of cellulosic ethanol production. They include *Clostridium acetobutylicum*, *Clostridium beijerinkii*, and *Clostridium saccharolyticum*, able to produce the superior biofuel butanol (3), as well as those cellulolytic, ethanol-producing species such as *Clostridium thermocellum*, *Clostridium cellulolyticum*, and *Clostridium phytofermentans* (4). Yet other species, or their products, are finding therapeutic applications in the treatment of cancer (*Clostridium sporogenes* and *Clostridium novyii*) (5), aberrant muscular dysfunctions (botulinum toxin of *C. botulinum*) (6) and Dupuytren's contracture (collagenase from *Clostridium histolyticum*) (7). A prerequisite for both the rational development of therapeutic countermeasures against pathogens and the full exploitation of beneficial properties is to better understand, and thereafter modify, essential biological processes at the molecular level. A pivotal requirement is the ability to ablate or modify gene function through "knock-out" and "knock-in".

Until recently, most efforts directed at clostridial gene modifications were concentrated on conventional, recombination-based procedures in which plasmids carrying homologous sequences of the target region are inserted into the host genome at the chosen site by homologous recombination (8). In many instances, procedures relied on insertion of the entire plasmid into the chromosome by a Campbell-like mechanism, resulting in an unstable single cross-over insertion. In certain instances, allelic exchange has been accomplished, generating double crossover, stable mutants. The general ineffectiveness of these mutagenic approaches can largely be attributed to low frequencies of DNA transfer, the absence of vectors conditional for replication and the lack of availability of negative selection markers. The latter impedes the effective deployment of allelic exchange procedures.

In recent years, clostridial researchers have turned to the use of group II intron retargeting methodologies. This has been made possible by the seminal work of the Alan Lambowitz laboratory who have pioneered the use of such technology in heterologous systems (9, 10). Through a series of elegant studies, they were able to define, and hence exploit, the retargeting specificity of the Ll.ltrB group II intron of *Lactobacillus lactis* (11, 12). Their findings were exploited in the development of the group II intron-based, pMTL007 ClosTron plasmid, which for the first time allowed the reproducible generation of mutants in a wide range of different clostridial species (13). Crucially, the ClosTron incorporated a retro-transposition-activated marker (RAM) based on an *ermB* gene. RAM elements essentially comprise an inactive copy of an antibiotic resistance gene which becomes activated during the process of retro-transposition (insertion of the intron into an alternative target site). Consequently, the successful insertion of the intron into its target site is accompanied by acquisition of resistance to erythromycin (Em), in the case of the ClosTron. More recently, the system has been refined and streamlined (14, 15) to minimise the labour-intensity and maximise the accessibility of the mutagenesis method, and to add facilities to make multiple mutations (14) and deliver small amounts of heterologous "cargo" DNA to the chromosome (14). Full details may be found at http://www.clostron.com. ClosTron technology is now in widespread use (16–24).

2. Materials

2.1. Culture Media

1. Clostridial Growth Medium (CGM): Dissolve 2 g $(NH_4)_2SO_4$, 0.5 g KH_2PO_4, 1 g K_2HPO_4, 0.1 g $MgSO_4{\cdot}7H_2O$, 2 g tryptone, 1 g yeast extract, and 50 g glucose in 800 mL distilled water. Then add 0.75 mL $FeSO_4{\cdot}7H_2O$, 0.5 mL $CaCl_2$, 0.5 mL $MnSO_4{\cdot}H_2O$, 0.1 mL $CoCl_2$, and 0.1 mL $ZnSO_4$ (all from 20 g/L stock solutions). Adjust the pH to 7.0 with NaOH, make up to 1 L with distilled water, add agar to 1.2%, and sterilise by autoclaving. CGM is routinely used for *C. acetobutylicum*.
2. 2× Yeast Extract Tryptone Glucose Medium (2× YTG): Dissolve 16 g tryptone, 10 g yeast extract, and 5 g NaCl in 800 mL of dH_2O. Adjust the solution to pH 5.2 (using HCl). Then adjust the volume to 900 mL using dH_2O and sterilise by autoclaving. Once cool, aseptically add 100 mL of sterile 20% glucose solution, giving a 2% final glucose concentration. Adjust the final volume to 1,000 mL (if necessary) using sterile dH_2O. This medium is routinely used for *C. beijerinckii*.
3. Brain Heart Infusion Supplement (BHIS) Medium: This medium is the commercially available BHI medium (Oxoid) supplemented with 10 mg/L of L-cysteine and 5 mg/mL of yeast extract, and is routinely used for *C. difficile*.
4. Tryptone Yeast Extract Thioglycollate Medium (TYG): Dissolve 30 g tryptone, 20 g yeast extract, and 1 g sodium thioglycollate in about 800 mL of dH_2O. Adjust final volume to 1 L. Sterilise by autoclaving. TYG medium is routinely used for *C. sporogenes* and group I strains of *C. botulinum*.
5. L Broth (LB) Medium: Add the following to about 800 mL of dH_2O; 10 g tryptone extract, 5 g yeast extract, and 5 g NaCl. Dissolve, and adjust final volume to 1 L. Sterilise by autoclaving. LB is routinely used for *Escherichia coli*.
6. 2× Yeast Extract Tryptone Medium (2× YT): Dissolve in approximately 800 mL of dH_2O, 16 g tryptone, 10 g yeast extract, and 5 g NaCl. Adjust pH to 7.0 with 5N NaOH, and make up the volume to 1 L with dH_2O. Sterilise by autoclaving. This medium can also be used for *E. coli*.

7. Clostridium Basal Medium (CBMS): This medium is used when recycling the ErmRAM in ClosTron mutants of *C. acetobutylicum*. It is prepared as follows. First make up all the stock solutions below in water. Filter sterilise all stocks except glucose, which can be autoclaved. Store $MnSO_4$, $FeSO_4$, and vitamin stocks at 4°C. The other stocks can be stored at room temperature. Mix 200 mg $MgSO_4 \cdot 7H_2O$, 7.58 mL of a 1 mg/mL stock $MnSO_4 \cdot H_2O$, 10 mL of a 1 mg/mL stock $FeSO_4 \cdot 7H_2O$, 1 mL of a 1 mg/mL *p*-aminobenzoic acid stock, 20 μL of a 0.1 mg/mL stock biotin, 1 mL of a 1 mg/mL stock thiamine-HCl, 4 g casein hydrolysate (enzymatic) and make up to 800 mL with water. Sterilise by autoclaving, and then add 100 mL of a 50% w/v stock glucose (5% final), 10 mL of a 50 mg/mL stock K_2HPO_4, 10 mL of a 50 mg/mL stock KH_2PO_4, and 20 mL of a 250 g/L stock $CaCO_3$. Make the final volume up to 1,000 mL with sterile water. For CBMS agar, add agar to 1% and exclude the $CaCO_3$.

2.2. Antibiotic Supplements

Antibiotics are used when appropriate at the following concentrations: 100 μg/mL Ampicillin (Ap) from a stock solution of 10 mg/mL in dH_2O, 25 μg/mL chloramphenicol (Cm) in agar and 12.5 μg/mL in liquid (from stock solution of 25 mg/mL in ethanol), 15 μg/mL thiamphenicol (Tm) (from stock solution of 15 mg/mL in a 1:1 mix of ethanol and dH_2O), 10 μg/mL tetracycline (Tc) (from a stock solution of 10 mg/mL in ethanol), 2.5 μg/mL erythromycin (Em) (from a stock solution of 10 mg/mL in ethanol), 20 μg/mL lincomycin (Lm) (from a stock solution of 20 mg/mL in dH_2O), 250 μg/mL cycloserine, and 8 μg/mL cefoxitin (from a combined stock solution of 25 mg/mL cycloserine and 0.8 mg/mL cefoxitin in dH_2O or cycloserine separate, as required). Ap, Cm, and Tc are stored at –20°C, all others at 4°C (see Note 1).

2.3. Electrocompetent Clostridial Cells

1. 0.4 cm gap electroporation cuvettes (Biorad).
2. 284 mM sucrose solution: 9.72 g sucrose in 100 mL dH_2O, filter sterilised.
3. 100 mM sodium phosphate buffer (pH 7.4): 1.140 g anhydrous monobasic sodium phosphate and 5.749 g anhydrous dibasic sodium phosphate in 500 mL dH_2O. Check the pH and adjust to 7.4 if necessary with NaOH or HCl. Sterilise the solution by autoclaving.
4. Electroporation buffer (EPB): Aseptically add 1 mL *of* sterile 100 mM sodium phosphate buffer (pH 7.4) to 19 mL of sterile 284 mM sucrose solution (final concentrations are 5 mM sodium phosphate (pH 7.4) and 270 mM sucrose).

2.4. Molecular Biology Reagents

1. *E. coli* strains are donor strain CA434 (*thi*-1, *hsd*S20 (rB$^-$, mB$^-$), *supE44, recAB, ara-14, leuB5, proA2, lacY1, galK, rpsL20, xyl-5, mtl-1* – R702 Tra$^+$, Mob$^+$, conjugative plasmid Tc^R) and cloning host TOP10 (F-mcr*A*, Δ(mrr-hsd*RMS*-mcr*BC*), Φ80lac*Z*Δ*M15*, Δlac*X74*, rec*A1*, ara*D139*, Δ(ara leu)*7697*, gal*U*, gal*K*, rps*L*, end*A1*, nup*G*).
2. Plasmids used are pMTL007C-E2 (is based on the pCB102 replicon, carries *catP* and confers resistance to Cm in *E. coli* and Tm in *Clostridium* spp.) (14); pAN-2 (is based on the p15a replicon, carries a *tet* gene and confers resistance to Tc) is compatible with pMTL007C-E2, and carries the φ3TI methyltransferase gene of *B. subtilis* phage φ*3t*I, which protects DNA from *C. acetobutylicum Cac*824I DNA restriction activity (13); pMTL85151-PPS-flp3 (based on the PIM13 replicon, carries *catP* and confers resistance to Cm in *E. coli* and Tm in *Clostridium* spp.) carries a yeast *flp* gene encoding FLP recombinase under the transcriptional control of the promoter from the *C. acetobutylicum* thiolase (*thl*) gene (14).
3. The following oligonucleotide primers are required: EBS Universal (5′-CGAAATTAGAAACTTGCGTTCAGTAAAC-3′); spofdx-seq-F1 (5′-GATGTAGATAGGATAATAGAATCC ATAGAAAATATAGG-3′); pMTL007-R1 (5′-AGGGTAT CCCCAGTTAGTGTTAAGTCTTGG-3′); RAM-F (5′-ACG CGTTATATTGATAAAAATAATAATAGTGGG-3′), and; RAM-R (5′-ACGCGTGCGACTCATAGAATTAT TTCCTCCCG-3′) and should be used at 10 μM.
4. PCR template for generation of the intron retargeting fragment can be obtained from the Sigma-Aldrich TargeTron Kit (TA0100) or by using the plasmids pMTL20IT1 and pMTL20IT2 (14). Both plasmids are based on the ColE1 replicon, carry an Ap resistance gene (*bla*) and confer resistance to Ap.
5. DNA modifying enzymes are Phusion DNA polymerase and DNA Ligase, and the restriction endonucelases *Hin*dIII, *Bsr*GI, *Bgl*II, *Nde*I, *Not*I, *Xho*I, and *Sal*I. All enzymes are purchased from New England Biolabs and used under the conditions recommended by the manufacturer.
6. The following Qiagen kits are employed, and used according to the manufacturer's instructions: PCR purification kit (Cat no. 28104), Plasmid miniprep kit (Cat. no. 27106), Gel purification kit (Cat. no. 28704).
7. All DNA samples are electrophoresed on standard agarose gels at either 1.0% or 1.2% (w/v) dependent on the protocol.

3. Methods

The basic principle of ClosTron technology is to make specific changes to the group II intron such that it preferentially inserts into the DNA region of interest. This is accomplished using a computer algorithm able to predict the alterations necessary for intron retargeting. Such an algorithm can be implemented locally using published data (12), through the TargeTron design site of Sigma-Aldrich (http://www.sigma-genosys.com/targetron/) or using the free access design tool at http://www.clostron.com. The predicted changes are then introduced into the ClosTron plasmid (pMTL007 series), by the substitution of a 350-bp fragment between the unique *Bsr*GI and *Hin*dIII restriction sites of the chosen ClosTron plasmid with the newly designed, retargeted fragment.

Conventionally, the generation of the fragment is accomplished using a single Splicing by Overlap Extension (SOE) PCR (25) in conjunction with an appropriate set of four primers (three of which are designed through the intron retargeting algorithm, with a fourth universal primer) and a special template. This template is supplied as part of the Sigma-Aldrich TargeTron kit. Alternatively, two specially constructed plasmids (14), can be used as a template for the SOE PCR. However, the most cost- and time-effective method of generating the retargeted plasmids is to have the 350 bp fragment re-synthesised and cloned into the chosen pMTL007 vector by an appropriate gene synthesis company, such as DNA2.0. Using this route, the required plasmids can be available within 2 weeks following their design.

Once the required retargeted pMTL007-series plasmid has been generated, it is simply introduced into the desired clostridial host, either by electroporation or by conjugation from an *E. coli* donor, with initial selection for the antibiotic resistance gene present on the vector backbone, i.e., thiamphenicol (Tm) in the case of pMTL007C-E2 (Fig. 1a). A single Tm resistant clone is then streaked onto medium containing erythromcyin (Em). All Em resistant clones represent clones in which the group II::RAM element of the ClosTron has inserted into the chromosome, resulting in the activation of the *ermB* gene of the RAM element. Finally, the isolated integrant needs to be screened by PCR to ascertain whether the intron has inserted into the intended location in the genome. Screening should ideally be undertaken with four primers, capable of amplifying the two insertion junctions as well as the entire insertion. The latter act as a positive control for the presence of wild-type cells, to give a characteristic smaller DNA product in which no insertion has occurred. The PCR products generated at the junction sites should be subjected to nucleotide sequencing. It is essential that selected clones are

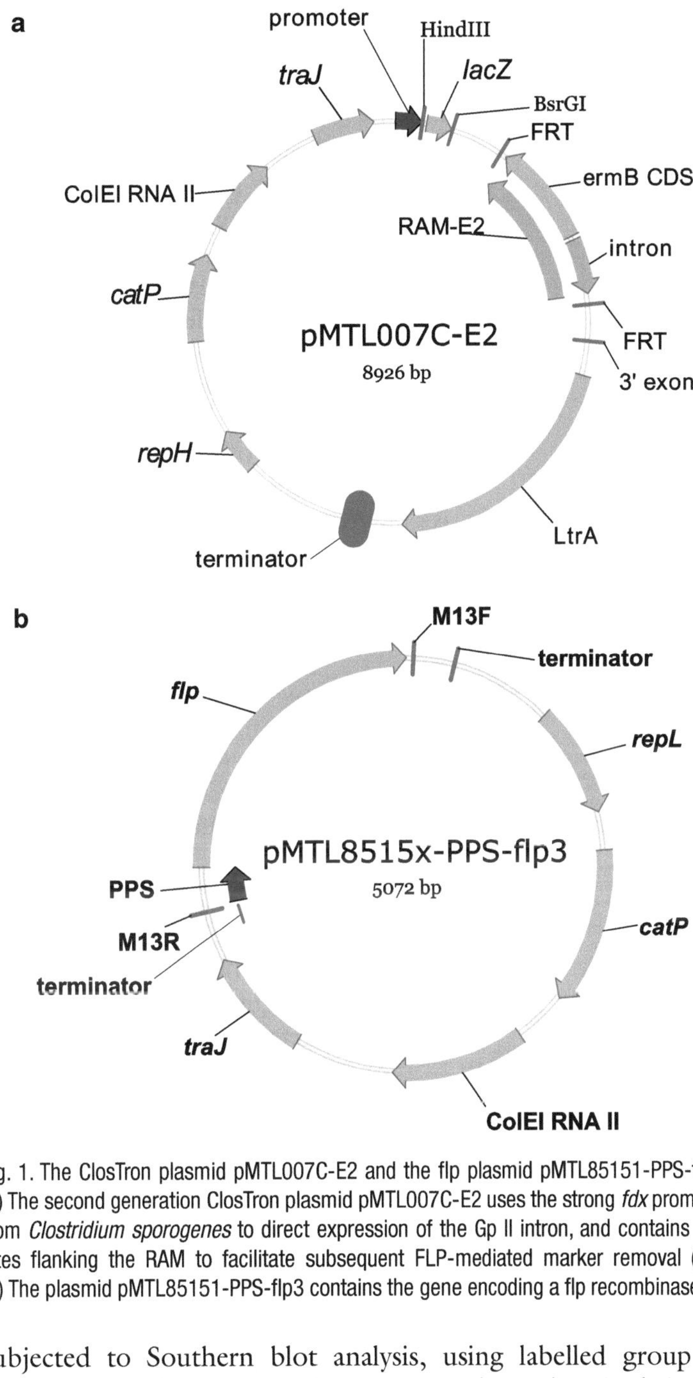

Fig. 1. The ClosTron plasmid pMTL007C-E2 and the flp plasmid pMTL85151-PPS-flp3. (**a**) The second generation ClosTron plasmid pMTL007C-E2 uses the strong *fdx* promoter from *Clostridium sporogenes* to direct expression of the Gp II intron, and contains FRT sites flanking the RAM to facilitate subsequent FLP-mediated marker removal (14). (**b**) The plasmid pMTL85151-PPS-flp3 contains the gene encoding a flp recombinase.

subjected to Southern blot analysis, using labelled group II intron-derived DNA as a probe, to ensure that only a single insertion of the element has occurred. All that remains is to check the integrants for loss of the pMTL007 delivery vehicle, easily determined by Tm sensitivity. This can be achieved through simple subculture in the absence of Tm selection. Loss of the pMTL007 plasmid, and its encoded *ltrA* gene, is essential, as the continued presence of LtrA protein in the cell may, if the intron is inserted into the sense strand, lead to splicing of the intron encoding sequence from the RNA. In this instance, the insertion would not be mutagenic.

Once an integrant has been isolated and verified a number of subsequent steps maybe instigated. Complementation of the mutant created is undertaken, using an appropriate vector carrying the wild-type gene. The pMTL80000 series of modular vectors are ideal for this purpose (15). In certain instances, further mutations may be desirable. In this case, the inserted *ermB* gene can be deliberately excised making use of directly repeated FLP Recognition Target (FRT) sites on some ClosTron plasmids. Following the introduction of a plasmid encoding FLP-recombinase (14), the transformant is subcultured and then screened for loss of Em resistance. Such a clone is then verified by PCR and nucleotide sequencing. Once a derivative lacking *ermB* is obtained, ClosTron technology can be used again to insertionally inactivate another gene (14).

3.1. Intron design

1. The easiest way to design the retargeted intron is to use the step-by-step guide found at http://www.clostron.com.
2. Choose whether to use the freely available tool at http://www.clostron.com based on the Perutka algorithm (12) or the equivalent tool available at the Sigma-Aldrich TargeTron Design Site. Access to the latter is secured by purchase of a Sigma-Aldrich TargeTron kit. Here, we assume you have chosen the former.
3. Follow the instructions on screen. Briefly, enter a name for your project, then paste in the sequence of your target gene. You will be offered a ranked order of target sites, together with the sequences of the primers EBS2 and EBS1d, specific for each retargeting region, and IBS primer required for splicing, necessary for generation of the 350-bp retargeting region by SOE PCR (25). You will also be given the entire sequence of the 350-bp retargeting region if you choose the DNA synthesis route. All this information can be downloaded.

3.2. Generation of Retargeted Plasmids by SOE PCR

1. Having secured the necessary oligonucleotides, make the following primer mixture: 12 μL dH_2O, 2 μL IBS (100 μM), EBS1d (100 μM), EBS2 (20 μM), and EBS Universal (20 μM) primers.
2. Assemble a PCR reaction using a proof reading polymerase, 1 μL of the above primer mixture and 1 μL template (either supplied with the Sigma-Aldrich TargeTron Design kit or a template made by mixing the two plasmids pMTL20IT1 and pMTL20IT2 in a ratio of 1:1 (14), ~100 ng). Prepare and perform the PCR in triplicate (see Note 2).

3. Use the following PCR cycle conditions: Denature at 94°C for 30 s followed by 30 cycles of 94°C for 15 s, 55°C for 30 s, and 72°C for 30 s with a final extension at 72°C for 2 min.
4. Visualise the PCR product on a 1% (w/v) agarose gel and cut the band (of all three reactions) that corresponds to the expected ~350-bp fragment. Purify the DNA, digest with *Bsr*GI and *Hin*dIII, and clean with a PCR clean-up kit (see Note 3).
5. Digest the ClosTron vector (here pMTL007C-E2, Fig. 1a) with the same pair of enzymes (*Bsr*GI and *Hin*dIII), ensuring complete digestion with both enzymes has occurred.
6. Check the vector is completely linearised on an agarose gel (load the entire digest), and excise the electrophoretic DNA band corresponding to the vector backbone. Additionally, you should see a ~250-bp fragment, corresponding to the unaltered targeting region. Purify the linearised vector DNA.
7. Ligate the retargeted region with the digested and purified vector. Perform also a control ligation, using only the vector DNA. Ligation can be performed for 30 min at room temperature.
8. The ligation mixture is then transformed by electroporation into *E. coli* TOP10. After recovery the cells are then plated onto medium containing 25 μg/mL Cm (when using pMTL007C-E2) to select for transformants and incubated at 37°C overnight. Additionally, "blue-white" selection using Xgal can be performed as the empty ClosTron vector carries a stuffer region encoding the LacZα fragment which will be replaced by the retargeting region.
9. After 24–48 h colonies big enough to be picked should appear. The presence of an insert with the correct size can either be verified by colony PCR, using the primers spofdx-seq-F1 and pMTL007-R1 (will yield a 548-bp fragment after retargeting and a 440-bp fragment from the parental plasmid), or by restriction digest. Make sure that there are no colonies on your control plate (see Note 4). Inoculate three separate 1-mL LB broth cultures from three independent colonies and incubate overnight at 37°C. In parallel, restreak each onto fresh agar medium, and incubate for 24–48 h, for safekeeping.
10. Extract the plasmid DNA from the purified transformant colonies using a standard Miniprep kit. To authenticate the clones, digest the isolated DNA using either *Sal*I or *Bgl*II. This should result in a linearised electrophoretic DNA fragment band when visualised on an agarose gel. The "empty" digested ClosTron vector should give two fragments as two *Sal*I/*Bgl*II sites are present, one of which disappears when the retargeted region is inserted. Clones that give the correct

restriction pattern must be sequenced as the retargeted region has a highly complex secondary structure which can easily cause mistakes when amplified by PCR. To sequence you can use the primers spofdx-seq-F1 and pMTL007-R1.

3.3. Generation of Retargeted Plasmids by Synthesis

1. The retargeting region is only 353-bp in size which can easily be synthesised by an appropriate DNA synthesis company and custom cloned into the ClosTron vector of choice, e.g., DNA2.0. On http://www.clostron.com you can find detailed instructions on how to order your ClosTron plasmids containing your retargeted region of choice.
2. Once received through the post (it will take about 2 weeks in total), transform the plasmid DNA into an *E. coli* TOP10 for safekeeping.

3.4. Plasmid Transfer by Conjugation

1. In the majority of clostridial species (*C. difficile*, *C. botulinum*, *C. sporogenes*, *C. beijerinckii*, *C. novyi*, and *Clostridium sordellii*), retargeted ClosTron plasmids are routinely introduced from an *E. coli* donor by conjugative plasmid transfer (26–29). Electrotransform the *E. coli* donor strain CA434 with your retargeted ClosTron plasmid and grow the resultant transformant overnight in 5 mL of LB medium supplemented with an appropriate antibiotic (here 12.5 μg/mL Cm) to ensure the plasmid is retained.
2. Also grow the recipient clostridial strain overnight at 37°C, in 1 mL of rich medium (BHIS for *C. difficile* and TYG for *C. botulinum* and *C. sporogenes*) under anaerobic conditions (see Note 5).
3. Pellet 1 mL of CA434 overnight culture harbouring your ClosTron plasmid at 1,500 × *g* in a bench top microfuge for 1 min, wash the pellet in 0.5 mL PBS, and spin as before. Take the pellet into the anaerobic cabinet.
4. In the anaerobic cabinet, resuspend the CA434 pellet in 200 μL of an overnight culture of the clostridial recipient. Then aliquot the mixture on one non-selective plate as eight drops of 25 μL. Do not invert the plate. Incubate at 37°C for between 8 and 24 h.
5. Using a disposable loop scrape the bacterial growth of the plate and resuspend in 1 mL of PBS. Plate the cells (200 μL per plate) on selective medium (selecting for the antibiotic resistance encoded on the ClosTron plasmid, for example, 15 μg/mL of Tm for pMTL007C-E2) including a counter-selection against the *E. coli* donor (usually cycloserine 250 μg/mL and for *C. difficile* 8 μg/mL cefoxitin in addition) and incubate for 1–3 days anaerobically, at 37°C.

3.5. Plasmid Transfer by Electroporation

1. Some clostridial strains (mainly *C. acetobutylicum*, *C. botulinum*, and *C. perfringens*) can be transformed by electroporation. The retargeted plasmid has to be purified from an *E. coli* strain and electroporated into electrocompetent clostridial cells. Methods for preparation and electroporation of electrocompetent clostridial cells have been described previously (30). The electroporation method outlined here for *C. acetobutylicum*, is essentially as described by Mermelstein et al. (31).
2. The retargeted plasmid needs to be purified from an *E. coli* host which also carries plasmid pAN-2, to ensure appropriate methylation of the ClosTron plasmid (13).
3. Place culture media, EPB, and agar plates in the anaerobic cabinet 1 day prior to use. Pre-chill electroporation cuvettes in the –20°C freezer.
4. Inoculate 10 mL of 2× YTG with a heavy loop of fresh *C. acetobutylicum* cells. Vortex to re-suspend thoroughly. Serially dilute the first 10 mL culture into 10^{-1}, 10^{-2}, and 10^{-3} 10 mL cultures. Incubate overnight anaerobically at 37°C.
5. Use all 10 mL of the most dilute overnight culture that still shows active growth to inoculate 60 mL of 2× YTG. Incubate at 37°C until $OD_{600} = 1.1$ (usually about 3.5 h).
6. Pour 35 mL of the culture into each of two 50 mL Falcon tubes and centrifuge at 4°C at 5,000 × *g* for 10 min.
7. Aliquot 20 mL of EPB on ice inside the anaerobic cabinet during this centrifugation.
8. Place the tubes on ice and carefully transfer into the anaerobic cabinet. Then carefully pour off the supernatants and re-suspend each pellet in 5 mL of EPB by vortexing. Return the tubes onto ice.
9. Then transfer them out of the anaerobic cabinet and centrifuge as before. Replace the tubes on ice and transfer them carefully back into the anaerobic cabinet.
10. Carefully pour off the supernatant and re-suspend each pellet in 1.15 mL EPB by vortexing. Transfer the cultures to one tube and return to ice. The cells are now ready to electroporate, and should be kept on ice and used promptly. There are enough cells for four transformations in total.
11. Add 20 μL (~100 ng) of plasmid DNA (methylated as appropriate) to each chilled 0.4 cm gap electroporation cuvette held on ice. Then transfer the cuvettes into the anaerobic cabinet.
12. Gently add 570 μL of competent cells to each cuvette, and incubate on ice for 2 min. Electroporate immediately using 2.0 kV, 25 μF, and ∞ Ω. Immediately add 1 mL of pre-warmed (37°C) anaerobic 2× YTG from a recovery tube to the cuvette

and mix gently. Transfer the entire transformation mixture back into the recovery tube and allow 1–3 h recovery (depending upon the antibiotic selection) (see Note 6).

13. Spin the cells at room temperature and discard the supernatant. Re-suspend the pellet in 1 mL of 2× YTG, plate transformants onto appropriate selective agar plates, and incubate for 1–3 days anaerobically at 37°C.

3.6. Screening for Intron Insertion

1. Once transformant/transconjugant colonies are big enough to pick (approx. 1 mm diameter) restreak two to four well-isolated colonies to purity.
2. Restreak a loop full on selective medium to screen for integrants or alternatively resuspend a loop full in PBS and plate on the appropriate medium (in the case of pMTL007C-E2 use 2.5 μg/mL Em as a spliced integrated RAM will confer Em resistance) (see Note 7).
3. Once colonies appear on the selective plates (1–3 days) restreak them to purity. Then inoculate 1 mL of liquid culture (use a medium appropriate to your clostridia species) and grow overnight. Extract the genomic DNA by a method of your choice and screen by PCR. Several different PCR reactions can be used: the RAM-F and RAM-R primers should give a band of ~900-bp for the integrant, which means a spliced RAM, otherwise the size should be ~1,300-bp. The most useful PCRs are across the intron–exon junction using a screening primer (which anneals outside of the retargeted region) and, depending on the sense of the insertion, the primers EBS universal and RAM-R. The different primers and combinations are outlined in Fig. 2.
4. To double check that your intron has inserted into your target gene it is advisable to sequence the PCR products from the intron–exon junctions (Fig. 2).
5. Finally, loss of the ClosTron plasmid needs to be confirmed. The integrant should not be able to grow on medium containing the antibiotic appropriate to the marker present on the vector backbone (here Tm consistent with the *catP* gene on pMTL007C-E2). Often the plasmid has already been lost during the preceding steps. If this is not the case, the integrant needs to be passaged several times on non-selective medium and then checked again for plasmid loss (on selective medium).

3.7. Southern Blot Analysis

1. It is always advisable (see Note 8) to establish that your isolated mutant contains a single intron insertion by Southern blot analysis using established methodology (see Note 9).
2. An intron-specific probe is best generated by PCR using EBS2 primer and a primer that anneals within the intron but

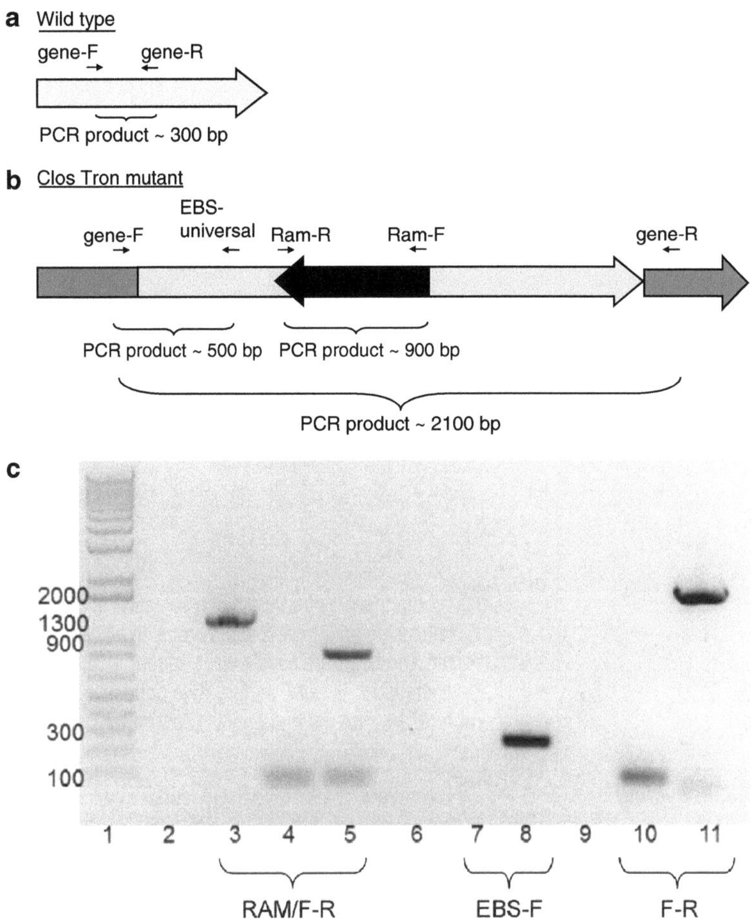

Fig. 2. Screening for a ClosTron mutant. (**a**) A schematic representation of the targeted wild type gene showing an appropriate position for the design of screening (flanking) primers. (**b**) The schematic representation of a ClosTron mutant showing annealing sites of the RAM-primers and possible PCR products that would be generated, including the exon–intron junction. (**c**) This agarose gel shows PCR products from a *tcdA* ClosTron mutant in *Clostridium difficile*. *Lanes*: 1, size makers; 2, empty; 3, pMTL007C-E2::*tcdA*; 4, wild type (wt); 5, ClosTron mutant (CT); 6, empty; 7, wt; 8, CT; 9, empty; 10, wt, and; 11, CT. The products in lanes 3–5 have been amplified with RAM-primers, in lanes 7 and 8 using EBS universal and a forward flanking primer, and in lanes 10 and 11, using forward and reverse flanking primers.

upstream of the *ermB* RAM. Avoidance of *ermB*-specific sequences in your probe is advisable as certain clostridial species can carry related sequences. Use genomic DNA of a ClosTron mutant as the PCR template (see Note 10).

3. The Southern blot analysis of a number of ClosTron mutants in *C. acetobutylicum* is shown in Fig. 3. One of them shows two bands (lane 6), indicative of the insertion of two intron elements. In this instance, examination of one of the other mutant replicates enabled the isolation of a strain containing only one intron insertion (see Note 11).

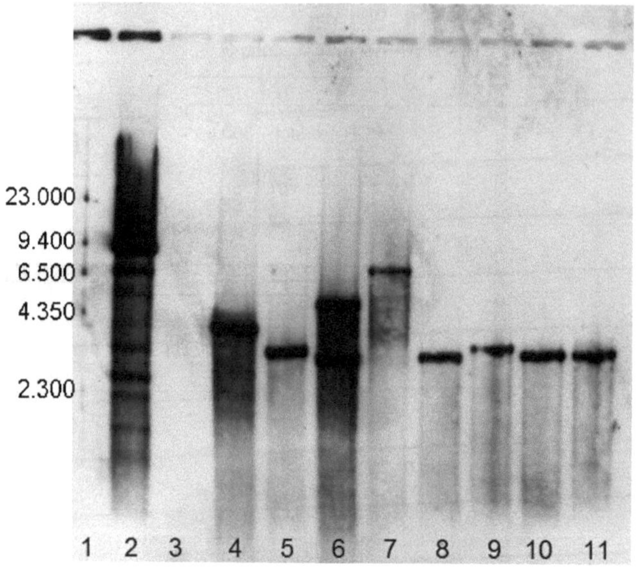

Fig. 3. Southern blot analysis of a selection of ClosTron mutants of *Clostridium acetobutylicum* using an intron-specific probe. In all cases, genomic DNA has been cleaved with *Hin*dIII. The *lanes* are as follows: 1, ladder; 2, plasmid control; 3, wild type; 4, *ptb* mutant; 5, *ack* mutant; 6, *adhE* mutant (two bands); 7, *adhE2* mutant; 8, *bdhA* mutant; 9, *bdhB* mutant; 10, *ctfA* mutant, and 11, *ctfB* mutant.

3.8. Recycling and Reuse of the ErmRAM

1. Sometimes it can be desirable to make multiple mutations in the same strain. ClosTron plasmids like pMTL007C-E2 contain flp recombinase target (FRT) sites flanking the ErmRAM. The recombinant expression of flp recombinase (supplied on a plasmid, Fig. 1b) in a ClosTron mutant background will lead to excision of the *ermB* marker from the chromosome. The ClosTron can then be used again to create another insertional mutation. This has been exemplified in *C. acetobutylicum* (14).
2. The flp plasmid pMTL85151-PPS-flp3 (prepared in an *E. coli* host additionally carrying pAN-2 when using *C. acetobutylicum*) is transferred into the clostridial ClosTron mutant by electroporation.
3. Pre-reduce one CBMS plate containing Em (10 μg/mL) and Tm (15 μg/mL) for each mutant that requires marker removal (adjust selection according to mutants).
4. Plate out clostridial mutant(s) containing the flp plasmid. Incubate anaerobically at 37°C for 2–3 days.
5. Inoculate four pre-reduced CBMS broth supplemented with Tm at 15 μg/mL, with each of the required clostridial mutants. Incubate anaerobically overnight at 30°C.
6. The next day, serially dilute each broth culture to 10^{-6} in PBS. Plate out 100 μL of the neat, 10^{-2}, 10^{-3}, 10^{-4}, and 10^{-6}

dilutions onto CBMS agar supplemented with Tm at a final concentration of 15 μg/mL. Incubate anaerobically at 30°C for 3 days.

7. Pick 50 colonies for each strain and grid plate onto CBMS agar containing Tm at a final concentration of 15 μg/mL. Incubate anaerobically at 30°C for 2–3 days.
8. Pick all 50 colonies and replica plate onto CBMS supplemented with either Em (10 μg/mL) or Tm (15 μg/mL). Incubate anaerobically at 30°C and check after 24 and 48 h, making a note of which colonies showing no growth on plates supplemented with Em, but growth in the presence of Tm.
9. Set up 1 mL cultures in CBMS broth of any clones which showed no growth on plates supplemented with Em. Also subculture these clones onto CBMS agar containing Tm (15 μg/mL) and onto CBMS agar containing no antibiotic.
10. The next day prepare genomic DNA from all broth cultures and carry out PCR screens to confirm marker removal, using primers that flank the insertion site of your target gene.

3.9. Complementation Studies

1. To definitely establish that the observed phenotype has been caused by the ClosTron-derived insertion, it is necessary to complement the mutation through the introduction of the wild-type gene. It is most simply achieved through the use of an autonomous vector, although, dependent on size, delivery of DNA to the chromosome is also possible using ClosTron technology (see Note 12).
2. Depending on the *Clostridium* spp. used, choose an appropriate vector from the pMTL80000 modular series, which provide an easily interchangeable selection of different replicons and antibiotic resistance markers (15). Choose for example pMTL84151.
3. If known, choose the promoter region of your gene (otherwise a suitable heterologous promoter) and amplify it by PCR with primers containing the restriction sites *Not*I and *Nde*I (CATATG, ATG is target gene start codon).
4. Also amplify your target gene, with the restriction sites *Nde*I (cloned together with the promoter this will result into an *Nde*I fusion exactly at the start codon), and for example *Xho*I.
5. Then clone the promoter fragment and the gene into the vector backbone, in this example digested with *Not*I and *Xho*I, and transfer the vector (after verifying and safekeeping in *E. coli*) using an appropriate method, into your ClosTron mutant.

6. Grow the transconjugants/transformants under antibiotic selection (it is advisable to transfer the empty plasmid into the wild type, in order to grow all strains under the same antibiotic pressure) and carry out the necessary experiments to establish the phenotype.

4. Notes

1. Chloramphenicol may generally not be used in *Clostridium* for selection of markers such as *catP*, due to its anaerobic reduction. In its place, thiamphenicol is routinely employed.
2. Although the PCR product is only ~350-bp in size, it contains extensive secondary structure which can lead to errors during PCR. This effect can be minimised by performing three independent PCR reactions in parallel and then combining the three PCR products and using the resultant mixture in the ligation reaction with cleaved ClosTron vector DNA.
3. The electrophoretic DNA band at ~350-bp should be the most intense. This is the DNA fragment that needs to be excised and gel purified. It is, however, not uncommon to see fainter electrophoretic DNA bands of 250- and ~100-bp.
4. If there are colonies on your control plate you should repeat the digestion of the vector, followed by a dephosphorylation step. Additionally, ligation time and reaction volume can be increased to improve the ligation efficiency.
5. The standard ClosTron vector uses a RAM element based on the *ermB* gene (ErmRAM) and, therefore, relies on positive selection for Em resistance. The *C. difficile* strain CD630 carries an *ermB* gene. Therefore, in order to make mutants in this strain, a specially selected Em-sensitive derivative (CD630Δerm) is utilised, in which the *ermB* gene has been deleted (32).
6. For maximum transformation efficiencies it is desirable to site either the electroporator, or more conveniently, the electroporation cuvette holder, inside the anaerobic cabinet. The latter can be achieved by fitting appropriate sockets to the housing of the cabinet, and siting the actual electroporator apparatus outside of the cabinet.
7. Other *C. difficile* strains, such as the PCR-Ribotype 017 strain R20291 are naturally Em resistance by a different mechanism (33), as they naturally lack the *ermB* gene. In this case, ClosTron mutants maybe selected on the basis of acquisition

of resistance to lincomycin (14). Alternatively, other modified ClosTron plasmids have been generated in which either a *catP* (34), or *aad9* (unpublished data) has been inserted into the group II intron in place of the ErmRAM. In these instances, insertion of the group II element into the target gene can be selected on the basis of acquisition of resistance to Tm or spectinomycin, respectively (34).

8. Mechanistically, the insertion of more than one group II intron into the chromosome of a single cell should be a rare event. However, it does occur. Thus far we have generated over 65 clostridial mutants in this laboratory, and in just two instances, insertion of more than one intron has occurred (see Fig. 3).
9. Genomic DNA isolation and Southern blot procedures are undertaken using standard procedures (35). Typically, DNA is cleaved with *Hin*dIII or *Eco*RV.
10. The EBS2 primer can be taken from one of your target genes, it does not matter which, and used to check all your ClosTron mutants in a Southern blot. The EBS2 primer sequence is given to you when you design your retargeting region, even if you get the fragment synthesised you can easily just order that primer.
11. It is advisable to always isolate and verify three independent ClosTron mutants.
12. In addition to gene "knock-out," the ClosTron has "knock-in" capability to deliver cargo DNA to the chromosome. This DNA is introduced into the domain IV region of the intron by its insertion into a specially created unique *Sal*I. In ClosTron plasmids containing the ErmRAM, the maximum size that can be delivered is approximately 1.0-kb (14). In variants lacking the ErmRAM, approximately 2.0-kb can be delivered, although in this case intron insertion cannot be selected through acquisition of resistance to Em.

Acknowledgments

The authors acknowledge the financial support of UK Medical Research Council (G0601176), the European Union (HEALTH-F3-2008-223585), the UK Biotechnology and Biological Sciences Research Council (BB/E021271/1, BB/D001498/1, BB/F003390/1, and BB/G016224/1), SysMO (Systems Biology of Microorganisms) and Morvus Technologies Ltd.

References

1. Dürre P. **(2005) Handbook** on **Clostridia,** CRC Press, New York. 2005.
2. Onderdonk A.B., and Allen S.D. (1994) *Clostridium.* Manual of clinical microbiology. In: Murray PR, Baron EJ, Pfaller MA, Tenover FC, Yolken RH, editors. 6 th ed. Washington, DC: ASM Press; p. 1210.
3. Dürre P. (2008) Fermentative butanol production: bulk chemical and biofuel. *Ann. N. Y. Acad. Sci.* **1125,** 353–362.
4. Demain A.L., Newcomb M., and Wu J.H.D. (2005) Cellulase, clostridia, and ethanol. *Microbiol. Mol. Biol. Rev.* **69,** 124–154.
5. Minton N.P. (2003) *Nature Rev. Microbiol.* **1,** 237–242.
6. Johnson E.A. (2005) *Clostridium botulinum* neurotoxins – applications in medicine and potential agents of bioterrorism. *Clin. Microbiol. News.* **27,** 147–151.
7. Hurst L.C., Badalamente M.A., Hentz V.R., Hotchkiss R.N., Kaplan T.T.D., Meals R.A., Smith T.M., and Rodzvilla J. (2009) Injectable collagenase *Clostridium histolyticum* for Dupuytren's contracture. *N Engl J Med* **361,** 968–979.
8. Heap J.T., Cartman S.T., Pennington O.J., Cooksley C.M., Scott J.C., Blount B., Burns D., and Minton N.P. (2008) Development of genetic knock-out systems for clostridia. In: Bruggermann, H, Gottschalk, G. (Eds), Clostridia: Molecular biology in the post-genomic era. Caister Academic Press. Norfolk, UK, pp. 179–198.
9. Karberg M., Guo H., Zhong J., Coon R., Perutka J., and Lambowitz A.M. (2001) Group II introns as controllable gene targeting vectors for genetic manipulation of bacteria. *Nature Biotechnol.* **19,** 1162–7.
10. Zhong J., Karberg M., and Lambowitz A.M. (2003) Targeted and random bacterial gene disruption using a group II intron (targetron) vector containing a retrotransposition-activated selectable marker. *Nucleic Acids Res.* **31,** 1656–64.
11. Mohr G., Smith D., Belfort M., and Lambowitz A.M. (2000) Rules for DNA target-site recognition by a lactococcal group II intron enable retargeting of the intron to specific DNA sequences. *Genes Dev.* **14,** 559–73.
12. Perutka J., Wang W., Goerlitz D., and Lambowitz A.M. (2004) Use of computer-designed group II introns to disrupt *Escherichia coli* DExH/D-box protein and DNA helicase genes. *J Mol Biol.* **336,** 421–39.
13. Heap J.T., Pennington O.J., Cartman S.T., Carter G.P., and Minton N.P. (2007) The ClosTron: a universal gene knock-out system for the genus *Clostridium. J. Microbiol. Methods.* **70,** 452–64.
14. Heap J.T., Kuehne S.A., Ehsaan M., Cartman S.T., Cooksley C.M., Scott J.C., and Minton N.P. (2010) The ClosTron: Mutagenesis in *Clostridium* refined and streamlined. *J. Microbiol Methods.* **80,** 49–55.
15. Heap T.J., Pennington O.J., Cartman S.T., and Minton N.P. (2009) "A modular system for *Clostridium* shuttle plasmids. *J. Microbiol. Methods* **78,** 79–85.
16. Bradshaw M., Marshall K.M., Heap J.T., Tepp W.H., Minton N.P., and Johnson E.A. (2010) Construction of a Nontoxigenic *Clostridium botulinum* Strain for Food Challenge Studies *Appl. Environ. Microbiol.* **76,** 387–93.
17. Burns D.A., Heap J.T., and Minton N.P. (2010) SleC is essential for germination of *Clostridium difficile* spores in nutrient-rich medium supplemented with the bile salt taurocholate. *J. Bacteriol.* **192,** 657–64.
18. Cooksley C.M., Davis I.J., Winzer K., Chan W.C., Peck M.W., and Minton N.P. (2010) Regulation of neurotoxin production and sporulation by a putative *agrBD* signaling system in proteolytic *Clostridium botulinum. Appl. Environ. Microbiol.* **76,** 4448–60.
19. Camiade E., Peltier J., Bourgeois I., Couture-Tosi E., Courtin P., Antunes A., Chapot-Chartier M.P., Dupuy B., and Pons J.L. (2010) Characterization of Acp, a peptidoglycan hydrolase of *Clostridium perfringens* with N-acetylglucosaminidase activity that is implicated in cell separation and stress-induced autolysis. *J Bacteriol.* **192,** 2373–84.
20. Dong H., Zhang Y., Dai Z., and Li Y. (2010) Engineering clostridium strain to accept unmethylated DNA. PLoS One. **5,** e9038.
21. Kirby J.M., Ahern H., Roberts A.K., Kumar V., Freeman Z., Acharya K.R., and Shone C.C. (2009) Cwp84, a surface-associated cysteine protease, plays a role in the maturation of the surface layer of *Clostridium difficile. J. Biol. Chem.* **284,** 34666–34673.
22. Underwood S., Guan S., Vijayasubhash V., Baines S.D., Graham L., Lewis R.J., Wilcox M.H., and Stephenson K. (2009) Characterization of the sporulation initiation pathway of *Clostridium difficile* and its role in toxin production. *J.Bacteriol.* **191,** 7296–7305.

23. Emerson J.E., Reynolds C.B., Fagan R.P., Shaw H.A., Goulding D., and Fairweather N.F. (2009) A novel genetic switch controls phase variable expression of CwpV, a *Clostridium difficile* cell wall protein. *Mol. Microbiol.* **74,** 541–556.
24. Twine S.M., Reid C.W., Aubry A., McMullin D.R., Fulton K.M., Austin J., and Logan S.M. (2009) Motility and flagellar glycosylation in *Clostridium difficile. J. Bacteriol.* **191,** 7050–7062.
25. Warrens A.N., Jones M.D., and Lechler R.I. (1997) Splicing by overlap extension by PCR using asymmetric amplification: an improved technique for the generation of hybrid proteins of immunological interest. *Gene.* **186,** 29–35.
26. Purdy D., O'Keeffe T.A., Elmore M., Herbert M., McLeod A., Bokori-Brown M., Ostrowski A., and Minton N.P. (2002) Conjugative transfer of clostridial shuttle vectors from *Escherichia coli* to *Clostridium difficile* through circumvention of the restriction barrier. *Mol Microbiol.* **46,** 439–452.
27. Williams D.R., Young D.I., and Young M. (1990) Conjugative plasmid transfer from *Escherichia coli* to *Clostridium acetobutylicum. J. Gen. Microbiol.* **136,** 819–26.
28. Theys J., Pennington O.P, Dubois L., Anlezark G., Vaughan T., Mengesha A., Landuyt W., Anné J., Burke P.J., Dûrre P., Wouters B.G., Minton N.P., and Lambin P. (2006) Repeated cycles of *Clostridium*-directed enzyme prodrug therapy result in sustained antitumour effects *in vivo. British J. Cancer* **95,** 1212–9.
29. Davis T.O., Henderson I., Brehm J.K., and Minton N.P. (2000) Development of a transformation and gene reporter system for group II, non-proteolytic *Clostridium botulinum* type B strains. *J. Mol. Microbiol. Biotechnol.* **2,** 59–69.
30. Mauchline M.L., Davis T.O., and Minton N.P. (1999) Clostridia. In: Manual of Industrial Microbiology and Biotechnology, Demain AL, Davies JE (eds), ASM Press, pp. 475–492.
31. Mermelstein L.D., and Papoutsakis E.T. (1993) *In vivo* methylation in *Escherichia coli* by the *Bacillus subtilis* phage φ*3t*I methyltransferase to protect plasmids from restriction upon transformation of *Clostridium acetobutylicum* ATCC 824. *Appl. Environ. Microbiol.* **59,** 1077–1081.
32. Hussain H.A., Roberts A.P., and Mullany P. (2005) Generation of an Em-sensitive derivative of *Clostridium difficile* strain 630 (630Deltaerm) and demonstration that the conjugative transposon Tn*916*DeltaE enters the genome of this strain at multiple sites. *J. Med. Microbiol.* **54,** 137–141.
33. Cartman S.T., Heap J.T., Kuehne S.A., Cockayne A., and Minton N.P. (2010) The Emergence of 'Hypervirulence' in *Clostridium difficile. Int. J. Med. Microbiol.* **300,** 387–395.
34. Kuehne S.A., Cartman S.T., Heap J.T., Kelly M.L., Cockayne A., and Minton N.P. (2010) The role of toxin A and toxin B in *Clostridium difficile* infection. *Nature* **467,** 711–713.
35. Cooksley C.M., Davis I.J., Winzer K., Cockayne A., Chan W.C., Peck M.W., and Minton N.P. (2010) A putative *agrBD* signalling system regulates toxin production and sporulation in proteolytic *Clostridium botulinum. J Bacteriol* 76, 4448–4460.

Chapter 24

High-Throughput Transposon Mutagenesis of *Corynebacterium glutamicum*

Nobuaki Suzuki, Masayuki Inui, and Hideaki Yukawa

Abstract

Construction of gene disruption mutants and analysis of the resultant phenotypes are an important strategy to study gene function. A simple and high-throughput method developed for microorganisms combines two different types of transposons, direct genomic DNA amplification and thermal asymmetric interlaced-PCR. The considerable utility of this approach is demonstrable in *Corynebacterium glutamicum*, where 18,000 transposon disruptants enabled the generation of an insertion library covering nearly 80% of the organism's 2,990 ORFs.

Key words: Transposon, Mutagenesis, Genome, Disruptant library, TAIL-PCR, *Corynebacterium glutamicum*, High throughput, Phi29, Rolling circle DNA amplification

1. Introduction

Transposon mutagenesis is a powerful tool for constructing large mutant pools. It nonetheless engenders inherent difficulties that demand sustained development of several techniques in order to avoid any undesirable issues that may otherwise arise. One difficulty is in identifying transposon insertion sites, consequently necessitating complex procedures, including extraction of genomic DNA, cloning of the inserted site and inverted PCR. The second difficulty arises from the fact that the position of transposon insertion is wholly dependent upon transposon characteristics.

Considerable progress in technique development has of late been realized. For instance, a rolling circle DNA amplification method recently developed using Phi29 DNA polymerase

James A. Williams (ed.), *Strain Engineering: Methods and Protocols*, Methods in Molecular Biology, vol. 765,
DOI 10.1007/978-1-61779-197-0_24, © Springer Science+Business Media, LLC 2011

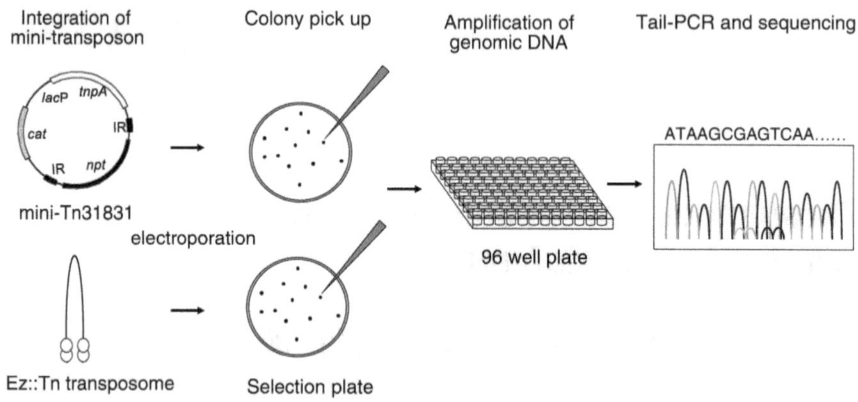

Fig. 1. Illustration of transposon mutagenesis and determination of insertion locations. Mutagenized cells are selected by plating on complex medium plates containing kanamycin. After colony pick up, mutants are arrayed into 96-well plates and analyzed (Copyright© American Society for Microbiology, ref. 11).

can amplify DNA templates 10,000-fold in a few hours, and consequently obviate numerous genomic DNA extraction steps (1). Thermal asymmetric interlaced (TAIL)-PCR can be used to determine the transposon insertion location instead of cloning or inverted PCR because it can amplify unknown DNA sequences adjacent to known sequences such as transposons (2, 3). Besides, utilization of a variety of transposons can assist in the generation of a variety of insertion mutants. An artificial transposon, miniTn*31831*, which was constructed using insertion sequence IS*31831*, exhibits no obvious target sequence specificity in *Corynebacterium glutamicum* (4, 5). A different Tn*5*-based mini-transposon is also randomly inserted into a host's genomic DNA (6–9). Favorite DNA sequences for insertion by miniTn*31831* are different from those by Tn*5*-based mini-transposons (5, 10).

By using the combination described above, direct genomic DNA amplification by Phi29 DNA polymerase, TAIL-PCR, and two kinds of transposons, we developed a new, high-throughput transposon mutagenesis method which can bypass some limitations of conventional transposon mutagenesis (Fig. 1). An insertion library covering nearly 80% of the 2,990 ORFs of *C. glutamicum* was generated by using this method (11). This approach is useful to construct gene disruptant library of microorganisms.

2. Materials

2.1. Transposon Mutagenesis

1. Complex liquid medium for *C. glutamicum*: 2.0 g urea, 7.0 g $(NH_4)_2SO_4$, 0.5 g K_2HPO_4, 0.5 g $MgSO_4$, 6 mg $FeSO_4/7H_2O$, 6 mg $MnSO_4/4–6H_2O$, 200 μg biotin, 200 μg thiamin/HCl, 1.0 g Yeast extract, 1.0 g casamino acids, 40.0 g glucose, H_2O to 1.0 l. After mixing, sterilize by autoclaving for 20 min.

For complex solid medium, autoclave the agar at 1.5% final concentration together with the ingredients of complex liquid medium. After cooling the medium to about 50°C, add kanamycin to 50 μg/ml and pour the medium into sterile disposable plates.

2. Kanamycin: 5 mg/ml. Use at 50 μg/ml final concentration in complex medium for transposon insertion mutant selection.
3. Transformation buffer: 10% glycerol. After mixing glycerol with deionized H_2O, sterilize by autoclaving for 20 min.
4. Gene pulserR (Bio-Rad, Richmond, CA).
5. Electroporation cuvettes: (0.1 cm gap, Bio-Rad), cool on ice until use.
6. miniTn*31831* transposon: (pMV23 plasmid) (4).
7. Tn*5*-based mini-transposon: (EZ::Tn<Kan2> transposome system, Epicentre, Madison, WI).

2.2. Whole Genome Amplification

1. GenomiPhi DNA Amplification Kit (GE Healthcare, NJ).
2. Tris–EDTA buffer: 10 mM Tris–HCl and 1 mM EDTA, pH 8.0. Mix with 10 ml of 1 M Tris–HCl, pH 8.0, and 2 ml of 500 mM EDTA, pH 8.0 in 1.0 l of sterile H_2O.
3. Complex media + kanamycin: *See* Subheading 2.1, item 1.
4. 50% glycerol: After mixing glycerol with deionized H_2O, sterilize by autoclaving for 20 min.

2.3. TAIL-PCR

1. ExTaq polymerase set (Takara, Shiga, Japan): containing ExTaq polymerase, 10× ExTaq buffer and dNTP mixture (2.5 mM each).
2. Transposon PCR primers: (see Note 1).

   *AP*1 primer (5′-NGTCGA(G/C)(A/T)GANA(A/T)GAA) (32 μM)

   *GSP*1 primer (5′-CTCCTTCATTACAGAAACGGC) (3.2 μM)

   *GSP*2 primer (5′-GCTGAGTTGAAGGATCAGATC) (3.2 μM)
3. Corresponding ORF PCR primers: (3.2 μM).

2.4. Sequencing

1. BigDye Terminator (Applied Biosystems, CA).
2. 5× sequencing buffer (Applied Biosystems).
3. ABI PRISM Genetic Analyzer (Applied Biosystems).
4. BigDye® XTerminator (Applied Biosystems).
5. Sequencing primers:

   miniTn*31831* (5′-AGGTTTCCGTAATTTGAACCACTAC ATT) (3.2 μM);

   Tn*5*-based mini-transposon (5′-ACAACAAAGCTCTCATC AACCGTGG9) (3.2 μM).

2.5. Cell Direct PCR

1. Sterile water: Autoclave.
2. ExTaq polymerase set (Takara, Shiga, Japan): containing ExTaq polymerase, 10× ExTaq buffer and dNTP mixture (2.5 mM each).

3. Methods

3.1. Transformation of *C. glutamicum*

1. A 2.5-ml overnight culture of *C. glutamicum* in complex medium is inoculated into 100 ml of new complex medium and grown with shaking (220 rpm) at 33°C.
2. Cells are harvested by centrifugation at 4°C for 10 min at 3,000 × *g* when OD_{610} of culture reaches 0.6 (2–2.5 h), followed by washing twice with 50 ml of ice-cold sterilized 10% glycerol.
3. Cells are suspended in 10% glycerol to a final volume of 1.5 ml, and 100 μl aliquots transferred to new tubes and stored at −80°C until transformation.
4. For transformation by transposon, cells prepared in step 3 above are thawed on ice and mixed with 0.05–0.1 μg miniTn*31831* transposon or Tn*5*-based mini-transposon DNA, and held on ice for 2 or 3 min (see Note 2).
5. The suspension is transferred to an ice-cold electroporation cuvette and electroporated by Gene pulserR with resistance, capacitance, and field strength settings at 200 Ω, 25 μF, and 19.5 kV/cm, respectively (see Note 3).
6. Immediately after the pulse, cells are gently diluted with 1.0 ml of complex medium and incubated at 33°C for 2 h.
7. Appropriate aliquots are spread on complex medium plates containing 50 μg/ml kanamycin. Transposon insertion mutants are obtained after 1–2 days of incubation at 33°C (transformation efficiency, see Notes 4 and 5).

3.2. Whole Genome Amplification

1. 4.5 μl of reaction buffer in GenomiPhi DNA Amplification Kit is mixed with 0.5 μl of enzyme mix on ice.
2. A small number of mutant cells (10^7–10^8) grown on selection plates is transferred by toothpick into 100 μl of each of (a) complex medium containing kanamycin and (b) Tris–EDTA buffer (see Note 6).
3. Complex medium samples are incubated at 33°C for 15 h, and cultures are mixed with an equal volume of 50% glycerol and stored at −80°C.
4. 0.5 μl of Tris–EDTA cell suspensions prepared in step 2 are mixed with 4.5 μl sample buffer in GenomiPhi DNA Amplification Kit, incubated at 95°C for 5 min, and chilled on ice for several minutes.

5. A 5.0-μl of reaction buffer with enzyme mix (step 1) is added to a 5-μl of sample (step 4) and reacted for 22 h at 30°C according to the manufacturer's protocol (genome amplification period was recently changed, see Note 7).
6. To stop the genome amplification reaction, samples are heated at 65°C for 10 min and then cooled to 4°C. After the reaction, amplified high molecular weight genomic DNAs (>10 kb) are observed in most samples.

3.3. TAIL-PCR

To identify transposon location on the genome, TAIL-PCR is employed. Basically, the procedure described by Liu et al. is followed, omitting the third PCR (3).

1. 1 μl of amplified genomic DNA, 1 μl ExTaq buffer, 0.8 μl dNTP mixture, 0.05 μl ExTaq, 0.8 μl AP1 primer, 0.8 μl GSP1 primer, and 5.6 μl H_2O are mixed.
2. Set primary thermal cycling as follows:
   (a) 1 cycle: 94°C for 1 min
   (b) 5 cycles: 94°C for 1 min; 65°C for 1 min; 72°C for 3 min
   (c) 1 cycle: 94°C for 1 min; 30°C for 3 min ramp to 72°C for 3 min; 72°C for 3 min
   (d) 15 cycles: 94°C for 30 s; 68°C for 1 min 72°C for 3 min; 94°C for 30 s; 68°C for 1 min 72°C for 3 min; 94°C for 30 s; 44°C for 1 min 72°C for 3 min
   (e) 1 cycle: 72°C for 1 min
3. PCR start (primary PCR). After reaction, smear bands are observed using 1% agarose gel electrophoresis.
4. 1 μl of 50 times diluted primary PCR products are mixed with 1 μl ExTaq buffer, 0.8 μl dNTP mixture, 0.05 μl ExTaq, 0.8 μl AP1 primer 0.8 μl GSP2 primer, and 5.6 μl H_2O.
5. Set second thermal cycling as follows:
   (a) 1 cycle: 94°C for 1 min
   (b) 12 cycles: 94°C for 30 s; 64°C for 1 min 72°C for 3 min; 94°C for 30 s; 64°C for 1 min 72°C for 3 min; 94°C for 30 s; 44°C for 1 min 72°C for 3 min
   (c) 1 cycle: 72°C for 5 min
6. After amplification, 2 μl aliquots of secondary PCR products are used for sequencing (see Notes 8 and 9).

3.4. Sequencing

Sequencing is performed on an ABI PRISM 3100 Genetic Analyzer.

1. Mix with 2 μl of second PCR product, 0.5 μl of sequencing primer, 2 μl of BigDye Terminator, 2 μl of 5× sequencing buffer, and 3.5 μl of H_2O.

2. Cycle sequencing is performed as follows: 94°C for 1 min, 96°C for 10 s, 50°C for 5 s, 60°C for 4 min for 40 cycles.
3. PCR products are purified by ethanol precipitation or BigDye® XTerminator method, and sequenced using a genetic analyzer (see Note 10).

3.5. Mapping of Insertion Sites on Genome by Cell-Direct PCR

Transposon insertion sites in mutants are identified by BLAST search (Fig. 2, see Note 11). After identification, transposon insertions within ORFs are confirmed by cell-direct PCR using custom primers flanking each annotated ORF. Since the length of transposed DNA of miniTn*31831* and Tn*5*-based mini-transposon are 1.8 kb and 1.2 kb, respectively, successful transposon insertion within an ORF causes an increase in the length of cell-direct PCR products. Amplified DNA fragments are compared with each corresponding ORF of wild type strain on agarose gels (see Note 12).

1. A small number of cells is picked up on a toothpick and suspended in 6.6 μl of sterilized water.
2. Cell suspensions are mixed with 0.8 μl of reverse and forward primers of a corresponding ORF, 1 μl ExTaq buffer, 0.8 μl dNTP mixture, and 0.05 μl ExTaq.
3. PCR is performed as follows:
   (a) 1 cycle: 98°C for 1 min
   (b) 30 cycles: 94°C for 30 s; Set at appropriate temperature for primers for 30 s; 72°C for appropriate period (normally calculate the time as 1 min need to amplify 1 kb DNA)
   (c) 1 cycle: 72°C for 7 min
4. After PCR, reaction products are compared on agarose gel.

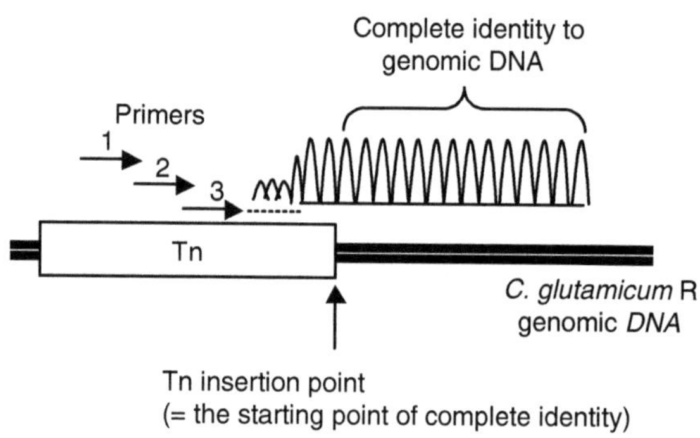

Fig. 2. Amplified DNA fragment and position of primers. By using BLAST search, insertion position is identified. Primers 1, 2, and 3 indicate GSP1, GSP2, and sequencing primer, respectively (Copyright© American Society for Microbiology, ref. 11).

4. Notes

1. GSP1 and GSP2 primers are designed to anneal with 910 and 726 bp upstream from C terminus of miniTn*31831* or with 345 and 161 bp upstream from C terminus of Tn*5*-based mini-transposon, respectively.
2. In the case of *C. glutamicum* R (GC content, 54.1%; 2,990 ORF), 55.3 and 87.6% of miniTn*31831* and Tn*5*-based mini-transposon, were inserted into coding sequences, respectively. miniTn*31831* tends to transpose more into AT-rich regions compared to Tn*5*-based mini-transposon (5), and the AT ratio of noncoding regions is relatively higher than that of coding regions on *C. glutamicum* genome. The distribution pattern of each transposon on the genome was also different (Figs. 3 and 4).
3. If preparation of competent cells is well done, resultant electroporation time constant values should be over 4.0.
4. Utilization of unmethylated pMV23 transposon plasmid DNA improves transformation efficiency. By using *dam*, *dcm* strain to prepare transposon DNA, for example, *Escherichia coli* JM110, the number of transposon insertion mutants was increased 100–1,000-fold in several *C. glutamicum* strains (12).
5. Transposon insertion mutants appeared at insertion efficiencies of 2.0×10^5 and 3.0×10^4 colony/μg DNA, respectively, for miniTn*31831* and Tn*5*-based mini-transposon in *C. glutamicum* R.

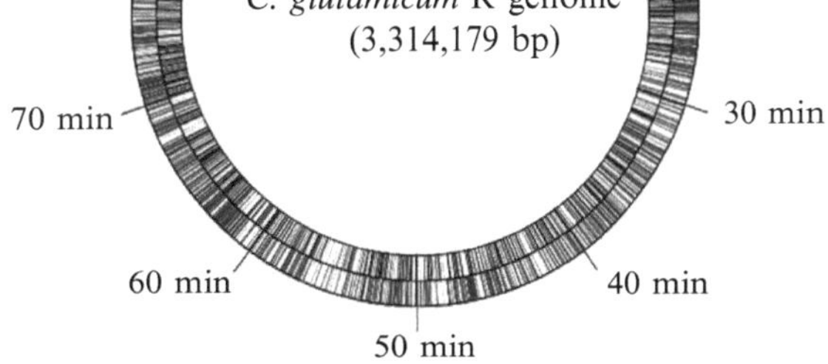

Fig. 3. Distribution of transposon insertions on *Corynebacterium glutamicum* genome. *Outermost* and *middle circles* indicate individual miniTn*31831* and Tn*5*-based mini-transposon insertions, respectively (Copyright© American Society for Microbiology, ref. 11).

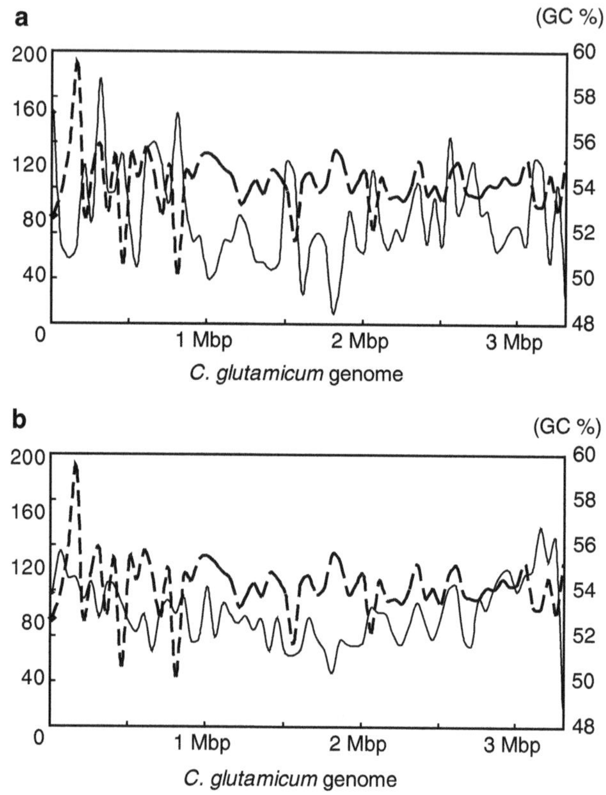

Fig. 4. Number of transposon insertions on *Corynebacterium glutamicum* genome. *Solid* and *dashed lines* mark the number of transposon insertions and GC content, respectively. (**a**) miniTn*31831* and (**b**)Tn*5*-based mini-transposon. The numbers were calculated per 50 kb (Copyright© American Society for Microbiology, ref. 11).

6. Successive utilization of 96-well plates of each step, cell growth, whole genome amplification reaction, primary, second PCR of TAIL PCR, and sequencing, is convenient in so far as it avoids many of the confusions of sample handling.
7. Whole genome amplification period by GenomiPhi DNA Amplification Kit is altered to 1.5 h in the GenomiPhi V2 Kit.
8. After the second PCR, one or two large bands can be observed by agarose gel electrophoresis. Most of the bands are smaller than 500 bases.
9. The sequenced lengths of TAIL-PCR products are less than 200 bases.
10. If second PCR products could not be obtained or sequenced, third PCR step described by Liu et al. (3) should be included in TAIL-PCR procedure.
11. To do high-throughput analysis of sequenced data, automated BLAST search is performed using an in-house Perl script.
12. Confirmation by cell-direct PCR should be performed because TAIL-PCR could amplify false DNAs.

Acknowledgments

We wish to thank Dr. C. Omumasaba for critical reading of the manuscript. This research was partly supported by New Energy and Industrial Technology Development Organization (NEDO), Japan.

References

1. Dean F. B., Nelson J. R., Giesler T. L., and Lasken R. S. (2001) Rapid amplification of plasmid and phage DNA using Phi29 DNA polymerase and 5 multiply-primed rolling circle amplification. *Genome Res.* **11**, 1095–1099.
2. Knobloch J. K., Nedelmann M., Kiel K., Bartscht K., Horstkotte M. A., Dobinsky S., Rohde H., and Mack D. (2003) Establishment of an arbitrary PCR for rapid identification of Tn*917* insertion sites in *Staphylococcus epidermidis*: characterization of biofilm-negative and nonmucoid mutants. *Appl. Environ. Microbiol.* **69**, 5812–5818.
3. Liu Y. G., Mitsukawa N., Oosumi T., and Whittier R. F. (1995) Efficient isolation and mapping of *Arabidopsis thaliana* T-DNA insert junctions by thermal asymmetric interlaced PCR. *Plant J.* **8**, 457–463.
4. Vertès A. A., Asai Y., Inui M., Kobayashi M., Kurusu Y., and Yukawa H. (1994) Transposon mutagenesis of coryneform bacteria. *Mol. Gen. Genet.* **245**, 397–405.
5. Vertès A. A., Inui M., Kobayashi M., Kurusu Y., and Yukawa H. (1994) Isolation and characterization of IS*31831*, a transposable element from *Corynebacterium glutamicum*. *Mol. Microbiol.* **11**, 739–746.
6. Goryshin I. Y. and Reznikoff W. S. (1998) Tn5 in vitro transposition. *J. Biol. Chem.* **273**, 7367–7374.
7. Goryshin I. Y., Miller J. A., Kil Y. V., Lanzov V. A., and Reznikoff W. S. (1998) Tn*5*/IS*50* target recognition. *Proc. Natl. Acad. Sci. U. S. A.* **95**, 10716–10721.
8. Oram D. M., Avdalovic A., and Holmes R. K. (2002) Construction and characterization of transposon insertion mutations in *Corynebacterium diphtheriae* that affect expression of the diphtheria toxin repressor (DtxR). *J. Bacteriol.* **184**, 5723–5732.
9. Goryshin I. Y., Jendrisak J. L., Hoffman M., Meis R., and Reznikoff W. S. (2000) Insertional transposon mutagenesis by electroporation of released Tn*5* transposition complexes. *Nat. Biotechnol.* **18**, 97–100.
10. Herron P. R., Hughes G., Chandra G., Fielding S., and Dyson P. J. (2004) Transposon Express, a software application to report the identity of insertions obtained by comprehensive transposon mutagenesis of sequenced genomes: analysis of the preference for in vitro Tn*5* transposition into GC-rich DNA. *Nucleic. Acids. Res.* **32**, e113.
11. Suzuki N., Okai N., Nonaka H., Tsuge Y., Inui M., and Yukawa H. (2006) High throughput transposon mutagenesis of *Corynebacterium glutamicum* and construction of a single-gene disruptant mutant library. *Appl. Environ. Microbiol.* **72**, 3750–3755.
12. Vertès A. A., Inui M., Kobayashi M., Kurusu Y., Yukawa H. (1993) Presence of *mrr*- and *mcr*-like restriction systems in coryneform bacteria. *Res. Microbiol.* **144**, 181–185.

Chapter 25

Mini-Mu Transposon Mutagenesis of Ethanologenic *Zymomonas mobilis*

Katherine M. Pappas

Abstract

Zymomonas mobilis is a facultatively anaerobic α-proteobacterium with a considerable potential for industrial ethanol production. An important tool in the generation of stable mutants for this organism is described in this chapter; it entails insertional mutagenesis with the help of the transposable element mini-Mu. The latter is delivered into *Z. mobilis* with the use of plasmid pULB113 (RP4::mini-Mu) that self-transfers in the organism at notable frequencies and remains highly stable even under nonselective conditions. Transposition of mini-Mu and subsequent mutagenesis occur readily in *Z. mobilis* pULB113 transconjugants and result in the generation of large numbers of random mutants. This can be demonstrated by the isolation of various auxotrophs with single or multiple nutritional requirements, the vast majority of which bears insertions at different chromosomal locations, even when exhibiting the same requirement. Therefore, transposon mutagenesis with the use of mini-Mu serves as a simple and effective tool for indiscriminate mutant production in *Z. mobilis*.

Key words: Transposon mutagenesis, Insertional inactivation, Mini-Mu, *Zymomonas mobilis*, Auxotroph production

1. Introduction

Zymomonas mobilis is currently considered the bacterial alternative to yeast, as it ferments sugar-rich substrates to ethanol at faster rates and comparable, if not higher, yields (1, 2). In doing so, *Z. mobilis* utilizes the Entner–Doudoroff glycolytic pathway which, compared to that of Embden–Meyerhof–Parnas, generates less energy for cellular purposes and results in minimal accumulation of biomass. Genetically engineered *Z. mobilis* strains that co-ferment pentoses and hexoses also stand as potent candidates for lignocellulosic ethanol production (3–5). Apart from ethanol, alterations in the robust sugar uptake and break-down routes of

James A. Williams (ed.), *Strain Engineering: Methods and Protocols*, Methods in Molecular Biology, vol. 765,
DOI 10.1007/978-1-61779-197-0_25,

Z. mobilis lead to synthesis of other products of economic importance, such as of sorbitol, phenylacetylcarbinol (PAC; precursor of ephedrine and pseudoephedrine) and levan, a polyfructan (2). Lastly, of additional biotechnological interest is the abundance of rare classes of fatty acids and bacterial steroids (hopanoids) in the *Z. mobilis* cell membranes, the unusual content of which is relevant to the ethanol stress imposed on this organism's physiology (2, 6). Current advances in the genomics front open up new perspectives in the study of *Z. mobilis* (7–9) and in this direction, of great advantage is the small size of its genome (ca. 2 Mb), which facilitates novel "omics" and systems biology approaches (10), and also traditional engineering and optimization.

Production of mutants is an integral part in functional analysis of organisms and strain enhancement. Methods that have been employed toward mutant production in *Z. mobilis* include conventional mutagenesis with the use of UV-irradiation or chemical mutagens (11–13), evolutionary adaptation (5) or specific gene knock-out via allele exchange (14). Except for the last, instability of mutants produced by most other methods has been repeatedly reported in literature and likely owes to the strong DNA-repair activity exhibited by wild-type *Zymomonas* (12).

An alternative route to the generation of mutants, stable and most often polar, involves transposon mutagenesis. The latter entails insertional gene inactivation that is usually nonleaky and also provides readily identifiable genetic markers carried by the transposon (15). Of several transposons that have been tested in *Z. mobilis* in order to mark or mutate the genome (i.e., Tn*5*, Tn*10*, Tn*951*, and Tn*1725*; (11, 16, 17)), a truncated version of the transposable phage Mu (Mu3A or mini-Mu; (18)) has proven to be the most successful in yielding random insertions into the genome (19). The last is exhibited by the high scores of auxotroph mutants obtained by use of plasmid pULB113 – a plasmid RP4 derivative carrying mini-Mu (20) – which range from 2.6% of independently isolated pULB113 transconjugants at direct screenings, to 30% obtained after an enrichment procedure (19). Given that pULB113-mediated transposition has been reported to occasionally result in R′ formation (20), its use in *Z. mobilis* may also provide an effective means toward chromosomal marker mobilization. The procedures by which *Z. mobilis* mutants are generated and detected with the use of pULB113 and mini-Mu are outlined below.

2. Materials

2.1. Bacterial Cultures, Conjugation, and Marker Selection

1. Strain MXR (pULB113) is kindly provided by Dr. F. Van Gijsegem, Laboratoire Interactions Plantes Pathogènes, UMR 217, INRA/AgroParisTech/UPMC.

2. *Z. mobilis* minimal medium (MM; (21)): 2% (w/v) glucose, 0.1% (w/v) $(NH_4)_2SO_4$, 0.1% (w/v) KH_2PO_4, 0.1% (w/v) K_2HPO_4, 0.05% (w/v) $Mg_2SO_4{\cdot}7H_2O$, and 0.05% (w/v) NaCl. After autoclaving and cooling to 50°C, pantothenic acid is added to a 0.5 μg/mL final concentration. Pantothenic acid stock solution is 0.5 mg/mL, filter-sterilized and kept at 4°C for at least a month (see Note 1).
3. *Z. mobilis* complete medium (CM; (21)): MM without sodium chloride and pantothenic acid, supplemented with 0.5% (w/v) yeast extract (Oxoid). To make solid media add 1.5% (w/v) agar prior to autoclaving.
4. Antibiotic stock solutions: 100 mg/mL kanamycin, 25 mg/mL tetracycline dissolved in 50% (v/v) ethanol, 100 mg/mL ampicillin-Na salt, 50 mg/mL rifampicin dissolved in DMSO, 10 mg/mL novobiocin, 100 mg/mL streptomycin, and 10 mg/mL nalidixic acid dissolved in water adding a couple of drops of 1N NaOH. Stocks are filter-sterilized and stored at −20°C; working aliquots can be kept at 4°C for 1 month. Tetracycline and rifampicin stock solutions or media containing them should be light-protected. Final concentrations of antibiotics for marker selection in *Z. mobilis* are 100 μg/mL kanamycin, 40 μg/mL tetracycline, 20 μg/mL rifampicin, 250 μg/mL ampicillin, and 40 μg/mL novobiocin.
5. Luria-Bertani (LB) medium: 1% (w/v) NaCl, 0.5% (w/v) yeast extract, 1% (w/v) Tryptone. M9 minimal medium and antibiotic concentrations for marker selection in *Escherichia coli* are as standardly described (22).
6. Filter holders (Swinnex-25, Millipore) and filters of 0.45 μm pore size, 25-mm diameter (MF-Millipore) are used for bacterial matings.
7. 10- or 20-mL syringes are used for bacterial culture filtering.
8. 96-Well plates serve for multiple isolate subculturings and viable cell counts.
9. A 48-pin replicator facilitates multiwell plate inoculations (Boekel Scientific, Festerville, PA).
10. Normal saline solution (NS): 0.9% (w/v) NaCl.

2.2. Auxotroph Determination and Enrichment

1. Amino acid and nucleotide stock solutions are 5 mg/mL (except for tyrosine and tryptophan, 0.5 mg/mL), filter-sterilized and kept at 4°C for at least a month. Vitamin stocks are 0.5–1 mg/mL. Final concentrations for nutritional requirement analysis are 10 μg/mL for amino acids and bases and 0.5 μg/mL for vitamins.
2. Amoxicillin and clavulanic acid are used for auxotroph enrichment at a 400 μg/mL and 100 μg/mL final concentration,

respectively. Stock solutions are 100 mg/mL for both, prepared and kept as above (see step 4 in Subheading 2.1).

2.3. DNA Isolation and Restriction

1. Cell resuspension buffer: TAE 1× (see step 1 in Subheading 2.4).
2. Lysis stock solution: 0.6% (w/v) Tris–base, 3% (w/v) SDS.
3. Working-lysis solution: 50 mM Tris–HCl, 3% (w/v) SDS, pH 12.6. Prepare fresh by adding 40 μL of 10N NaOH to 10 mL of the lysis stock solution (step 2; store in a plastic vial).
4. Lysis neutralization solution: 3 M potassium acetate, pH 4.8 (adjusted with glacial acetic acid; (23)). Store at 4°C.
5. Deproteinization mix: 25:24:1 (v/v/v) phenol:chloroform: isoamyl alcohol saturated with 10 mM Tris–HCl, pH 8.0 (22). Store at 4°C.
6. Isopropanol and ethanol (analytical grade) and 70% (v/v) ethanol for DNA precipitation.
7. TE buffer (1×): 10 mM Tris–HCl, 1 mM EDTA, pH 7.6. Prepare from stocks of 2 M Tris–HCl, pH 7.6 (Tris–base pH adjusted with HCl) and 0.5 M EDTA, pH 8.0 (pH adjusted with NaOH pellets).
8. RNase stock solution: 10 mg/mL DNase-free pancreatic RNase A in 10 mM Tris–HCl, 15 mM NaCl, pH 8.0. Stock is heat-treated at 100°C for 10 min to inactive DNase, aliquoted and stored at −20°C (22). Working concentration is 20 μg/mL.
9. Standard SDS/alkaline-lysis materials (22) are used for plasmid DNA preparations from *E. coli* and *Z. mobilis*, with the exception that for *Z. mobilis* the regular alkaline lysis solution II (1% (w/v) SDS, 0.2 N NaOH) contains 5 mM EDTA (19). Organic reagents for extractions and precipitations as in steps 4 and 5.
10. Restriction enzymes and buffers according to manufacturer.

2.4. DNA Electrophoresis

1. TAE Electrophoresis buffer (1×): 40 mM Tris–acetate, 1 mM EDTA, pH 7.9. Prepare from a 50× stock: 242 g/L Tris–base, 57.1 mL/L glacial acetic acid, 100 mL/L 0.5 M EDTA, pH 8.0.
2. Agarose gels: 0.8% (w/v) ultra pure agarose in 1× TAE. The suspension is brought to boil to dissolve the agarose, cooled down to 50°C and poured onto 20- or 30-well horizontal gel casting plates. 0.5 μg/mL ethidium bromide is incorporated in the gel prior to casting (add appropriate quantity from a 10-mg/mL stock) or used for postelectrophoresis staining of the gel. In this case prepare ~500 mL of 0.5 μg/mL ethidium bromide in 1× TAE and store in a dark container (attention: ethidium bromide is light-sensitive and carcinogenic).

3. Sample loading dye (6×): 30% (w/v) glycerol, 0.25% (w/v) xylene cyanol, 0.25% (w/v) bromophenol blue.
4. DNA fragment size markers: λ/*Hind*III or 1-kb ladder (New England Biolabs).
5. A gel imaging system serves for gel documentation. A UV-transparent ruler may be optionally photographed besides the gel, to act as size reference.

2.5. Southern Blotting of DNA

1. A vacuum system is preferably used for transfer and blotting of DNA (VacuGene XL; GE Healthcare, formerly Amersham Biosciences).
2. Nylon membranes are used for DNA blotting (Hybond™-N, GE Healthcare, or other brands guaranteeing low backgrounds in nonradioactive hybridizations).
3. Depurination solution: 0.25N HCl.
4. Denaturation solution: 0.5N NaOH, 1.5 M NaCl (store in a plastic bottle).
5. Neutralization solution: 1 M Tris–HCl (pH 8.0), 1.5 M NaCl.
6. SSC (20×) for transfer: 8.82% (w/v) Na-citrate, 17.53% (w/v) NaCl. Also prepare a tenfold dilution to wash membranes after transfer completion.
7. A UV transilluminator is used for DNA cross-linking on membranes at 254 nm.

2.6. Nonradioactive DNA Labeling and Hybridization

1. A gel extraction kit (i.e., QIAquick Gel Extraction; Qiagen, Valencia, CA) aids in the purification of the DNA fragment to be labeled.
2. The non-radioactive DIG DNA Labeling Kit (Roche) is used for DNA labeling with digoxigenin-UTP, according to the manufacturer.
3. Hybridizationsolution:5×SSC,0.1%(v/v) *N*-laurylsarcosine-Na salt, 0.02% (w/v) SDS, 1% (w/v) blocking reagent (Roche). Heat to near-boiling while stirring, to dissolve the blocking reagent. Store at –20°C.
4. Posthybridization washing solution 1: 2× SSC, 0.1% (w/v) SDS. Prepare fresh from 20× SSC, 10% (w/v) SDS stocks.
5. Posthybridization washing solution 2: 0.5× SSC, 0.1% (w/v) SDS (stringent wash; prepare as above). Warm to hybridization temperature before use.
6. Detection buffer 1: 100 mM Tris–HCl, 150 mM NaCl, pH 7.5 (see Note 2). Prepare from 2 M Tris–Cl (pH 7.8), 5 M NaCl stocks. Adjust pH if necessary.
7. Detection buffer 2: as above, with 1% (w/v) blocking reagent (Roche). Store at –20°C.

8. Detection buffer 3: 100 mM Tris–HCl, 100 mM NaCl, pH 9.5. Prepare from 1 M Tris–Cl (pH 9.5), 5 M NaCl stocks.
9. Anti-Digoxigenin – AP, Fab Fragments (Roche). Store at 4°C.
10. Colorimetric detection reagents NBT/BCIP (DIG DNA Labeling and Detection Kit or DIG Nucleic Acid Detection Kit, Roche). Light-protect, store at −20°C.
11. *N,N*-Dimethylformamide (DMF) to decolorize the filter for rehybridization (*optional*).
12. Alkaline probe stripping solution: 0.2N NaOH, 0.1% (w/v) SDS. Prepare fresh from 10N NaOH, 10% (w/v) SDS stocks (*optional*).
13. SSC (2×): 300 mM NaCl, 30 mM sodium citrate (dilute from a 20× SSC stock 1:10; for stock preparation see step 6 in Subheading 2.5) (*optional*).
14. Whatman 3 MM filter paper is used throughout for filter protection and storage.

3. Methods

Z. mobilis is hardly transformed via chemical transformation methods; it can be transformed via electroporation, although at low yields and solely with vectors that exhibit high stability in the organism. By far, the most successful means for gene transfer into *Z. mobilis* is direct or helped (triparental) conjugation with the aid of the IncP1 transfer system (13).

Employing bacterial conjugation in order to introduce foreign genes in *Z. mobilis*, cross-inhibition of the mating partner – a frequent complication in bacterial co-cultures – should be taken into account. For example, *Zymomonas* inhibits *Pseudomonas* to complete mating obstruction (19). *E. coli* is also inhibited by *Z. mobilis* – often to an order of magnitude population decrease – although it can still act as DNA donor. Different *Z. mobilis* strains vary in their performance as recipients, with the most robust ethanol producers (ATCC 31821 variant strains ZM4 and CP4) proving fortuitously also the best recipients. Lastly, specific strain derivatives (i.e., of ATCC 10988 that have been studied) may exhibit a drop in conjugal receptiveness at late growth stages, contrarily to others or the parental strain. Overall, conjugal performance depends largely on the mating parents and tests should be run to determine optimal conditions in each case.

When gene integration is pursued by means of transposition or targeted recombination, the stability of the plasmid vehicle used is of less consequence inasmuch as its entry and minimal sustenance are achieved (i.e., such that will allow for subsequent recombinational events to take place). The broad-host range plasmid RP4 is lost in 40 generations from Z. *mobilis*, yet transfer of pULB113 (RP4::mini-Mu) leads to the generation of extremely stable transconjugants, most certainly due to spontaneous integration events involving mini-Mu. Notably, not all transposons, despite their reported promiscuity or that of the vehicle they are hosted on, are as successful in yielding multiple and stable insertions in *Z. mobilis* (19).

The success of pULB113 in randomly fusing into the *Z. mobilis* genome hints to the possibility that other Mu-based insertional tools, such as Mud*lac* used for expressional studies (23), or cointegrates involving pULB113 of R′ nature that act to create in vivo gene banks or mobilize the chromosome (20), may also prove effective in the study and engineering of *Z. mobilis*. Current knowledge of the bacterium's genome sequence will certainly aid in future mapping of Mu-based insertions and, thus, in full extrapolation of pathways and regulons indicated by loss-of-function mutants.

3.1. Conjugal Transfer of pULB113 into Z. mobilis

1. *E. coli* (LB, optionally containing 50 μg/mL ampicillin, 12.5 μg/mL tetracycline and/or 50 μg/mL kanamycin for pULB113 selection) and *Z. mobilis* (MM or CM) cultures to be used in matings are prepared from fresh precultures and are harvested at mid-log phase (should not exceed 5×10^8 cells/mL, see Note 3). *E. coli* grows best at 37°C, shaken (22); *Z. mobilis* grows at 30°C, standing, in screw-cap bottles filled to 80–90% of their volume (caps should be left slightly loose upon tightening, to avoid excessive CO_2 pressure build-up in the bottle).
2. Matings between *E. coli* donors carrying pULB113 – i.e., strain MXR (pULB113) – and *Z. mobilis*, are performed on nitrocellulose filters at donor/recipient cell ratios of 1:3 to 1:5 (other *E. coli* strains, preferably *recA*–, can also be used). For this, culture aliquots are drawn with syringe and filtered with the use of a Swinnex filter holder – filter system (that can be disassembled). 2–3 mL sterile NS are then passed through the filter to wash the cells, and the filter is removed from the holder and placed onto solid CM medium (bacteria side up), for 5–6 h at 30°C (see Notes 4, 5). Viable cell counts from donor and recipient cultures should be conducted at this stage in order to monitor premating bacterial population numbers (see step 3 below).
3. At the end of the mating period, cells are recovered from filters in 1 mL NS (the cell paste can be completely resuspended if the filter is placed inside an empty Petri plate and rinsed/scraped off with the NS several times). The mating cell resuspension is used for postmating donor and recipient cell counts, as well as transconjugant counts and platings.

Viable cell counts are conducted by doing serial tenfold dilutions of the mating resuspension in NS and spotting 20-μL from each dilution onto plates selective for either the donor or the recipient markers. Natural resistance of *Z. mobilis* to nalidixic acid (50 μg/mL) and streptomycin (100 μg/mL) serves to counter-select *E. coli*. Alternatively for this purpose, *Z. mobilis* spontaneous mutants resistant to rifampicin and/or novobiocin can be isolated and used as mating recipients throughout. *E. coli* cell counts are monitored on LB plates at 37°C, at which conditions *Z. mobilis* does not grow.

4. *Z. mobilis* transconjugants for pULB113 can be isolated by plating 200 μL aliquots of the mating resuspension on CM plates, selective for both recipient (see step 3 above) and pULB113 markers (amp, kan and tet; see Note 6). Transconjugant colonies appear in 4–5 days (at a frequency of ca. 10^{-3} per CP4 recipient cells plated; (20)) and can be transferred to fresh selective media in order to verify marker expression. Cultures of isolates determined to be kept at this stage can be stored in 20% (v/v) glycerol at −20°C for 3–5 years or at −70°C permanently.

3.2. Screening for Mini-Mu Insertions: Identification of Auxotrophs

1. *Z. mobilis* pULB113 transconjugants are subcultured for 16–30 generations in selective liquid medium to allow for transposition events to occur (approximately eight to nine cell divisions are estimated to take place in 24 h of growth, starting with a 0.5% (v/v) late-log inoculum into fresh medium). Multiple isolate subculturings can be carried out in multiwell plates with the use of a pin replicator. Slow-growing isolates are often observed at this stage; these most likely carry mutations affecting growth traits.
2. Auxotrophy is one of the most frequent traits to be encountered via random mutagenesis and can be tested by mini-streaking isolates from step 1 on solid minimal medium with the use of a toothpick, which minimizes rich medium carry-over onto the MM plate (25–50 streaks per Petri plate). Nonreversible auxotrophs as well as leaky mutants are obtained in such screenings. Further auxotyping of isolates is carried out in subsequent steps, starting with broad characterization of nutritional requirements (e.g., amino acids, vitamins or bases), followed by specific compound determination for each of the isolates.
3. For auxotroph enrichment, newly emerged transconjugant colonies are pooled together and used to inoculate liquid MM supplemented with antibiotics that inhibit cell wall formation (i.e., 400 μg/mL amoxicillin and 100 μg/mL clavulanic acid) to a final cell suspension that does not exceed 10^6–10^7 bacteria per mL. The suspension is grown at 30°C for ca.

20 h (or as long as it takes for a control culture without antibiotics to start duplicating), at which period, dividing wild-type cells are killed and nongrowing mutants survive. Cells are then harvested, plated, and screened for auxotrophy. Following this procedure, several orders of magnitude lethality is observed; however, a tenfold increase in auxotroph frequencies is also achieved.

3.3. DNA Isolation and Restriction from Mini-Mu Insertion Mutants

1. Small scale total DNA is prepared from *Z. mobilis* mutants (or the parental strain) as follows: cells from 3 mL late-log cultures (see Note 7) are harvested and the pellet resuspended in 200 μL 1× TAE. Two volumes (400 μL) of working-lysis solution are added, the suspension briefly mixed by inversion and incubated for 15 min at 65°C (at this period the lysate clears). 50 μL of ice-cold neutralization solution is added to the lysate, followed by an equal volume (700 μL) of the deproteinization mix. The mixture is vortexed to complete emulsification for 5–7 s and kept on ice for 10 min. The emulsion is centrifuged in a microfuge (13,523 × *g*; 10 min) and the upper aqueous phase removed to a fresh tube and extracted again with deproteinization mix. DNA is precipitated from the cleared aqueous phase by adding NaCl to 0.5 M, and 0.8 volumes of isopropanol or 2 volumes of ethanol, mixing and centrifuging as before. The DNA pellet is washed with 70% (v/v) ethanol, dried and resuspended in 50 μL 1× TE buffer (preferably supplemented with RNase to a final concentration of 20 μg/mL), at 65°C for 20 min (this step is important in order to inactivate nucleases; see Note 7).
2. One tenth of the total DNA preparation is digested with the use of a frequent cutter for *Z. mobilis* (i.e., *Eco*RI or *Hind*III, 1–2 units/μg DNA, for 2 h at 37°C) to test for quality and quantity of the prepared DNA.

3.4. DNA Electrophoresis

1. DNAs from *Z. mobilis* mutants likely to bear mini-Mu insertions (see step 1 in Subheading 3.3) are digested with a restriction enzyme chosen such as to (1) not cut into mini-Mu, (2) cut pULB113 close to mini-Mu, and (3) cut the *Z. mobilis* chromosome relatively frequently (i.e., *Pvu*II or *Kpn*I; digestions as in step 2, Subheading 3.3). This serves to detect restriction fragment length polymorphisms brought about by the insertions (see Note 8, Fig. 1).
2. Digested DNAs aiming to be Southern-blotted are electrophoresed overnight in a 0.7–0.8% (w/v) agarose gel. The gel is then submerged in ethidium bromide staining solution for 20 min, rinsed with distilled water and documented under UV-exposure (preferably at 300–365 nm to limit DNA damage) with the use of a gel imaging system.

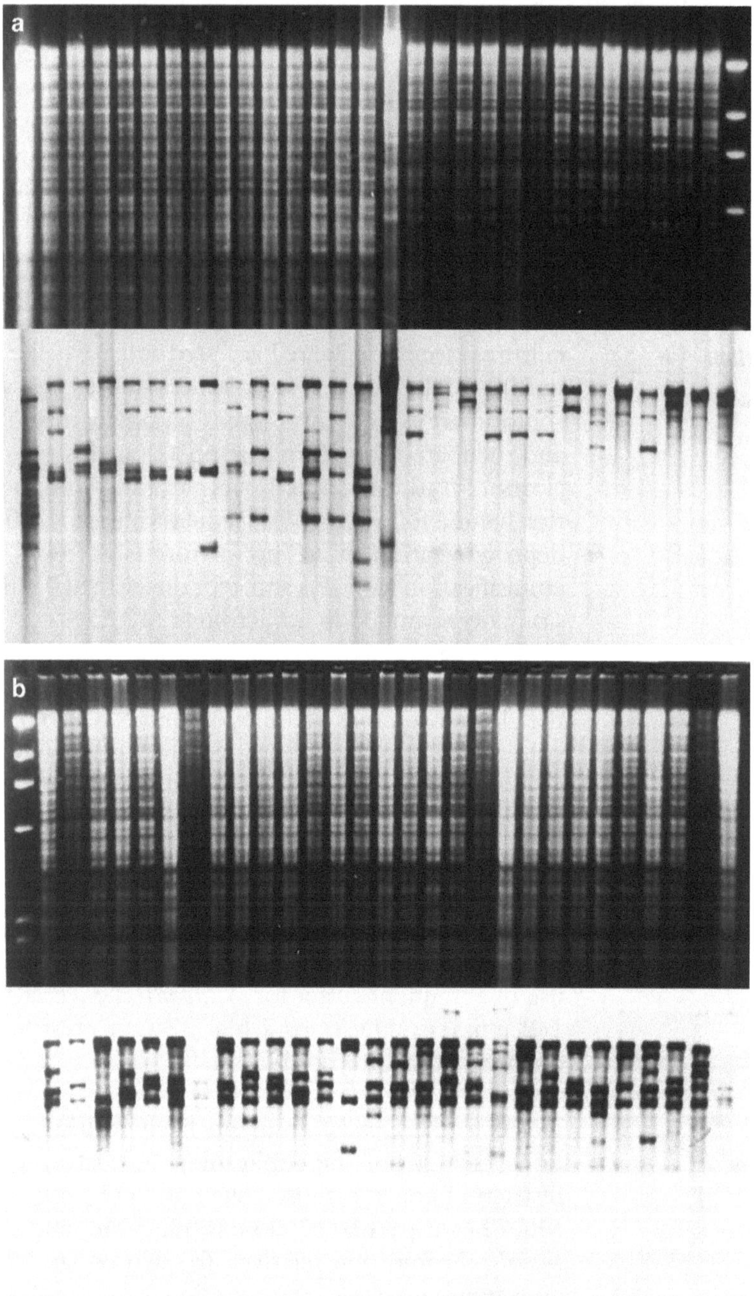

Fig. 1. Detection of mini-Mu transposition by Southern hybridization. In each composite image, ethidium bromide-stained agarose gels are shown on *top* and the corresponding hybridization results with the labeled 2.9-kb *Pst*I fragment of mini-Mu, on *bottom*. (**a**) Chromosomal DNA digestions from independent methionine-requiring *Zymomonas mobilis* CP4 mutants. *Lanes* at the *left half* of the gel are digested with *Pvu*II (negative control DNA from parental CP4 at *left-most lane*) and at the *right half* with *Kpn*I. (**b**) *Pvu*II chromosomal digestions from cysteine-requiring *Z. mobilis* CP4 mutants (*right-most lane* carries parental strain DNA). Only two pairs of isolates in each of gels (**a**) and (**b**) show identical patterns (reproduced with kind permission from John Wiley and Sons (19)). (**c**) The majority of samples electrophoresed on (**b**) are cut in this gel with *Eco*RI. The lack of polymorphism in this case underlines the importance of choice of enzyme in insertional profiling.

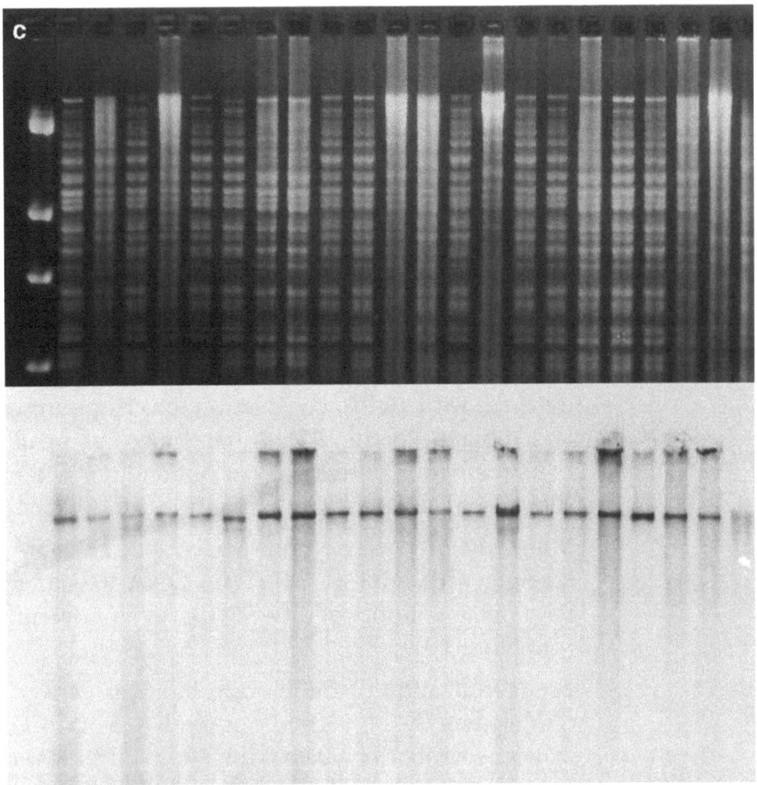

Fig. 1. (continued)

3.5. Southern Blotting of DNA

1. Blotting of the gel is carried out on nylon membrane filters by use of vacuum (50–55 mbar), applying conditions for high-molecular weight DNA transfer, according to the manufacturer (VacuGene XL, GE Healthcare). Depurination, denaturation, and neutralization solutions are subsequently poured on top of the gel and, by end of each treatment, removed by a 10-mL pipette to be replaced by the next solution, in a manner such that the gel always remains covered. Each treatment lasts as long as it takes for respective solutions to completely saturate the gel (an indication for this is given by the time it takes for the bromophenol blue front to turn into yellow at HCl-mediated depurination; the same time interval – usually 10–20 min – applies to following steps). Final DNA transfer is carried out with the use of 20× SSC – poured onto and covering the gel as before – and is complete within an hour (ethidium bromide-staining of the gel after transfer verifies that no residual DNA is usually left behind). At transfer completion, the electrophoresis origin for each lane is marked with pencil on the filter (by piercing the gel at the sites of wells), the vacuum is turned off, the gel removed, and the filter collected and washed briefly in 2× SSC. The filter is then

air-dried, cross-linked under UV at 254 nm (face-down on a transilluminator) for 3 min (see Note 9), and stored protected in Whatmann paper.

3.6. Detection of Mini-Mu Insertions by Southern Hybridization

1. A mini-Mu – specific probe for use in Southern hybridizations may be generated by PCR-amplification of Mu-specific sequences or, alternatively, by isolation of an internal 2.9-kb *Pst*I fragment of mini-Mu from pULB113. For this, large-scale plasmid preparation of pULB113 is required (pULB113 is a low-copy plasmid; at least a liter of the *E. coli* host should be treated according to regular SDS/alkaline-lysis maxi preparation procedures – (23)). The isolated plasmid DNA is digested with *Pst*I to completion, electrophoresed overnight, and the appropriate 2.9-kb band (second smallest in a five-band restriction pattern) is cut out and purified from the agarose slice (QIAquick Gel Extraction Kit, Qiagen).
2. 0.5–1 μg of purified mini-Mu – specific DNA is used for labeling with DIG-dUTP (DIG DNA Labeling kit, Roche), to a final volume of 20–30 μL in an overnight reaction (see Note 10).
3. Prehybridization of the blotted filter (as derived at step 3 in Subheading 3.4) is carried out for 1–2 h at 68°C in a hybridization chamber (Hybaid Mini Oven). Filter(s) are placed in rotisserie bottles filled with ca. 20 mL of hybridization solution per 100 cm^2 of membrane. Prehybridization is followed by overnight hybridization at same conditions; for this, the hybridization solution used to equilibrate and block the membrane is replaced with ca. 5–7 mL of fresh solution containing the probe (4–6 μL of labeled DNA). The probe solution is denatured for 10 min at 95–100°C prior to use, while care is taken to avoid letting the membrane dry amidst solution change. After use, the probe solution can be stored at –20°C for future reuse (see Note 11).
4. Posthybridization washing of unbound probe requires two 5-min washes in solution 1 at room temperature and two 15-min washes in prewarmed solution 2 at hybridization temperature (68°C; stringent washes). These can be carried out in the hybridization chamber, using at least 25 mL of washing solution per bottle for each treatment (see Note 12), at top rotational speed.
5. For signal detection, the filter is removed from the hybridization bottle and treated in a glass Tupperware-type of container on a shaking platform, following the recommended protocol materials and steps (DIG-dUTP label detection by anti-DIG antibody/alkaline phosphatase (AP) conjugate; DIG DNA Labeling and Detection Kit or DIG Nucleic Acid Detection Kit, Roche). Care is taken at all stages not to let the filter dry (see Notes 13, 14).

6. For final signal development, NBT/BCIP is used as AP substrate (Roche); the filter is treated in 10–15 mL of freshly prepared color detection solution in a plastic bag (sealed with the use of a regular bag-sealer) and stored in the dark at room temperature or 37°C (for faster results). The color precipitate reaches saturation from anywhere between a few minutes to 24 h, according to probe strength.
7. To stop the color reaction, the filter is washed thoroughly in distilled water. It is then documented with the use of a PC-scanner or other imaging device while wet (color is then more intense), air-dried, and stored permanently in the dark.
8. To rehybridize with a second probe, the filter should be decolorized and the probe removed. For this, the filter is treated in 50–100 mL DMF at 50–52°C, under hood (attention: the DMF flash point is 58°C). As soon as the signals fade, the filter is rinsed in water and then shaken gently in alkaline probe stripping solution for 30 min at 37°C. The released probe is washed away in three consecutive washes with 50 mL 2× SSC, 15 min each, and the filter stored dry for future use (*optional*).

4. Notes

1. Unless otherwise mentioned, solutions are prepared in deionized H_2O and kept at room temperature. Stock and working solutions as well as consumables for bacteriological or molecular use are sterilized by autoclaving or, where indicated, by 0.2 μM filtration. Aseptic conditions are employed when handling bacteria.
2. Tris-based detection buffer 1 for posthybridization washes and anti-DIG antibody binding is as reported in the first digoxigenin-labeling protocol released by Boehringer Mannheim (GmbH); it performs equally well to the currently recommended maleic acid buffer.
3. For mating experiments it is recommended that *Z. mobilis* cultures are harvested at mid-exponential phase. Over-grown *Z. mobilis* yields strong backgrounds on selective medium and for some strains the recipient performance drops.
4. In conjugation experiments where highest transconjugant numbers are not a requisite, mild centrifugation of donor and recipient culture mixtures (i.e., 4,602 × *g*, 1 min in a microfuge) and spotting onto a filter can substitute for syringe-filtering.
5. *E. coli* strains carrying plasmids of the IncP1 incompatibility group conjugate best on solid surfaces. Additionally, nonselective medium used to nourish the bacteria at mating intervals

should preferably favor the recipient (i.e., CM used instead of LB in matings between *E. coli* donors and *Z. mobilis* recipients).

6. *Z. mobilis* is inherently resistant to a wide range of antibiotics. Selection of true transconjugants (or electrotransformants) is safer when multiple markers of foreign DNA are used.
7. DNA should be prepared from *Z. mobilis* cultures at a phase no later than mid- to late-exponential (particularly plasmid DNA). At more progressed growth stages deproteinization gets to be more difficult and, additionally, plasmid yields drop (unpublished observation). It should be noted here that *Z. mobilis* exhibits strong nucleolytic activity; it is important that the concentration of EDTA is kept to at least 5 mM throughout lysis and the final DNA resuspension is heat-treated at 65°C.
8. In order to detect transposon-mediated RFLPs (caused by Mu or other elements), various digests of chromosomal DNA and test hybridizations have to be carried out, in order to resort to the enzymes yielding optimal pattern differences (compare images in Fig. 1).
9. Alternative capillary or electrophoretic transfer methods (22) may be substituted for the described vacuum transfer. For multiple strippings and reprobings, optimal retention of blotted DNA on nylon filter (particularly neutral nylon) is required; this is best achieved by both UV cross-linking and baking at 80°C in a dry oven, for 2 h.
10. Regardless of the high DIG DNA protocol sensitivity, it is recommended that at least 0.5 μg DNA is labeled, since a stronger probe leads to faster specific signal development against nonspecific background. High quality of DNA to be labeled (pure from contaminants and least degraded) is also important for background elimination.
11. Alternative heat-sealable bag-based hybridization methods (22) may be substituted for the described hybridization chamber. DIG-labeled DNA can be used for at least 15 years (that has been tested) without loss in performance; it is also stable in the hybridization solution.
12. Proposed solution volumes for hybridization and signal detection steps apply to 100 cm^2 filters.
13. Throughout Southern blotting, hybridization and detection procedures, nylon filters should be treated with care and stored protected in Whatmann paper. If not for harsh filter treatment, backgrounds also rise due to bad quality of probe or the filter drying at any stage prior to adding the probe, the anti-DIG antibody conjugate or the AP substrate(s). Good equilibration and constant wetness of filter at those stages is important. Glass tupperware instead of plastic is also preferred.

14. If chemilluminescence is used for signal development (i.e., CSPD), X-ray film or a phosphorimager screen will be required to record the signal. Higher sensitivity is achieved this way; nevertheless in most cases it is more practical to monitor the emergence of signal(s) on the membrane itself.

Acknowledgments

The author sincerely wishes to thank Professor M. A. Typas for his generous guidance in the study of *Z. mobilis*. Special thanks to Dr. I. Galani for her contribution in data described in this work. The NKUA Research Committee is acknowledged for grant no. 70/4/9824.

References

1. Gunasekaran P., and Raj C. (1999) Ethanol fermentation technology - *Zymomonas mobilis*. *Current Science* 77, 56–68.
2. Rogers P.L., Jeon Y.J., Lee K.J., and Lawford H.G. (2007) *Zymomonas mobilis* for fuel ethanol and higher value products. *Adv. Biochem. Eng. Biotechnol.* **108**, 263–288.
3. Zhang M., Eddy C., Deanda K., Finkelstein M., and Picataggio S. (1995) Metabolic engineering of a pentose metabolism pathway in ethanologenic *Zymomonas mobilis*. *Science* **267**, 240–243.
4. Deanda K., Zhang M., Eddy C., and Picataggio S. (1996) Development of an arabinose-fermenting *Zymomonas mobilis* strain by metabolic pathway engineering. *Appl. Environ. Microbiol.* **62**, 4465–4470.
5. Dien B.S., Cotta M.A., Jeffries T.W. (2003) Bacteria engineered for fuel ethanol production: current status. *Appl. Microbiol. Biotechnol.* **63**, 258–266.
6. Moreau R.A., Powell M.J., Osman S.F., Whitaker B.D., Fett W.F., Roth L., and O'Brien D.J. (1995) Analysis of intact hopanoids and other lipids from the bacterium *Zymomonas mobilis* by high-performance liquid chromatography. *Anal. Biochem.* **224**, 293–301.
7. Seo J.S., Chong H., Park H.S., Yoon K.O., Jung C., Kim J.J. et al. (2005) The genome sequence of the ethanologenic bacterium *Zymomonas mobilis* ZM4. *Nat. Biotechnol.* **23**, 63–68.
8. Yang S., Pappas K.M., Hauser L.J., Land M.L., Chen G.L., Hurst G.B. et al. (2009) Improved genome annotation for *Zymomonas mobilis*. *Nat. Biotechnol.* **27**, 893–894.
9. Kouvelis V.N., Saunders E., Brettin T.S., Bruce D., Detter C., Han C. et al. (2009) Complete genome sequence of the ethanol producer Zymomonas mobilis NCIMB 11163. J. Bacteriol. **191**, 7140–7141.
10. Yang S., Tschaplinski T.J., Engle N.L., Carroll S.L., Martin S.L., Davison B.H. et al. (2009) Transcriptomic and metabolomic profiling of *Zymomonas mobilis* during aerobic and anaerobic fermentations. *BMC Genomics* **10**, 34.
11. Skotnicki M.L., Lee K.J., Tribe D.E., and Rogers P.L. (1982) Genetic alteration of *Zymomonas mobilis* for ethanol production. In: Hollender A et al (ed) Genetic engineering of microorganisms for chemicals, Plenum Press, NY
12. Typas M.A., and Galani I. (1992) Chemical and UV mutagenesis in *Zymomonas mobilis*. *Genetica* **87**, 37–45.
13. Sprenger G.A., Typas M.A., and Drainas C. (1993) Genetics and genetic engineering of *Zymomonas mobilis*. *World J. Microbiol. Biotechnol.* **9**, 17–24.
14. Kalnenieks U., Galinina N., Toma M.M., Pickford J.L., Rutkis R., and Poole R.K. (2006) Respiratory behaviour of a *Zymomonas mobilis adhB::kan*(r) mutant supports the hypothesis of two alcohol dehydrogenase isoenzymes catalysing opposite reactions. *FEBS Lett.* **580**, 5084–5088.
15. Reznikoff W.S., and Winterberg K.M. (2008) Transposon-based strategies for the identification of essential bacterial genes. *Methods Mol. Biol.* **416**, 13–26.
16. Carey T., Sewell G.M., Osman Y.A., and Ingram L.O. (1987) Expression of the lactose

transposon (Tn*951*) in *Zymomonas mobilis*. *Appl. Environ. Microbiol.* **45**, 1163–1168.

17. Walker M.J., and Pemberton J.M. (1988) Construction of transposons encoding genes for b-glucosidase, amylase and poly-galacturonate trans-eliminase from *Klebsiella oxytoca* and their expression in a range of gram-negative bacteria. *Curr. Microbiol.* **17**, 69–75.
18. Faelen M., Resibois A., and Toussaint A. (1978) Mini-Mu: an insertion element derived from temperate phage Mu-1. *Cold Spring Harbor Symp. Quant. Biol.* **43**, 1169–1177.
19. Pappas K.M., Galani I., and Typas M.A. (1997) Transposon mutagenesis and strain construction in *Zymomonas mobilis*. *J. Appl. Microbiol.* **82**, 379–388.
20. Van Gijsengem F., and Toussaint A. (1982) Chromosome transfer and R-prime formation by an RP4::mini-Mu derivative in *Escherichia coli*, *Salmonella typhimurium*, *Klebsiella pneumoniae* and *Proteus mirabilis*. *Plasmid* 7, 30–44.
21. Galani I., and Typas M.A. (1985) Growth requirements and the establishment of a chemically defined minimal medium in *Zymomonas mobilis*. *Biotechnol. Lett.* 7, 673–678.
22. Sambrook J., and Russell D.W. (2001) Molecular Cloning: Laboratory Manual. Cold Spring Harbor Laboratory Press, Cold Spring Harbor, New York.
23. Castilho B.A., Olfson P., and Casadaban M.J. (1984) Plasmid insertion mutagenesis and *lac* gene fusion with mini-Mu bacteriophage transposons. *J. Bacteriol.* **158**, 488–495.

Chapter 26

Engineering Thermoacidophilic Archaea using Linear DNA Recombination

Yukari Maezato, Karl Dana, and Paul Blum

Abstract

Thermoacidophilic archaea comprise one of the major classes of extremophiles. Most belong to the family Sulfolobales within the phylum Crenarchaeota. They are of applied interest as sources of hyperstable enzymes, for biomining of base and precious metals, and for evolutionary studies because of their use of eukaryotic-like subcellular mechanisms. Genetic methods are available for several species particularly *Sulfolobus solfataricus*. This organism has a considerable number of methods available for the construction of novel cell lines with unique functions. This chapter presents recent developments in the use of homologous recombination and linear DNA for the engineering of site-specific changes in the genome of *S. solfataricus*.

Key words: Archaeal recombineering, Linear DNA recombination, Cell line construction, Crenarchaeota, Hyperthermophiles

1. Introduction

Extremophiles are organisms typically unicellular that thrive in otherwise hostile conditions. Many of these are of great fundamental and applied interest as sources of robust hyperstable enzymes particularly suitable for analysis by protein crystallography (1), for biofuels production (2, 3), and in the recovery of base or semiprecious metals (4). These organisms are also of great interest because of their evolutionary relationships to eukaryotes. As archaea, they utilize eukaryotic-like mechanisms for the synthesis, modification, and turnover of macromolecules including DNA, RNA, and protein (5, 6).

Recombination-based cell line engineering (recombineering) has become highly developed for use in the bacterium *Escherichia*

James A. Williams (ed.), *Strain Engineering: Methods and Protocols*, Methods in Molecular Biology, vol. 765,
DOI 10.1007/978-1-61779-197-0_26,

coli because of the addition of phage recombination components that enhance homologous recombination between linear DNA molecules (7). This capacity has greatly improved efforts to create large complex DNA molecules with diverse applications. Interestingly, homologous linear DNA recombination occurs at a high frequency in natural isolates of *Sulfolobus solfataricus* obviating the need for modification of the endogenous recombination system. The ability to use this method was unexpectedly efficient and likely reflects two features of *S. solfataricus*. The first is a lack of a restriction modification barrier that often plagues bacterial genetic systems. The second is the presence of a eukaryotic-like recombination system. For example, recombination components endogenous to *S. solfataricus* include *Rad*A, an ortholog of eukaryotic *Rad*51; *Mre*11 and *Rad*50, orthologs of eukaryotic *Mre*11/*Rad*50; and *Topoisomerase VI*, an ortholog of eukaryotic *Spo*11 (8–10). Collectively, these recombination proteins mediate efficient and site-specific recombination between homologous linear DNA sequences.

A recent compilation of methods have been described for the general cultivation and storage of *S. solfataricus* (11) and are reiterated here in a condensed form. Methods have also been described for genetic manipulation involving successive single crossover events using nonreplicating (suicide) gene delivery systems (11–14). The purpose of this chapter is to expand and update the current methods used for the genetic manipulation of *S. solfataricus* that are based on linear DNA recombination.

2. Materials

2.1. Cultivation

Cells are grown aerobically in glass Erlenmeyer flasks with polypropylene screw top caps at 80°C in rotary water bath shaker filled with glycerol. Cultivation uses Allen's medium as modified by Brock (15, 16).

1. Brock salts medium (Modified Allen's medium) (1×): 1.3 g/L $(NH4)_2SO_4$, 0.28 g/L KH_2PO_4, 0.12 g/L $MgSO_4$, 0.072 g/L $CaCl_2{\cdot}2H_2O$, 0.02 g/L $FeCl_3{\cdot}6H_2O$, and 0.0045 g/L $Na_2B_4O_7{\cdot}10H_2O$ (see Note 1). The minor components of the basal salts solution consist of trace elements that are prepared fresh and added volumetrically. The concentration of these trace elements and the amounts added per liter of medium are 100 μL of 18 mg/mL $MnCl_2{\cdot}4H_2O$, 10 μL of 22 mg/mL $ZnSO_4{\cdot}7H_2O$, 10 μL of 5.0 mg/mL $CuCl_2{\cdot}2H_2O$, 10 μL of 3.0 mg/mL $NaMoO_4{\cdot}2H_2O$, 10 μL of 3.0 mg/mL $VOSO_4{\cdot}H_2O$, and 10 μL 1.0 mg/mL $CoSO_4{\cdot}7H_2O$. The final volume is adjusted to 1 L and the pH is adjusted to 3.0

by adding concentrated sulfuric acid followed by sterilization by autoclaving. Carbon sources including tryptone and sugars are added at 0.2% (w/v).

2. Gelrite: Dissolve 6 g gelrite (Kelco) in 500 mL water and autoclave (see Note 2).
3. Complex medium: 0.2% (w/v) Difco tryptone in Brock salts (see Note 3).
4. Defined medium: 0.2% (v/v) lactose in Brock salts (see Note 4).
5. Complex medium plates: A solid medium is prepared by adding a sterilized solution containing 1.2% (w/v) gelrite (Kelco) to equal parts of sterilized 2× basal salts with 16 mM magnesium chloride previously adjusted to a pH of 3.0 with sulfuric acid. The resulting solid medium contains 0.6% (w/v) gelrite and 8 mM magnesium chloride.

2.2. Transformation

1. 20 mM Sucrose: Dissolve 1.37 g of sucrose in 200 mL water, filter (0.45 μm) sterilize, and store at room temperature.
2. X-gal solution: Dissolve 10 mg/mL bromo-chloro-indoly-galactopyranoside in *N,N*-dimethylformamide, store at -20°C.
3. Electroporation: MidSci electroporation cuvette (0.1-cm electrode).

3. Methods

3.1. Chromosomal Recombination Using Single Linear Molecules

3.1.1. Disruption of aldhT

A single linear DNA molecule for directed chromosomal recombination is the simplest engineering method to perform. The molecule must contain a selectable marker (gene) usually placed at the center of the targeted open reading frame. This is accomplished either by cloning the selectable marker gene into the correct location via preexisting or constructed compatible restriction sites or, by overlap extension PCR to create appropriate fusion joints between the various molecules. A plasmid encoded, gene disruption construct, is then amplified by PCR and the resulting linear DNA is used for transformation of appropriate *S. solfataricus* cell lines (Fig. 1a).

1. Linear DNA molecules of a target locus (*aldhT::lacS* in this case) from plasmid DNA is obtained by PCR (17).
2. PCR amplicons are purified as described in manufacturer's PCR product clean up kit (see Note 5).
3. Purified linear DNA is used for transformation (see Subheading 3.4).

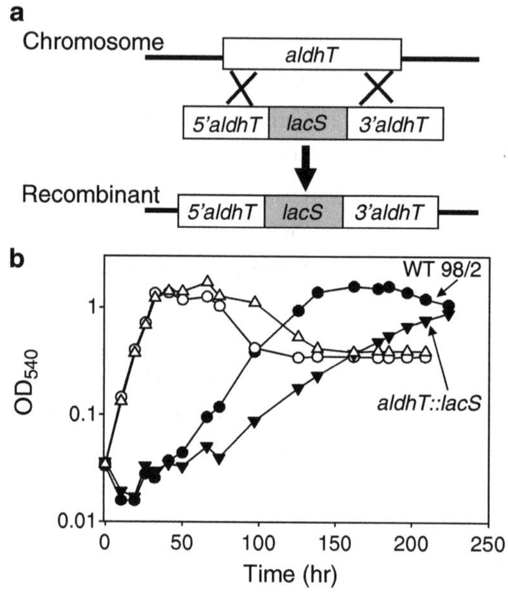

Fig. 1. Linear DNA transformation and recombination. (**a**) A single linear DNA fragment is transformed into ***Sulfolobus***. Homologous recombination between linear DNA fragment and chromosome results in allele replacement. (**b**) Phenotypic analysis of ***aldhT::lacS*** strain showing ***aldhT*** is not required for arabinose utilization. ***Solid circle***=wild type in 0.2% arabinose, ***open circle***=wild type in 0.2% tryptone, ***solid triangle***=***aldhT::lacS*** strain in 0.2% arabinose, and ***open triangle***=***aldhT::lacS*** strain in 0.2% tryptone.

An example provided here concerns the construction of a cell line lacking the *aldhT* gene. Aldehyde dehydrogenase (*aldhT*) is a key enzyme for conversion of aldehydes to carboxylic acids via oxidation. An aldehyde dehydrogenase gene (*aldhT*, SSO3117) had been proposed as part of aldehyde metabolism (18, 19), and in pentose catabolism based on enzymatic activity of the recombinant enzyme and its ability to catalyze conversion of 2,5-dioxopentanoic acid to 2-oxoglutaric acid (20). To further clarify the in vivo function of AldhT, the corresponding gene, *aldhT*, was inactivated by disruption through insertion of the *lacS* gene. The *lacS* gene encodes a beta-glycosidase and confers the ability to utilize lactose as the sole carbon and energy source on *S. solfataricus* strains lacking this gene (13). The linear DNA construct was then integrated by homologous recombination into the chromosome following DNA transformation (11). The resulting mutant was then genotyped using PCR, restriction analysis, and DNA sequencing, combined with phenotypic analysis, to assess the mutant's ability to catabolize pentose sugars (Fig. 1b). Interestingly, the mutant lacking *aldhT* retained the ability to catabolize the pentose arabinose. This finding indicates that *aldhT* is not required for arabinose catabolism or, that *S. solfataricus* harbors additional pathways for utilization of this sugar.

A second use of single linear DNA recombination is for the transfer of mutated genes between cell lines. This requires the use of appropriate host strains that allow for the selection of the gene disruption through recovery of the disrupting genetic marker. In the following example, a disruption mutation that inactivated a DNA polymerase is described. Y-family DNA polymerases are error-prone enzymes conserved in all three domains of life that bypass bulky DNA lesions in a process referred to as translesion DNA synthesis. Three Y-family DNA polymerases have been identified in humans, two in bacteria, and only one in *S. solfataricus*. Loss of the human Y-family DNA polymerase polη causes Xeroderma pigmentosum (a skin abnormality leading to cancer) and causes resistance to the cancer drug cisplatin (21). One of the intensively studied Y-family DNA polymerases is Dpo4 from *S. solfataricus* (22). While initial studies of a disruption mutant demonstrated Dpo4 deficiency elicited a general stress response and drug sensitivity (1), subsequent studies required the mutation be placed in an alternative cell line to enable more directed studies, including complementation. This was accomplished first by amplifying the disrupted allele using PCR, purifying the resulting linear DNA and then transforming it into an alternative host strain followed by selection for the disrupting genetic marker. The *dpo4::lacS* mutation was moved from strain PBL2025 (14) into strain PBL2069 (Fig. 2). While PBL2025 is a *lacS* deletion

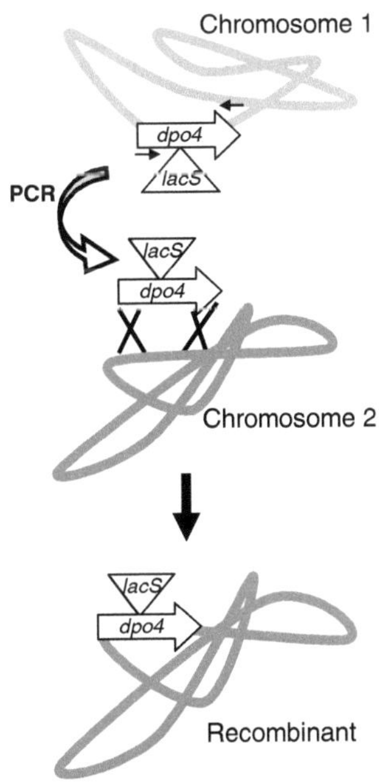

Fig. 2. Allele transfer between chromosomes. Target allele from existing strain (chromosome 1) is amplified by PCR. Amplified allele from chromosome 1 is then transformed and integrated into another strain by homologous recombination (chromosome 2).

mutant allowing for the use of *lacS* as a genetic marker, PBL2069 harbors additional mutations expanding its genetic utility. PBL2069 is deleted for *lacS* and three additional loci encoding alpha-glucosidases resulting in its inability to use maltose as the sole carbon and energy source. This catabolic deficiency is restored using *malA* an alpha-glucosidase as a genetic marker (Maezato, Dana, and Blum, unpublished).

1. Linear DNA encoding the target locus (*aldhT::lacS* in this case) is obtained by PCR (17) from the starting strain genome.
2. PCR amplicons are purified using and following a protocol in a PCR product clean up kit (see Note 5).
3. Purified linear DNA is used for transformation (see Subheading 3.4).

3.2. Chromosomal Recombination Using Two Linear Molecules: Construction and Characterization of an S. solfataricus dps Mutant

PCR methods can replace the use of conventional cloning using unique restriction sites to juxtapose genes in proper orientations that enable construction of mutations. In the example described next, a disruption mutation was constructed using two linear DNA molecules each the product of an overlap extension PCR reaction that undergo recombination between each other and flanking homologous chromosomal sequences. A *dps* loss of function mutation generated by *lacS* insertion was constructed in *S. solfataricus* strain PBL2025 (12) using linear recombination (23). To simplify the process, a new strategy was employed requiring three simultaneous crossovers between two PCR products and the homologous region of the chromosome (Fig. 3a). The PCR products were produced by overlap extension PCR fusing either the 5′ or 3′ end of *dps* and its flanking sequences together with the *lacS* gene (SSO3019) resulting in fragments of about 1.5 kb. The *lacS* insert was placed 50 nucleotides (nt) into the *dps* open reading frame. The two PCR products were then cotransformed as described (11) and homologous recombinants were recovered by enrichment in a miminal lactose medium as described (13). Clonal recombinant cultures were established by colony purification on a solid complex medium containing tryptone (0.2%w/v). The *dps* allele was examined in three purified isolates by PCR using primers complementary to regions located 5′ and 3′ to the *dps* coding region. The uninterrupted allele produced an amplicon of 1 kb while the *lacS* disrupted allele produced an amplicon of 2.8 kb.

3.3. Chromosomal Recombination Using Multiple Linear DNAs: Inactivation of S-layer Domain Genes

The construction of mutations can be further simplified through the use of more than two linear DNA molecules. In the final example, three PCR products are used to construct a disruption mutation in an *S. solfataricus* protein (SSO0011) annotated as having an S-layer domain. S-layers are protein monolayers that surround the exterior of diverse organisms including archaea.

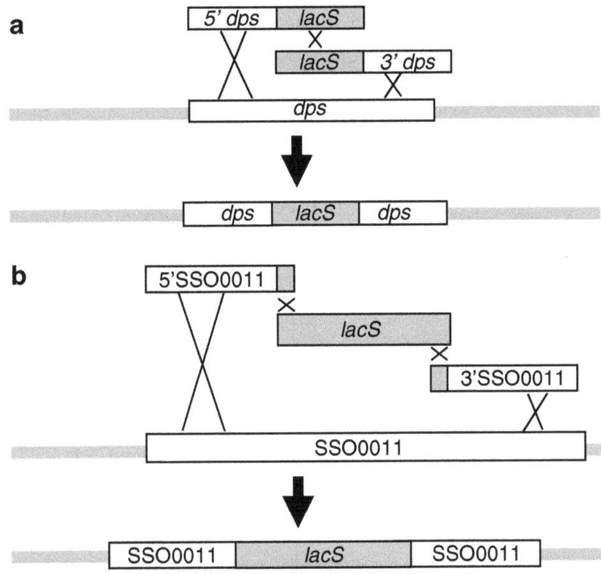

Fig. 3. Advanced linear DNA transformation and recombination. (**a**) Two linear DNA fragment recombination. Using linear DNA fragments composed of 5′ end of target DNA (***dps***) and 3′ end of target gene each fused with a marker gene (created by overlapping extension PCR). A disruption of the target gene (***dps***) is created when two linear DNA molecules undergo homologous recombination with the target gene on a chromosome. (**b**) Multiple linear DNA fragment recombination. 20–30 bp of nucleotide complementary to ***lacS*** were added to the reverse and forward primers of 5′SSO0011 and 3′SSO0011 fragment (respectively), thus creating SSO0011 DNA fragments with a short complementary region to ***lacS***. Unlike two fragment linear DNA recombination (**a**), no overlap extension PCR is necessary for this multiple linear DNA transformation method.

Proteins annotated as having S-layer domains can be the primary structural component of the S-layer or, associated with this structure through an interaction mediated by the S-layer domain-containing region. To understand the relative importance of SSO0011, PCR products constituting the final disruption construct were made in vitro using overlap extension PCR. The resulting DNAs encompassed: (1) the 5′ end and flanking region of SSO0011 fused to the 5′ end of *lacS*; (2) the central portion of *lacS*; and (3) the 3′ end of *lacS* fused to the 3′ end of SSO0011 combined with some of its flanking region (Fig. 3b). The three molecules are then cotransformed into a suitable host strain and recombinants are selected. The resulting mutation consists of a disruption of SSO0011 resulting from multiple recombination events including two between the three linear DNA molecules and another two recombination events between the homologous regions of the chromosome and the 5′ and 3′ flanking regions of SSO0011.

1. Linear DNA molecules of a target locus (*aldhT::lacS* in this case) are created by overlapping extension (OLE) PCR (24) (see Note 6).
2. PCR amplicons are purified by gel electrophoresis, using the manufacturer's protocol for a gel extraction kit (see Note 7).
3. Purified linear DNA is used for transformation (see Subheading 3.4).

3.4. Preparation of Electrocompetent Cells for Transformation

Transformation of *S. solfataricus* by electroporation (see Notes 8 and 9).

1. Prepare a cell pellet from 25 mL of mid-log phase culture growing on complex medium by centrifugation for 15 min at 3,000 × *g* in a Sorval F21-8x50Y rotor or equivalent at room temperature.
2. Add 1 mL of 20 mM sucrose and resuspend cells by gently shaking or tapping tube.
3. Add additional 4 mL of 20 mM sucrose and recover the cells as described in step 1.
4. Repeat steps 2 and 3, twice.
5. Resuspend the cell pellet in 20 mM sucrose to a final volume of 1 mL, these electrocompetent cells are enough for up to 20 transformations.
6. For each transformation, use 0.05 mL of electrocompetent cells (prepared above). Transfer the electrocompetent cells to a sterile 0.5-mL polypropylene microcentrifuge tube and preheat at 50°C for 10 min in a heating block.
7. Add DNA (~1 μg) to preheated electrocompetent cells and incubate at 50°C for an additional 3 min.
8. Transfer the cell-DNA mixture into a prewarmed (50°C) electroporation cuvette, and electroporate at 100 Ω, 2.0 kV, 25 μF (around 2.0–2.5 ms).
9. Transfer electroporated mixture into a flask containing preheated (80°C) defined medium and incubate at 80°C with shaking.
10. When growth is apparent (typically 6–7 days), process culture.

3.5. Recombinant Cell Line Screening

The following text explains the procedure for the identification of *lacS* recombinants derived from enrichment cultures produced as described above. This procedure can be applied to screen for other types of recombinants.

1. Perform tenfold serial dilutions of the enrichment culture on complex medium (0.2% tryptone, w/v) plates and incubate at 80°C until colonies form (typically 5–7 days).

2. Recombinant strains are detected by spraying colonies with X-gal solution. Incubate plates for an additional 1 h at 80°C until blue colored colonies are observed.
3. Pick and patch blue colonies on complex medium plates, incubate at 80°C for 3 days, and retest the LacS phenotype (blue color) by applying X-gal.
4. Inoculate confirmed patches into 25-mL shake flask cultures in complex medium and grow to mid-log phase (i.e., $OD_{540} = 0.5$).
5. Take 5 mL of culture; isolate genomic DNA for PCR genotyping and sequencing analysis. The remainder of the culture is used to prepare a frozen permanent stock (see below).

3.6. Storage of Cell Lines

Cell lines can be stored for long periods through the preparation of a frozen permanent.

1. Prepare 20 mL of mid-log phase culture growing in complex medium (cell density of 0.5 OD_{540}) and collect a cell pellet by centrifugation (3,000 × g) at room temperature.
2. Discard the supernatant and resuspend the cell pellet in 0.93 mL of Brock Salts medium.
3. Transfer the resuspended cells into a sterile 1.5 mL polypropylene microcentrifuge tube and label appropriately.
4. Add 0.07 mL of dimethyl sulfoxide (DMSO) and mix well.
5. Flash freeze in an ethanol-dry ice bath for 10 min. Store it at -80°C.

4. Notes

1. All media components and chemicals should be prepared in water and sterilized by autoclaving, unless it is specified in the text.
2. Gelrite needs to be completely dissolved in water by boiling before the autoclaving.
3. Tryptone could be added to Brock salts medium and autoclaved together. For 0.2% complex medium, add 1 g of Difco tryptone powder to 1 L Brock salts medium, and autoclave to sterilize.
4. To prevent hydrolysis of lactose by high heat, a stock solution of lactose (i.e., 10% stock solution, filter sterilized) should be prepared and added to the autoclaved Brock salts medium to make 0.2% (v/v) lactose defined medium.
5. To clean PCR product, MinElute® PCR purification kit (Qiagen #28004) is used.

6. The primers for OLE PCR primers for three linear DNA fragment transformation contain 20–24 nt homologous to the 5′ end(or 3′ end) of the target genes and 20–24 nt of the genetic marker gene.
7. To purify DNA from gel, QIAquick® Gel Extraction Kit (Qiagen #28704) is used.
8. *Recombinogenic strains*: For some as yet unknown reason, only *S. solfataricus* strain 98/2 (25) (NCBI accession number: NZ_ACUK00000000) has been found to be useful for directed chromosomal recombination. Importantly, other strains of this organism notably *S. solfataricus* strain P2 have been reported not to be recombinogenic (26). Therefore, all work regarding this type of recombineering has been conducted with this particular strain of *S. solfataricus.*
9. *Transformation*: Transformation of *S. solfataricus* is accomplished by electroporation. The amount of DNA, electroporation field strengths, pre- and postelectroporation conditions, and the total number of cells transformed are factors that can influence the transformation efficiency.

References

1. Wong J. H., Brown J. A., Suo Z., Blum P., Nohmi T., Ling H. (2010) Structural insight into dynamic bypass of the major cisplatin-DNA adduct by Y-family polymerase Dpo4. *EMBO J* **29**, 2059–2069.
2. Miller P. S., Blum P. H. (2010) Extremophile-inspired strategies for enzymatic biomass saccharification. *Environ Technol* **31**, 1005–1015.
3. Verhaart M. R., Bielen A. A., van der Oost J., Stams A. J., Kengen S. W. (2010) Hydrogen production by hyperthermophilic and extremely thermophilic bacteria and archaea: mechanisms for reductant disposal. *Environ Technol* **31**, 993–1003.
4. Orell A., Navarro C. A., Arancibia R., Mobarec J. C., Jerez C. A. (2010) Life in blue: Copper resistance mechanisms of bacteria and Archaea used in industrial biomining of minerals. *Biotechnol Adv doi:10.1016/j.biotechadv.2010.07.003.*
5. Blum P. (2001) *Archaea, ancient microbes, extreme environments and the origin of life.*, Vol. 50, Academic Press, New York.
6. Blum P. (2008) *Archaea, new models for prokaryotic biology* Horizon Press, Norwich UK.
7. Sawitzke J. A., Thomason L. C., Costantino N., Bubunenko M., Datta S., Court D. L. (2007) Recombineering: *in vivo* genetic engineering in *E. coli*, *S. enterica*, and beyond. *Methods Enzymol* **421**, 171–199.
8. Constantinesco F., Forterre P., Elie C. (2002) *NurA*, a novel 5'-3' nuclease gene linked to *rad50* and *mre11* homologs of thermophilic Archaea. *EMBO Rep* **3**, 537–542.
9. Nichols M. D., DeAngelis K., Keck J. L., Berger J. M. (1999) Structure and function of an archaeal topoisomerase VI subunit with homology to the meiotic recombination factor Spo11. *EMBO J* **18**, 6177–6188.
10. Seitz E. M., Brockman J. P., Sandler S. J., Clark A. J., Kowalczykowski S. C. (1998) RadA protein is an archaeal RecA protein homolog that catalyzes DNA strand exchange. *Genes Dev* **12**, 1248–1253.
11. Sowers K. R., Blum P. H. DasSarma S. (2007) Gene transfer in archaea, in *Methods for General and Molecular Microbiology* (C. A. Reddy, T. J. B., J. A. Breznak, G. A. Marzluf, and T. M. Schmidt Ed.), American Society for Microbiology Press, Washington D.C.
12. Schelert J., Dixit V., Hoang V., Simbahan J., Drozda M., Blum P. (2004) Occurrence and characterization of mercury resistance in the hyperthermophilic archaeon *Sulfolobus solfataricus* by use of gene disruption. *J Bacteriol* **186**, 427–437.

13. Worthington P., Hoang V., Perez-Pomares F., Blum P. (2003) Targeted disruption of the alpha-amylase gene in the hyperthermophilic archaeon *Sulfolobus solfataricus*. *J Bacteriol* **185**, 482–488.
14. Schelert J., Drozda M., Dixit V., Dillman A., Blum P. (2006) Regulation of mercury resistance in the crenarchaeote *Sulfolobus solfataricus*. *J Bacteriol* **188**, 7141–7150.
15. Allen M. B. (1959) Studies with *Cyanidium caldarium*, an anomalously pigmented chlorophyte. *Arch Mikrobiol* **32**, 270–277.
16. Brock T. D., Brock K. M., Belly R. T., Weiss R. L. (1972) Sulfolobus: a new genus of sulfur-oxidizing bacteria living at low pH and high temperature. *Arch Mikrobiol* **84**, 54–68.
17. Sambrook J., Russell D. W. (2001) Molecular Cloning: A Laboratory Manual 3rd ed., vol. 2. *2*.
18. Chong P. K., Burja A. M., Radianingtyas H., Fazeli A., Wright P. C. (2007) Proteome and transcriptional analysis of ethanol-grown *Sulfolobus solfataricus* P2 reveals ADH2, a potential alcohol dehydrogenase. *J Proteome Res* **6**, 3985–3994.
19. Chong P. K., Burja A. M., Radianingtyas H., Fazeli A., Wright P. C. (2007) Proteome analysis of *Sulfolobus solfataricus* P2 propanol metabolism. *J Proteome Res* **6**, 1430–1439.
20. Brouns S. J., Walther J., Snijders A. P., van de Werken H. J., Willemen H. L., Worm P., de Vos M. G., Andersson A., Lundgren M., Mazon H. F., van den Heuvel R. H., Nilsson P., Salmon L., de Vos W. M., Wright P. C., Bernander R., van der Oost J. (2006) Identification of the missing links in prokaryotic pentose oxidation pathways: evidence for enzyme recruitment. *J Biol Chem* **281**, 27378–27388.
21. Gratchev A., Strein P., Utikal J., Sergij G. (2003) Molecular genetics of Xeroderma pigmentosum variant. *Exp Dermatol* **12**, 529–536.
22. Zhang H., Guengerich F. P. (2010) Effect of N2-guanyl modifications on early steps in catalysis of polymerization by *Sulfolobus solfataricus* P2 DNA polymerase Dpo4 T239W. *J Mol Biol* **395**, 1007–1018.
23. Maaty W. S., Wiedenheft B., Tarlykov P., Schaff N., Heinemann J., Robison-Cox J., Valenzuela J., Dougherty A., Blum P., Lawrence C. M., Douglas T., Young M. J., Bothner B. (2009) Something old, something new, something borrowed; how the thermoacidophilic archaeon *Sulfolobus solfataricus* responds to oxidative stress. *PLoS One* **4**, e6964.
24. Higuchi R., Krummel B., Saiki R. K. (1988) A general method of *in vitro* preparation and specific mutagenesis of DNA fragments: study of protein and DNA interactions. *Nucleic Acids Res* **16**, 7351–7367.
25. Rolfsmeier M., Blum P. (1995) Purification and characterization of a maltase from the extremely thermophilic crenarchaeote *Sulfolobus solfataricus*. *J Bacteriol* **177**, 482–485.
26. Jonuscheit M., Martusewitsch E., Stedman K. M., Schleper C. (2003) A reporter gene system for the hyperthermophilic archaeon *Sulfolobus solfataricus* based on a selectable and integrative shuttle vector. *Mol Microbiol* **48**, 1241–1252.

Chapter 27

Targeted Gene Disruption in Koji Mold *Aspergillus oryzae*

Jun-ichi Maruyama and Katsuhiko Kitamoto

Abstract

Filamentous fungi have received attentions as hosts for heterologous protein production because of their high secretion capability and eukaryotic post-translational modifications. One of the safest hosts for heterologous protein production is Koji mold *Aspergillus oryzae* since it has been used in the production of Japanese fermented foods for over 1,000 years. The production levels of proteins from higher eukaryotes are much lower than those of homologous (fungal) proteins. Bottlenecks in the heterologous protein production are suggested to be proteolytic degradation of the produced protein in the medium and the secretory pathway. For construction of excellent host strains, many genes causing the bottlenecks should be disrupted rapidly and efficiently. We developed a marker recycling system with the highly efficient gene-targeting background in *A. oryzae*. By employing this technique, we performed multiple gene disruption of the ten protease genes. The decuple protease gene disruptant showed fourfold production level of a heterologous protein compared with the wild-type strain.

Key words: *Aspergillus oryzae*, Filamentous fungi, Multiple gene disruptions, Heterologous protein production, Highly efficient gene-targeting, Marker recycling

1. Introduction

Koji mold *Aspergillus oryzae* is a filamentous fungus that is one of the excellent hosts for heterologous protein production due to its high protein productivity and the safety guaranteed by its use in the manufacture of Japanese fermented foods for over 1,000 years (1). In general, the production level of proteins from animals and plants is much lower than the homologous (fungal) proteins (2–5). In the heterologous protein production of filamentous fungi, proteolytic degradation of the produced protein is one of the bottlenecks limiting the yields (6, 7). For example, *A. oryzae* has 134 protease genes (8), many of which might cause degradation of the heterologous protein. To enhance the ability of protein

James A. Williams (ed.), *Strain Engineering: Methods and Protocols*, Methods in Molecular Biology, vol. 765, DOI 10.1007/978-1-61779-197-0_27,

production, it is important to generate a host applicable to multiple rounds of genetic manipulations. However, compared with other microorganisms such as bacteria and yeasts, only a few attempts to manipulate many genes such as multiple disruptions of protease genes had been performed for breeding of industrial strains in filamentous fungi.

In order to carry out multiple gene disruptions rapidly and efficiently, we have developed a marker recycling system with the highly efficient gene-targeting background (9). The *pyrG* gene encoding orotidine-5′-phosphate (OMP) decarboxylase is used for marker recycling, which allows multiple gene disruptions in *A. oryzae*, since *pyrG*-excised strains can be positively selected by using 5-fluoro-orotic acid (5-FOA) that is converted to the toxic intermediate 5-fluoro-UMP by the enzyme (10). In each disruption process, the *pyrG* marker is excised by the direct repeats of ~300 bp upstream flanking region of the target gene, resulting in no residual ectopic/foreign DNA fragments in the genome. For the highly efficient gene-targeting background, *A. oryzae ligD* gene homologous to *Neurospora crassa mus-53* gene involved in nonhomologous chromosomal integration was disrupted, resulting in ~90% gene disruption efficiency that is much higher than the wild type (~40%) (9, 11). By using this system, we could generate a decuple protease gene disruptant showing fourfold higher level production of a heterologous protein than the wild-type strain (11).

2. Materials

2.1. Construction of Gene Disruption Fragments

1. MultiSite Gateway system (Invitrogen, San Diego, CA).
2. PrimeSTAR HS DNA Polymerase (TaKaRa, Otsu, Japan).
3. pDONR™P4-P1R (Invitrogen).
4. pDONR™P2R-P3 (Invitrogen).
5. pgEpG containing the *pyrG* gene (9).
6. pDEST™R4-R3 (Invitrogen).
7. *A. oryzae* RIB40 strain (wild type) (8).

2.2. Transformation of *A. oryzae*

1. NSPlD1 strain (Δ*ligD* Δ*pyrG* strain) [*niaD*⁻ *sC*⁻ *adeA*⁻ Δ*argB*::*adeA*⁻ Δ*ligD*::*argB* Δ*pyrG*::*adeA*] (9).
2. DPY medium: 2% dextrin, 1% polypeptone, 0.5% yeast extract, 0.5% KH_2PO_4, 0.05% $MgSO_4 \cdot 7H_2O$, pH 5.5. Mix all together and autoclave.
3. DPY liquid medium containing 20 mM uridine and 0.2% uracil. Uridine and uracil can be added before autoclaving.

4. Sterilized miracloth (Calbiochem, Darmstadt, Germany).
5. TF Solution I: 50 mM maleic acid (pH 5.5), 1% Yatalase (TaKaRa), 0.6 M $(NH_4)_2SO_4$. Prepare immediately before use and ultrafiltrate.
6. TF Solution II: 1.2 M sorbitol, 50 mM $CaCl_2{\cdot}2H_2O$, 35 mM NaCl, 10 mM Tris–HCl (pH 7.5). Mix all together and autoclave.
7. TF Solution III: 60% PEG 4000, 50 mM $CaCl_2{\cdot}2H_2O$, 10 mM Tris–HCl (pH 7.5). Mix all together and autoclave.
8. M+Met medium: 0.2% NH_4Cl, 0.1% $(NH_4)_2SO_4$, 0.05% KCl, 0.05% NaCl, 0.1% KH_2PO_4, 0.05% $MgSO_4{\cdot}7H_2O$, 0.002% $FeSO_4{\cdot}7H_2O$, 2% glucose, 0.15% methionine, pH 5.5. Mix all together and autoclave.
9. Top agar: M+Met medium including 1.2 M sorbitol and 0.8% agar. Autoclave.
10. M+Met agar medium containing 1.2 M sorbitol and 1.5% agar. Autoclave.
11. M+Met agar medium containing 1.5% agar. Autoclave.

2.3. Colony PCR for A. oryzae Transformants

Colony PCR Master Mix: 2.8 μl sterilized distilled water, 10 μl 2× PCR Buffer for KOD FX, 4 μl 2 mM dNTPs, 0.4 μl 10 pmol/μl primers, 0.4 μl KOD FX (1 U/μl; TOYOBO, Kyoto, Japan).

2.4. Genomic DNA Extraction

1. DPY liquid medium (see item 2 in Subheading 2.2).
2. Liquid nitrogen.
3. Sterilized miracloth (see item 4 in Subheading 2.2).
4. Metal corn (Yasui Kikai, Osaka, Japan): Bullet-shaped metal to break cells.
5. Multi-Beads Shocker (Yasui Kikai).
6. GE Solution: 50 mM EDTA (pH 8.0), 0.5% SDS, 0.1 mg/ml Proteinase K. Prepare immediately before use.
7. Ethanol precipitation solution: 40 ml ethanol, 1.6 ml 3 M sodium acetate (pH 5.2). Prepare immediately before use.
8. RNase TE: 5 ml TE (10 mM Tris–HCl (pH 8.0), 1 mM EDTA (pH 8.0)), 5 μl 20 mg/ml RNase A solution. Prepare immediately before use.
9. PCI: phenol/chloroform/isoamyl alcohol (25:24:1).
10. CI: chloroform/isoamyl alcohol (24:1).
11. 70% ethanol.
12. TE: 10 mM Tris–HCl (pH 8.0), 1 mM EDTA (pH 8.0).

2.5. Southern Analysis

1. Agarose gel electrophoresis equipment and agarose gel.
2. Restriction enzymes and 10× buffers.

3. Hybond N+ membrane (GE Healthcare, Piscataway, NJ).
4. ECL (enhanced chemiluminescence) direct nucleic acid labeling and detection system (GE Healthcare).
5. LAS-100plus luminescent image analyzer (Fuji Photo Film, Tokyo, Japan).

2.6. Positive Selection of pyrG-Excised Strains by Using 5-FOA

1. 1.6 mg/ml 5-FOA solution. Dissolve in distilled water at 55°C with shading and then ultrafiltrate.
2. 1 M uridine solution. Ultrafiltrate.
3. PD agar medium containing 0.8 mg/ml 5-FOA and 20 mM uridine/0.2% uracil. Autoclave 100 ml 2× PD (Potato/dextrose) agar (Nissui Phamaceutical, Tokyo, Japan) containing 0.4% uracil. After cooled, add 100 ml 1.6 mg/ml 5-FOA solution and 4 ml 1 M uridine. Pour into plates and store in dark.

3. Methods

For high gene disruption frequency, we previously generated a disruptant of the *ligD* gene encoding DNA ligase IV homolog involved in the final step of nonhomologous end joining (9) with the selective marker *argB* in the *A. oryzae* quadruple auxotrophic strain, NSAR1 (*niaD*⁻ *sC*⁻ *adeA*⁻ Δ*argB*::*adeA*⁻) (12). The Δ*ligD* strains grow and conidiate comparably to the nondisrupted transformants (Fig. 1a), suggesting that it can be used in experiments such as heterologous protein production. On the other hand, the Δ*ligD* strain reduces the growth in the presence of methyl methanesulfonate (MMS), a chemical mutagen.

In order to add uridine/uracil auxotrophy that is applicable to marker recycling and multiple gene disruption, the *pyrG* gene was disrupted with the selective marker *adeA* in the Δ*ligD* strain (NSR-ΔlD2). The Δ*ligD* Δ*pyrG* strain (NSPlD1) can be transformed with the plasmid harboring the wild-type *pyrG* gene. The transformants are able to grow in the absence of uridine/uracil while the Δ*ligD* Δ*pyrG* strain does not form colonies on the same medium (Fig. 1b). However, only the Δ*ligD* Δ*pyrG* strain shows resistance against 5-FOA. These results demonstrate that positive selection using 5-FOA for uridine/uracil auxotrophs can be applied to *pyrG* marker recycling in *A. oryzae*.

By using 1.3–1.5 kb flanking regions of the target gene, the disruption efficiency is very high (~90%) in the Δ*ligD* background (9, 11). Transformants derived from the Δ*ligD* Δ*pyrG* strain produce a comparable level of heterologous proteins with the relevant wild-type strain. We further reported that decuple protease gene disruption increased heterologous protein yields (11).

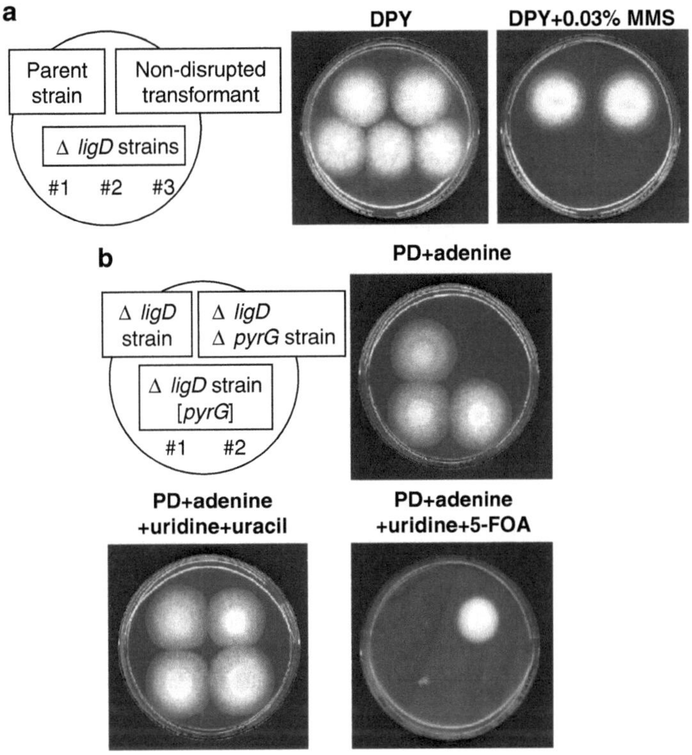

Fig. 1. Growth of the ***A. oryzae*** **Δ*ligD* Δ*pyrG*** strain. (**a**) Sensitivity of the **Δ*ligD*** strain to a chemical mutagen MMS. Conidia (~100 conidia/5 μl) were spotted on the DPY agar medium and incubated at 30°C for 3 and 5 days in the absence and presence of MMS, respectively. (**b**) Resistance of the **Δ*ligD* Δ*pyrG*** strain to 5-FOA. Conidia (~600 conidia/5 μl) were spotted on PD medium with indicated supplements and 5-FOA (0.8 mg/ml). The agar plates were incubated at 30°C for 3 days.

3.1. Construction of DNA Fragments for Gene Disruption in A. oryzae

Plasmid construction for gene disruption fragments is done by the MultiSite Gateway system as instructed by the manufacturer. PCR is performed with the PrimeSTAR HS DNA Polymerase that has a high fidelity.

1. Upstream flanking region (1.3–1.5 kb) of the target gene ORF is amplified with the genomic DNA of *A. oryzae* RIB40 strain as template, and inserted into pDONR™P4-P1R by the BP recombination reaction, generating a 5′ entry clone.
2. The upstream (0.3 kb) and downstream (1.3–1.5 kb) flanking regions of the gene are amplified. The two fragments are connected by fusion PCR (see Note 1), and inserted into pDONR™P2R-P3 by the BP recombination reaction, generating a 3′ entry clone.
3. The obtained 5′ and 3′ entry clones together with a center entry clone plasmid, pgEpG containing the *pyrG* gene (9), are mixed for the LR recombination reaction with the destination

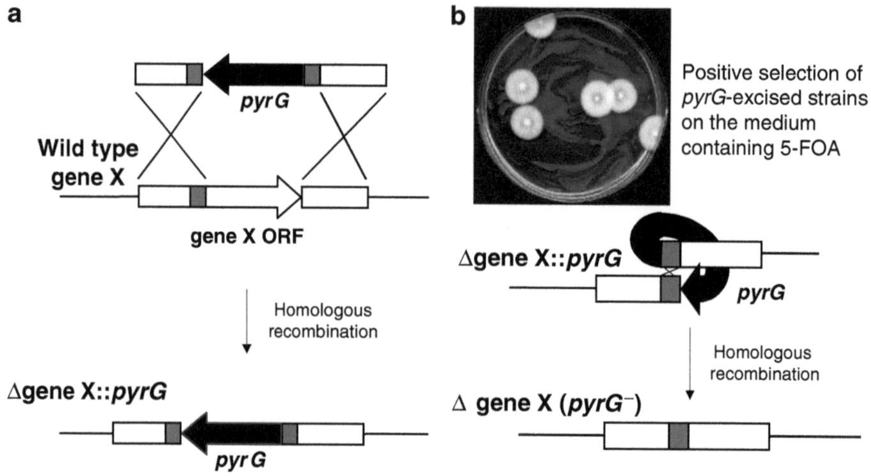

Fig. 2. Overview of multiple gene disruption by *pyrG* marker recycling in *A. oryzae*. (**a**) Targeted gene disruption with the *pyrG* marker. The *boxes* (1.3–1.5 kb) are the flanking regions used for disruption of the target gene. The 0.3-kb upstream flanking region of the target gene (*boxed in gray*) is attached at 5′-end of the downstream flanking regions, introducing direct repeats. (**b**) Excision of the *pyrG* marker targeted at the disrupted locus. By homologous recombination of the direct repeats consisting of the 0.3 kb upstream flanking region of the target gene (*boxed in gray*), the *pyrG* gene targeted at the disrupted locus is excised, and then the upstream and downstream flanking region are directly connected. Note that no ectopic/foreign DNA fragments are left in the genome after excision of the *pyrG* marker.

vector, pDEST™R4-R3, generating a plasmid including the gene disruption fragment.

4. The gene disruption fragment is amplified with the resultant plasmid as template. In this construct, 3′-end of the upstream flanking region of the ORF (~300 bp) is fused with the downstream flanking region of the ORF so that the *pyrG* marker is flanked by the ~300 bp directed repeats (Fig. 2a; gray box).

3.2. Transformation of the A. oryzae ΔligD ΔpyrG Strain

This procedure is a modified version according to the method of Kitamoto (1).

1. Inoculate the *ΔligD ΔpyrG* strain in 100 ml DPY liquid medium containing 20 mM uridine and 0.2% uracil. Shake the culture for ~20 h at 30°C.
2. Collect mycelia by filtration with a sterilized miracloth in funnel, and wash them with sterilized distilled water.
3. Incubate the mycelia in 10 ml TF Solution I at 30°C by mild agitation (50 strokes/min) for 3 h. Protoplast formation is checked by microscopic observation.
4. Separate protoplasts from mycelia by filtration through sterilized miracloth. Dilute the protoplast suspension with an equal volume of TF Solution II.

5. Gently precipitate the protoplasts by centrifugation (700 × *g*, 8 min, 4°C; see Note 2) and wash twice with 5–10 ml TF Solution II (see Note 3). Finally, protoplasts are resuspended in TF Solution II with the concentration at 1×10^7–5×10^7/ml.
6. Mix a gene disruption DNA fragment (~3 μg/10 μl; see Note 4) with the protoplast suspension (200 μl) and incubate on ice for 30 min.
7. Mix, in three serial steps, 250, 250, and 850 μl of TF Solution III with the DNA-protoplast mixture and keep at room temperature for 20 min.
8. Dilute the PEG-treated protoplast suspension with 5–10 ml TF Solution II, and centrifuge at 700 × *g* for 8 min at 4°C.
9. Resuspend the protoplasts in 500 μl of TF Solution II.
10. Add aliquots of the protoplast suspension in 4 ml Top agar, and pour the mixture onto M+Met agar medium containing 1.2 M sorbitol that is an osmotic stabilizer.
11. After 3–4 days cultivation at 30°C, transformants are visible on the agar medium. Inoculate them onto new M+Met agar medium (without sorbitol) for a single colony (see Note 5).

3.3. Colony PCR of A. oryzae Transformants

1. Design a forward primer annealing to the region immediately upstream of the gene disruption fragment, and a reverse primer annealing to the downstream flanking region of the ORF. Colony PCR using these primers reveals disruption of the target gene and *pyrG* marker excision.
2. Pick up mycelia and conidia from a single colony (see Note 6), and suspend them in 50 μl TE buffer.
3. Add 2 μl of the mycelia/conidial suspension in 18 μl of the Colony PCR Master Mix.
4. Run the PCR program and check the amplification by electrophoresis. Transformants showing disruption of the target gene are taken for Genomic DNA extraction and Southern analysis (see Note 7).

3.4. Genomic DNA Extraction of A. oryzae Transformants

1. Transformants are grown in 10 ml DPY liquid medium at 30°C for 16–18 h.
2. Mycelia are harvested by filtration with miracloth or filter paper, and washed with distilled water.
3. Freeze the mycelia (~250 mg in wet weight) in liquid nitrogen together with a metal corn, and break them using Multi-Beads Shocker at 2,000 rpm for 10 s.
4. Freeze them in liquid nitrogen and break the mycelia again.
5. After removing the metal corn, add 600 μl GE Solution and shake gently at 60°C for 30–60 min.

6. Add 700 μl PCI and mix well. Centrifuge at 20,000×g at 4°C for 10 min and take the supernatant (see Note 8).
7. Add 600 μl PCI and mix well. Centrifuge at 20,000×g at 4°C for 5 min and take the supernatant.
8. Add 550 μl CI and mix well. Centrifuge at 20,000×g at 4°C for 5 min and take the supernatant.
9. Add 900 μl ethanol precipitation solution and rotate gently. Centrifuge it at 20,000×g at 4°C for 10 min and remove the supernatant.
10. Dry briefly the pellet and add 400 μl RNase TE. After dissolving, incubate at 37°C for 30 min.
11. Add 400 μl PCI and mix well. Centrifuge at 20,000×g at 4°C for 5 min and take the supernatant.
12. Add 350 μl CI and mix well. Centrifuge at 20,000×g at 4°C for 5 min and take the supernatant.
13. Add 1 ml ethanol precipitation solution and invert several times. Centrifuge at 20,000×g at 4°C for 10 min, and remove the supernatant.
14. Add 1 ml 70% ethanol. Centrifuge at 20,000×g at 4°C for 10 min, and remove the supernatant.
15. Dissolve the pellet with 100 μl TE and store the solution at 4°C (see Note 9).

3.5. Southern Analysis of A. oryzae *Transformants*

1. Digest genomic DNAs with restriction enzymes and load them for agarose electrophoresis.
2. Transfer the digested genomic DNAs onto Hybond N+ membrane.
3. Use ECL direct nucleic acid labeling and detection system and an instrument such as LAS-100plus luminescent image analyzer for labeling and detection.

3.6. Positive Selection of pyrG*-Excised Strains by Using 5-FOA*

This process is performed for positive selection of *pyrG*-excised strains by using agar medium containing 5-FOA (Fig. 2b). The *pyrG* marker inserted at the target locus is excised out by homologous recombination with the direct repeats, in which the flanking regions of the target ORF are directly connected without leaving any ectopic/foreign DNA fragments.

1. Conidia (1×10^6–5×10^6/plate) of the gene disruptants with the *pyrG* marker are spread on PD agar medium containing 0.8 mg/ml 5-FOA and 20 mM uridine/0.2% uracil and then incubated at 30°C.
2. After 4–5 days cultivation, growing colonies are transferred onto another 5-FOA agar medium supplemented with uridine/uracil (see Note 10).

3. The strains are confirmed for *pyrG* marker excision by colony PCR and Southern analysis (see Subheadings 3.3–3.5 and Note 11).

3.7. Successive Round of Gene Disruption and Marker Recycling

Resultant 5-FOA resistant strains are uridine/uracil auxotroph that is therefore applicable to successive rounds of gene disruptions using the *pyrG* marker as instructed above (see Note 12).

4. Notes

1. For fusion PCR, the reverse and forward primers of the upstream (0.3 kb) and downstream flanking regions of the gene should be overlapped.
2. We centrifuge protoplast suspensions without brake.
3. When protoplasts are suspended and mixed, we use wide-mouthed pipettes such as sterile transfer pipettes (Sarstedt, Nümbrecht, Germany).
4. The amount of a gene disruption fragment is enough to obtain ~20 transformants.
5. Since filamentous fungi such as *A. oryzae* are multinuclei, transformants are inoculated onto another selective medium for a single colony to make them homokaryotic. One step of this inoculation process is sufficient to obtain homokaryotic transformants.
6. The region with newly formed conidia on 3-day culture is taken for colony PCR.
7. If colony PCR indicates heterokaryotic (with both bands showing wild type and disrupted gene loci), transformants should be inoculated on another agar plate with selective medium.
8. When genomic DNA solutions are taken, 1,000 μl pipette tips should be used to avoid shearing the genomic DNA.
9. The amount of genomic DNA is enough to be used ~5 times for Southern analysis.
10. With this inoculation condition, six to eight colonies appear per one plate.
11. For genomic DNA extraction, 20 mM uridine and 0.2% uracil are added in DPY medium of the *pyrG*-excised strains.
12. In experiments for heterologous protein production, we transform non-*pyrG*-excised strains with an expression plasmid (11).

Acknowledgments

We thank Dr. Jaewoo Yoon and Yukiko Oshima for experimental help. This study was supported by a Grant-in-Aid for Scientific Research (S) from the Ministry of Education, Culture, Sports, Science, and Technology of Japan and by the Program for the Promotion of Basic Research Activities for Innovative Biosciences (PROBRAIN) of Japan.

References

1. Kitamoto K. (2002) Molecular biology of the Koji molds. *Adv. Appl. Microbiol.* **51**, 129–153.
2. Tsuchiya K., Nagashima T., Yamamoto Y., Gomi K., Kitamoto K., and Kumagai C. (1994) High level secretion of calf chymosin using a glucoamylase prochymosin fusion gene in *Aspergillus oryzae*. *Biosci. Biotechnol. Biochem.* **58**, 895–899.
3. Nakajima K., Asakura T., Maruyama J., Morita Y., Oike H., Shimizu-Ibuka A., Misaka T., Sorimachi H., Arai S., Kitamoto K., and Abe K. (2006) Extracellular production of neoculin, a sweet-tasting heterodimeric protein with taste-modifying activity, by *Aspergillus oryzae*. *Appl. Environ. Microbiol.* **72**, 3716–3723.
4. Jin F.J., Watanabe T., Juvvadi P.R., Maruyama J., Arioka M., and Kitamoto K. (2007) Double disruption of the proteinase genes, *tppA* and *pepE*, increases the production level of human lysozyme by *Aspergillus oryzae*. *Appl. Microbiol. Biotechnol.* **76**, 1059–1068.
5. Ito K., Asakura T., Morita Y., Nakajima K., Koizumi A., Shimizu-Ibuka A., Masuda K., Ishiguro M., Terada T., Maruyama J., Kitamoto K., Misaka T., and Abe K. (2007) Microbial production of sensory-active miraculin. *Biochem. Biophys. Res. Commun.* **360**, 407–411.
6. Archer D.B., and Peberdy J.F. (1997) The molecular biology of secreted enzyme production by fungi. *Crit. Rev. Biotechnol.* **17**, 273–306.
7. van den Hombergh J.P., van de Vondervoort P.J., Fraissinet-Tachet L., and Visser J. (1997) *Aspergillus* as a host for heterologous protein production: the problem of proteases. *Trends Biotechnol.* **15**, 256–263.
8. Machida M., Asai K., Sano M., Tanaka T., Kumagai T., Terai G., Kusumoto K., Arima T., Akita O., Kashiwagi Y., Abe K., Gomi K., Horiuchi H., Kitamoto K., Kobayashi T., Takeuchi M., Denning D.W., Galagan J.E., Nierman W.C., Yu J., Archer D.B., Bennett J.W., Bhatnagar D., Cleveland T.E., Fedorova N.D., Gotoh O., Horikawa H., Hosoyama A., Ichinomiya M., Igarashi R., Iwashita K., Juvvadi P.R., Kato M., Kato Y., Kin T., Kokubun A., Maeda H., Maeyama N., Maruyama J., Nagasaki H., Nakajima T., Oda K., Okada K., Paulsen I., Sakamoto K., Sawano T., Takahashi M., Takase K., Terabayashi Y., Wortman J.R., Yamada O., Yamagata Y., Anazawa H., Hata Y., Koide Y., Komori T., Koyama Y., Minetoki T., Suharnan S., Tanaka A., Isono K., Kuhara S., Ogasawara N., and Kikuchi H. (2005) Genome sequencing and analysis of *Aspergillus oryzae*. *Nature* **438**, 1157–1161.
9. Maruyama J., and Kitamoto K. (2008) Multiple gene disruptions by marker recycling with highly efficient gene-targeting background (Δ*ligD*) in *Aspergillus oryzae*. *Biotechnol. Lett.* **30**, 1811–1817.
10. Boeke J.D., LaCroute F., and Fink G.R. (1984) A positive selection for mutants lacking orotidine-5-phosphate decarboxylase activity in yeast: 5-fluoro-orotic acid resistance. *Mol. Gen. Genet.* **197**, 345–346.
11. Yoon J., Maruyama J., and Kitamoto K. (2011) Disruption of ten protease genes in the filamentous fungus *Aspergillus oryzae* highly improves production of heterologous proteins. *Appl. Microbiol. Biotechnol.* **89**, 747–759.
12. Jin F.J., Maruyama J., Juvvadi P.R., Arioka M., and Kitamoto K. (2004) Development of a novel quadruple auxotrophic host transformation system by *argB* gene disruption using *adeA* gene and exploiting adenine auxotrophy in *Aspergillus oryzae*. *FEMS Microbiol. Lett.* **239**, 79–85.

Chapter 28

Selectable and Inheritable Gene Silencing through RNA Interference in the Unicellular Alga *Chlamydomonas reinhardtii*

Karin van Dijk and Nandita Sarkar

Abstract

Reverse genetic approaches have become invaluable tools to tap into the wealth of information provided by sequenced genomes. In 2007, sequencing of the *Chlamydomonas reinhardtii* genome was completed, and with this an increased demand for the development of reverse genetic strategies for gene analysis. In a variety of organisms, including *Chlamydomonas*, inverted repeat transgenes have been used to produce strains silenced for a specific gene due to the production of double stranded RNA (dsRNA). Here, we describe a tandem inverted repeat system designed to overcome some of the typical challenges that arise when transgenes are used to trigger gene silencing including the lack of a screenable phenotype, unpredictable levels of silencing, silencing of the transgene itself and thus loss of target gene silencing, and finally silencing of unintended genes (off-target genes). The described strategy allows selection of target gene silencing by inducing co-silencing of the target gene and a gene, *MAA7*, silencing of which produces a selectable RNAi-induced phenotype. This selection, therefore, precludes extensive molecular screening for transgenic strains exhibiting target gene silencing, and also ensures heritable silencing through many generations.

Key words: RNA interference, siRNA, Reverse genetics, Inverted repeat transgenes, Gene silencing, *Chlamydomonas*, Functional genomics, Algae

1. Introduction

It has been over a decade since the discovery of the conserved cellular mechanism RNA silencing through a double-stranded RNA (dsRNA) intermediate, originally coined RNA interference (RNAi) (1). Although the RNAi machinery is involved in various mechanisms of silencing, the most intensively studied is gene transcript silencing through small interfering RNAs (siRNAs)

James A. Williams (ed.), *Strain Engineering: Methods and Protocols*, Methods in Molecular Biology, vol. 765,
DOI 10.1007/978-1-61779-197-0_28, © Springer Science+Business Media, LLC 2011

which are produced from long dsRNAs. Ultimately, the RNAi machinery uses one of the siRNA strands as a guide to target homologous RNA molecules for degradation or translation silencing (2, 3). Since the discovery scientists working on a wide variety of organisms have capitalized on this biological phenomenon and employed it into a powerful tool that has revolutionized reverse genetic approaches particularly in organisms where other reliable reverse genetics strategies such as transposon tagging and insertional mutagenesis are not available. *Chlamydomonas reinhardtii* is one of these organisms. This alga has been a model organism for a variety of biological processes, including eukaryotic photosynthesis and flagellar assembly, and more recently for the production of biofuels (4, 5). As the *Chlamydomonas* genome is now fully sequenced (4), the demand for reverse genetic tools for gene analysis and genome manipulation has become pivotal.

A variety of RNA silencing techniques have been developed for *Chlamydomonas*, including the use of transgenes producing dsRNA or antisense RNA (6–9) and strains producing artificial miRNAs (amiRNAs) (10–13). Although stable RNA silencing strains can be engineered through the use of inverted repeat (IR) transgenes, there are a variety of challenges researchers typically face. First, since silencing of most genes does not typically provide for a screenable phenotype, and the level of silencing is unpredictable, generation of a silenced strain requires that at least a few if not many transgenic strains have to be analyzed at the molecular level to determine the extent of gene silencing. Second, the transgene itself can be silenced, most commonly at the transcriptional level (7), resulting in loss of dsRNA production targeting the gene of interest, and thus loss of gene silencing. Finally, with the production of dsRNA, and ultimately the production of small-interfering RNAs (siRNAs), there is always the chance that unintended genes (off-target genes) can be silenced if a subset of the produced siRNAs has sequence identity to those genes.

In this chapter, we describe a RNA silencing approach, using tandem IR transgenes (TIR), designed to overcome some of these challenges (6, 7). One of the key features of this system is that silencing of the gene is selectable. To enable selection, the transgene is designed such that an IR sequence targeting the gene of interest is within an IR targeting the tryptophan synthase β-subunit gene, *MAA7*. Silencing of *MAA7* can easily be selected for as strains that lack tryptophan synthase β-subunit can grow on a cytotoxic medium containing 5-fluoroindole (5-FI) (see Fig. 1) (14). Because the production of target gene dsRNA is tied to *MAA7* dsRNA production, selection for *MAA7* silencing ultimately results in target gene silencing. In short, this selection therefore precludes extensive molecular screening for transgenic strains exhibiting target gene silencing, and also ensures heritable silencing through many generations.

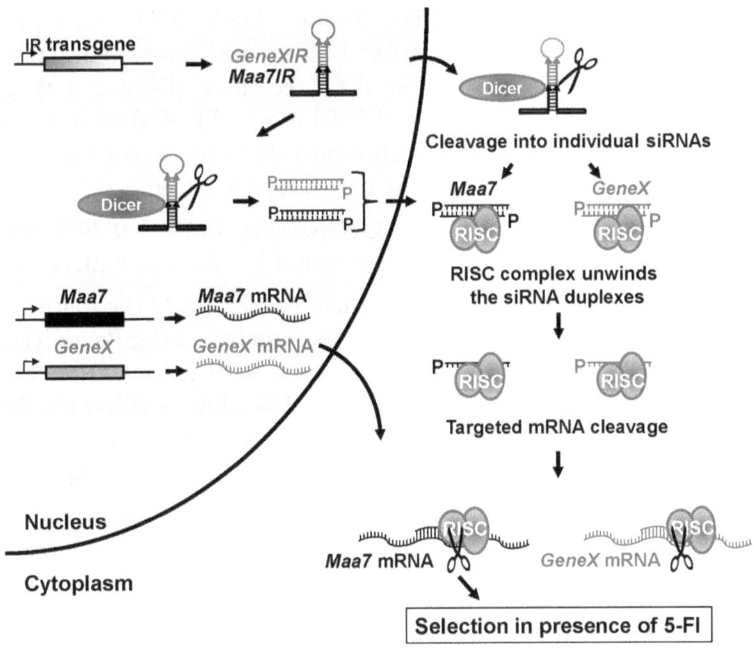

Fig. 1. Model showing the selectable RNAi downregulation of the gene of interest (*GeneX*) and *Maa7* transcripts using the TIR system. The TIR construct produces an mRNA sequence containing the inverted repeats of *Maa7* and GeneX which fold into a double-stranded hairpin structure. Dicer processes the dsRNA into siRNAs which can target both *Maa7* and GeneX transcripts. The potential nuclear location of Dicer is hypothetical. In the cytoplasm, the siRNAs gets unwound and incorporated into the RISC complex and these siRNA-RISC target *Maa7* and GeneX mRNA for degradation.

2. Materials (see Note 1)

2.1. Crude Genomic DNA Extraction

1. RHDB buffer: 10 mM Tris–HCl (pH 8.3), 2.5 mM $MgCl_2$, 50 mM KCl, 0.15% NP 40, 0.45% Tween 20.
2. Proteinase K: Dissolve proteinase K to 10 mg/ml in water or 10 mM Tris–HCl (pH 7) right before use.

2.2. Primer Design and Polymerase Chain Reaction Amplification of the 3′ UTR and the Spacer Region

1. rTth DNA Polymerase, XL (Applied Biosystems) with 3.3× XL buffer, 25 mM $Mg(OAc)_2$ and 1.25 mM dNTPs.
2. Primers for IR amplification (LIR1, RIR1, LIR2, RIR2).
3. Primers for spacer amplification: SL: GCATCCTCAA GCATCCTTCTATTC SR: GTAGGAGGCACAGGAAGA GCAAA.
4. *MAA7/X* IR vector (AY710294) (7).
5. 10× TE buffer: 100 mM Tris–HCl (pH 8.0), 10 mM EDTA (pH 8.0). Sterilize solution by autoclaving before storage at room temperature.

2.3. Coldfusion Cloning of PCR Fragments in Vector

1. Tris Borate EDTA (TBE) gel running buffer: Make a 10× stock solution and dilute that 1:10 for a working solution. For the 10× stock dissolve 108 g Tris–HCl and 7.45 g Na_2EDTA in roughly 800 ml water prior to adding 55 g boric acid. Adjust the volume to 1 l and store at room temperature. The pH should be around 8.3.
2. 6× Gel loading buffer: 0.24% bromophenol blue, 0.25% xylene cyanol FF, 30% w/v glycerol in H_2O. Store at 4°C.
3. SOC medium (15): Add the following components to 980 ml water and autoclave: 20 g Bacto Tryptone, 5 g Yeast Extract, 0.58 g NaCl, 0.19 g KCl. After the mixture is cooled, add 10 ml/l of 20% glucose (filter sterilized) and 10 ml/l of 2 M Mg stock solution and store at room temperature.
4. 2 M Mg stock solution for SOC medium: Make a 2 M stock of Mg^{2+} comprised of 1 M $MgCl_2$ and 1 M $MgSO_4$. Filter-sterilize and store at room temperature.
5. LB + ampicillin plates: Add the following to 1 l water: 10 g Tryptone, 5 g Yeast extract, 10 g NaCl and 18 g Bacto Agar. Mix contents, autoclave, cool mixture to 60°C, and add ampicillin to 100 mg/l and pour plates.
6. 100× Ampicillin stock: Dissolve 1 g of ampicillin sodium salt in 10 ml of water. Filter sterilize, dispense into 1 ml aliquots, and store at −20°C.
7. QIAquick Gel Extraction Kit (Qiagen).
8. Cold fusion cloning kit (System Biosciences).
9. *Eco*RI and buffer (New England Biolabs or other suppliers).
10. 1 kb DNA ladder (Invitrogen or other suppliers).
11. 10 mg/ml ethidium bromide.

2.4. Confirmation of Correct Plasmid Constructs

1. Terrific broth (TB): Mix the following components in a final volume of 900 ml: 12 g Bacto Tryptone, 24 g Bacto Yeast Extract, 5 ml glycerol. Autoclave, let the mixture cool and add 100 ml sterile phosphate solution.
2. Phosphate solution for TB: Mix the following components in a final volume of 250 ml: 5.78 g KH_2PO_4 and 31.35 g K_2HPO_4. Filter-sterilize the solution and store at room temperature.
3. QIAprep Spin Miniprep Kit (Qiagen).

2.5. Autolysin Preparation

1. HS Medium for autolysin preparation: To prepare HS medium, mix 7.5 ml of Beyjerinck's salts, 7.5 ml of phosphate salts (solution 2 for TAP medium), and 1.5 ml of Trace elements (Solution 3 for TAP medium) in 1,485 ml of water and autoclave.
2. Beyjerinck's salt: Dissolve 100 g NH_4Cl, 4 g $MgSO_4{\cdot}7H_2O$, and 2 g $CaCl_2{\cdot}2H_2O$ in a final volume of 1 l with water.

3. Nitrogen-free HS medium for autolysin preparation: Prepare nitrogen-free Beyjerinck's salts (omit NH_4Cl) and prepare HS medium as described above with N-free Beyjerinck's salts.
4. *Chlamydomonas* mating type strains CC620 and CC621.
5. Lugol's iodine (Fisher Scientific or other company): mix 6% KI and 4% I in water by stirring overnight protected from light by wrapping container in foil. Store at room temperature covered in aluminum foil.

2.6. Chlamydomonas Glass Bead Transformations

1. TAP Medium (liquid and plates): To prepare the final TAP medium mix 25 ml solution 1, 0.375 ml solution 2, 1 ml solution 3, 1 ml glacial acetic acid, and 2.42 g Tris base in 1 l water and autoclave. To prepare solid medium add 15 g of Bacto Agar.
   (a) Solution 1 (TAP salts): Dissolve 15 g NH_4Cl, 4 g $MgSO_4 \cdot 7H_2O$, and 2 g $CaCl_2 \cdot 2H_2O$ in a final volume of 1 l with water.
   (b) Solution 2 (Phosphate salts): Dissolve 28 g K_2HPO_4 and 14.4 g KH_2PO_4 in a final volume of 100 ml with water.
   (c) Solution 3 (Hutner's trace element): Mix solution A (50 g acid free EDTA in 550 ml water) with Solution B (11.40 g H_3BO_3, 22 g $ZnSO_4 \cdot 7H_2O$, 5.06 g $MnCl_2 \cdot 4H_2O$, 4.99 g $FeSO_4 \cdot 7H_2O$, 1.61 g $CoCl_2 \cdot 6H_2O$, 1.57 g $CuSO_4 \cdot 5H_2O$, and 1.1 g ammonium molybdate in 550 ml of water). Heat the mixture to 100°C, cool to 90°C ,and adjust the pH between 6.5 and 6.8 with 20% KOH. Adjust the volume to 1 l with water and let the mixture stand until color changes from green to purple. Store in dark at 4°C. This solution can be frozen in aliquots at –20°C.
2. TAP selection medium: TAP medium with 5-fluoroindole (5-FI) and paromomycin should be made 1–2 days before plating the transformants. After autoclaving the TAP medium with Bacto agar, the medium should be cooled to about 60°C prior to the addition of 5-FI and paromomycin to final concentrations of 7 mM 5-FI and 5 mM paromomycin. The plates should be poured immediately, covered in foil, and allowed to dry overnight at room temperature.
3. Paromomycin stock: Dissolve 1 g of paromomycin (Sigma) in 10 ml of water. Filter sterilize, dispense into 1 ml aliquots, and store at –20°C.
4. 5-Fluoroindole (5-FI) stock: make 100 mM 5-FI (Sigma) in ethanol, dispense into 1 ml aliquots, and store at –20°C in the dark.
5. Glass bead preparation for transformations: Acid wash 0.5 mm diameter glass beads (Sigma) to remove impurities and neutralize the pH by multiple rinses in distilled water. Next, air

dry the beads and bake at 400°F for 2 h. Place 300 mg of the beads in 15 ml conical tubes and autoclave the tubes.

6. 20% PEG (polyethylene glycol): Make a 20% PEG 6000 (Sigma p-5413) solution in water, autoclave, and store at −20°C for further use.

2.7. RNA Isolation and cDNA Synthesis

1. Trizol reagent (Invitrogen).
2. Chloroform.
3. Phenol:chloroform:isoamyl alcohol (25:24:1).
4. Isopropanol.
5. 75% ethanol.
6. DEPC water (Ambion).
7. Superscript III kit (Invitrogen).
8. RNase-free microcentrifuge tubes and pipette tips.
9. RNase-free TE: 10 mM Tris–HCl (pH 8.0), 10 mM EDTA (pH 8.0). Make in DEPC water and sterilize solution by autoclaving before storage at room temperature.

2.8. Quantitative PCR Analysis to Detect Co-silencing of the MAA7 and Gene X

1. SsoFast EvaGreen Supermix Kit (BioRad).
2. Primers for detection of gene of interest transcript: custom designed.
3. Primers for detection of MAA7 transcript (optional): *MAA7*-F CGCAAGTCTACTGTTCGCATT and *MAA7*-R TCGCCT CGTTGTAGTCCTTCT.
4. Primers for detection of constitutively expressed CBLP transcript (optional): CBLP-F TGCTGTGGGACCTGGCTGA and CBLP-R GCCTTCTTGCTGGTGATGTTG.

3. Methods

In the system we describe here, we use a TIR transgene, where the IR of the gene to be downregulated is positioned within another IR, targeting expression of tryptophan synthase β-subunit encoded by *MAA7*. Downregulation of *MAA7* can be selected for by incubating cells on media containing 5-FI. Tryptophan synthase β-subunit converts 5-FI to the cytotoxic compound 5-fluorotryptophan, hence the only surviving cells on such media are those expressing no or very little tryptophan synthase β-subunit. This occurs in those cells that effectively downregulate *MAA7* expression through the production of dsRNA produced from the IR transgene transcripts, and since downregulation of *MAA7* expression occurs via dsRNA produced from the TIR transcript, production of dsRNA targeting the gene of interest is ensured.

The transgene is constructed in the plasmid *MAA7/X* IR (Fig. 2a, AY710294) (7). As is displayed in Fig. 2a, this vector contains a spacer region (Sp) which can be replaced by an IR of the gene of interest. Once inserted, the IR is positioned within

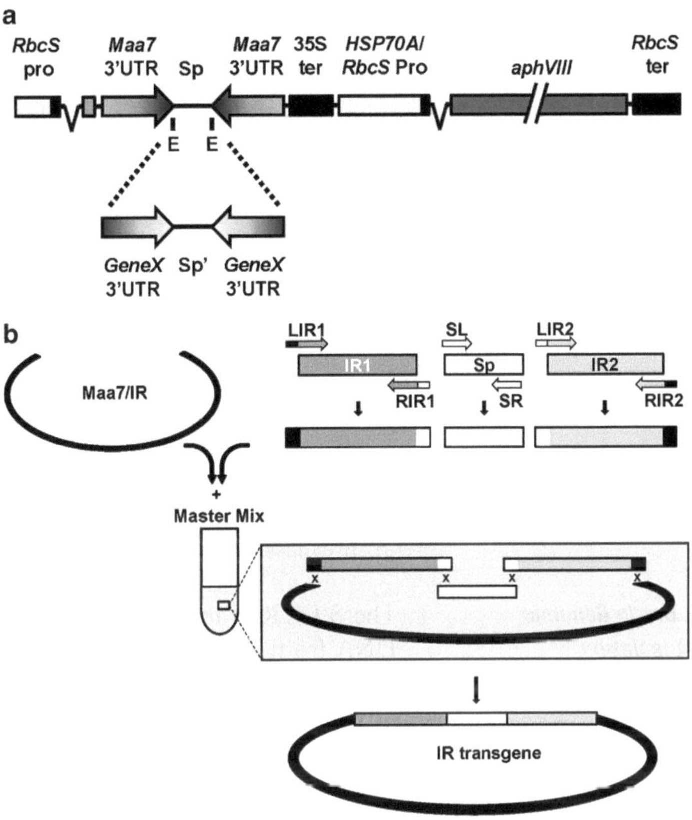

Fig. 2. A diagram of the vector (*MAA7/X* IR, [7]) (**a**) and the cloning strategy used to generate the TIR transgene (**b**). (**a**) A spacer (Sp) flanked by ***Eco***RI sites (E) and the *Maa7* 3′UTR in sense and antisense orientation can be replaced by a DNA fragment containing sense and antisense 3′UTR sequences from gene of interest (X) flanked by a spacer region (Sp′). This inverted repeat construct is under the transcriptional control of the *Rubisco* promoter (*RbcS* pro) and *Cauliflower Mosaic Virus* 35S terminator (35S ter), and will produce a tandem inverted repeat (as shown in Fig. 1) when transcribed. Downstream of the construct is the selectable marker, *aphVIII,* encoding for paromomycin resistance, under the transcriptional control of the *Hsp70/Rubisco* promoter (*Hsp70A/RbcS* pro) and *Rubisco* terminator (*RbcS* ter). (**b**) This schematic represents the use of the coldfusion kit to engineer the transgene in *MAA7/X* IR. The *MAA7/X* IR vector is linearized by *Eco*R1 and three PCR reactions are performed to amplify the two inverted repeat fragments (IRs) and the spacer (Sp). The primers for the two IR fragments are designed such that each of the primers contains extensions of sequences that are homologous to the vector or the spacer (indicated by the *white* and *black boxes* added to the primers). Next, the linearized vector and the PCR fragments are mixed with the coldfusion master mix to allow stitching together of the homologous sequences. *IR1* IR in the sense orientation, *IR2* IR in the antisense orientation, *LIR1* Forward primer for IR1, *RIR1* Reverse primer for IR1, *SL* Forward primer for Sp, *RL* right primer for S, *LIR2* Forward primer for IR2, *RIR2* Reverse primer for IR2.

the IR targeting *MAA7*, ensuring selectable production of the dsRNA. Transcription of the transgene is under the control of the constitutive *RbcS* promoter and 35S terminator. Flanking this transgene is the selectable marker, *aphVIII*, encoding for paromomycin resistance (16–18) which allows for selection of plasmid integration in the *Chlamydomonas* genome during the transformation process (see below). The coldfusion cloning kit (System Biosciences) is used for the construction of the TIR constructs. It provides fast and efficient assembly and cloning of various PCR products in a specific order in any plasmid. In short, 300–1,000 bp of the 3′UTR of the gene of interest is cloned in both directions flanking a spacer region of 210 bp. The benefit of using the 3′UTR is that it is unique for each gene, thus ensuring specificity of gene downregulation. For the spacer region we use the spacer present in the *MAA7/X* IR plasmid.

Below we describe detailed instructions on the design of the transgene, introduction of this vector in *Chlamydomonas*, and finally the selection of strains downregulated for the target gene. Unless noted otherwise all *Chlamydomonas* cultures are incubated at room temperature (22–24°C) under continuous light (110 kmol/m^2 s photosynthetically active radiation). When liquid cultures are used they are incubated with constant agitation of 250 rpm unless noted otherwise.

3.1. Crude Genomic DNA Isolation

1. The 3′UTR of interest can be amplified using crude genomic DNA from a *Chlamydomonas* culture grown in TAP medium (*see* Subheading 3.5). Spin down 100 μl of an overnight culture in a 1.5-ml microcentrifuge tube at 13800 × *g* for 30 s.
2. Remove supernatant, resuspend the pellet in 100 μl water, and spin down cells for 30 s.
3. Resuspend the pellet in 100 μl RHDB buffer.
4. Add 1 μl Proteinase K and incubate mixture at 56°C for 60 min.
5. Next, transfer tube to 95–100°C for 15–20 min to inactivate proteinase K.
6. Vortex the tube for 5 s and centrifuge for 30 s at 13800 × *g*.
7. Remove the supernatant and use 5 μl in a 25-μl PCR reaction.

3.2. Primer Design and PCR Amplification of the 3 UTR and the Spacer Region

1. These instructions are for the use of the primer design program provided by IDT, but can be adapted for any primer design program. Obtain the genomic sequence of your gene of interest. This can be obtained from the *Chlamydomonas* Center website, http://www.chlamy.org/#. Click on the genome browser icon, followed by the advanced search and type in the name of the gene. A screen will display your gene of interest and several links. Click on the P link and in that

screen there is an option to view the genomic sequence, which if the gene is annotated will identify the introns, exons, and translations, allowing you to determine the 3′UTR sequence.

2. Copy the 3′UTR sequence and paste it in a primer design program. We typically use the primer design program on the IDT website, which can be accessed using the following URL http://www.idtdna.com/scitools/scitools.aspx. Click on the PrimerQuest icon, paste your sequence in the available screen and select for PCR primers. In the advanced settings select 300–1,000 bp for a target size. The melting temperature (Tm) should be between 58°C and 65°C (see Note 2).
3. Once you obtain the primer sequences you have to modify them for the coldfusion cloning kit (see Fig. 2b). The coldfusion kit requires that there is a 15-bp overlap between sequences that need to be stitched together. In addition, the ends that will be stitched to the vector need an additional number of bases to ensure regeneration of the enzyme site the construct will be cloned in. Add the following sequences to you primers:
   (a) To the 5′ end of the LIR1 primer add AATTGATCCAG AATTC (16 bp of the vector).
   (b) To the 5′ end of the RIR1 primer add GATGCTTGAG GATGC (to allow for recombination with the spacer sequence).
   (c) To the 5′ end of the LIR2 primer add TCCTGTGCC TCCTAC (15 bp overlap with the 3′ end of the spacer).
   (d) To the 5′ end of the RIR2 primer add ACGAAT TGACGAATTC for recombination with the vector.

   The following primer combination can be used to PCR amplify the 210 spacer region from *MAA7/X* IR vector: GCATCCTCAAGCATCCTTCTATTC/ GTAGGAGGCACAGGAAGAGCAAA.
4. Primers should be ordered desalted, resuspended in 1× TE to a concentration of 50 pmol/μl and stored at −20°C until use.
5. The primers can next be used to amplify the IR segments and spacer region, using genomic DNA and *MAA7/X* IR vector as a template, respectively. We routinely use rTth DNA polymerase (Applied Biosystems) as it has worked well in our hands for amplification of genomic DNA from *Chlamydomonas*. Turn on the PCR machine and preheat PCR block and lid to 93°C for a hot start.
6. Set up a 25-μl PCR reaction on ice, by adding the following components in the listed order to a 200-μl PCR tube: x μl of water (to make final volume 50 μl), 15 μl 3.3× XL buffer, 3 μl 25 mM Mg $(OAc)_2$, 8 μl 1.25 mM dNTPs, 100 ng genomic DNA or *MAA7/X* IR DNA (10–100 ng), 0.6 μl of each

primer (at 50 pmol/µl), and finally 1 µl rTth Polymerase. Water volumes should be adjusted to accommodate different concentrations of template DNA.

7. Quickly mix components by tapping tube and spin down contents in a centrifuge for 5 s. Place the PCR tube in a PCR machine and start cycling using the following parameters: 93°C for 30 s, 55°C (or temperature 2–3°C below the primer Tm which should be calculated based upon the gene-specific end of the primer) for 30 s and 72°C for 1 min (use an extension time of roughly 2 min/kb) for 30 cycles. Include an initial melting step of 93°C for 3 min and a final 72°C extension step for 5 min and allow reaction to cool to 4°C (see Note 3).
8. PCR products should be run on an agarose gel and gel purified (see below) before use in cloning reactions.

3.3. Coldfusion Cloning of PCR Fragments in Vector

1. The *MAA7/X* IR vector can be obtained from the *Chlamydomonas* Center and should be linearized by digesting it with *Eco*RI. Set up a restriction digest by combining about 2 µg of *MAA7/X* IR DNA in a 30–50 µl reaction with 1–2 µl *Eco*RI in 1× buffer. Incubate this mixture for at least 3 h at 37°C. It is very important that the vector is completely linearized, and thus gel purification (as described below) is highly recommended (see Note 4).
2. These instructions assume the use of a BIORAD mini gel apparatus, but can be adapted to any gel system. Make a 1% agarose gel (percentage depends on size of DNA fragment) by mixing 0.5 g agarose with 50 ml 1× TBE buffer and boiling the mixture in a microwave until all agarose particles have dissolved (be careful not to mix vigorously when removing the flask from the microwave, as the solution can super-boil). After cooling the mixture to about 50–60°C (warm to the touch), add 2 µl 10 mg/ml ethidium bromide, mix by swirling, pour entire contents in gel tray with a comb in place, and let it solidify for 20–30 min, until set.
3. After the gel has set, remove the comb and place the gel in gel box containing 1× TBE buffer.
4. Add 10 µl 6x gel loading buffer to the 50 µl PCR reaction and place entire contents in a well. Add a DNA ladder in one of the wells, and run gel for approximately an hour at 90–100 V.
5. Use a Gel documentation system to take a picture of the gel and identify the correctly sized DNA bands of PCR products.
6. Place gel on a UV illuminator and excise out DNA fragment using a clean razor blade. Make sure exposure of DNA to UV light is as short as possible to prevent nicking of DNA.

7. Place gel piece in a 1.5-ml microcentrifuge tube, weigh it, and use a gel purification kit to gel purify DNA. We routinely use the Qiaquick gel extraction kit to purify DNA. After the final elution measure DNA concentration using a spectrophotometer (see Note 5).
8. Once the PCR fragments and linearized vector are gel purified and DNA concentrations are known, set up the recombination reactions as follows and suggested by the manufacturer (System Biosciences).
9. The recommended molar ratio for insert versus vector is 2:1, with the vector concentration at 150–200 ng for a 6-kb vector, roughly the size of *MAA7/X* IR. In general, the reactions tend to be more successful when at least 50 ng of the insert is used. However, since the PCR products (the spacer, and the IR sequences) are relatively small, 210 bp and 300–1,000 bp, respectively, a higher insert versus vector ratio could be used. For example use 400 ng vector and 50 ng insert.

   Set up the reaction by mixing 400 ng vector and 50 ng of each PCR fragment in a 10 μl reaction volume that contains the buffer and coldfusion master mix. If the total volume for vector and inserts exceeds 7 μl, adjust the master mix and water to a total volume of 20 μl.
10. Briefly spin mixture and incubate at room temperature for 5 min followed by a 10-min incubation on ice. This mixture is now ready for transformation in *Escherichia coli* competent cells.
11. The cold fusion kit comes with *E. coli* competent cells which can be used to transform mixtures. Alternatively other high efficiency *E. coli* competent cells could be used.
12. Thaw competent cells on ice, add 10 μl of coldfusion reaction mixture to thawed cells, mix gently, and place the mixture on ice for 30 min.
13. Heat-shock mixture by placing tube in a 42°C waterbath for 50 s.
14. Transfer tube to ice for 2 min.
15. Add 250 ml SOC medium to tube, transfer entire contents to a 15-ml tube (optional), and place in a 37°C shaker at 190 rpm for 1 h.
16. Spread 25, 100 μl and the rest of the tube on three separate LB+ ampicillin plates.
17. Incubate plates upside down in a 37°C incubator overnight. Colonies should appear within 24 h.

3.4. Confirmation of Correct Plasmid Constructs

1. The next step is to determine if the plasmids contain the correct insert. One can use colony PCR to determine which colonies contain the correct construct (see Note 6), but we recommend

isolating plasmids and performing restriction enzyme digest analysis to ensure correct plasmid construction.

2. Pick about ten colonies to prepare plasmids from. We commonly use the QIAprep Spin Miniprep Kit to isolate plasmids.
3. Inoculate 5 ml of Terrific Broth containing tetracycline with a needle point of cells from a colony and incubate tubes in a 37°C shaker at 220–250 rpm overnight, and follow instructions provided by manufacturer (Qiagen) to isolate plasmids.
4. Once plasmids have been isolated, confirmation of the correct insert can be established by digestion with *Eco*RI. Using this enzyme one should be able to cut out the entire inserted fragment from the plasmid, since this cloning procedure should have regenerated the *Eco*RI sites. If there is an introduced *Eco*RI site present in the IRs, extra bands will appear during gel analysis of the digests. Include a digestion of the *MAA7/X* IR plasmid as a control.
5. Set up a 10-μl enzyme digest by adding 100–200 ng plasmid DNA to a 1.5-ml microcentrifuge tube containing 1× *Eco*RI buffer and 0.2–0.3 μl *Eco*RI enzyme, mix contents, quickly spin in a centrifuge, and incubate at 37°C for 1 h. Next, run digests on a 0.8% agarose gel (see Subheading 3.3 for agarose gel electrophoresis) and analyze DNA bands, using a gel documentation system.
6. If there are no introduced *Eco*RI sites in the construct, you should see two bands: one representing the vector and the other the inserted IR.

3.5. Autolysin Preparation

Before transforming the linearized plasmid into *Chlamydomonas*, the cell wall needs to be digested by autolysin. Autolysin is a cell wall digesting enzyme produced by mating *Chlamydomonas* gametes. Production of high quality autolysin is critical for successful transformation of *Chlamydomonas* as poor quality autolysin commonly results in failures of transformation.

1. Use roughly half of confluent overnight TAP plate cultures of the mating type strains CC620 and CC621 to inoculate separate 500 ml flasks containing 150 ml HS medium and incubate them under continuous light with continuous agitation for 3–4 days.
2. Collect cells by centrifugation at 3,000 × *g* for 18 min and resuspend the pellet in 100 ml nitrogen-free HS medium.
3. Add 10 μl Lugol's iodine to a 100-μl sample and count cells using a hemocytometer. Next, resuspend the cells in nitrogen-free HS medium to a final density of 1×10^6 cells/ml.
4. Transfer 250 ml of each cell type in separate 500-ml flasks. Multiple flasks can be used at this point.

5. Induce gamete formation by gently shaking the cells at 65 rpm for 24 h in continuous light.
6. Collect cells by centrifugation at 3,000 × *g* for 18 min and resuspend in nitrogen-free HS medium to a final density of 2×10^7 cells.
7. Transfer the cells to a 500-ml flask and allow the cells to recover for 1 h in continuous light without agitation.
8. Mix equal volumes of mating type cells in a 500-ml Erlenmeyer flask and let it stand in light for 1–2 h.
9. Centrifuge the cells at 3,000 × *g* for 5 min at 4°C to pellet the cells.
10. Transfer the supernatant to a sterile microcentrifuge tube and centrifuge at 18,000 × *g* for 15 min at 4°C to remove all cellular debris.
11. Collect the supernatant in a sterile microcentrifuge tube (see Note 7). The supernatant is the autolysin and aliquots of autolysin can be stored at −20°C for several months. Before freezing, one can also test the efficiency of cell wall digestion by following step 7 in Subheading 3.6.

3.6. Chlamydomonas Glass Bead Transformations

One of the most efficient ways of introducing the TIR construct in *Chlamydomonas* is by glass bead transformation. Higher efficiency of transformation can be achieved when the plasmid is linearized. Here, we provide a protocol adapted from Kindle (19).

1. Use a few loops full of an overnight TAP plate culture of wild-type cells (or the strain that needs to be transformed with the TIR construct) to inoculate 150 ml TAP medium in a 500-ml flask (see Note 8). Incubate the culture under continuous light and constant agitation until it reaches a cell density of $2–8 \times 10^6$ cells/ml.
2. Harvest cells at 3,000 × *g* for 15 min and resuspend in 10 ml liquid TAP medium.
3. Count the cells and collect 4×10^7 cells for each transformation. We typically do about 13 transformations for one construct.
4. Centrifuge the cells at 3,000 × *g* for 15 min and resuspend in desired volume of autolysin to remove the cell wall (typically 2 ml of autolysin is required to remove the cell wall of 4×10^7 cells).
5. Incubated the cells resuspended in autolysin under continuous light with constant shaking at low speed (65 rpm) for 80 min.
6. Treatment of autolysin for 80 min should remove the cell wall of about 40–60% of the cells. It is critical to check the efficiency of cell wall digestion with autolysin because the efficiency of transformation directly correlates with it.

7. To check for the efficiency of cell wall digestion add 1 ml triton X-100 to 100 μl of the cells and count the cells that are not lysed (A). Also count the total number of cells without the Triton X treatment (B). Estimate the efficiency of cell wall digestion by using the following formula: $(1 - A/B) \times 100$. The efficiency should be 60% or more (see Note 9).
8. After autolysin treatment, harvest the cells at $1,500 \times g$ for 5 min at 4°C, carefully remove the supernatant and resuspend cells in TAP medium to a concentration of 13.3×10^7 cells/ml.
9. For each glass bead transformation, take a sterile 15 ml conical tube and add the following in the order specified: (1) 300 mg sterile glass beads, (2) 300 μl cells, (3) 1 μl linearized plasmid at 1–1.5 μg/μl, and (4) 100 μl of PEG 6000.
10. Mix the contents, vortex the tube at high speed for exactly 30 s, and immediately add 700 μl TAP medium.
11. Transfer the contents into a 250-ml flask, add 10 ml TAP medium, and incubate for 2 days under dim light (50 μM/m^2 s photosynthetically active radiations) without agitation. This allows the cells to recover and permits the induction of RNAi construct expression.
12. Next, harvest cells by centrifugation at $3,000 \times g$ for 5 min, resuspend in 1 ml TAP, and spread onto four plates containing TAP medium supplemented with 5-FI and paromomycin (see Note 10).
13. Incubate the plates inverted under dim light with a layer of paper towels on top of them for 10–14 days until colonies appear.
14. Once colonies appear they should be maintained under constant selection of both paromomycin and 5 μM 5-FI to prevent silencing of the IR constructs. To select for cells with the strongest phenotype (indicating a stronger *MAA7* down-regulation), prepare TAP plates containing paromomycin and two different concentrations of 5-FI (5 and 7 μM). Serially dilute ten transformants and spot plate them on three different plates: TAP medium and TAP medium containing 5 or 7 μM of 5-FI. Incubate the plates under dim light (50 μM/m^2 s photosynthetically active radiations) with a paper towel on top for 8–10 days. Colonies that survive the best on both plates can be picked for further molecular analysis.

3.7. RNA Isolation and cDNA Synthesis

Cells that grow on 5-FI and paromomycin should have down-regulated expression levels of both *MAA7* and the gene of interest. A variety of methods can be used to detect down regulation of *MAA7* and the gene of interest. Our method of choice is reverse transcription (RT) followed by quantitative PCR (qPCR) but one can use more traditional methods like semi-Quantitative PCR or

Northern blot analysis. One can also determine protein levels of the expressed gene of interest using immunoblot analysis, provided there are available antibodies to detect the protein of interest. However, for the scope of this chapter, we discuss only the detection method using qPCR. To measure expression of the transcripts, preparation of high quality total RNA is essential. Only use RNase-free tips and tubes and clean bench surface and pipettors with 70% isopropanol or RNAZap (Ambion). Changing gloves often will also prevent contamination with RNase. Unless indicated, centrifugation should be performed at 13,000 × *g*.

1. Inoculate a loop full of overnight plate cultures of transgenic and wild-type *Chlamydomonas* cells in 150 ml TAP medium with 2.5 μM 5-FI and grow under continuous light and shaking until the cells reach a density of $2–4 \times 10^6$ cells/ml.
2. For each culture, Pellet 10^8 cells by centrifugation, resuspend in 1 ml TAP medium, and transfer to a microcentrifuge tube (10^8 cells will provide approximately 50–60 μg of total RNA).
3. Centrifuge the cells for 1.5 min at maximum speed and discard the supernatant. Care should be taken to remove all the supernatant without disturbing the pellet.
4. Using a pipette tip spread the cell pellet on the microcentrifuge tube walls and freeze the tube in liquid nitrogen for at least 10 min.
5. Remove the tube from liquid nitrogen, immediately add 1 ml Trizol, and vortex at maximum speed for 10 min. Let the tube stand at room temperature for 5 min to ensure complete dissociation of the nucleoprotein complexes.
6. Centrifuge the tube for 10 min at 4°C to remove insoluble debris and transfer the supernatant to a clean microcentrifuge tube.
7. Add 0.25 ml chloroform and mix well by inverting the tube for 30 s followed by vortexing at maximum speed for 2 min. This mixing is essential for the separation of the two phases. Allow the tube to stand at room temperature for 2 min and centrifuge for 15 min at 4°C.
8. Transfer the top aqueous phase to a clean microcentrifuge tube. Care should be taken not to remove the bottom Trizol/chloroform phase or the interphase.
9. Add an equal volume of phenol:chloroform:isoamyl alcohol (25:24:1) and vortex for 2 min. Let the mixture stand for 5 min at room temperature and centrifuge for 15 min at 4°C.
10. Transfer the top aqueous phase to a clean microcentrifuge tube and add one volume of isopropanol. Incubate the tube

on ice for an hour and precipitate the RNA by centrifugation for 20 min at 4°C.

11. Remove the isopropanol, add 1 ml 75% ethanol, and mix the sample by vortexing for 15–20 min at room temperature.
12. Centrifuge for 10 min at 4°C and carefully remove the supernatant. We recommend a second 75% ethanol wash to remove all the residual salt.
13. Dry the pellet for 10–20 min. If higher quality RNA is desired the pellet can be resuspended in 500 ml Trizol and steps 6–12 be repeated. Once the RNA pellet is obtained, resuspend the pellet in 100 μl of RNase-free TE. To ensure complete resuspension of the pellet the tube can be heated at 55°C for 5–10 min. Aliquot RNA and store samples at -20°C. RNA can be stored in this manner for several months or at -80°C for a year. Repeated freeze thaw cycles of RNA samples should be avoided.
14. Here, we have outlined procedures for cDNA synthesis using the Superscript III kit (Invitrogen) but other cDNA kits could be used. About 5 μg of RNA produces sufficient cDNA for the qPCR analysis. As per manufacturer's recommendation, place 5 μg of total RNA in a PCR tube, add 1 μl of 50 μM oligo-dT adapter and 1 μl of 10 mM dNTP in a total volume of 10 μl. Incubate the mixture at 65°C for 5 min to allow the adapter to attach to the polyA tail of the RNA followed by 1–2 min incubation at 4°C.
15. Next, the cDNA is synthesized by mixing the oligo-dT primed RNA with 2 μl of 10× RT buffer, 4 μl of 10 mM $MgCl_2$, 2 μl 0.1 M DTT, 1 μl of RNase OUT, and 1 μl of Superscript III RT.
16. Incubate the mixture at 50°C for 55 min and stop the reaction by heating the mixture at 85°C for 5 min followed by 4°C for a few minutes.
17. To remove any traces of RNA treat the mixture with RNase H for 20 min at room temperature.
18. The cDNA is stable at 4°C for 1–2 weeks but can be stored for a couple of months at -20°C. Avoid repeated freeze thaw cycles of cDNA samples. One of the most common reasons for quantitative PCR failure is the degradation of cDNA samples.

3.8. Quantitative PCR Analysis to Detect Co-silencing of the MAA7 and Gene X

Once the cDNA is synthesized, transcript levels of *MAA7* and the gene of interest can be measured by a variety of methods. Here, we describe the use of quantitative PCR (qPCR). The success of qPCR relies on good primer design. Poor primer design can result in nonspecific products or primer dimers. Primer design software (http://www.quantprime.de) can be used to design qPCR primers (20), but primers can also be designed using any standard primer design algorithms. For the optimal performance of the primer set, the specific reaction conditions, the annealing temperatures, and $MgCl_2$ conditions should be optimized. In addition, prior to using primers the primer efficiency needs to be determined (see Note 11).

1. The key parameters for primer design are as follows:
   - The amplicon size should be between 60 and 150 bps.
   - The optimal melting temperature (Tm) of the primers should be in the range of 58–65°C.
   - The annealing temperatures of both the primers should be similar.
   - The last base at 3′ end of the primers should have no more than two Gs or Cs.
   - Primers should be designed such that there are minimal primer hairpin structures and/or primer dimer formations.
   - To avoid getting signals from contaminating genomic DNA, primers can be designed in the 3′UTR or in regions that span an intron.
2. For the detection of *MAA7* transcripts, one can use the primers described in (21): *MAA7*-F and *MAA7*-R.
3. To normalize the levels of *MAA7* transcripts primers designed to detect the constitutively expressed CBLP gene can be used. CBLP is a gene that encodes the G protein β subunit in *Chlamydomonas*. The primers sets that have been described in (21) for CBLP transcripts are CBLP-F and CBLP-R.
4. Once the cDNA is obtained and the primers are optimized, transcript abundance can be measured by qPCR. For a 20-μl reaction, add 10 μl of 2× SsoFast EvaGreen Supermix (BioRad), 400 nM of each primer and 5 μl of 1:50 diluted cDNA, and water to a final volume of 20 μl. The parameters on the thermocyler are adjusted based on the optimizations of the primers. The abundance of the transcripts is measured via the $2^{-(\Delta\Delta)CT}$ method.

4. Notes

1. Unless stated otherwise, all solutions should be prepared in water that has a resistivity of 18.2 MΩ-cm and total organic content of less than five parts per billion. This standard is referred to as “water” in this text.
2. To ensure that the silencing phenotype produced by the IR construct is not due to silencing of off-target sites, we recom-

mend making two independent silencing constructs by designing the primers to amplify nonoverlapping regions of the genomic sequence. Moreover, primer melting temperatures are important. If they are below 58°C increase the length of the primer until they are at least 58°C.

3. PCR conditions should be optimized before proceeding to qPCR analysis. Once primers are designed use the primers in an optimization test where a gradient of primer melting temperatures is used. Also, if needed one can perform an optimization test using a primer dilution series. Use conditions under which the Ct values are the lowest.
4. To ensure that the vector is completely linearized, use 5–10 ng of the linearized and purified vector and transform into competent cells. There should be none or very few colonies growing on the transformation plates.
5. It is critical that during gel extraction of DNA bands all ethanol remaining from wash steps is removed prior to eluting DNA. To ensure ethanol removal, lengthen the centrifugation times after the wash steps to 5 min.
6. Colony PCR can be used to determine if transformed *E. coli* cells contain the TIR construct. To do so, use a needle or toothpick to transfer a small amount of bacteria to water in a PCR tube and use primers designed to one of the IRs to amplify DNA. Analyze products using gel electrophoresis.
7. Autolysin solutions tend to get contaminated very easily; hence care should be taken to perform all the steps in a sterile environment.
8. It is important to use overnight grown plate cultures as inoculum for the transformations because older cultures will lead to lower transformation efficiency.
9. If the treatment with autolysin gives a cell wall removal efficiency of 40% or lower, incubate the cells for a longer period up to 4 h. If after 4 h the efficiency of cell wall digestion is still low, we suggest discontinuing the transformation and redoing it with another batch of autolysin.
10. When using 5-FI in media (including liquid media) protect the media from light by wrapping plates or flasks in aluminum foil.
11. One way to determine primer efficiency is to make a standard curve, using a real-time PCR machine as described below. We typically use the SsoFast EvaGreen Supermix Kit (BIORAD) but other SYBR green mixes would work.
    (a) Make tenfold serial dilutions of the target template and amplify these inputs using PCR conditions as suggested by the manufacturer (BIORAD).

(b) Plot the Ct values against the concentration and calculate the slope. It is important to keep the primer concentration the same. We therefore suggest making a mastermix that includes everything except the template input and aliquot that in the PCR tubes, followed by template addition. The final primer concentration in the mix should be between 100 and 500 nM.

(c) The efficiency (E) of primers can be obtained from $10^{(-1/\text{slope})}$. The amplification efficiency can be calculated using the following formula: $\%E = (E - 1) \times 100$. The amplification efficiency should be between 90 and 110%.

Acknowledgments

Research in the van Dijk laboratory is supported by Grant Number P20 RR16469 from the National Center for Research Resources (NCRR), a component of the National Institutes of Health (NIH), Grant number 0940177 from the National Science Foundation (NSF) and Grant number EPS-1004094 from NSF.

References

1. Fire A., et al. (1998) Potent and specific genetic interference by double-stranded RNA in *Caenorhabditis elegans*. *Nature.* **391**, 806–811.
2. Carthew R.W., Sontheimer E.J. (2009) Origins and Mechanisms of miRNAs and siRNAs. *Cell.* **136**, 642–655.
3. Ghildiyal M., Zamore P.D. (2009) Small silencing RNAs: an expanding universe. *Nat Rev Genet.* **10**, 94–108.
4. Merchant S.S., et al. (2007) The *Chlamydomonas* genome reveals the evolution of key animal and plant functions. *Science.* **318**, 245–250.
5. Scott S.A., et al. (2010) Biodiesel from algae: challenges and prospects. *Curr Opin Biotechnol.* **21**, 277–286.
6. Kim E.J., Cerutti H. (2009) Targeted gene silencing by RNA interference in *Chlamydomonas*. *Methods Cell Biol.* **93**, 99–110.
7. Rohr J., et al. (2004) Tandem inverted repeat system for selection of effective transgenic RNAi strains in *Chlamydomonas*. *Plant J.* **40**, 611–621.
8. Schroda M. (2006) RNA silencing in *Chlamydomonas*: mechanisms and tools. *Curr Genet.* **49**, 69–84.
9. Waterhouse P.M., Helliwell C.A. (2003) Exploring plant genomes by RNA-induced gene silencing. *Nat Rev Genet.* **4**, 29–38.
10. Grimm D. (2009) Small silencing RNAs: state-of-the-art. *Adv Drug Deliv Rev.* **61**, 672–703.
11. Molnar A., et al. (2009) Highly specific gene silencing by artificial microRNAs in the unicellular alga *Chlamydomonas reinhardtii*. *Plant J.* **58** 165–174
12. Zeng Y., Wagner E.J., Cullen B.R. (2002) Both natural and designed micro RNAs can inhibit the expression of cognate mRNAs when expressed in human cells. *Mol Cell.* **9**, 1327–1333.
13. Zhao T., et al. (2009) Gene silencing by artificial microRNAs in *Chlamydomonas*. *Plant Journal.* **58**, 157–164.
14. Dutcher S.K., et al. (1992) Tryptophan analog resistance mutations in *Chlamydomonas reinhardtii*. *Genetics.* **131**, 593–607.
15. Hanahan D. (1983) Studies on transformation of *Escherichia coli* with plasmids. *J. Mol. Biol.* **166**, 557–580.
16. Cerutti H., et al. (1997) A eubacterial gene conferring spectinomycin resistance on *Chlamydomonas reinhardtii*: integration into the nuclear genome and gene expression. *Genetics.* **145**, 97–110.

17. Sizova I., Fuhrmann M., Hegemann P. (2001) A *Streptomyces rimosus aphVIII* gene coding for a new type phosphotransferase provides stable antibiotic resistance to *Chlamydomonas reinhardtii*. *Gene*. **277**, 221–229.
18. Sizova I.A., et al. (1996) Stable nuclear transformation of *Chlamydomonas reinhardtii* with a *Streptomyces rimosus* gene as the selective marker. *Gene*. **181**, 13–18.
19. Kindle K.L. (1990) High-frequency nuclear transformation of *Chlamydomonas reinhardtii*. *Proc Natl Acad Sci U S A*. **87**, 1228–1232.
20. Arvidsson S., et al. (2008) QuantPrime--a flexible tool for reliable high-throughput primer design for quantitative PCR. *BMC Bioinformatics*. **9**, 465.
21. Zhao T., et al. (2007) A complex system of small RNAs in the unicellular green alga *Chlamydomonas reinhardtii*. *Genes Dev*. **21**, 1190–1203.

Erratum

Selectable and Inheritable Gene Silencing through RNA Interference in the Unicellular Alga *Chlamydomonas reinhardtii*

Karin van Dijk and Nandita Sarkar

James A. Williams (ed.), *Strain Engineering: Methods and Protocols*, Methods in Molecular Biology, vol. 765,
DOI 10.1007/978-1-61779-197-0_28, © Springer Science+Business Media, LLC 2011

DOI 10.1007/978-1-61779-197-0_28

The publisher regrets that in the online and print versions of this book, on page 470, in item 7, "1 μL of trition-X" incorrectly reads "1 ml of trition-X"."

The online version of the original chapter can be found at
http://dx.doi.org/10.1007//978-1-61779-197-0_28

Index

James A. Williams (ed.), *Strain Engineering: Methods and Protocols*, Methods in Molecular Biology, vol. 765, DOI 10.1007/978-1-61779-197-0, © Springer Science+Business Media, LLC 2011

MIX
Papier aus verantwortungsvollen Quellen
Paper from responsible sources
FSC® C105338

If you have any concerns about our products,
you can contact us on
ProductSafety@springernature.com

In case Publisher is established outside the EU,
the EU authorized representative is:
Springer Nature Customer Service Center GmbH
Europaplatz 3, 69115 Heidelberg, Germany

Printed by Libri Plureos GmbH
in Hamburg, Germany